# Technologies for the Treatment and Recovery of Nutrients from Industrial Wastewater

Ángeles Val del Río
*University of Santiago de Compostela, Spain*

José Luis Campos Gómez
*University Adolfo Ibáñez, Chile*

Anuska Mosquera Corral
*University of Santiago de Compostela, Spain*

A volume in the Advances in Environmental Engineering and Green Technologies (AEEGT) Book Series

IGI GLOBAL
DISSEMINATOR OF KNOWLEDGE

www.igi-global.com

Published in the United States of America by
IGI Global
Information Science Reference (an imprint of IGI Global)
701 E. Chocolate Avenue
Hershey PA, USA 17033
Tel: 717-533-8845
Fax: 717-533-8661
E-mail: cust@igi-global.com
Web site: http://www.igi-global.com



Library of Congress Cataloging-in-Publication Data

Names: Val del Rio, Angeles, 1980- editor. | Campos Gomez, Jose Luis, 1971- editor. | Mosquera Corral, Anuska, 1969- editor.
Title: Technologies for the treatment and recovery of nutrients from industrial wastewater / Angeles Val del Rio, Jose Luis Campos Gomez, and Anuska Mosquera Corral, editors.
Description: Hershey PA : Information Science Reference, [2017] | Series: Advances in environmental engineering and green technologies | Includes bibliographical references and index.
Identifiers: LCCN 2016033126| ISBN 9781522510376 (hardcover) | ISBN 9781522510383 (ebook)
Subjects: LCSH: Sewage--Purification--Nutrient removal. | Factory and trade waste--Purification. | Nutrient pollution of water
Classification: LCC TD758.5.N87 T43 2017 | DDC 631.8/69--dc23 LC record available at https://lccn.loc.gov/2016033126

This book is published in the IGI Global book series Advances in Environmental Engineering and Green Technologies (AEEGT) (ISSN: 2326-9162; eISSN: 2326-9170)

British Cataloguing in Publication Data
A Cataloguing in Publication record for this book is available from the British Library.

# Advances in Environmental Engineering and Green Technologies (AEEGT) Book Series

ISSN: 2326-9162
EISSN: 2326-9170

Mission

Growing awareness and an increased focus on environmental issues such as climate change, energy use, and loss of non-renewable resources have brought about a greater need for research that provides potential solutions to these problems. Research in environmental science and engineering continues to play a vital role in uncovering new opportunities for a "green" future.

The **Advances in Environmental Engineering and Green Technologies (AEEGT)** book series is a mouthpiece for research in all aspects of environmental science, earth science, and green initiatives. This series supports the ongoing research in this field through publishing books that discuss topics within environmental engineering or that deal with the interdisciplinary field of green technologies.

Coverage

- Electric Vehicles
- Biofilters and Biofiltration
- Industrial Waste Management and Minimization
- Waste Management
- Green Technology
- Alternative Power Sources
- Pollution Management
- Renewable Energy
- Cleantech
- Water Supply and Treatment

IGI Global is currently accepting manuscripts for publication within this series. To submit a proposal for a volume in this series, please contact our Acquisition Editors at Acquisitions@igi-global.com or visit: http://www.igi-global.com/publish/.

The Advances in Environmental Engineering and Green Technologies (AEEGT) Book Series (ISSN 2326-9162) is published by IGI Global, 701 E. Chocolate Avenue, Hershey, PA 17033-1240, USA, www.igi-global.com. This series is composed of titles available for purchase individually; each title is edited to be contextually exclusive from any other title within the series. For pricing and ordering information please visit http://www.igi-global.com/book-series/advances-environmental-engineering-green-technologies/73679. Postmaster: Send all address changes to above address. 

## Titles in this Series

*For a list of additional titles in this series, please visit: www.igi-global.com*

*Oceanographic and Marine Cross-Domain Data Management for Sustainable Development*
Paolo Diviacco (Istituto Nazionale di Oceanografia e di Geofisica Sperimentale (OGS), Italy) Adam Leadbetter (Marine Institute, Ireland) and Helen Glaves (British Geological Survey, UK)
Information Science Reference • copyright 2017 • 425pp • H/C (ISBN: 9781522507000) • US $205.00 (our price)

*Examining the Development, Regulation, and Consumption of Functional Foods*
Sailas Benjamin (University of Calicut, India)
Medical Information Science Reference • copyright 2017 • 344pp • H/C (ISBN: 9781522506072) • US $185.00 (our price)

*Applying Nanotechnology for Environmental Sustainability*
Sung Hee Joo (University of Miami, USA)
Information Science Reference • copyright 2017 • 558pp • H/C (ISBN: 9781522505853) • US $225.00 (our price)

*Exploring the Nutrition and Health Benefits of Functional Foods*
Hossain Uddin Shekhar (University of Dhaka, Bangladesh) Zakir Hossain Howlader (University of Dhaka, Bangladesh) and Yearul Kabir (University of Dhaka, Bangladesh)
Medical Information Science Reference • copyright 2017 • 523pp • H/C (ISBN: 9781522505914) • US $200.00 (our price)

*Smart Cities as a Solution for Reducing Urban Waste and Pollution*
Goh Bee Hua (National University of Singapore, Singapore)
Information Science Reference • copyright 2016 • 362pp • H/C (ISBN: 9781522503026) • US $190.00 (our price)

*Food Science, Production, and Engineering in Contemporary Economies*
Andrei Jean-Vasile (Petroleum-Gas University of Ploiesti, Romania)
Information Science Reference • copyright 2016 • 473pp • H/C (ISBN: 9781522503415) • US $240.00 (our price)

*Smart Grid as a Solution for Renewable and Efficient Energy*
Ayaz Ahmad (COMSATS Institute of Information Technology, Pakistan) and Naveed Ul Hassan (LUMS School of Science & Engineering, Pakistan)
Information Science Reference • copyright 2016 • 415pp • H/C (ISBN: 9781522500728) • US $220.00 (our price)

*Biologically-Inspired Energy Harvesting through Wireless Sensor Technologies*
Vasaki Ponnusamy (Universiti Tunku Abdul Rahman, Malaysia) Noor Zaman (King Faisal University, Saudi Arabia) Tang Jung Low (Universiti Teknologi Petronas, Malaysia) and Anang Hudaya Muhamad Amin (Multimedia University, Malaysia)

# Table of Contents

# Detailed Table of Contents

High strength waste streams, namely rejected water from a wastewater treatment plant, livestock slurry, and agro-food wastewater, are characterized by its high organic matter and nutrient content which favours processes aiming to recover energy and nutrients, instead of removing them. In this regard physical-chemical processes are suitable technologies to attain these objectives. Among others, stripping coupled with absorption, struvite precipitation, membrane separation, and vacuum evaporation, are all physical-chemical processes aiming to concentrate nutrients in a stream that can later be reused as fertilizer. In this chapter the main physical-chemical processes will be defined and described in terms of the objective of each process technique, their theoretical fundamentals, environmental effects (air, water and soil emissions, resource depletion), technical indicators (efficiencies, energy consumption, etc.), and by-product characteristics.



This chapter gives an overview on the main technologies for nutrient removal from industrial wastewater by focusing on principles and operational parameters of real applications. A plethora of technologies can achieve the nutrients removal from wastewater depending mainly on their concentration and forms; however, biological nitrification and denitrification and chemical precipitation are the most common processes used today to remove nitrogen and phosphorus, respectively. Stripping, adsorption and membrane based processes for nutrients recovery can be economically viable only when nitrogen concentration is higher than 1.5-2 gN/L. On the other hand, phosphorus recovery should always be pursued and struvite crystalization is the most common option that should be evaluated together with biological phosphorus accumulation in sludge or plants for the following post-processing and valorization.



In this chapter the fundamental aspects involved in biological treatment processes applied for the removal of nitrogen from wastewaters are described. A comprehensive review of the literature is provided including kinetic, microbiological and biochemical aspects of nitrification, denitrification and anammox, as well as key operational parameters affecting the processes. This information is relevant for designing wastewater treatment processes applied for the removal of nitrogen from wastewaters. The information is also essential for predicting and controlling the performance of these treatment processes.

Constructed Wetlands (CWs) are low-cost and sustainable systems for wastewater treatment. Traditionally they have been used for urban and domestic wastewater treatment, but in the last two decades, the applications for industrial wastewater treatment increased due to the evolution of the technology and the extended research on the field. Nowadays, CWs have been applied to the treatment of different kind of wastewaters as such as refinery and petrochemical industry effluents, food industry effluents including abattoir, dairy, meat, fruit and vegetables processing industries, distillery and winery effluents, pulp and paper, textile, tannery, aquaculture, steel and mixed industrial effluents. In this chapter, the authors present the main types of CWs, explain how they work and the expected performances, and describe the principal applications of CWs for industrial wastewater treatment with particular attention to suspended solids, organic matter and nutrient removal. A review of these applications as well as some case studies will be discussed.



Aerobic Granular Sludge (AGS) has been successfully applied for carbon, nitrogen and phosphorous removal from wastewaters, in a single tank, reducing the space and energy requirements. This is especially beneficial for, often space restricted, industrial facilities. Moreover, AGS holds a promise for the toxic pollutants removal, due to its layered and compact structure and the bacteria embedding in a protective extracellular polymeric matrix. These outstanding features contribute to AGS tolerance to toxicity and stability. Strategies available to deal with toxic compounds, namely granulation with effluents containing toxics and bioaugmentation, are addressed here. Different applications for the toxics/micropollutants removal through biosorption and/or biodegradation are presented, illustrating the technology versatility. The anthropogenic substances effects on system performance and bacterial populations established within AGS are also addressed. Combination of contaminants removal to allow water discharge, and simultaneous valuable products recovery are presented as final remark.

Application of anammox based processes is nowadays an efficient way to remove nitrogen from wastewaters, being good alternative to the conventional nitrification-denitrification process. This chapter reviews the possible configurations to apply the anammox process, being special attention to the previous partial nitritation, necessary to obtain the adequate substrates for anammox bacteria. Furthermore a description of the main technologies developed and patented by different companies was performed, with focus on the advantages and bottlenecks of them. These technologies are classified in the chapter based on the type of biomass: suspended, granular and biofilm. Also a review is presented for the industrial applications (food industry, agricultural wastes, landfill leachates, electronic industry, etc.), taking into account full scale experiences and laboratory results, as well as microbiology aspects respect to the anammox bacteria genera involved. Finally the possibility to couple nitrogen removal, by anammox, with phosphorus recovery, by struvite precipitation, is also evaluated.



In most cases, the anammox process is used for nitrogen removal from reject water coming from dewatering of digested sludge. However, there are more industrial streams suitable for treatment by partial nitritation/anammox process. The landfill leachate may be a good example of such wastewater. Generally, landfilling is the most used solution for treatment of urban solid wastes. The problem with landfill leachate production and management is one of the most important issues associated with the sanitary landfills. These streams are highly contaminated wastewater with a complex mixture of organic and inorganic compounds and characterized by a high ammonia content and low biodegradable organic fraction matter. The objective of this chapter is the short characteristic of landfill leachate and a short review of its treatment methods with special focus on nitrogen removal by partial nitritation/anammox process.

# Foreword

Sustainable development has become a global theme. Under this theme, industrial, agricultural and municipal sectors should develop so-called sustainable technologies. Anuska and her colleagues have been working towards sustainable wastewater/wastes treatment technologies for many years, including the application of anammox based processes, aerobic granules, nutrient removal and recovery, modeling, etc. So I must say that they are qualified to write this book.

Compared to sustainable technologies, conventional wastewater/waste treatment processes consume energy for aeration for COD oxidation and nitrification, and thereby result in considerable loss of chemical energy by COD oxidation (about 14 MJ/kg COD as metabolic heat) and greenhouse gases (GHG) emissions into the atmosphere. Moreover, COD requirement for denitrification and biological phosphate removal and sludge production are also key problems for conventional processes. Facing increasingly stricter effluent standards, wastewater treatment plants (WWTPs) will cause more energy and organic carbon consumption and thus higher GHG emissions and larger sludge production if conventional processes have to be used.

Today, water problems have to be solved in a more integrated and innovative way and sustainability has become a major concern, as mentioned above. Therefore, developing more sustainable processes is becoming essential. What is pursued in sustainable processes is to minimize COD oxidation and GHG emissions, to recover nutrients (especially phosphate) in wastewater/wastes and organic energy in excess sludge, and finally to reuse effluents.

The book, "Technologies for the Treatment and Recovery of Nutrients from Industrial Wastewater", focuses on industrial nutrient recovery, which is indeed a hot topic in wastewater treatment. It is estimated that the remaining accessible reserves of phosphate rock will run out in 50 years, if the growth of demand for fertilizers remains at 3% per year. Reducing usage will help the reserves last longer, but the biggest gains will probably be made from the recovery of phosphates, both from wastewaters and livestock waste. With regards to wastewater, such recovery would have the additional benefits of minimizing eutrophication and alleviating the scaling of process equipment at wastewater treatment plants. For these reasons, nutrient removal from wastewater has recently become integrated with phosphate recovery.

The book summarized the state of the art of both nitrogen and phosphorous removal & recovery from industrial effluents. Chapter 1 provides a fundament of biogeochemical cycles of nitrogen and phosphorus, which is important to understand the circulation of elements through aquatic, aerial and terrestrial ecosystems. In Chapter 2, eutrofication caused by N and P is elucidated, and related laws and regulations on local and global scales are reviewed as they are key measures to promote technical progresses for controlling eutrofication. Following these fundaments, Chapter 3 introduces all developed key physical-chemical processes for recovering nutrients from wastewater treatment, livestock slurry, and agro-food wastewater. Besides physical-chemical processes, chemical and biological processes play an important role in recovering nutrient as biological nitrification and denitrification and chemical precipitation are the most common processes used today to remove and recover nitrogen and phosphorus, which are emphasized in Chapter 4. Regarding nitrogen removal, nitrification and denitrification as conventional processes, and newly developed processes like anammox are explained in detail in Chapter 5. Other autotrophic processes based on the use of sulphur compounds or methane are described in Chapter 6. As a powerful tool, modelling of related biological nutrient removal in Chapter 7 is described with case studies. Moreover, some conventional and novel technologies like constructed wetlands and aerobic granular sludge are also introduced in Chapters 8 and 9, in which another point of views for nutrient removal are elucidated. Finally, Chapters 10 and 11 demonstrate applications of anammox technologies (ANAMMOX®, CANON®, DEMON®, etc.), which allow the sustainable treatment of industrial effluents like landfill leachates, and can deepen an impression on practical applications of anammox.

Transformation from removal to recovery is a global trend as wastewater and/or wastes are resources to be regained, rather than real wastes. Future technologies of wastewater treatment will definitely focus on sustainable approaches to recover all useful resources. In this respect, as an example, STOWA in the Netherlands launched a new framework for future wastewater treatment, called as "NEWs", i.e. nutrients + energy + water factories. The framework of NEWs emphasizes energy/resource recovery from wastewater. Clearly, this book is a right orientation for this purpose. The book could also serve as a source of information for professionals of the wastewater sector and academics to explain wastewater treatment subjects from an applied point of view, especially useful for process design and, therefore for students, from both degree and doctorate level.

*Xiaodi Hao*

*Beijing University of Civil Engineering and Architecture, China*

# Preface

Wastewater produced from human activities, services and many industrial sectors contains as major contaminants organic matter, nitrogen and phosphorous. The disposal of this type of wastewater into the environment, without adequate treatment, provokes a negative impact into the receiving water. Among the different negative consequences the eutrophication, produced by an excess of nutrients (N and P), is a worldwide problem. At this respect, the Environmental Protection Agency (EPA) of the United States of America recognizes the nutrient pollution in water as one of most widespread, costly and challenging environmental problems of the U.S. Similarly, the European Environment Agency (EEA) has identified the misuse of fertilizers, for agriculture purposes, as a source of the poor water quality of the freshwater bodies in Europe.

The U.S. and the EU are promoting the development of appropriated technologies able to either restore the quality of the water bodies or to treat the produced wastewater to avoid the disposal of N and P in significant amounts to the environment. These actions are expected to mitigate the spread of the eutrophication impact.

Among the types of wastewater, those from the industrial origin are in general the most loaded sources of N and P. Furthermore; the industry sector is interested in closing the water cycle to reuse the wastewater coming from the production process. Both aspects make the evaluation of existing and new treatment systems for N and P removal/recovery necessary.

The operation of treatment systems for nutrients (N and P) removal from wastewater to meet the discharge limits imposed by the authorities is known to be costly, especially regarding energy consumption. For this reason, the development of innovative technologies able to eliminate these pollutants efficiently regarding high effluent quality and low operational costs is of interest.

However, in some cases the best alternative consists of recovering these nutrients present in wastes (like wastewater) for further use, for example, as fertilizers. In this way, the current trend of moving towards a circular economy is accomplished. This alternative is of particular concern for the phosphorous, as its sources in the world are expected to be limited shortly. Up to date the use of organic fertilizers from waste-based sources in Europe has not been regulated yet, making these "green" fertilizers not competitive with the traditional chemical ones. Nevertheless, The Commission published recently (March 2016) the first deliverable of the Circular Economy Package with new rules on organic and waste-based fertilizers in the EU, which opens the door for the nutrients recovery technologies being competitive.

In this world situation, where the environment preservation targets towards the elimination of nutrients pollution from water bodies and their recovery for further use, the present book is proposed as an excellent tool to provide an understandable approach to those technologies already available for nutrients recovery and removal, with particular emphasis on the industrial wastewater. The primary objective of this book focuses on providing the state of the art of the nitrogen and phosphorus removal and recovery from different kinds of industrial effluents from an applied point of view. The impact expected from this publication is to serve as a guide for those interested in the application of innovative solutions, compared to existing ones, to remove and recover nutrients from industrial wastewater. Taking into account its approach this publication is of relevance to the scientific, academic and industrial community since it can be a document of reference providing useful information from the basics of the applied processes to the application of emerging technologies to the wastewater above.

For these reasons, the target audience includes professionals from the industrial and wastewater treatment sectors, and scientists working in the research and development of new technologies for nutrients treatment and recovery. The document could also serve as a source of information for academics to explain wastewater treatment subjects from an applied point of view, especially useful for process design and, therefore for students, from both degree and doctorate level.

The book comprises eleven chapters, including the general aspects related to the nutrients pollution (Chapters 1 and 2), the basis of the removal/recovery processes (Chapters 3, 4, 5, and 6), the concepts of modelling and control to be applied to the wastewater treatment plants (WWTPs) (Chapter 7), and the description of particular treatments and technologies (Chapters 8, 9, 10, and 11). The first chapter explains the circulation of N and P in the environment throughout the biogeochemical cycles and underlines the effects of anthropogenic activities in their alteration. Because of the release of an excess of nutrients coming from industrial processes to the water bodies, provoking significant problems (e.g. eutrophication), to impose laws and regulations able to stop and/or minimize this issue emerges. Chapter 2 describes the efforts of the European Union and the United States of America on the development of regulations. These laws mainly impose the necessity to perform an adequate treatment of the wastewater to meet the discharge limits. Chapter 3 describes the leading technologies able to cope with this treatment by physico-chemical processes and Chapter 4 those based on chemical and biological processes, focused on the particular case of the wastewater generated in the industrial sector. The next two chapters go in depth on the fundamentals of biological nitrogen removal, with a particular emphasis on the autotrophic processes, which are the subject of extensive research at the moment to achieve the energy self-sufficiency of the WWTPs (Chapters 5 and 6). Chapter 7 presents relevant information on the modelling and control of nutrients removal in WWTPs. Because the effluents produced by the different industrial sectors are characterized by the presence of various pollutants and ranges of concentrations, a broad type of technologies, have to be considered for their treatment. Factors like the nitrogen and/or phosphorous levels, the presence of organic matter, the presence of specific toxic pollutants, the temperature, the pH, etc. will determine the best process/technology to be used. A revision of these applicable technologies for industrial wastewater treatment is summarized in the following chapters: constructed wetlands (Chapter 8), aerobic granular sludge (Chapter 9) and anammox process (Chapter 10). Finally, a case study of a particular type of wastewater (landfill leachate) is presented in Chapter 11. A short description of each of the elaborated chapters is provided below.

The Chapter 1 titled "Biogeochemical Cycles of Nitrogen and Phosphorus: Implications of Anthropogenic Activities" intends to describe the natural circulation of the elements (N and P) in the environment, and the way the different human activities can alter it, with special attention to the role of the metals. Therefore, this is an introductory chapter which helps to focus the necessity to take protective actions to avoid the nutrients pollution in the water bodies. The comprehension of the biogeochemical cycles is crucial to know the circulation of an individual element through aquatic, aerial and terrestrial ecosystems. Then the main biological processes for nitrogen removal involved in the WWTPs are concisely defined.

This chapter underlines the relationship between the main anthropogenic implications and the input of exogenous compounds altering the natural cycles. In this way, the presence of N and P compounds generated by different industrial processes together with the addition of other inorganic chemicals, like metals, are affecting the biodiversity of the ecosystems, by disrupting the naturally closed balance of the biogeochemical cycles.

To mitigate these negative impacts is the responsibility of the different governments to develop regulations on this subject. In this line Chapter 2, titled "Nutrients Pollution in Water Bodies: Related Legislation in Europe and the United States of America," deals with the problematic of N and P presence in excess in natural environments and presents the different produced normative to reduce and/or prevent their effects on the environment. The countries promulgate these documents, sometimes based on the results from international conventions and programmes. In this chapter, a review of the laws and regulations that affect the discharge of nitrogen and phosphorus, in Europe and the United States of America, are addressed. Finally, a brief explanation about several international initiatives is provided to understand the global framework concerning nutrients pollution. Frequently the limits imposed for nitrogen and phosphorus discharge to water bodies are based on environmental criteria. However in the case of industrial activities certain "exceptions" to these limits are considered. Despite the different legal frameworks around the world set the limits for nutrients discharges into water bodies, the basis for the requirements imposed on industrial activities are similar either in the European Union or the United States of America. In both cases, the limits that the industries must fulfill concerning pollutants discharges are fixed by the efficiency of the technologies available to reduce them. Furthermore, a distinction is performed between the direct discharge to natural water bodies and the discharge to sewer systems. In general, the recommended techniques for nitrogen removal are based on the biological nitrification/denitrification process. For phosphorus removal, the most selected one is usually the chemical precipitation. Nowadays, the action of different social partners (scientific community, NGOs, environmental programs, etc.) is increasing the pressure for the establishment of nutrients limits criteria more protective for the environment.

Having in mind the importance of the applied technologies, in the next chapters of the book a review is provided of the main features of the existing and under development ones.

To begin the Chapter 3, titled "Systems Based on Physical-Chemical Processes: Nutrient Recovery for Cycle Closure," describes the main used physical-chemical processes. It goes in depth in their objectives, theoretical fundamentals, environmental effects (air, water and soil emissions, resource depletion), technical indicators (efficiencies, energy consumption, etc.), and by-product characteristics. The chapter summarizes the application of these technologies to the recovery of nutrients from high strength waste streams, like the rejected water from WWTPs, livestock slurry, and agro-food wastewater. These effluents contain high organic matter and nutrient concentrations, which favor the utilization of processes aiming at the recovery of energy and nutrients, instead of only removing them. Among others, stripping coupled to absorption, struvite precipitation, membrane separation, and vacuum evaporation, are all physical-chemical processes aiming at concentrating nutrients in a stream, for further valorization as fertilizer. In this way the valorization of organic wastes (e.g. livestock manure and other industrial wastewaters) to recover renewable energy or to obtain added value products (e.g. nutrient fertilizers) is possible, reducing the environmental impacts of their management. Furthermore, in this way the increase of scarcity of raw materials and energy sources, which is currently of concern in the world, diminishes. Having all this in mind in this chapter, the leading process technologies already in operation, as well as the new ones under development, have been reviewed. Their degree of readiness as well as the hindering issues, which should be overcome before they can be worldwide spread, is shown. Among them, the operational costs, the quality of the recovered products and the need to open the market for this kind of products are the main limiting factors. Nevertheless, these recovering technologies present the proper potential for their implementation in a near future.

Although physical-chemical processes are undoubtedly useful, the biological processes have also their importance. The Chapter 4 titled "Chemical and Biological Processes for Nutrients Removal and Recovery" provides an overview on the main chemical and biologically based technologies for nutrients treatment from industrial wastewater. Here these technologies are presented focused on practical principles like applicability and operational parameters of full scale implementations. The fundamentals of these reviewed processes are also described. A comparative analysis based on energy consumption between nutrients removal and recovery is also provided. Despite there is a significant amount of technologies suitable for nutrients removal from wastewater, this chapter points at the combined biological nitrification and denitrification and the chemical precipitation as the most common processes used today to remove nitrogen and phosphorus, respectively. In the case of phosphorus, its recovery should always be pursued to recycle a critical raw material with limited resources. Although struvite recovery is a robust technology for P recovery, other possible technologies, including enhanced biological P removal or novel ion exchange resins should be promoted especially for low concentrated streams. On the other hand, nitrogen recovery is not a concern, but it should be evaluated to identify those cases which are economical and environmentally sustainable in comparison with its removal. Furthermore, restrictions from the legislation have to be taken into account when processes as ammonia stripping and recovery as ammonium sulphate are applied, even if this is the most traditional and reliable system. In those cases where the biological processes can be implemented, the nitrogen deammonification and its recovery via the Haber-Bosh process are energetically favorable. Considering only biological processes, the nitrogen removal via-nitrite pathway represents an advance over conventional activated sludge systems and enables the optimization of denitritation process with the concurrent anoxic phosphorus uptake.

Due to the relevance of the biological processes to understand the key features to guarantee their optimum performance is necessary. For this the Chapter 5, titled "Fundamentals of the Biological Processes for Nitrogen Removal," describes the kinetic, microbiological and biochemical aspects of the three most applied biological nitrogen removal processes: nitrification, denitrification and anammox. The key operational parameters affecting those processes are also identified here. This information is relevant to define the values of these parameters allowing for the optimal performance of the processes and to use them during the design of nitrogen removal systems in wastewater treatment plants. This information is also essential for predicting and controlling the operation of these treatment processes. Although these processes are implemented at full-scale, several challenges remain unsolved, such as the development of novel treatment concepts with the capacity to mitigate the emission of greenhouse gases (methane and nitrous oxide) and to recover nitrogenous compounds as valued by-products from industrial waste streams.

In particular, those biological processes for nitrogen removal based on autotrophic organisms are under extensive study as they allow saving the organic matter contained in the wastewater for further valorization by methane production. Chapter 6, titled "Autotrophic Denitrification Processes," intends to cover this concept. This chapter presents a review of autotrophic biological processes to reduce nitrogen by the use of sulphur compounds, hydrogen or methane. These processes perform satisfactorily for the treatment of low C/N ratio wastewater. Although not all of them are in the same application stage, in this chapter, their description and evaluation of their application from an economical point of view has been elaborated. From this analysis, the combination of the conventional nitrification with the autotrophic denitrification, using methane or hydrogen, arises as the most efficient and cost-effective alternatives. The large cost of these options is mainly due to the electricity consumption associated to the aeration to oxidize ammonium to nitrite or nitrate (nitrification) needed for the posterior autotrophic denitrification.

The comprehension of the fundamentals corresponding to the studied processes enables the establishment of the optimal conditions to be imposed during the operation of the technologies for nutrients removal. The use of models helps the selection of this conditions and the development of control systems. In this line the Chapter 7, titled "Modelling and Control of Nitrogen and Phosphorus Removing Systems," examines the existing models (based on the ASM) developed for nutrient (N and P) removal in biological treatment systems. Not only models including biological processes but also physicochemical ones have been discussed. Particular attention has been paid to those elaborated to fulfil the needs created by some emerging or novel technologies and to comply with the objectives to be evaluated for each one. Among them, the level of GHG production or the possibilities for phosphorous recovery are considered, which increase the models complexity. Models can also help with the control design of these processes providing information about process dynamics and potential loop interactions. This information constitutes a useful tool to be able to design, manage, operate, optimize and benchmark these novel technologies with potentially existing standard technologies. Therefore detailed modelling of nutrient removal becomes essential.

A wide range of technologies is available for nutrients removal. As an example in Chapter 8, titled "Constructed Wetlands for Industrial Wastewater Treatment and Removal of Nutrients," the evaluation of the operation and expected the performance of the main types of Constructed wetlands (CWs) is provided. Several examples of application to treat industrial wastewater containing suspended solids, organic matter and nutrient removal are described. The CWs are low-cost and sustainable engineered systems, which use natural processes involving wetland vegetation, soils, and their associated microbial assemblages to assist in treating wastewater. In CWs pollutants are removed by physical, chemical and microbial processes (i.e. nitrification, denitrification, sulphate reduction, carbon metabolization, etc.). Traditionally, they have been applied for municipal and domestic wastewater treatment. However, in the last two decades, this technology evolved and spread its application to treat industrial wastewater. In this chapter case studies focused on the performance of several pilot or full scale CWs in operation in industrial locations has been discussed.

Nowadays, the specific needs of the industry are leading the development of new technologies. In this sense the Chapter 9, titled "Aerobic Granular Sludge: Treatment of Wastewaters Containing Toxic Compounds," reviews the application of technologies based on aerobic granular sludge (AGS) for the treatment of industrial wastewaters containing very specific toxic compounds. Among others, the industry faces the challenges of the reduction of energy consumption and space required for implantation, which are features of the AGS based systems. Furthermore, they can cope simultaneously with the wastewater treatment and value added products for recovery. In this way not only the sustainability of the process is promoted, but also its economic value increases, thus benefiting its implementation in industrial facilities. Up to date, existing developments at full scale have limited application for the treatment of industrial effluents. In this chapter the state of the art of the different studies regarding the applicability of aerobic granular sludge to remove nitrogen and phosphorus simultaneously from wastewater containing toxic compounds is presented. For these particular cases, AGS based technologies are presented as a promising alternative, which is gaining interest in the last years. Different applications for the removal of toxic compounds through biosorption and/or biodegradation are presented, illustrating the versatility of the technology. The effects of anthropogenic substances on the system performance and the bacterial populations established within AGS are also addressed. The combination of contaminants removal to allow water discharge and simultaneous recovery of valuable products are presented in the final remarks.

The discovery of new biological processes for nitrogen is expected to transform the wastewater treatment plants. The Chapter 10, titled "Anammox Process: Technologies and Application to Industrial Effluents," provides an overview of the existing configurations to apply the anammox process in combination to the previous partial nitritation, necessary to obtain the adequate mixture of substrates for the anammox bacteria. At the moment the application of the anammox process is considered essential to achieve the energy self-sufficiency in sewage treatment plants. Although its implementation at industrial scale is also feasible, its application for the treatment of industrial wastewater is not so common. The low number of industrial applications can be attributed to the lack of a compendium of successful experiences. In this chapter a summarized description of the leading technologies developed and patented by different companies is presented, including their advantages and bottlenecks, if any. A classification has been arranged based on the type of biomass grown in the system: suspended, granular and biofilm. Furthermore, full scale experiences for the treatment of industrial wastewater (food industry, agricultural wastes, landfill leachates, electronic industry, etc.) are summarized. Finally, new approaches where the nitrogen removal, by anammox, is coupled to phosphorus recovery, by struvite precipitation, are also evaluated. From the content of the chapter it is clear that despite the anammox process was discovered only around 25 years ago, its application at full scale is nowadays a success as reflects the significant number of different technologies developed by the companies in the water sector.

Finally, the Chapter 11, titled "Treatment of Landfill Leachate by the Anammox Process," describes a case of study where the anammox process is used for the treatment of landfill leachate. The production and further management of this kind of wastewater are one of the most important issues associated with the sanitary landfills. Although the EU has restricted the disposal of biodegradable compounds to landfills, existing ones continue to produce polluted effluents. These streams are characterized by their high pollution content and complex composition due to the presence of a mixture of organic and inorganic compounds, high ammonia content and low biodegradable organic fraction matter. So far, there has not been developed a standard treatment method for this type of wastewater. Therefore, the objective of this chapter is to define the characteristic composition of landfill leachates and it possibilities of treatment. In this chapter, a review of the different processes applied to this type of effluents is performed including chemical, physical and biological processes described in the previous chapters. Special focus has been put on the nitrogen removal by the combined partial nitritation/anammox process. Due to the complex composition of landfill leachate and the presence of inhibitory substances like heavy metals, organic carbon or salts, there is a tendency to use combined treatment methods with some pretreatment, especially for organic carbon removal, and also with some post-treatment methods like constructed wetlands or with AOP-based technologies. The use of partial nitritation in combination with the anammox process gives a cost effective alternative for biological removal of nitrogen from landfill leachate.

Summarizing the collection of chapters presented in this book represents a useful contribution to the field of nutrients pollution and treatment from wastewater of industrial origin. The contents go from the fundamentals of the studied to the case studies where results from full scale experiences are provided, both useful either for academics or professionals of the sector. Environmental implications and regulations are also presented to contextualize the problem of anthropogenic accumulation of nutrients. The differently described technologies can help to understand, with a complete perspective, the real scenario for their application to the problems of industrial effluents treatment. The information provided by this book was elaborated to present the evaluated solutions considering their economic viability and compromise with the environment and the sustainable development and by the concept of "circular economy". Furthermore, this book comprises relevant information regarding industrial wastewater and nutrient removal and recovery which is not available at the moment in existing publications. Finally, these contents are expected to be useful for a broad target audience.

# Acknowledgment

The editors would like to acknowledge the help of all the people involved in this project and, more specifically, to the authors, reviewers and editorial advisory board that took part in the review process. Without their support, this book would not have become a reality.

First, the editors would like to thank each one of the authors for their contributions which significantly improved the quality of the book. Our sincere gratitude goes to the authors of the chapters who contributed their time and expertise to this book.

Second, the editors wish to acknowledge the valuable contributions of the reviewers regarding the improvement of quality, coherence, and content presentation of chapters. Most of the authors also served as referees; we highly appreciate their double task. The contribution of the editorial advisory board constituted an important help.

The editors want to express also their gratitude to Prof. Dr. Xiaodi Hao for his commitment in the elaboration of the preface of this book.

To finish, the editors want to acknowledge the Cost Action Water2020 for providing the contact network to identify the authors of some of the elaborated chapters.

*Ángeles Val del Río*
*Universidade de Santiago de Compostela, Spain*

*José Luis Campos Gómez*
*University Adolfo Ibáñez, Chile*

*Anuska Mosquera-Corral*
*Universidade de Santiago de Compostela, Spain*

# Chapter 1
# Biogeochemical Cycles of Nitrogen and Phosphorus:
## Implications of Anthropogenic Activities

**Andrea Fra-Vázquez**
*University of Santiago de Compostela, Spain*

**Daniel Valenzuela-Heredia**
*University Adolfo Ibáñez, Chile*

## ABSTRACT

*The study of biogeochemical cycles is important to understand the circulation of elements through aquatic, aerial and terrestrial ecosystems. Among others, N and P are considered as the limiting nutrients that determine the productivity of organisms. Since human activity is increasing its influence over these cycles at a global scale, it is important to analyze the implications of anthropogenic variations in order to detect the sources and try to prevent or reduce their impact. Moreover, not only the increasing or diminished concentration of nitrogen and phosphorus in nature by human resource exploitations is needed to account today. As metals play a crucial role in the dynamics of these essential elements, their presence in the environment also requires a significant attention.*

## INTRODUCTION

Many chemical elements such as carbon (C), nitrogen (N), phosphorus (P) and sulphur (S) are essential to form the fundamental molecules which organisms depend on. The circulation of these elements at local and global scale, known as biogeochemical cycles, has been widely studied. This concept includes the interactions through both biotic ("bio-") and abiotic ("geo-") parts of an ecosystem of different chemical elements and compounds ("chemical-") along with their fluxes ("cycles"). The abiotic factors comprise water (hydrosphere), land (lithosphere) and air (atmosphere), whereas the biotic part is referred as the biosphere. Elements and compounds can be held for extended periods of time in certain places (reservoirs) or transported (fluxes). Biogeochemical cycles of different elements are closely linked, creating interactions that maintain order in the ecosystem with fluxes from highly reduced to highly oxidized forms

DOI: 10.4018/978-1-5225-1037-6.ch001

and back (Madsen, 2011). However, as human population continues increasing, its impact on Earth's environment becomes more and more significant. Many of these anthropogenic actions may change the global cycles of different chemical elements (Butcher, Charlson, Oriand, & Wolfe, 1992). Thus, the knowledge of these cycles is essential to understand and predict the behavior of certain compounds that may be hazardous to human health (Pepper, Gerba, & Gentry, 2015).

The addition of limiting nutrients may dramatically change the dominant species in both aquatic and terrestrial ecosystems, which can originate a decrease in the biological diversity. The eutrophication is one of the best-documented consequences of anthropogenic alterations on nitrogen and phosphorus cycles in aquatic environments (Vitousek et al., 1997). It consists in the increase of algae growth on the surface of the water and the development of anoxic environments, also known as "algal blooms". The availability of oxygen in depth decreases due to the high algae growth on the surface of the water, which leads to the death of fishes and changes the composition of the biota. The main organisms involved in this "blooms" are cyanobacteria (*Anabaena*, *Aphanizomenon*, *Occillatoria*) and algae (*Asterionella*, *Dinophysis*) (Smil, 2000). Therefore, the main consequences of eutrophication are changes in animal species composition and oxygen fluxes, along with the increase of sulfide concentrations by sulfate reduction under anoxic conditions.

Even though there are other important biogeochemical cycles, this chapter is focused in two main nutrients: N and P. They are essential for organisms to live because they are considered as the nutrients that most often limit the productivity and functioning of terrestrial and aquatic ecosystems (Wang, Houlton, & Field, 2007). The transformations of reactive species throughout the cycles in different ecosystems will be analyzed for both chemical elements. Moreover, human activity may provoke an unbalance in the biogeochemical cycles by the accumulation of chemical at toxic concentrations or the scarcity of essential compounds for organisms. Therefore, the implications of anthropogenic activities that break the equilibrium in those biogeochemical cycles will be discussed, with special attention to the role of the metals.

Biological reactions involved in nitrogen and phosphorus cycles require metals for the activity of specific enzymes that attend reactions across which different oxidation states of nitrogen and phosphorus help to cell-sustaining. The cell has a fine-tuning mechanism to handle the uptake and storing metals (e.g. Fe, Cu, Zn, Ag, and Cd). Many human activities have a significant impact on the release of metals to the environments, causing that the metal concentrations can be lethal for organisms, due that the mechanism to homeostasis is not sufficient to maintain the cellular stability. In this section, we review metals involved in biological reactions in the N and P cycles and the concentrations that can delay the biogeochemical move of these important elements for ecosystems functions. A microbiological and proteomic point view is used to illustrate the enzymes involved in each step of biological transformation of nitrogen and phosphorous compounds.

## NITROGEN

### Nitrogen Cycle

Nitrogen is present in a wide variety of chemical forms: organic compounds (urea, amines, proteins, nucleic acids), reduced inorganic forms ($NH_4^+$ and $NH_3$), and oxidized inorganic forms ($N_2O$, NO, $NO_2^-$, $NO_3^-$). The chemical form of nitrogen present in a certain ecosystem depends on the reservoir. The principal form of nitrogen in terrestrial systems and oceans is organic nitrogen, nitrate ($NO_3^-$) and ammonium ($NH_4^+$), whereas in the atmosphere it mainly occurs as nitrogen gas ($N_2$) (Butcher, et al., 1992). Nitrogen forms are continuously transported from one reservoir to another and transformed by different processes (Schlesinger & Bernhardt, 2013). The gases generated in land as a result of microbial activity or anthropogenic processes (biomass and fossil fuel combustion) are transferred to the atmosphere. Nitrogen can also be emitted from the ocean through volatilization of $NH_3$, $N_2$ and $N_2O$. Wet deposition (of inorganic and organic forms by rain or snow) and dry deposition (gases like $HNO_3$ and $NH_3$) occur from the atmosphere to land and ocean surfaces. Finally, both organic and inorganic nitrogen compounds are discharged from continents to coastal by rivers and groundwater.

The nitrogen cycle is divided into different processes: nitrogen fixation, assimilation, mineralization, nitrification, denitrification and anammox (Figure 1). These transformations may be chemical but they are mainly controlled by microorganisms, such as bacteria, archaea and fungi (Bernhard, 2010).

## Nitrogen Fixation

Even though the composition of the atmosphere is 78% nitrogen, biological systems are N-limited because most of the organisms are not able to use $N_2$ (Galloway, Schlesinger, Levy, Michaels, & Schnoor, 1995). The biological nitrogen fixation (BNF) is the process in which the conversion of stable $N_2$ to $NH_3$ or $NH_4^+$ is carried out by microorganisms (Equation 1). Before the human activity, BNF was the most important process converting $N_2$ into reactive nitrogen.

$$N_2 + 8\,H^+ + 8\,e^- \rightarrow 2\,NH_3 + H_2 \qquad (1)$$

There is a great physiological and phylogenetic diversity among the nitrogen-fixing organisms involved in BNF (Bernhard, 2010). They can be free-living (*Cyanobacteria*) or symbiotic (*Rhizobium*); aerobic (*Pseudomonas*) or anaerobic (*Clostridium*); phototrophic (*Cyanobacteria*) or chemotrophic (*Azotobacter*). A good example of a symbiotic association in terrestrial ecosystems is the nitrogen-fixing bacteria from genus *Rhizobium* found on the roots of many leguminous plants (e.g. peas, clover, soybeans). In this case, the root exudates from the plant serve as a signal to attract the bacteria. Then, bacteria initiate the nitrogen fixation in nodules formed on the roots. The plant supplies organic nutrients to microorganisms, and they produce nitrogen in a usable form for the plant.

Despite the great diversity, all nitrogen-fixing organisms have in common the presence of an enzymatic complex called nitrogenase (or dinitrogenase), which catalyzes the reduction of $N_2$ to $NH_3$. This enzyme is extremely oxygen sensitive, and aerobic nitrogen-fixing bacteria have developed different protective mechanisms (Stein & Klotz, 2016), such as the spatial decoupling. For instance, in aquatic systems, *Cyanobacteria* have heterocysts, which are structures that protect the enzyme by creating a low-oxygen environment.

Moreover, some nitrogen can be fixed abiotically by lightning or certain industrial processes like fossil fuel combustion and the Haber-Bosch process (Schlesinger & Bernhardt, 2013). The lighting is a process that occurs in nature and that consists in the conversion of $N_2$ and molecular oxygen into NO by electrical discharges at high temperature in the atmosphere. Then, NO is oxidized to $NO_2$ and $HNO_3$, which are introduced in the ecosystem by wet and dry deposition. The anthropogenic processes will be discussed later.

## Assimilation/Mineralization

Once nitrogen has been fixed in the soil or aquatic systems, organisms may take up nitrogen compounds by absorption through their roots. They directly incorporate $NH_3$ or $NH_4^+$ as part of their biomass in the form of organic nitrogen compounds in the N-assimilation process. When plants and animals die, the general decomposition of organic material takes place through the mineralization. This process consists in the conversion of organic nitrogen into $NH_4^+$ by heterotrophic microorganisms, which use organic matter as an energy source.

## Nitrification

Nitrification is a two-step process to convert ammonia to nitrite and then to nitrate. Nitrifying microorganisms are chemolithoautotrophs that fix carbon dioxide to produce organic carbon and obtain energy through the oxidation of reduced inorganic nitrogen compounds ($NH_3$ or $NH_4^+$ to $NO_2^-$ or $NO_3^-$). However the energy generated from the oxidation is very small and, as a result, their growth yields are very low. This process occurs under aerobic conditions in soil and aquatic environments and two microbial groups are involved. The oxidation of ammonia to nitrite (Equation 2) is mainly catalyzed by ammonia oxidizing bacteria (AOB), such as genera *Nitrosomonas, Nitrosococcus* and *Nitrosospira*, but also by ammonia oxidizing archaea (AOA). Nitrite oxidizing bacteria (NOB), like *Nitrospira* and *Nitrobacter*, oxidize nitrite to nitrate (Equation 3).

$$NH_4^+ + 3/2\ O_2 \rightarrow NO_2^- + H_2O + 2H^+ \quad (2)$$

$$NO_2^- + 1/2\ O_2 \rightarrow NO_3^- \quad (3)$$

Even though both groups are phylogenetically separated, there are interactions among both microbial groups and frequently observed together. However, a new microorganism from genus *Nitrospira* able to carry out a complete nitrification has been recently discovered, and it is known as "comammox" (complete ammonia oxidizer) (Daims et al., 2015).

## Dissimilatory Nitrate Reduction

Once $NO_3^-$ is released, it can be reduced to $NH_4^+$ by dissimilatory nitrate reduction to ammonium (DNRA) and then incorporated into biomass as organic nitrogen compound in the nitrogen assimilation process (Equation 4). It is an important process for many plants and organisms which occurs in the absence of oxygen.

$$NO_3^- \rightarrow NO_2^- \rightarrow NH_4^+ \quad (4)$$

## Denitrification

The denitrification process consists in the conversion of nitrate to produce nitrogen gas and returning it to the atmosphere (Equation 5). This process occurs under anaerobic conditions and microorganisms use nitrate as a final electron acceptor instead of oxygen. It occurs in certain soils and sediments, or anoxic zones in lakes and oceans (Bernhard, 2010). It is mainly performed by denitrifying bacteria, such as genera *Bacillus*, *Pseudomonas* and *Paracoccus*, which are heterotrophic and need external organic carbon. They can directly couple all the reactions and obtain $N_2$ as the final product. However, many bacteria and archaea are not able to encode all the enzymes and the incomplete pathway may lead to release intermediates, like nitric oxide (NO) and nitrous oxide ($N_2O$), which is considered a greenhouse gas (Stein & Klotz, 2016).

$$NO_3^- \rightarrow NO_2^- \rightarrow NO \rightarrow N_2O \rightarrow N_2 \quad (5)$$

Denitrification is an important process because it allows the removal of nitrate from the ecosystem in order to return nitrogen to the atmosphere as nitrogen gas. In this way, it is a very important process commonly used for nitrogen removal in wastewater treatments.

## Anammox

An alternative biological route has been recently discovered, representing an important process in the global nitrogen biogeochemical cycle. This process, known as anaerobic ammonium oxidation (anammox), consists in the $NH_4^+$ oxidation coupled to $NO_2^-$ reduction to form $N_2$ gas and takes place under anoxic environments (Equation 6) (Strous et al., 2006). It is carried out by bacteria from phylum *Planctomycetes*, like *Brocadia anammoxidans*, which show a specialized organelle called anammoxosome (Stein & Klotz, 2016).

$$NH_4^+ + NO_2^- \rightarrow N_2 + 2\,H_2O \quad (6)$$

The main advantage of this process is that removes simultaneously nitrite and ammonia without contributing to greenhouse gas emissions because $N_2O$ is not produced.

## Anthropogenic Perturbations

The complexity of the nitrogen cycle makes it very sensitive to disturbances (Wollast, Mackenzie, & Chou, 1993). Without anthropogenic perturbations, the amount of N fixed is approximately balanced by the amount of N converted back to $N_2$. In the absence of human influences, BNF and lightning were the only sources of new nitrogen forms in the environment (Fowler et al., 2013). However, humans have developed new processes, like Haber-Bosch process and fossil fuel and biomass combustion that change the fluxes and pathways of the global nitrogen cycle (Galloway, et al., 1995). The global nitrogen cycle has changed to be dominated by anthropogenic processes.

As the global human population continues growing, nitrogen supply is essential to provide enough food. It is known that around 75% of the nitrogen created by anthropogenic processes is added to agro-systems to sustain food production (Schlesinger & Bernhardt, 2013). Therefore, an increase in the nitrogen supply will be traduced in an increase of the yield productivity of crops. Most of the nitrogen contained in fertilizers is fixed industrially by the Haber-Bosch process, which uses high pressure and temperature with metallic catalysis to produce $NH_3$ from $N_2$ and $H_2$. This process became the dominant source of fertilizer nitrogen since 1930 (Schlesinger & Bernhardt, 2013).

The combustion of biomass and fossil fuels creates energy but also waste gases are generated, like NO by the reaction of $N_2$ and $O_2$. In this way, nitrogen fixed by plants (fossil nitrogen) becomes reactive again when fossil fuels are extracted and burned (Schlesinger & Bernhardt, 2013). The end product of NO oxidation is nitric acid ($HNO_3$), the main component of the acid rain.

The $N_2O$ emissions are mostly produced by natural biological processes like nitrification-denitrification reactions in terrestrial and aquatic systems. However, this GHG has increased its concentration in the atmosphere since pre-industrial times, and it is considered as one of the causes of the climate change. This increase is due to anthropogenic emissions from the use of fossil fuels as an energy source and the application of nitrogenous fertilizers in agricultural activities. It is chemically inert in the troposphere but, it can catalyze the stratospheric ozone depletion, which implies an important impact on human health (Wollast, et al., 1993).

The addition of fertilizers increases the amount of nitrogen available for loss. $NH_4^+$ can be converted by nitrifying bacteria to $NO_3^-$, which can reach aquatic systems such as rivers, lakes and aquifers, and provoke environmental problems (Canfield, Glazer, & Falkowski, 2010). Nitrate excess may contribute to groundwater pollution, which can lead to a human health risk by contamination of drinking water (Schlesinger & Bernhardt, 2013). Moreover, elevated amounts of nitrate can cause environmental problems like eutrophication. This phenomenon has consequences for N cycle (Howarth et al., 2011). As a result of lower oxygen availability, nitrification activity is reduced, the nitrate supply is decreased and, consequently, denitrification also slows. Moreover, the high sulphide levels favor the dissimilatory nitrate reduction to ammonium (DNRA), which competes with denitrification for nitrate. Finally, the anaerobic ammonium oxidation process is also stopped by the high sulphide concentration. All of these consequences result in more nitrogen in the aquatic system and less converted to $N_2$.

Despite the benefits that supplies in the nitrogen cycle provide in food production, there are important consequences due to the human modification of the global nitrogen cycle, like impact on human health, biodiversity loss and climate change (Fowler, et al., 2013).

The levels of inorganic nutrients in terrestrial and aquatic environments, specifically trace metals, can be strongly augmented by anthropogenic derived activities, such as land-use modification, wetland degradation and the excessive use of pesticides and mining. All of these activities contribute to unusually high concentrations of metals in the environment in relation to the normal background levels (Prasad, 2001). For example, soils in North Rhine-Westphalia, an area with the highest population density in Germany, were contaminated with Cd, Cr, Cu, Ni, Pb, and Zn (Kim et al., 2015). In marine ecosystems, trace metals are in the order of nanomolar and picomolar concentrations, a well-documented environmental impaired is the mining dumping deposited on the coast of Chañaral, a mining city of Chile (Correa et al., 1999).

Metals are essential in the cycles since they act as cofactors or redox intermediaries (Cavet, Borrelly, & Robinson, 2003). The great redox potential of metals to obtain energy or nutrients in biological metabolism involves the necessity of accounting with specific mechanisms to deal with their potential toxicity. This causes a detriment in the velocity of biogeochemical reactions that leads to the accumulation

*Table 1. Oxidation states of molecules involved in nitrogen cycle*

| Molecule | Name | Oxidation State of N |
|---|---|---|
| $NH_3$ | Ammonia | -3 |
| $N_2H_4$ | Hydrazine | -2 |
| $NH_2OH$ | Hydroxylamine | -1 |
| $N_2$ | Nitrogen | 0 |
| $N_2O$ | Nitrous oxide | +1 |
| NO | Nitric oxide | +2 |
| $NO_2^-$ | Nitrite | +3 |
| $NO_3^-$ | Nitrate | +5 |

of chemicals at concentrations that are toxic and also the shortage of some substances that are needed due to the physiological requirements of organisms (Gleeson, Clipson, Melville, Gadd, & McDermott, 2005). The implications of the presence of these metals in the nitrogen cycle will be discussed in the next paragraphs. An enzymatic analysis is used to understand the direct influence of metals over biological reactions that involve changes in the redox status of nitrogen species (Table 1).

## Roles of Metals in Nitrogen Cycle

The $N_2$ reduction (Equation 1) involves the activity of nitrogenase enzyme that catalyzes the three successive two-electron transfer reactions: from nitrogen to diimine, then to hydrazine and finally to two molecules of ammonia. Its activity requires $Mg^{2+}$. In bacterial genera *Anabaena, Azospirillum, Azotobacter, Chlorobium, Chromatium, Desulfovibrio, Frankia, Gloeothec, Rhizobium* and *Rhodopseudomonas*, the enzyme is composed of 2 metalloproteins: component I is MoFe protein and component II is Fe protein (Haaker & Klugkist, 1987). In some cases, the molybdenum may be replaced by vanadium or iron, like in *Azotobacter chroococcum* (Fay et al., 2010) or in *Rhodobacter capsulatus* (Petersen, Fisher, & Lowe, 2008; Petersen, Mitchell, Fisher, & Lowe, 2008; Siemann, Schneider, Drottboom, & Muller, 2002).

Ammonia oxidation is the first and rate-limiting step of chemoautotrophic nitrification (Beman, Roberts, Wegley, Rohwer, & Francis, 2007), catalyzed by ammonia monooxygenase (*amo*) enzyme. A copper center is present in the catalytic domain of the enzyme, and iron and zinc are also necessary. $Cu^{+2}$ stimulates the ammonia-dependent $O_2$ consumption and the hydrazine-dependent ethane oxidation (Ensign, Hyman, & Arp, 1993). Moreover, iron is a catalytic component of *amo* enzyme in *Nitrosomonas europaea*, possibly as part of the oxygen activating center (Zahn, Arciero, Hooper, & DiSpirito, 1996).

In denitrification process, oxidizing forms of nitrogen are used by anaerobic bacteria as alternative electron acceptors during respiration. This process involves a group of enzymes, all of them requiring metals as cofactors: nitrate reductase, nitrite reductase, nitric oxide reductase and nitrous oxide reductase. The nitrate reductase enzyme is activated by $Ca^{+2}$ and $Mg^{+2}$, but concentrations of 5 mM cause an inhibition higher than 50% of activity in *Cucumis sativus* and *Cucurbita maxima*. $Mo^{6+}$ and $Ni^{+2}$ are necessary for nitrate reduction, but an excess of $Ni^{+2}$ causes inhibition of denitrification in *Aspergillus niger*. Additionally, a high sensitivity of the nitrate reductase enzyme to other metals was observed, where micromolar concentrations of $Cd^{+2}$, $Pb^{+2}$, $Zn^{+2}$, and $Cu^{+2}$ caused inhibition (Aiken, Peyton, Apel, & Petersen, 2003). In the case of $Mn^{+2}$, Cr and Co, the inhibition is caused by millimolar concentrations

(Oji et al., 1985; Panda & Choudhury, 2005). The reduction of nitrite to nitric oxide is catalyzed by two proteins that are negatively affected by high concentrations of metals involved in their catalytic activity. The first type is a copper-nitrite reductase enzyme that catalyzes the electron transfer. The second type is involved in the conventional denitrification and it is a heme protein, a dimeric enzyme with two heme groups per monomer. In *Neurospora crassa*, the nitrite reductase enzyme has iron atoms, but also Cu, Mg and Zn are necessary for the enzymatic activity (Prodouz & Garrett, 1981). Similar requirement are exhibit by *Paracoccus denitrificans*, *Achromobacter xylosoxidans* (Lam & Nicholas, 1969), *Alcaligenes faecalis*, *Fusarium oxysporum*, *Haloarcula marismortui*, *Haloferax denitrificans*, *Hyphomicrobium denitrificans, Pseudomonas chlororaphis*, *Pseudoalteromonas haloplanktis* and *Rhodobacter sphaeroides* (Aiken, et al., 2003; Bueno, Gomez-Hernandez, Girard, Bedmar, & Delgado, 2005; Deligeer et al., 2002; Yousafzai & Eady, 2002).

Another pathway in which nitrite is reduced is through dissimilatory nitrite reduction to ammonia (DNRA). The enzyme involved in this reaction is a ferricytochrome oxidoreductase. In many bacteria, the flux of electrons necessary for nitrite reduction to ammonia is catalyzed by a multiheme cytochrome (Almeida et al., 2007; Lukat et al., 2008; Tikhonova et al., 2006). $Ca^{2+}$ is necessary for the activity of ammonia-forming nitrite reductase (Bykov & Neese, 2012; Cunha et al., 2003). Copper and Zinc have noxious effects on the reduction of nitrite to ammonia (Kajie & Anraku, 1986). Assimilation of nitrite in algae and higher plants is mediated by a ferredoxin-nitrite reductase. Another assimilatory nitrite pathway is catalyzed by NADH-nitrite reductase enzyme or by NADPH-nitrite reductase. In the first case, an iron-sulfur flavoprotein (FAD) containing a siroheme group is the responsible of mediates the 6-electron reduction of nitrite to ammonia. The NADPH-dependent enzyme acts as a detoxifying mechanism for the nitrite generated by nitrate respiration (Tyson, Busby, & Cole, 1997).

The reduction of nitric oxide to oxide nitrous is catalyzed by a nitric oxide reductase, which is a heme-thiolate protein. It is present, for example, in *Aspergillus oryzae, Fusarium lichenicola, Fusarium oxysporum, Trichosporon cutaneum* and *Histoplasma capsulatum* (Watsuji, Takaya, Nakamura, & Shoun, 2003; Daiber, Shoun, & Ullrich, 2005). The nitric oxide reductase in *Thermus thermophilus* and *Bacillus azotoformans* is a heme-copper oxidase (Blomberg, Blomberg, & Siegbahn, 2006; Suharti, Heering, & de Vries, 2004). In *Pyrobaculum aerophilum*, the nitric oxide reductase contains two heme iron and one non-heme iron (de Vries, Strampraad, Lu, Moenne-Loccoz, & Schroder, 2003). Another type of nitric oxide reductase uses the menaquinol pathway for electrons transference. This fact is considered as a defense mechanism in *Bacillus azotoformans* against the toxicity of NO, although it is possible to be used as serving the bioenergetics needs of the organism. The nitric oxide reductase in this organism contains two copper atoms and two heme centers.

The reduction of $N_2O$ by nitrous-oxide reductase (*nor*) enzyme plays a critical environmental role in preventing the release of this greenhouse gas into the atmosphere. Nitrous-oxide reductase is a cupredoxin that contains copper centers (Savelieff et al., 2008; Snyder & Hollocher, 1987). *Paracoccus pantotrophus* harbors a nitrous-oxide reductase composed by a binuclear copper center, which serves to receive electrons from cytochrome c552 (Rasmussen, Brittain, Berks, Watmough, & Thomson, 2005). The nitrous-oxide reductase of *Pyrobaculum aerophilum* is an efficient metallo-oxidase that catalyzes the oxidation of cuprous and ferrous ions to reduce $N_2O$ to $N_2$ (Fernandes et al., 2010). In *Rhodobacter sphaeroides* a concentration of 0.001 mM $Cu^{+2}$ is necessary for full activity (Sato, Okubo, & Yamazaki, 1998). The *nor* activity in *Pyrobaculum aerophilum* requires at least 1 $Fe^{+2}$ atom per nitrous-oxide reductase molecule, and $Ca^{+2}$ is essential for the enzymatic activity in *Wolinella succinogenes* (Teraguchi & Hollocher, 1989). In *Rhodobacter sphaeroides* Ni is present in the structure of the enzyme but in a concentration of 0.1 mM of $Ni^{+2}$ the activity decreases 50%. This species also has zinc requirements for $N_2O$ reduction (2 atoms per enzyme). However, a concentration of 0.1 mM of $Zn^{+2}$ causes complete inhibition of nitrous-oxide reductase activity (Sato, et al., 1998).

The enzyme involved in the anammox pathway is a hydrazine oxidoreductase (*hao*). This enzyme is a metalloenzyme that requires iron in different states of oxidation. In Candidatus *Maribeggiatoa* and *Planctomycete* KSU-1, the *hao* enzyme contains eight heme centres (MacGregor et al., 2013; Shimamura et al., 2007). The anammox reaction occurs in the anammoxosome. This organelle is a cytoplasmic membrane-bounded region, which makes up about 30–60% of the cell volume. Some species like Candidatus *Brocadia anammoxidans*, Candidatus *Kuenenia stuttgartiensis* and Candidatus *Scalindua brodae*, have this organelle surrounded by ladderane lipids and exclusively contains the hydrazine oxidoreductase as the major protein to combine nitrite and ammonia in a one-to-one fashion (Jetten et al., 2001).

## PHOSPHORUS

### Phosphorus Cycle

Phosphorus plays an essential role in biological systems as one of the main components in all organisms (Filippelli, 2008). The phosphorus biogeochemical cycle is considerably different from those of other chemical elements because the occurrence of phosphorus compounds is usually in the solid phase and the atmosphere plays a minor role in the global cycle. Therefore, the P cycle has three main reservoirs in the lithosphere, hydrosphere and biosphere (Wollast, et al., 1993). Lithospheric storage of P is dominated by marine and freshwater sediments (Smil, 2000). Moreover, the oxidation-reduction reactions are not significant, which implies that P mainly occurs in the biosphere as inorganic phosphate ($PO_4^{3-}$) or organic forms of phosphate. $PO_4^{3-}$ is involved in the structural composition of many minerals. Phosphate is required for the formation of many molecules, and it is essential for life as a constituent of genetic material (deoxyribonucleic acid (DNA) and ribonucleic acid (RNA)), energy transference (adenosine triphosphate (ATP)), and structural support (phospholipids) (Butcher, et al., 1992).

A scheme of the phosphorus biogeochemical cycle is shown in Figure 2. Most of the phosphorus is naturally present in terrestrial systems forming part of rocks as certain minerals like apatite [$Ca_5(PO_4)_3OH$], the most abundant P-mineral. The main movement of P is due to the constant but slow erosion of rocks to form soil, which gradually introduces particulate and dissolved forms of P into groundwater. This biogeochemical cycle includes interactions with biological (organic) and mineralogical (inorganic) systems but, unlike other biogeochemical cycles, biota does not dominate this one. Plants can incorporate phosphorus to form biomolecules: it is taken up by plants and then converted into an organic form by the assimilation process. However, a small amount of P present in soils is available to plants as $PO_4^{3-}$, which commonly makes this element as a growth-limiting nutrient in terrestrial ecosystems. For this reason, plants have developed specific strategies to increase the amount of P. For example, organic acids produced by plant roots may dissolve apatite minerals (Filippelli, 2002) and the secretion of phosphatase enzyme by plant roots allow to release inorganic P from organic matter. Another tactic is the symbiotic affiliation with certain fungi that excretes phosphatase enzyme and, in exchange, the plant provides carbohydrates to the fungi. Decomposition of dead biomass eventually releases phosphates back into the soil-groundwater system. Dissolution/precipitation reactions of P-containing minerals along with adsorption/desorption of P on mineral surfaces may influence dissolved P concentration in groundwater.

P is transported to rivers and finally reaches the oceanic reservoir. In the marine system, dissolved P is the nutrient limiting the biological productivity. Dissolved $PO_4^{3-}$ is taken up by photosynthetic organisms at the base of the marine food chain. The marine P cycle is linked to the marine C and N cycles through the photosynthetic fixation of these elements by phytoplankton (Schlesinger & Bernhardt, 2013). When aquatic organisms die, they end as marine sediments by deposition in the ocean and eventually, they are converted into sedimentary rocks by geologic pressure. On a very large timescale (hundreds of millions of years), these sediments are uplifted to form new rocks, completing the global cycle (Liu & Chen, 2014).

## Anthropogenic Perturbations

Nowadays, human activity has a direct impact on the phosphorus biogeochemical cycle. The main interferences are deforestation, application of fertilizers, recycling of organic wastes, increase of cultivation, and urban and industrial waste disposal (Smil, 2000). Deforestation after burning selective tree harvesting increases the erosion and leads to soil loss, which releases P to land and it is eventually transported from terrestrial to aquatic ecosystems. Due to phosphorus is an important nutrient for plants, the application of synthetic fertilizers containing P, N and other nutrients directly to the field has increased. The recycling of organic wastes is related to the application of animal manure to land due to its high nutrient content, which has increased in the last decades. However, plants are not able to use all the phosphate released by erosion or fertilizers (both, synthetic and manure) (Filippelli, 2008) and, as a consequence, P can reach the soil-groundwater system. Apart from agriculture, untreated urban and industrial wastewaters are also anthropogenic contributors to P cycle (Filippelli, 2002).

There are important environmental consequences of anthropogenic activity in the P-cycle that must be taken into account (Liu & Chen, 2014). Regarding mineral conservation, today the phosphate rocks are non-renewable resources due their highly intensive extraction. Moreover, the intensive livestock activities have become larger and spatially concentrated in specific areas, which leads to an excess of phosphorus in the manure applied to land that can reach aquatic ecoystems. The main environmental consequence of an increase of phosphate in aquatic ecosystems is an excess growth of certain organisms like algae. As it was previously explained, this phenomenon known as eutrophication is linked with a loss of biodiversity of the ecosystem. Phosphates themselves are toxic to people and animals only in very high concentrations. However, eutrophication may have consequences such as the bad taste and odor of drinking water, the formation of trihalomethanes during chlorination and the serious health problem by ingesting toxins released by certain species of algal blooms (Smil, 2000).

Microorganisms like *Bacteria*, dispose of inclusion bodies to accumulate P under adverse conditions that not permit the development of their physiologic fitness. The retention of the P in more competitive microorganism blocks their use by other; this involved in some cases a detriment in the diversity and the ecosystem balance.

## Roles of Metals in Phosphorous Cycle

The dynamic and complex diagram of interactions that take place between metals in the enzymatic regulation and the enzymatic activity are reflected in all organisms. Regarding metals that interfere with the enzymatic reactions in the P cycle, the concentrations that imposed a limit of activation and inhibition of the enzymatic reaction depend on the type of metabolism and the concentration of other metals. Phytase is a phosphatase enzyme that catalyzes the hydrolysis of the phytate. Its activity is essential to release inorganic phosphate and make it available to cell metabolism. The reaction is likely to proceed through a direct attack of a water molecule over the phytate and the subsequent stabilization of the pentavalent transition state by the bound of calcium ions. Then, the release of phosphates and inositol occurs. Rapid equilibrium ordered a mechanism in which the binding of $Ca^{+2}$ to the active site is necessary for the activation of the enzyme and the release of phosphate. $Ca^{+2}$ is also required as a substrate because the enzyme is only able to hydrolyze the calcium-phytate complex. Other metals involved in the catalytic action of phytase are exposed in Table 2.

Alkaline phosphatase is a metalloenzyme that catalyzes the dephosphorylation of molecules that contain phosphate (nucleotides, proteins, lipids). In general, its transcription is upregulated during phosphate starvation (Luo et al., 2010; Yeung et al., 2009). In *Haloarcula marismortui* the highest specific activity is obtained when the medium is completely devoid of phosphate and enzymatic activity cannot be detected when phosphate concentrations are higher than 0.1 mM (Goldman, Hecht, Eisenberg, & Mevarech, 1990). In *Pseudomonas aeruginosa* low levels of high-molecular weight alkaline phosphatase

*Table 2. Metals involved in activity of enzymes of the phosphorous cycle*

| Enzyme | Organisms Analyzed | | Metal Required | | Metal Inhibitor | |
|---|---|---|---|---|---|---|
| Phytase | *Aspergillus fumigatus* | *Lentinula edodes* | $Ba^{+2}$ | $Mn^{+2}$ | $Al^{+3}$ | $Fe^{+3}$ |
| | *Aspergillus niger* | *Pectobacterium wasabiae* | $Ca^{+2}$ | $Mg^{+2}$ | $Ba^{+2}$ | $Li^{+}$ |
| | *Bacillus amyloliquefaciens* | *Penicillium parvum* | $Co^{+2}$ | $Hg^{+2}$ | $Cd^{+2}$ | $Mg^{+2}$ |
| | *Bacillus licheniformis* | *Pseudomonas syringae* | $Cu^{+2}$ | $Ni^{+2}$ | $Ca^{+2}$ | $Mn^{+2}$ |
| | *Bacillus nealsonii* | *Raoultella terrigena* | $Fe^{+2}$ | $K^{+}$ | $Co^{+2}$ | $Hg^{+2}$ |
| | *Cryptococcus laurentii* | *Rhizopus microsporus* | $Li^{+}$ | $Zn^{+2}$ | $Cu^{+2}$ | $Ni^{+2}$ |
| | *Enterococcus hirae* | *Schizophyllum commune* | | | $Cr^{+3}$ | $K^{+}$ |
| | *Janthinobacterium sp.* | *Schwanniomyces capriottii* | | | $Fe^{+2}$ | |
| | *Kodamaea ohmeri* | *Selenomonas ruminantium* | | | | |
| | | *Serratia sp.* | | | | |
| Alkaline phosphatase | *Aeropyrum pernix* | *Helicoverpa armigera* | $Ba^{+2}$ | $Mg^{+2}$ | $Ag^{+2}$ | $Hg^{+2}$ |
| | *Anabaena sp.* | *Phaeodactylum tricornutum* | $Ca^{+2}$ | $Mn^{+2}$ | $Ba^{+2}$ | $Mn^{+2}$ |
| | *Aspergillus caespitosus* | *Pseudomonas sp.* | $Co^{+2}$ | $Ni^{+2}$ | $Cd^{+2}$ | $Ni^{+2}$ |
| | *Aspergillus oryzae* | *Pyrococcus abyssi* | $Sr^{+2}$ | $K^{+}$ | $Ca^{+2}$ | $Sr^{+2}$ |
| | *Cobetia marina* | *Saccharomyces cerevisiae* | $Fe^{+2}$ | $Zn^{+2}$ | $Co^{+2}$ | $Zn^{+2}$ |
| | *Cyberlindnera jadinii* | *Shewanella sp.* | | | $Cu^{+2}$ | |
| | *Haloarcula marismortui* | *Thermus sp.* | | | | |
| | *Halomonas sp.* | *Thermus thermophilus* | | | | |

Data obtained from BRENDA protein data bank.

*Figure 1. Global nitrogen cycle*

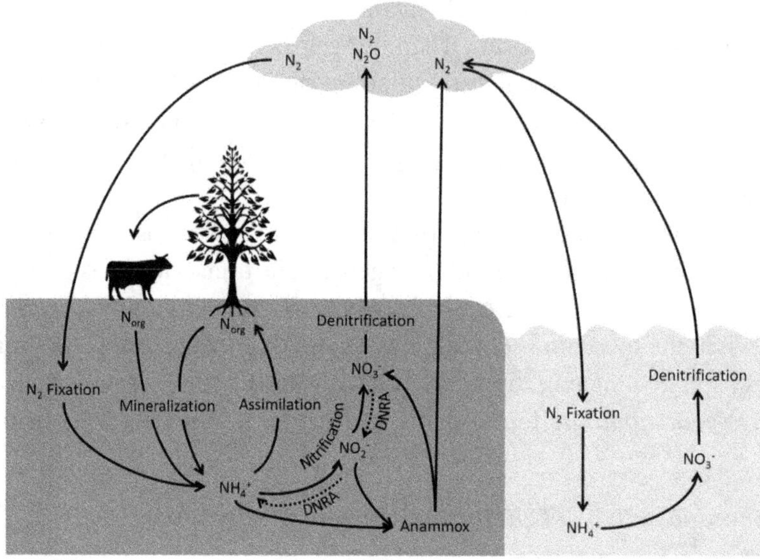


are constitutively produced, but low-molecular weight alkaline phosphatase is induced by phosphate deprivation (Tan & Worobec, 1993). The cell localization of this enzyme varies among species and also depends on the environmental phosphorus conditions. In *Pseudomonas aeruginosa* and *Synechococcus elongatus* the enzyme is associated with the outer layer of the cell wall (Block & Grossman, 1988; Day & Ingram, 1975), whereas in *Bacillus licheniformis* it appears as particulate form associated with the membrane (Schaffel & Hulett, 1978). During phosphate starvation, *Haloarcula marismortui* and *Vibrio* sp. secrete the enzyme to the extracellular matrix (Goldman, et al., 1990; Hauksson, Andresson, & Asgeirsson, 2000). Calcium is essential for the alkaline phosphatase activity. Certain organisms such as *Aeropyrum pernix, Pasteurella multocida, Campylobacter jejuni, Pyrococcus furiosus, Phaeodactylum tricornutum, Haloarcula marismortui* require $Ca^{2+}$ ions to activate the enzyme. Optimal concentration for enzymatic activity in *Haloarcula marismortui* is 3.4 mM (Goldman, et al., 1990). *Aspergillus oryzae* shows different isoenzymes of the enzyme, which are differentially regulated by $Ca^{2+}$: the isoenzyme I is activated by $Ca^{2+}$, whereas the isoenzyme III is weakly activated (Sakurai, Toda, & Hideo, 1981). The catalytic activities of the enzyme are also metal ion dependent (Table 2), and $Mg^{+2}$, $Zn^{+2}$ and $Co^{+2}$ are the main activators. The alkaline phosphatase in *Escherichia coli* contains an active center with three distinct metal binding sites in each subunit (Wyckoff, Handschumacher, Murthy, & Sowadski, 1983). It contains one $Mg^{+2}$ ion per monomer, which is located near the active site and stabilizes the transferred phosphoryl group in the transition state (Zalatan, Fenn, & Herschlag, 2008). Moreover, the enzyme employs a binuclear metallocenter of two $Zn^{+2}$ ions, which facilitates the phosphate hydrolysis (Nikolic-Hughes, O'Brien P, & Herschlag, 2005). In *Shewanella* sp., the activity is greatly enhanced by simultaneous addition of $Mg^{+2}$ and $Zn^{+2}$. The maximum specific activity was observed with 20 mM $MgCl_2$ and 5 mM $ZnCl_2$ (Suzuki et al., 2005). Moreover, $Mg^{2+}$ protects the enzyme against thermal inactivation in *Serratia marcescens* (Bhatti, Alvi, & Chaudhry, 2000). Both *Thermotoga maritime* and *Thermotoga neapolitana* require $Co^{+2}$ for the enzymatic activity (Savchenko, Wang, Vieille, & Zeikus, 2001; Wojciechowski, Cardia, & Kantrowitz, 2002) and in *Thermus yunnanensis* it is involved in the

*Figure 2. Global phosphorus cycle*

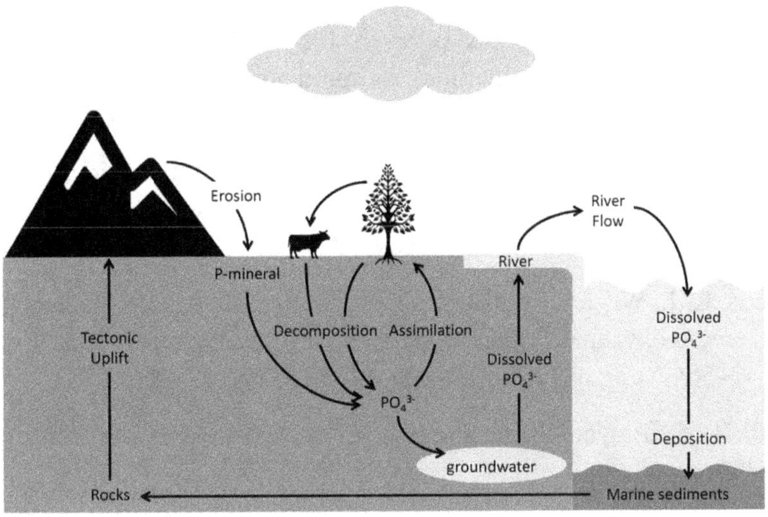


thermo stability of the enzyme (Gong et al., 2005). In *Halobacterium salinarum* the kinetic behavior of dephosphorylation is determined by the ratio between the concentration of $Mn^{2+}$ and $Na^+$ or $K^+$. When the $Mn^{2+}$ increases and $Na^+$ or $K^+$ decreases, the kinetics show cooperative behavior (Marhuenda-Egea, Piera-Velazquez, Cadenas, & Cadenas, 2001).

## CONCLUSION

The biogeochemical cycles remain in equilibrium due to the continuously balanced transformation of chemical elements. However, with the development of industrial processes, an unbalance of these cycles has occurred. The main anthropogenic implications are related to the addition of external compounds that alter the cycles. The increase of N and P compounds generated by different industrial processes and the addition of other inorganic chemicals, like metals, are affecting the biodiversity of the ecosystems, due the break of the closed balance of the biogeochemical cycles.

## ACKNOWLEDGMENT

This work was supported by FONDECYT (1150285) of the Chilean Government and by the FISHPOL (CTQ2014-55021- R) of the Spanish Government. A. Fra-Vázquez belongs to the Galician Competitive Research Group (GRC 2013-032) and CRETUS Strategic Partnership (AGRUP2015/02), financed by the Galician Government.

## REFERENCES

Aiken, A. M., Peyton, B. M., Apel, W. A., & Petersen, J. N. (2003). Heavy metal-induced inhibition of Aspergillus niger nitrate reductase: Applications for rapid contaminant detection in aqueous samples. *Analytica Chimica Acta*, *480*(1), 131–142. doi:10.1016/S0003-2670(02)01598-2

Almeida, M. G., Silveira, C. M., Guigliarelli, B., Bertrand, P., Moura, J. J., Moura, I., & Leger, C. (2007). A needle in a haystack: The active site of the membrane-bound complex cytochrome c nitrite reductase. *FEBS Letters*, *581*(2), 284–288. doi:10.1016/j.febslet.2006.12.023 PMID:17207484

Beman, J. M., Roberts, K. J., Wegley, L., Rohwer, F., & Francis, C. A. (2007). Distribution and diversity of archaeal ammonia monooxygenase genes associated with corals. *Applied and Environmental Microbiology*, *73*(17), 5642–5647. doi:10.1128/AEM.00461-07 PMID:17586663

Bernhard, A. (2010). The Nitrogen Cycle: Processes, Players, and Human Impact. *Nature Education Knowledge*, *3*(10), 25.

Bhatti, A. R., Alvi, A., & Chaudhry, G. R. (2000). Evidence on the presence of two distinct alkaline phosphatases in Serratia marcescens. *FEMS Microbiology Letters*, *182*(1), 131–135. doi:10.1111/j.1574-6968.2000.tb08886.x PMID:10612744

Block, M. A., & Grossman, A. R. (1988). Identification and Purification of a Derepressible Alkaline Phosphatase from Anacystis nidulans R2. *Plant Physiology*, *86*(4), 1179–1184. doi:10.1104/pp.86.4.1179 PMID:16666051

Blomberg, L. M., Blomberg, M. R., & Siegbahn, P. E. (2006). A theoretical study on nitric oxide reductase activity in a ba(3)-type heme-copper oxidase. *Biochimica et Biophysica Acta*, *1757*(1), 31–46. doi:10.1016/j.bbabio.2005.11.004 PMID:16375849

Bueno, E., Gomez-Hernandez, N., Girard, L., Bedmar, E. J., & Delgado, M. J. (2005). Function of the Rhizobium etli CFN42 nirK gene in nitrite metabolism. *Biochemical Society Transactions*, *33*(Pt 1), 162–163. doi:10.1042/BST0330162 PMID:15667294

Butcher, S. S., Charlson, R. J., Oriand, G. H., & Wolfe, G. V. (1992). Global Biogeochemical Cycles. London: Academic Press (Harcourt Brace Jovanovich).

Bykov, D., & Neese, F. (2012). Reductive activation of the heme iron-nitrosyl intermediate in the reaction mechanism of cytochrome c nitrite reductase: A theoretical study. *Journal of Biological Inorganic Chemistry*, *17*(5), 741–760. doi:10.1007/s00775-012-0893-0 PMID:22454108

Canfield, D. E., Glazer, A. N., & Falkowski, P. G. (2010). The evolution and future of Earth's nitrogen cycle. *Science*, *330*(6001), 192–196. doi:10.1126/science.1186120 PMID:20929768

Cavet, J. S., Borrelly, G. P., & Robinson, N. J. (2003). Zn, Cu and Co in cyanobacteria: Selective control of metal availability. *FEMS Microbiology Reviews*, *27*(2-3), 165–181. doi:10.1016/S0168-6445(03)00050-0 PMID:12829266

Correa, J. A., Castilla, J. C., Ramírez, M., Varas, M., Lagos, N., Vergara, S., & Brown, M. T. et al. (1999). Copper, copper mine tailings and their effect on marine algae in Northern Chile.[journal article]. *Journal of Applied Phycology*, *11*(1), 57–67. doi:10.1023/A:1008027610826

Cunha, C. A., Macieira, S., Dias, J. M., Almeida, G., Goncalves, L. L., Costa, C., & Romao, M. J. et al. (2003). Cytochrome c nitrite reductase from Desulfovibrio desulfuricans ATCC 27774. The relevance of the two calcium sites in the structure of the catalytic subunit (NrfA). *The Journal of Biological Chemistry*, *278*(19), 17455–17465. doi:10.1074/jbc.M211777200 PMID:12618432

Daiber, A., Shoun, H., & Ullrich, V. (2005). Nitric oxide reductase (P450nor) from Fusarium oxysporum. *Journal of Inorganic Biochemistry*, *99*(1), 185–193. doi:10.1016/j.jinorgbio.2004.09.018 PMID:15598501

Daims, H., Lebedeva, E. V., Pjevac, P., Han, P., Herbold, C., Albertsen, M., & Wagner, M. et al. (2015). Complete nitrification by Nitrospira bacteria. *Nature*, *528*(7583), 504–509. http://www.nature.com/nature/journal/v528/n7583/abs/nature16461.html#supplementary-information. doi:10.1038/nature16461 PMID:26610024

Day, D. F., & Ingram, J. M. (1975). In vitro studies of an alkaline phosphatase-cell wall complex from Pseudomonas aeruginosa. *Canadian Journal of Microbiology*, *21*(1), 9–16. doi:10.1139/m75-002 PMID:803858

de Vries, S., Strampraad, M. J., Lu, S., Moenne-Loccoz, P., & Schroder, I. (2003). Purification and characterization of the MQH2:NO oxidoreductase from the hyperthermophilic archaeon Pyrobaculum aerophilum. *The Journal of Biological Chemistry*, *278*(38), 35861–35868. doi:10.1074/jbc.M300857200 PMID:12799376

Deligeer, F., Fukunaga, R., Kataoka, K., Yamaguchi, K., Kobayashi, K., Tagawa, S., & Suzuki, S. (2002). Spectroscopic and functional characterization of Cu-containing nitrite reductase from Hyphomicrobium denitrificans A3151. *Journal of Inorganic Biochemistry*, *91*(1), 132–138. doi:10.1016/S0162-0134(02)00442-7 PMID:12121770

Ensign, S. A., Hyman, M. R., & Arp, D. J. (1993). In vitro activation of ammonia monooxygenase from Nitrosomonas europaea by copper. *Journal of Bacteriology*, *175*(7), 1971–1980. PMID:8458839

Fay, A. W., Blank, M. A., Lee, C. C., Hu, Y., Hodgson, K. O., Hedman, B., & Ribbe, M. W. (2010). Characterization of isolated nitrogenase FeVco. *Journal of the American Chemical Society*, *132*(36), 12612–12618. doi:10.1021/ja1019657 PMID:20718463

Fernandes, A. T., Damas, J. M., Todorovic, S., Huber, R., Baratto, M. C., Pogni, R., & Martins, L. O. et al. (2010). The multicopper oxidase from the archaeon Pyrobaculum aerophilum shows nitrous oxide reductase activity. *The FEBS Journal*, *277*(15), 3176–3189. doi:10.1111/j.1742-4658.2010.07725.x PMID:20597980

Filippelli, G. M. (2002). The Global Phosphorus Cycle. *Reviews in Mineralogy and Geochemistry*, *48*(1), 391–425. doi:10.2138/rmg.2002.48.10

Filippelli, G. M. (2008). The Global Phosphorus Cycle: Past, Present, and Future. *Elements*, *4*(2), 89–95. doi:10.2113/GSELEMENTS.4.2.89

Fowler, D., Coyle, M., Skiba, U., Sutton, M. A., Cape, J. N., Reis, S., & Voss, M. et al. (2013). The global nitrogen cycle in the twenty-first century. *Philosophical Transactions of the Royal Society of London. Series B, Biological Sciences*, *368*(1621). doi:10.1098/rstb.2013.0164

Galloway, J. N., Schlesinger, W. H., Levy, H. II, Michaels, A., & Schnoor, J. L. (1995). Nitrogen fixation: Anthropogenic enhancement-environmental response. *Global Biogeochemical Cycles*, *9*(2), 235–252. doi:10.1029/95GB00158

Gleeson, D. B., Clipson, N., Melville, K., Gadd, G. M., & McDermott, F. P. (2005). Characterization of fungal community structure on a weathered pegmatitic granite. *Microbial Ecology*, *50*(3), 360–368. doi:10.1007/s00248-005-0198-8 PMID:16328654

Goldman, S., Hecht, K., Eisenberg, H., & Mevarech, M. (1990). Extracellular Ca2(+)-dependent inducible alkaline phosphatase from extremely halophilic archaebacterium Haloarcula marismortui. *Journal of Bacteriology*, *172*(12), 7065–7070. PMID:2123861

Gong, N., Chen, C., Xie, L., Chen, H., Lin, X., & Zhang, R. (2005). Characterization of a thermostable alkaline phosphatase from a novel species Thermus yunnanensis sp. nov. and investigation of its cobalt activation at high temperature. *Biochimica et Biophysica Acta*, *1750*(2), 103–111. doi:10.1016/j.bbapap.2005.05.007 PMID:15955749

Haaker, H., & Klugkist, J. (1987). The bioenergetics of electron transport to nitrogenase. *FEMS Microbiology Letters*, *46*(1), 57–71. doi:10.1111/j.1574-6968.1987.tb02452.x

Hauksson, J. B., Andresson, O. S., & Asgeirsson, B. (2000). Heat-labile bacterial alkaline phosphatase from a marine Vibrio sp. *Enzyme and Microbial Technology*, *27*(1-2), 66–73. doi:10.1016/S0141-0229(00)00152-6 PMID:10862903

Howarth, R., Chan, F., Conley, D. J., Garnier, J., Doney, S. C., Marino, R., & Billen, G. (2011). Coupled biogeochemical cycles: Eutrophication and hypoxia in temperate estuaries and coastal marine ecosystems. *Frontiers in Ecology and the Environment*, *9*(1), 18–26. doi:10.1890/100008

Jetten, M. S., Wagner, M., Fuerst, J., van Loosdrecht, M., Kuenen, G., & Strous, M. (2001). Microbiology and application of the anaerobic ammonium oxidation ('anammox') process. *Current Opinion in Biotechnology*, *12*(3), 283–288. doi:10.1016/S0958-1669(00)00211-1 PMID:11404106

Kajie, S., & Anraku, Y. (1986). Purification of a hexaheme cytochrome c552 from Escherichia coli K 12 and its properties as a nitrite reductase. *European Journal of Biochemistry*, *154*(2), 457–463. doi:10.1111/j.1432-1033.1986.tb09419.x PMID:3002798

Kim, R. Y., Yoon, J. K., Kim, T. S., Yang, J. E., Owens, G., & Kim, K. R. (2015). Bioavailability of heavy metals in soils: Definitions and practical implementation--a critical review. *Environmental Geochemistry and Health*, *37*(6), 1041–1061. doi:10.1007/s10653-015-9695-y PMID:25841357

Lam, Y., & Nicholas, D. J. (1969). A nitrite reductase with cytochrome oxidase activity from Micrococcus denitrificans. *Biochimica et Biophysica Acta*, *180*(3), 459–472. doi:10.1016/0005-2728(69)90025-5 PMID:4309366

Liu, Y., & Chen, J. (2014). *Phosphorus Cycle Reference Module in Earth Systems and Environmental Sciences*. Elsevier.

Lukat, P., Rudolf, M., Stach, P., Messerschmidt, A., Kroneck, P. M., Simon, J., & Einsle, O. (2008). Binding and reduction of sulfite by cytochrome c nitrite reductase. *Biochemistry*, *47*(7), 2080–2086. doi:10.1021/bi7021415 PMID:18201106

Luo, M., Guo, Y. C., Deng, J. Y., Wei, H. P., Zhang, Z. P., Leng, Y., & Zhou, Y. F. et al. (2010). Characterization of a monomeric heat-labile classical alkaline phosphatase from Anabaena sp. PCC7120. *Biochemistry (Moscow)*, *75*(5), 655–664. doi:10.1134/S0006297910050172 PMID:20632947

MacGregor, B. J., Biddle, J. F., Siebert, J. R., Staunton, E., Hegg, E. L., Matthysse, A. G., & Teske, A. (2013). Why orange Guaymas Basin Beggiatoa spp. are orange: Single-filament-genome-enabled identification of an abundant octaheme cytochrome with hydroxylamine oxidase, hydrazine oxidase, and nitrite reductase activities. *Applied and Environmental Microbiology*, *79*(4), 1183–1190. doi:10.1128/AEM.02538-12 PMID:23220958

Madsen, E. L. (2011). Microorganisms and their roles in fundamental biogeochemical cycles. *Current Opinion in Biotechnology*, *22*(3), 456–464. doi:10.1016/j.copbio.2011.01.008 PMID:21333523

Marhuenda-Egea, F. C., Piera-Velazquez, S., Cadenas, C., & Cadenas, E. (2001). Kinetic regulation of an alkaline p-nitrophenylphosphate phosphatase from Halobacterium salinarum in low water system by Mn2+ and monovalent cations. *FEMS Microbiology Letters*, *198*(2), 111–115. doi:10.1111/j.1574-6968.2001.tb10627.x PMID:11430399

Nikolic-Hughes, I., O'Brien, P. J., & Herschlag, D. (2005). Alkaline phosphatase catalysis is ultrasensitive to charge sequestered between the active site zinc ions. *Journal of the American Chemical Society*, *127*(26), 9314–9315. doi:10.1021/ja051603j PMID:15984827

Oji, Y., Hamano, T., Ryoma, Y., Miki, Y., Wakiuchi, N., & Okamoto, S. (1985). Purification and Characterization of NADH-Nitrate Reductase from Leaves of 2-Row Barley, and Its Activity as Affected by Some Metabolites. *Journal of Plant Physiology*, *119*(3), 247–256. doi:10.1016/S0176-1617(85)80184-X

Panda, S. K., & Choudhury, S. (2005). Changes in nitrate reductase activity and oxidative stress response in the moss Polytrichum commune subjected to chromium, copper and zinc phytotoxicity. *Brazilian Journal of Plant Physiology*, *17*(2), 191–197. doi:10.1590/S1677-04202005000200001

Pepper, I., Gerba, C., & Gentry, T. (2015). *Environmental Microbiology* (3rd ed.). San Diego, CA: Academic Press.

Petersen, J., Fisher, K., & Lowe, D. J. (2008). Structural basis for VO2+ inhibition of nitrogenase activity (A): 31P and 23Na interactions with the metal at the nucleotide binding site of the nitrogenase Fe protein identified by ENDOR spectroscopy. *Journal of Biological Inorganic Chemistry*, *13*(4), 623–635. doi:10.1007/s00775-008-0360-0 PMID:18351402

Petersen, J., Mitchell, C. J., Fisher, K., & Lowe, D. J. (2008). Structural basis for VO(2+)-inhibition of nitrogenase activity: (B) pH-sensitive inner-sphere rearrangements in the 1H-environment of the metal coordination site of the nitrogenase Fe-protein identified by ENDOR spectroscopy. *Journal of Biological Inorganic Chemistry*, *13*(4), 637–650. doi:10.1007/s00775-008-0364-9 PMID:18386081

Prasad, M. N. V. (2001). *Metals in the Environment: Analysis by Biodiversity*. Taylor & Francis.

Prodouz, K. N., & Garrett, R. H. (1981). Neurospora crassa NAD(P)H-nitrite reductase. Studies on its composition and structure. *The Journal of Biological Chemistry*, *256*(18), 9711–9717. PMID:6457037

Rasmussen, T., Brittain, T., Berks, B. C., Watmough, N. J., & Thomson, A. J. (2005). Formation of a cytochrome c-nitrous oxide reductase complex is obligatory for N2O reduction by Paracoccus pantotrophus. *Dalton Transactions (Cambridge, England)*, (21): 3501–3506. doi:10.1039/b501846c PMID:16234931

Sakurai, Y., Toda, K., & Hideo, S. (1981). Multiple Forms and Some Properties of Alkaline Phosphatase Produced by Aspergillus oryzae on Solid Medium. *Agricultural and Biological Chemistry*, *45*(9), 1959–1967. doi:10.1080/00021369.1981.10864831

Sato, K., Okubo, A., & Yamazaki, S. (1998). Characterization of a multi-copper enzyme, nitrous oxide reductase, from Rhodobacter sphaeroides f. sp. denitrificans. *Journal of Biochemistry*, *124*(1), 51–54. doi:10.1093/oxfordjournals.jbchem.a022096 PMID:9644245

Savchenko, A., Wang, W., Vieille, C., & Zeikus, J. G. (2001). Alkaline phosphatase from Thermotoga neapolitana. *Methods in Enzymology*, *331*, 298–305. doi:10.1016/S0076-6879(01)31067-4 PMID:11265472

Savelieff, M. G., Wilson, T. D., Elias, Y., Nilges, M. J., Garner, D. K., & Lu, Y. (2008). Experimental evidence for a link among cupredoxins: Red, blue, and purple copper transformations in nitrous oxide reductase. *Proceedings of the National Academy of Sciences of the United States of America*, *105*(23), 7919–7924. doi:10.1073/pnas.0711316105 PMID:18535143

Schaffel, S. D., & Hulett, F. M. (1978). Alkaline phosphatase from Bacillus licheniformis. Solubility dependent on magnesium, purification and characterization. *Biochimica et Biophysica Acta*, *526*(2), 457–467. doi:10.1016/0005-2744(78)90137-7 PMID:718947

Schlesinger, W. H., & Bernhardt, E. S. (2013). Biogeochemistry. In E. S. Bernhardt (Ed.), *Biogeochemistry* (3rd ed., pp. i–ii). Boston: Academic Press.

Shimamura, M., Nishiyama, T., Shigetomo, H., Toyomoto, T., Kawahara, Y., Furukawa, K., & Fujii, T. (2007). Isolation of a multiheme protein with features of a hydrazine-oxidizing enzyme from an anaerobic ammonium-oxidizing enrichment culture. *Applied and Environmental Microbiology*, *73*(4), 1065–1072. doi:10.1128/AEM.01978-06 PMID:17172456

Siemann, S., Schneider, K., Drottboom, M., & Muller, A. (2002). The Fe-only nitrogenase and the Mo nitrogenase from Rhodobacter capsulatus: A comparative study on the redox properties of the metal clusters present in the dinitrogenase components. *European Journal of Biochemistry*, *269*(6), 1650–1661. doi:10.1046/j.1432-1327.2002.02804.x PMID:11895435

Smil, V. (2000). Phosphorus in the environment: Natural flows and human interferences. *Annual Review of Energy and the Environment*, *25*(1), 53–88. doi:10.1146/annurev.energy.25.1.53

Snyder, S. W., & Hollocher, T. C. (1987). Purification and some characteristics of nitrous oxide reductase from Paracoccus denitrificans. *The Journal of Biological Chemistry*, *262*(14), 6515–6525. PMID:3032972

Stein, L. Y., & Klotz, M. G. (2016). The nitrogen cycle. *Current Biology*, *26*(3), R94–R98. doi:10.1016/j.cub.2015.12.021 PMID:26859274

Strous, M., Pelletier, E., Mangenot, S., Rattei, T., Lehner, A., Taylor, M. W., & Le Paslier, D. et al. (2006). Deciphering the evolution and metabolism of an anammox bacterium from a community genome. *Nature*, *440*(7085), 790–794. doi:10.1038/nature04647 PMID:16598256

Suharti, H., Heering, H. A., & de Vries, S. (2004). NO reductase from Bacillus azotoformans is a bifunctional enzyme accepting electrons from menaquinol and a specific endogenous membrane-bound cytochrome c551. *Biochemistry*, *43*(42), 13487–13495. doi:10.1021/bi0488101 PMID:15491156

Suzuki, Y., Mizutani, Y., Tsuji, T., Ohtani, N., Takano, K., Haruki, M., & Kanaya, S. et al. (2005). Gene cloning, overproduction, and characterization of thermolabile alkaline phosphatase from a psychrotrophic bacterium. *Bioscience, Biotechnology, and Biochemistry*, *69*(2), 364–373. doi:10.1271/bbb.69.364 PMID:15725663

Tan, A. S., & Worobec, E. A. (1993). Isolation and characterization of two immunochemically distinct alkaline phosphatases from Pseudomonas aeruginosa. *FEMS Microbiology Letters*, *106*(3), 281–286. doi:10.1111/j.1574-6968.1993.tb05977.x PMID:8454193

Teraguchi, S., & Hollocher, T. C. (1989). Purification and some characteristics of a cytochrome c-containing nitrous oxide reductase from Wolinella succinogenes. *The Journal of Biological Chemistry*, *264*(4), 1972–1979. PMID:2536696

Tikhonova, T. V., Slutsky, A., Antipov, A. N., Boyko, K. M., Polyakov, K. M., Sorokin, D. Y., & Popov, V. O. et al. (2006). Molecular and catalytic properties of a novel cytochrome c nitrite reductase from nitrate-reducing haloalkaliphilic sulfur-oxidizing bacterium Thioalkalivibrio nitratireducens. *Biochimica et Biophysica Acta*, *1764*(4), 715–723. doi:10.1016/j.bbapap.2005.12.021 PMID:16500161

Tyson, K., Busby, S., & Cole, J. (1997). Catabolite regulation of two Escherichia coli operons encoding nitrite reductases: Role of the Cra protein. *Archives of Microbiology*, *168*(3), 240–244. doi:10.1007/s002030050494 PMID:9382703

Vitousek, P. M., Aber, J. D., Howarth, R. W., Likens, G. E., Matson, P. A., Schindler, D. W., & Tilman, D. G. et al. (1997). Human alteration of the global nitrogen cycle: Sources and consequences. *Ecological Applications*, *7*(3), 737–750. doi:10.1890/1051-0761(1997)007[0737:haotgn]2.0.co;2

Wang, Y. P., Houlton, B. Z., & Field, C. B. (2007). A model of biogeochemical cycles of carbon, nitrogen, and phosphorus including symbiotic nitrogen fixation and phosphatase production. *Global Biogeochemical Cycles, 21*(1). doi: 10.1029/2006gb002797

Watsuji, T. O., Takaya, N., Nakamura, A., & Shoun, H. (2003). A possible role of NADPH-dependent cytochrome P450nor isozyme in glycolysis under denitrifying conditions. *Bioscience, Biotechnology, and Biochemistry*, *67*(5), 1109–1114. doi:10.1271/bbb.67.1109 PMID:12834289

Wojciechowski, C. L., Cardia, J. P., & Kantrowitz, E. R. (2002). Alkaline phosphatase from the hyperthermophilic bacterium T. maritima requires cobalt for activity. *Protein Science*, *11*(4), 903–911. doi:10.1110/ps.4260102 PMID:11910033

Wollast, R., Mackenzie, F. T., & Chou, L. (1993). *Interactions of C, N, P and S. Biogeochemical Cycles and Global Change*. Springer-Verlag Berlin Heidelberg. doi:10.1007/978-3-642-76064-8

Wyckoff, H. W., Handschumacher, M., Murthy, H. M., & Sowadski, J. M. (1983). The three dimensional structure of alkaline phosphatase from E. coli. *Advances in Enzymology and Related Areas of Molecular Biology*, *55*, 453–480. PMID:6353889

Yeung, S. L., Cheng, C., Lui, T. K., Tsang, J. S., Chan, W. T., & Lim, B. L. (2009). Purple acid phosphatase-like sequences in prokaryotic genomes and the characterization of an atypical purple alkaline phosphatase from Burkholderia cenocepacia J2315. *Gene*, *440*(1-2), 1–8. doi:10.1016/j.gene.2009.04.002 PMID:19376213

Yousafzai, F. K., & Eady, R. R. (2002). Dithionite reduction kinetics of the dissimilatory copper-containing nitrite reductase of Alcalegenes xylosoxidans. The SO(2)(.-) radical binds to the substrate binding type 2 copper site before the type 2 copper is reduced. *The Journal of Biological Chemistry*, *277*(37), 34067–34073. doi:10.1074/jbc.M204305200 PMID:12082116

Zahn, J. A., Arciero, D. M., Hooper, A. B., & DiSpirito, A. A. (1996). Evidence for an iron center in the ammonia monooxygenase from Nitrosomonas europaea. *FEBS Letters*, *397*(1), 35–38. doi:10.1016/S0014-5793(96)01116-7 PMID:8941709

Zalatan, J. G., Fenn, T. D., & Herschlag, D. (2008). Comparative enzymology in the alkaline phosphatase superfamily to determine the catalytic role of an active-site metal ion. *Journal of Molecular Biology*, *384*(5), 1174–1189. doi:10.1016/j.jmb.2008.09.059 PMID:18851975

# Chapter 2
# Nutrients Pollution in Water Bodies:
## Related Legislation in Europe and the United States

**Ángeles Val del Río**
*Universidade de Santiago de Compostela, Spain*

**José Luis Campos Gómez**
*University Adolfo Ibáñez, Chile*

**Paula Carrera Fernández**
*Universidade de Santiago de Compostela, Spain*

**Anuska Mosquera-Corral**
*Universidade de Santiago de Compostela, Spain*

## ABSTRACT

*The pollution of water bodies by an excess of nutrients (N and P) is a worldwide problem with effects on the human health, ecosystems status, climate change, etc. To face with this important issue different regulations were promulgated by the countries, sometimes based on the results from international conventions and programmes. In this chapter, a review of the laws and regulations that affect the discharge of nitrogen and phosphorus is addressed, focused in the case of Europe and the United States. Finally, a brief explanation about international initiatives was performed to understand the global framework concerning nutrients pollution.*

## INTRODUCTION

The water resources are subjected to anthropogenic pressures derived from human activities (discharge of municipal and industrial wastewater, agriculture and farming, hydromorphological alterations, etc.). These activities are frequently responsible for the presence of an excess of nitrogen (N) and phosphorous (P) in natural environments provoking nutrient pollution of water bodies. This presence of large nutrients loads causes problems of eutrophication, reduction of biodiversity in aquatic ecosystems and deterioration of water quality. For these reasons in the last decades the problem of nutrients pollution has received much attention to avoid its potentially harmful effects and reduce nutrient losses to the aquatic environment.

The efforts to reduce nitrogen and phosphorus pollution have been focused not only on the development of treatment systems, to guarantee the adequate discharge of treated wastewater from industry and

DOI: 10.4018/978-1-5225-1037-6.ch002

municipal activities, but also on the improvement of the agriculture practices, principally concerning the use of fertilizers. An efficient use of these elements has clear environmental, socioeconomic, and national security benefits and contributes to the reduction of some of the effects of the global climate warming.

Nowadays attention is also paid to the ecosystems capacity to absorb the excessive and growing levels of nitrogen and phosphorous. In a recent study (Karabulut, Bouraoui, Grizzetti, Bidoglio, & Pistocchi, 2014) the production of reactive nitrogen in the world has been estimated. This study indicates that it increased more than 20 times from 1860 to 2005, and it currently amounts to around 187 tonnes annually, representing around 1.5-2.0 times the natural rate for the planet. From this reactive nitrogen, produced by anthropogenic activities, 90 tonnes have been estimated to end up in the oceans. With respect to the phosphorous the same study estimates that nearly half of the amount mined (10 Mtonnes annually) ends up in the oceans, which represents eight times the natural rate of input.

The nutrients are natural resources essential for life, but when they are released to the environment in excessive amounts severe problems arise. For this reason, the authorities and governments of the different world regions have the responsibility to develop regulations and directives that deal with these significant impacts. The developed policies need to be considered within an integrated and holistic approach to addressing the nutrient cycle including anthropogenic activities. Despite the establishment of common emission value limits (ELV) by international directives, each country established different limit values depending on their particular situation. Therefore, the laws and regulations for nitrogen and phosphorous effluent discharges differ around the world. This chapter will focus on reviewing regulations referred to nutrients for two regions in the world: Europe Union (EU) and the United States of America (commonly U.S.)

Among the different laws that concern water allocated for water uses or sources like bathing, drinking water, surface waters, inland and coastal waters, etc.; this chapter will focus especially on those related to water discharged from industrial activities. The composition of the wastewater originated from industrial activities varies widely according to the type of process taking place. For this reason, certain types of industrial wastewater contain higher concentrations of nutrients than others. Therefore, the discharge limits imposed by the regulations need to take into account not only this variability but also the geographic region and the class of water body receiving the effluent (lake, river, ocean, etc.). To fulfill the discharge limits each industrial facility needs to implement the corresponding treatment, if the wastewater is directly discharged into a water body, or a simpler pre-treatment if the effluent goes to a municipal wastewater treatment plant (WWTP). In general, an enhanced aeration treatment process is used to remove the nitrogen content, while the chemical phosphorus precipitation process is used to remove the phosphorus content from wastewater.

## EUROPEAN UNION

The European Union (EU) has set up some policy instruments (summarized in Table 1) to protect water bodies from nutrients pollution provoked by nitrogen and phosphorus contained in effluents discharge. A specific directive is devoted regulating nitrogen discharges (e.g. Nitrates Directive (91/676/EEC)) and although a particular directive is not available for phosphorous discharges, some European Member States address the phosphorus environmental detrimental effects by the implementation of the Nitrates Directive (91/676/EEC), the Water Framework Directive (2000/60/EC) and the Industrial Emissions Directive (2010/75/EU).

*Table 1. Overview of main EU Directives related to N and P emissions and concentrations in water bodies*

| Directive | Requirements/Limit values |
|---|---|
| Bathing Water Directive (76/160/EEC), amended by (2006/7/EC) | • Ammonia, nitrogen Kjeldahl, nitrates and phosphates must be checked when there is a tendency towards the eutrophication of the water.<br>• The directive established the method of analysis and inspection for these parameters. |
| Directive on Dangerous Substances (76/464/EEC), amended by (2006/11/EC) | • The Member States shall apply a system of zero-emission to discharges into the groundwater of specified listed substances, including organophosphorus compounds.<br>• The Member States shall apply to the groundwater the provisions of this Directive related to the substances belonging to specific families and groups of substances including: inorganic compounds of phosphorus and elemental phosphorus; and substances that can affect the oxygen balance, particularly: ammonia and nitrite/nitrate.<br>• Limit values for industrial effluents established according to the sector and type of product. |
| Urban Wastewater Treatment Directive (91/271/EEC), amended by (98/15/EC) | • All agglomerations must be provided with collecting systems for urban wastewater.<br>• Identification of sensitive areas.<br>• Requirements for discharges to sensitive areas:<br>- a reduction for total N of 70%-80% and total P of 80% of the influent<br>- maximum annual mean total N and P concentrations of 10-15 mg N/L and 1-2 mg P/L, depending on the size of the urban area. |
| Nitrates Directive (91/676/EEC) | • Establishment of a code of good agricultural practices.<br>• Designation of Nitrate Vulnerable Zones (NVZ).<br>• Establishment of action programmes with mandatory measures in vulnerable zones, including the application of N limits.<br>• Water quality criteria:<br>- 50 mg N/L of nitrate in groundwater and surface waters<br>- eutrophic status of surface waters<br>• Application limit for nitrogen from animal manure: 170 kg/ha/yr. |
| Water Framework Directive (2000/60/EC) | • Maintaining/establishing good ecological status in surface water bodies and good chemical and quantitative status in groundwater bodies<br>• Establishment of river basement management plans (RBMP).<br>• Designation of "protected areas".<br>• For "limit values" and "measures required" reference is made to other Directives.<br>• Estimation and identification of significant point and diffuse sources of pollution from urban, industrial, agricultural and other installations and activities, listing a series of substances including "substances which contribute to eutrophication (in particular, nitrates and phosphates)". |
| Groundwater Directive (2006/118/EC) | • Groundwater quality standards for nitrate and active substances in pesticides, including their relevant metabolites, degradation and reaction products.<br>• Threshold values for all pollutants and indicators of pollution which characterise groundwater as being at risk of failing to achieve good groundwater chemical status.<br>• Establishes 50 mg/L for nitrate as a binding maximum quality threshold. |
| Marine Strategy Framework Directive (2008/56/EC) | • Establishes a framework to take the necessary measures to achieve or maintain good environmental status in the marine environment by the year 2020.<br>• Specific descriptor for eutrophication based on: nutrients levels, direct effects and indirect effects. |
| Waste Framework Directive (2008/98/EC) | • The "wastewater" as waste is out of the scope of this Directive.<br>• Member States shall take the necessary measures to ensure that waste management is carried out without endangering human health, without harming the environment and, in particular without risk to water, air, soil, plants or animals. |
| Industrial Emissions Directive (2010/75/EU) | • Includes all the types of wastes generated by the industries spread to air, soil and water.<br>• Establishes the use of Best Available Techniques (BATs) documents to set the permit limit values.<br>• Values of emission limits for listed pollutants including "substances which contribute to eutrophication (in particular, nitrates and phosphates)". |

(adapted from (van Eekert, Verdoes, & de Buisonjé, 2012) and (Oenema et al., 2011))

## Summary of EU Directives and Regulations

The first European Directive that dealt with the pollution in water bodies was the Bathing Water Directive (76/160/EEC), posteriori amended in 2006 (2006/7/EC). This instruction was developed with the aim to protect the environment and public health by reducing the pollution of bathing water and by protecting it from further deterioration. Concerning the nutrients pollution the Directive stated that "ammonia, nitrogen Kjeldahl, nitrates and phosphates must be checked by the competent authorities when there is a tendency towards the eutrophication of the water". This Directive also established those methods of analysis and inspection used for these parameters. However over time, different controversies appeared among researchers regarding the most appropriate indicators and the numerical standards postulated in the directive (Figueras, Polo, Inza, & Guarro, 1997).

In 1976 the European Union also promulgated the Directive on Dangerous Substances (76/464/EEC), amended in 2006 (2006/11/EC). According to this directive, the Member States shall apply a system of zero-emission to discharges into the groundwater of specified listed substances, including among them organophosphorus compounds, inorganic compounds of phosphorus and elemental phosphorus, and substances provoking an adverse effect on the oxygen balance, particularly: ammonia and nitrites. This directive indicated that the limit values applicable to industrial effluents shall be established according to the sector and the type of product.

In 1991 two important directives were approved by the European Union regarding nutrients pollution: the Urban Wastewater Treatment (UWWT) Directive (91/271/EEC) and the Nitrates Directive (91/676/EEC). The main difference between them is that the UWWT Directive aims at controlling emissions of pollutants from point sources while the Nitrates Directive intends to protect water resources against nitrates from agriculture sources. Currently, both directives are the base of the programs: Good Ecological Status for Inland Water in EU and Good Environmental Status of European Regional Seas.

The UWWT Directive (91/271/EEC) concerns the collection, treatment and discharge of urban wastewater and the treatment and discharge of wastewater from certain industrial sectors. This Directive establishes some obligations regarding proper collection systems, appropriate treatment of collected wastewater and reinforced treatment in areas sensitive to eutrophication. It sets the secondary treatment as the normal standard but requires tertiary treatment where discharges affect sensitive areas identified in the Directive. More specifically the Directive requires:

- The collection and treatment of wastewater in all agglomerations of >2,000 population equivalents (p.e.)
- The secondary treatment of all discharges from agglomerations of > 2000 p.e., and more advanced treatment for agglomerations of >10,000 p.e. in designated sensitive areas and their catchments.
- A pre-authorization of all discharges of urban wastewater, of discharges from the food-processing industry and of industrial discharges into urban wastewater collection systems.
- The monitoring of the performance of treatment plants and receiving waters.
- The control of sewage sludge disposal and reuse, and treated wastewater reuse whenever it is appropriate.

In 1998, it was amended by the Directive (98/15/EC) due to problems of interpretation. This amendment clarifies the requirements for discharges from urban wastewater treatment plants in sensitive areas subjected to eutrophication. It establishes the limits for discharges to sensitive areas for total nitrogen and total phosphorous depending on the size of the communities:

- For populations between 10,000-100,000 p.e.: 2 mg P/L and 15 mg N/L
- For populations > 100,000 p.e.: 1 mg P/L and 10 mg N/L
- Furthermore, the Directive indicates that the removal efficiency of total N and P needs to be at least the 70-80% of the total N and 80% of the total P present in the untreated wastewater.

These limits serve as a guide for the Member States, but each country can set more restrictive limits in the transposition of the Directive. For example, Ireland set a limit for total ammonia of 10 mg N/L for discharges to water bodies. In Germany, the requirements for discharge are 25 mg/L of total N (including ammonia, nitrite and nitrate nitrogen); and 1.5 mg/L of total P. The permit of total N discharge is higher as long as the nitrogen load is reduced by at least 75% in a central WWTP. In Italy the emission limit values are 15 mg/L, 0.6 mg/L and 20 mg/L into surface water bodies and 30 mg/L, 0.6 mg/L and 30 mg/L into municipal collecting systems for ammonia (as $NH_4^+$), nitrite (as N) and nitrate (as N), respectively (CWW BREF, 2003).

The Nitrates Directive (91/676/EEC) aims at protecting water quality across Europe by preventing nitrates release from agricultural sources and by promoting the use of good farming practices. Therefore, it is one of the key instruments on the protection of waters against agricultural pressures. The Member States must monitor the quality of the water bodies, identify polluted water (or water sources at risk of pollution), identify Nitrate Vulnerable Zones (NVZ) and establish Codes of Good Agricultural Practice and Action Programs. In the Directive manure application in NVZs is restricted by the limit of 170 kg N/ha/yr. Considering that the ratio between N and P in these effluents ranges from 1.5 to 7 g N/g P, the phosphorus application by manure is indirectly limited by the directive. There are also other restrictions for manure in NVZs, e.g. regarding the procedure and periods for land use. The Nitrates Directive also indicates the necessity to do a monitoring and reporting every four years by the Member States. This report needs to include the following information: nitrate concentrations in ground water and surface water bodies; eutrophication of surface water; assessment of the impact of action programs on water quality and agricultural practices; revision of NVZs and action programs; estimation of future trends in water quality. As indicated before although phosphorus is not mentioned in the Nitrates Directive, its implementation by several member states includes the application of phosphorus regulations. For example the Nitrates Action Programme (NAP) 2015-2018 and Phosphorus Regulations from Northern Ireland (NAP-2015-2018) establish that "organic manure with more than 0.25 kg TP/kg TN (e.g. some anaerobic digestates) can only be applied where soil analysis shows there is a crop requirement for phosphorus."

Around ten years later the Water Framework Directive (2000/60/EC) was approved based on a new water policy approach more focused on the sources of nutrients pollution instead of the receiving water bodies. The main environmental objectives are to achieve and maintain a good status for all surface and ground water bodies throughout the EU territory and to prevent deterioration and ensure the conservation of high water quality. The Directive established the necessity to estimate and identify the significant points and diffuse sources of pollution from urban, industrial, agricultural and other installations and activities origin, listing a series of substances including substances which contribute to eutrophication (in particular, nitrates and phosphates). To achieve the objectives, River Basin Management Plans (RBMPs) have to be implemented. These plans should provide a clear indication that the objectives set for the river basin are fulfilled in the expected time. The first RBMPs were published in 2009 and they are updated every six years. A report from the European Commission (Karabulut et al., 2014) about the implementation of Good Ecological Status by 2015 indicated the established goals will not be reached for approximately 50% of EU water bodies, being hydromorphologial pressures and diffuse nutrient pollution the most significant pressures. In the mentioned report the agriculture is identified as a significant pressure in more than 90% of the RBMPs.

The Groundwater Directive (2006/118/EC) indicates that the groundwater is the most sensitive and the largest body of freshwater in the EU and, in particular, also the main source of public drinking water supplies in many regions. This Directive establishes a regime which sets underground water quality standards and introduces measures to prevent or limit inputs of pollutants into groundwater. In this sense, the Directive specifies groundwater quality standards for nitrate and active substances in pesticides, including their relevant metabolites, degradation and reaction products.

The Marine Strategy Framework Directive (2008/56/EC) sets the objective of achieving Good Environmental Status for marine waters by 2020 using coherent approaches across sea basins. Good Environmental Status is determined based on qualitative descriptors, which include one related to human-induced eutrophication. In this sense, the Directive established the necessity to diminish the eutrophication, especially its adverse effects such as losses in biodiversity, ecosystem degradation, harmful algal blooms and oxygen deficiency in bottom waters. The Directive stated that the evaluation of the eutrophication in marine waters needs to consider the evaluation of coastal and transitional waters under Water Framework Directive (2000/60/EC), also taking into consideration the information and knowledge of regional sea conventions. The assessment needs to combine information on:

1. Nutrient levels,
2. Direct effects and
3. Indirect effects of eutrophication.

The Directive 2008/56/EC indicates the necessity of cooperation between the Member States considering that the concentration of nutrients is related to nutrient loads from rivers in the catchment area.

The Waste Framework Directive (2008/98/EC) sets the basic concepts and definitions related to waste management. Although "wastewater" is not included because it is addressed in another specific regulation (i.e. UWWT Directive 91/271/EEC), the Directive lays down some basic waste management principles. It requires the management of waste without endangering human health and harming the environment, and in particular without risk to water, air, soil, plants or animals, without causing a nuisance through noise or odours, and without adversely affecting the countryside or places of particular interest. Therefore, it deals with the necessity to guarantee an efficient waste management to avoid the pollution of water bodies by these wastes, including indirectly the eutrophication problem.

Finally, the Industrial Emissions Directive (2010/75/EU), which replaces IPPC Directive 96/61/EC, will be paid special attention in this chapter to focus the problematic of nutrients pollution by industrial activities.

## Industrial Emissions Directive (2010/75/EU)

The Industrial Emission Directive (IED) (2010/75/EU) is the main EU instrument that regulates the pollutant emissions from industrial installations. The IED has the aim to prevent, reduce and as far as possible eliminate pollution arising from industrial activities. The Directive states the necessity of a general framework for the control of the main industrial activities and is based on five pillars:

1. **Integrated Approach**: Take into account the whole environmental performance of the plant, covering emissions to air, water and land, generation of waste, use of raw materials, energy efficiency, noise, prevention of accidents, and restoration of the site upon closure.
2. **Use of Best Available Techniques (BAT):** The BATs are defined taking into account the information provided by experts from the Member States, industry and environmental organisations. The conclusions obtained from the BATs are the base to set the permit conditions including emission limit values.
3. **Flexibility**: In particular cases the competent authorities have some flexibility to set less strict emission (with a justified document). For example, when the assessment shows that achieving the emission levels (indicated by the BAT) could be very expensive compared to the environmental benefits, based on geographical location, local environmental conditions or technical characteristics of the installation.
4. **Inspections**: The Member States shall coordinate investigations, at least, every 1 to 3 years, using risk-based criteria.
5. **Public Participation**: The Directive specifies the right of the public to: participate in the decision-making process, be informed of its consequences, access to the permits and know the results of the monitored activities.

Following this directive the European Pollutant Release and Transfer Register (E-PRTR) was created (EEA, 2016a). The E-PRTR is a public register to provide environmental information on major industrial activities which contains data by more than 30,000 industrial facilities covering 65 economic activities across Europe. In this register information regarding diffuse emissions of N and P (as kg/ha) to water is provided including atmospheric deposition, agriculture, UWWTPs, un-connected households and inland navigation.

The last report from the EU (Kourti et al., 2015) that reviews the IED implementation by BATs is based on the analysis of seven industrial sectors. These sectors were selected due to the availability of data for them (cement and lime, glass, iron and steel, chlor-alkali, tanneries, refineries and pulp and paper sector). In this report, the nitrogen is considered as a key water pollutant for five industries (glass, iron and steel, tanneries, refineries and pulp and paper) while the phosphorus is considered in only one sector (pulp and paper). Table 2 presents a summary extracted from this report for these sectors about the key water pollutants and BAT-AELs (Best Available Technology-Associated Emission Levels). The corresponding Best Available Techniques Reference Document (BREF) for each sector indicates precisely the BAT to accomplish with these BAT-AELs.

For the iron and steel sector, the IS BREF document (Remus, Aguado, Roudier, & Delgado, 2013) indicates for wastewater from the coking process and coke oven gas cleaning a pre-treatment based on ammonia stripping by using alkaline and steam. The BAT for this pre-treated wastewater is to use biological wastewater treatment (aerobic system with activated sludge) with integrated denitrification/nitrification stages. The BREF document preferably indicates the application of pre-denitrification/nitrification treatment to achieve values lower than 35 mg N/L for the sum of nitrogen as ammonia, nitrite and nitrate. In the case of the tannery sector, the BREF document (Black et al., 2013) suggests as BAT for nitrogen removal a biological process based on "nitrification of ammoniacal nitrogen compounds

*Table 2. Overview of main wastewater pollutants by sectors in the report of IED implementation by BATs –AELs*

| Industrial Sector | Key Water Pollutants |
|---|---|
| Cement and lime | No key water pollutants were identified. Emissions to water are considered insignificant and there are not BAT-AELs. |
| Glass | The BAT-AELs cover the following pollutants: ammonia, arsenic, cadmium, copper, fluorides, lead, nickel, phenols and zinc.<br>For ammonia WW discharge to surface water the BAT-AELs is: $NH_4^+ < 10$ mg/L |
| Iron and steel | Kjeldahl Nitrogen <45 mg N/L<br>Sum of ammonia, nitrite and nitrate <15-50 mg N/L |
| Tanneries | For direct discharges of WW after treatment (in mg/L):<br>COD=200-500 mg/L, $BOD_5$=15-25 mg/L, suspended solids < 35 mg/L, nitrogen as ammonia <10 mg/L, Cr < 0.3-1.0 mg/L, sulphide < 1 mg/L.<br>Indirect discharges of WW in urban WWTP:<br>Cr < 0.3-1.0 mg/L, sulphide < 1 mg/L |
| Chlor-alkali | Free chlorine<br>Hg |
| Refineries | Total N between 1-25 mg N/L<br>Heavy metals (Pb, Cd, Ni and Hg) |
| Pulp and paper | Total Phosphorus (TP) 0.001-0.07 kg P/t<br>Total Nitrogen (TN) 0.01-0.4 kg N/t<br>Absorbable Organic Halides (AOX) |

(Kourti et al., 2015)

to nitrates, followed by the reduction of nitrates to gaseous nitrogen". The document also specifies that the biological nitrogen removal has difficult implementation into existing plants where there are space limitations, indicating as emerging technique the use of biofilms. The BAT for refinery sector is also a biological process based on nitrification/denitrification for nitrogen removal. The REF BREF document (Barthe, Chaugny, Roudier, & Delgado, 2015) says that "wastewater treatment techniques are mature techniques, and emphasis has now shifted to prevention and reduction of contaminated wastewater streams to the final treatment units". The PP BREF document for pulp and paper sector (Suhr et al., 2015) mentioned different BAT-AELs for total N and total P depending on the type of the process used and the raw material. With respect to the BATs the document states the use of biological nitrification/denitrification process for nitrogen removal and tertiary treatment by chemical precipitation for phosphorus removal.

## Impact of Regulations Over Nutrients Pollution

The European Environmental Agency (EEA) is the organism responsible for the monitoring of the impact of the different policies concerning the environment. Regarding the nutrients pollution, the EEA performed several studies in the last years that suggest the occurrence of a "decoupling" between the excess of nitrogen and phosphorus production and the population growth. The monitoring studies (EEA, 2008) indicate that phosphorus concentrations decreased in rivers and to a lesser extent in lakes during the 1990s, reflecting the improvement in the wastewater treatment over this period. In the same period, the nitrate concentrations in rivers suffered a minor decrease while there is no evidence of changes in the groundwater and lakes nitrate levels. This positive change was also confirmed in a technical report elaborated by the European Commission (Karabulut et al., 2014), which indicates that nitrogen and phosphorus inputs to the EU environment have decreased considerably over the past 20 years. However, the same report underlined that an excess of nutrients continues affecting air and water quality, with negative impacts on the ecosystems and human health.

Regarding the industrial activities the emission to water of nutrients was evaluated by the European Environment Agency in a document (EEA, 2016b). It was observed that the manufacturing industry in 11 countries (Austria, Czech Republic, Germany, Greece, Hungary, Lithuania, Netherlands, Norway, Portugal, Spain and Sweden) has achieved absolute decoupling of nutrient emissions from economic growth. A decrease in emissions coupled with a decrease in gross value added (GVA) occurred in the United Kingdom, France, Italy, Belgium and Finland. However, in all cases (except Finland), the rate of emissions decrease was greater than that of GVA. An increase in nutrient emissions, accompanying the growth in GVA, was observed in Slovakia and Poland.

Therefore, Europe policies seem to be effective, although the problem is not solved yet and the eutrophication is still a significant issue. Further efforts are needed to manage the nutrients cycles in a more cost-effective, sustainable and resource efficient way, gaining importance the concept of Circular Economy.

### Circular Economy

Recently, on December 2015, the European Commission presented the new circular economy package to promote the water reuse at EU level. Among the action plans and initiatives that the Commission intends to develop in the period 2016-2017 the "water reuse in industrial activities" is included. Although the water reuse in many sectors is already a common practice, under the scope of the Industrial Emissions Directive (2010/75/EU) the Commission want to enhance the role of water reuse reviewing the BREFs for relevant industrial sectors.

Regarding the nutrients recovery from wastewater, the main route is their use as fertilizer in the soil. However, the existing Fertilisers Regulation in the EU only ensures free movement for traditional fertilisers typically made of mined or synthetic raw materials, not including a clear procedure for organic fertilisers. For this reason in the concept of the Circular Economy, the EU also includes a new regulation to boost the use of organic and waste-based fertilisers. This initiative will help to take market advantages for these fertilizers, which are not economically competitive with conventional fertilizers. This means that in the future the technologies for nutrients recovery, already developed but not sufficiently exploited, will increase interest.

## THE UNITED STATES OF AMERICA

In U.S. is the Environmental Protection Agency (EPA) the organism responsible for the control and prevention of the environment, including air, water and land, from pollution. Regarding the contamination of the water bodies by an excess of nutrients, the EPA states that "nutrient pollution is one of America's most widespread, costly and challenging environmental problems". A report of this agency (EPA-820-F-15-096, 2015) evaluated the costs related to nutrients pollution and classified them into two categories. The first category is costs associated with nutrient pollution reduction at its sources: municipal wastewater treatment plants, decentralized wastewater treatment systems, industrial wastewater treatment, urban and residential runoff. The second type is external costs derived from the impacts of the nutrients pollution: tourism and recreation, commercial fishing, property values, human health, drinking water treatment costs, mitigation, restoration. To reduce and mitigate these costs different programs are developed by the EPA in the frame of the major law in U.S. about water pollution: the Clean Water Act (CWA-1972).

## The Clean Water Act (CWA)

In 1948, the United States of America promulgated their first major law concerning the water pollution (The Federal Water Pollution Control Act). Later in 1972, it was amended, and the law became commonly known as the Clean Water Act (CWA). The main points of the CWA directive are:

- To establish the basic structure for regulating pollutant discharges into the water bodies of the United States.
- To give the EPA the authority to implement the pollution control programs such as setting wastewater standards for the industry.
- To maintain existing requirements to set water quality standards for all contaminants in surface waters.
- To made it unlawful for any person to discharge any pollutant from a point source into navigable water bodies, unless a permit was obtained under its provisions.
- To fund the construction of sewage treatment plants under the construction grants program.
- To recognize the need for planning to address the critical problems posed by nonpoint source pollution.

According to the CWA directive the different states and authorities are responsible for: establishing water quality standards according to the designated use and criteria to protect those uses, developing antidegradation policies and implementation methods, and promoting the protection of downstream waters. In the frame of the CWA two important programs were developed:

- **The National Pollutant Discharge Elimination System (NPDES):** This program is included in the CWA and its administration is performed by the EPA and the states. The NPDES establishes the permits that reflect federal standards for discharges. If the regulatory agency determines that a particular water body is not meeting water quality standards, discharge permits to those water sources are imposed. NPDES permits to regulate urban and industrial wastes that are collected in sewers and treated at municipal WWTPs. It also permits to control the direct discharge into water bodies of industrial point sources and concentrated animal feeding operations. The aims of these permits are to protect public health and aquatic life and to ensure that every facility treats its produced wastewater. The pollutants regulated by NPDES are classified as conventional pollutants (faecal coliforms or oil and grease), toxic pollutants (including pesticides, solvents, polychlorinated biphenyls, dioxins and heavy metals) and nonconventional pollutants (additional substances that may require regulation such as nitrogen and phosphorus).
- **The Total Maximum Daily Load (TMDL):** According to the CWA each state must develop TMDLs for all the water bodies. After developing the TMDLs, they are submitted by the states to EPA for approval. The TMDL program establishes the maximum pollutant level that can be present in a water body maintaining it quality standards. States must list waters that do not fulfill the

water quality standards, even after the application of required control, and then establish TMDLs for these listed water bodies. States are responsible for the development, achievement and maintenance of TMDL by allocating reductions among all sources.

## Nutrient Pollution Policy and Data: Numeric Nutrient Criteria

In the frame of the CWA the EPA is working to combat the nitrogen and phosphorus pollution problems in the U.S. throughout the Nutrient Pollution Policy and Data (EPA, 2016). One of the primary goals is try to develop numeric nutrient criteria for nitrogen and phosphorus pollution including values and parameters, to serve as a tool for protecting and restoring the designated uses of a water body with regards to nutrients pollution. The development of numeric nutrient criteria requires a great effort from the states and for this reason different guidance documents are proposed by the EPA:

1. **Stressor-Response Relationships:** This document describes a four-step approach for estimating and interpreting nutrient stressor-response relationships to derive numeric nutrient criteria and water quality standards under the CWA.
2. **Estuarine and Coastal Waters:** This document is to provide scientifically defensible technical guidance to develop quantitative nutrients and algal criteria for estuaries and coastal waters.
3. **Lakes and Reservoirs:** This document presents a general consensus about freshwater lake responses to nutrient additions. Essentially stated that an ambient of total phosphorus (TP) concentration greater than about 0.01 mg/L and/or total nitrogen (TN) of about 0.15 mg/L is able to predict the formation of algal bloom problems during the growing season. However due to the particularities of some regions of the country, with different soil compositions and precipitation regimes, the application of that rule has to be adjusted by zones. For this reason, the continental United States has been divided into 14 separate eco-regions of similar geographical characteristics and criteria will be developed for each of them.
4. **Rivers and Streams:** This document provides background information on classifying rivers and streams, selecting criteria variables, designing monitoring programs, building a database analysing nutrient and algal data, deriving regional criteria, and implementing management practices.
5. **Wetlands:** This document presents EPA's scientific recommendations on logical approaches for developing regional nutrient criteria for wetlands. The wetlands document has information on classifying wetlands, selecting criteria variables, designing monitoring programs, building a database analysing nutrient and algal data, deriving local rules, and implementing management practices.

The numeric nutrient criteria are still under development, and the states are classified in various levels according to the number of data for different types of water (Table 3). At the moment, the maximum level (Level 5), which corresponds to a complete set of N and P criteria for all water types (lakes/reservoirs, rivers/streams and estuaries), is only achieved by four states. The most of the states (28) are at the minimum level (Level 1) without N and/or P numeric criteria. Therefore, there is still a huge amount of work to do in the collection of data to set adequate policies about nutrient pollution. The numeric values established for N and P as adequate to avoid problems of nutrient pollution depends on each state and the type of water body. Examples of numeric nutrient criteria for different levels are showed in Table 4.

*Table 3. Summary of numeric total nitrogen and total phosphorus criteria based on state-provided information to EPA*

| Number of Level | Description* | Number of States Affected | States |
|---|---|---|---|
| Level 1 | No N and/or P criteria | 28 | Alabama, Alaska, Arkansas, Connecticut, Delaware, District of Columbia, Idaho, Indiana, Iowa, Kansas, Kentucky, Louisiana, Maine, Maryland, Michigan, Mississippi, New Hampshire, New York, North Carolina, North Dakota, Ohio, Pennsylvania, South Dakota, Tennessee, Texas, Utah, Washington, Wyoming |
| Level 2 | Some water bodies with N and/or P criteria | 16 | Arizona, California, Colorado, Georgia, Illinois, Massachusetts, Missouri, Montana, Nebraska, Nevada, New Mexico, Oklahoma, Oregon, South Carolina, Vermont, Virginia |
| Level 3 | 1 water type with N and/or P criteria | 3 | Rhode Island, U.S. Virgin Islands, West Virginia |
| Level 4 | 2 or more water types with N and/ or P criteria | 5 | Florida, Minnesota, New Jersey, Puerto Rico, Wisconsin |
| Level 5 | Complete set of N and P criteria for all water types | 4 | American Samoa, Guam, Hawaii, Northern Marianas Islands |

*( EPA, 2016 )* *Types of water: lakes/reservoirs, rivers/streams, estuaries.

*Table 4. Examples of numeric nutrient criteria for N and P in states from U.S. with different level classification*

| State | Criteria for N | Criteria for P |
|---|---|---|
| New York (Level 1) | For lakes/reservoirs, rivers/streams and wetlands the nitrite as N:<br>100 μg/L (aquatic life chronic standard)<br>20 μg/L (trout water) | --- |
| Arizona (Level 2) | For lakes/reservoirs: 1.6-1.9 mg/L as TN<br>For rivers/streams: 0.1-3 mg/L as TN | For lakes/reservoirs: 115-160 μg/L as TP<br>For rivers/streams: 0.05-0.8 mg/L as TP |
| West Virginia (Level 3) | --- | For lakes/reservoirs: 30-40 μg/L as TP |
| New Jersey (Level 4) | For lakes/reservoirs: 2 mg/L as nitrate<br>For rivers/streams: 2 mg/L as nitrate | For lakes/reservoirs: 0.05 mg/L as TP<br>For rivers/streams: 0.1 mg/L as TP |
| Hawaii (Level 5) | Open coastal water: 110-350 μg/L as TN and 3.5-25 μg/L as nitrite+nitrate<br>Ocean water: 50-100 μg/L as TN and 1.5-3.5 μg/L as nitrite+nitrate<br>Estuaries: 200-750 μg/L as TN and 8-70 μg/L as nitrite+nitrate<br>All embayments: 150-500 μg/L as TN and 5-35 μg/L as nitrite+nitrate | Open coastal water: 16-60 μg/L as TP<br>Ocean water: 10-25 μg/L as TP<br>Estuaries: 25-200 μg/L as TP<br>All embayments: 20-75 μg/L as TP |

The numeric nutrient criteria are a crucial tool to establish the values concerning nitrogen and phosphorus in the two important programs respect to the water pollution: the National Pollutant Discharge Elimination System (NPDES) and the Total Maximum Daily Loads (TMDLs).

## Industrial Wastewater

Wastewater discharges from industrial activities may contain pollutants at levels that could affect the quality of receiving waters or interfere with publicly owned treatment works (POTW) that receive those discharges. For this reason, industries must obtain permits if their discharges go directly to surface water or to perform a pre-treatment program if the discharges go to municipal sewage treatment plants. In this sense the industrial wastewater discharges need to consider the following points:

- **NPDES Permit Program:** It establishes specific limitations based on the type of facility/activity generating the discharge. The limits are based on both: the technology available to control the pollutants (i.e., technology-based effluent limits) and the protective limits for the receiving water (i.e., water quality-based effluent limits).
- **National Pretreatment Program:** It establishes the standards that need to accomplish the non-domestic wastewaters (i.e. industrial) to discharge into a POTW, which provides benefits to both, the POTWs and the industrial users, because reduces/eliminates pollution at the point source and prevents the introduction of contaminants in the POTW that can disturb its normal operation.
- **Effluent Limitations Guidelines and Standards:** Established for non-municipal effluents, like industry ones. These standards are based on the degree of pollutant reduction achieved by an industrial category through the application of pollutant control technologies. These are technology-based regulations (they are not based on risk or impacts upon receiving water bodies) and represent the compromise between the greatest pollutant reduction and the assumable cost for the industry. The standards for direct dischargers are included in the NPDES permits or other control mechanisms for indirect dischargers (i.e. Pretreatment Program).

As an example, the limits for the industry of fertilizer manufacturing regarding nitrogen and phosphorus values is presented in Table 5, regulated by the code federal regulation (CFR) number 418. The CFR 418 states that "the effluent limitations and guidelines representing the degree of effluent reduction attainable by the application of the best practicable control technology currently available".

To summarize industrial plants can either be connected to wastewater treatment plants or discharge directly into receiving waters. Discharges to wastewater treatment plants must comply with certain pre-treatment requirements. Direct discharges must have a NPDES permit which establishes the limits of pollutants in the effluent. The permits are based on the use of best available technology, if it is insufficient to meet water quality standards, further action may be required. The EPA recognizes that, although the

*Table 5. Limits of discharge for fertilizer manufacturing industry according to CFR 418*

| Parameter | Maximum for One Day | Average of Daily Values for 30 Consecutive Days | Discharge to a Publicly Owned Treatment Work |
|---|---|---|---|
| Total phosphorus (mg P/L) | 105 | 35 | 35 |
| Ammonia (mg N/L) | 50.0-187.5* | 25.0-62.5* | 30 |

* Depended of the discharge: point source or a new source.

NPDES program and pre-treatment requirements have made significant advances avoiding industrial pollution, the industrial activities are still a major cause of environmental degradation in some areas.

Respect to the treatment technologies recommended for the reduction of nutrients concentration there is information from the EPA regarding municipal facilities, which could be useful for the industrial wastewaters. A report that reviews different biological nitrogen removal technologies and their costs (EPA-823-R-07-002, 2007) indicates that the technology needs to be selected for each facility depending on the target effluent quality, operator experience, influent quality, existing treatment processes and possibility to remove phosphorus simultaneously. The technologies suggested in this report for biological nitrogen removal includes: Modified Ludzack-Ettinger (MLE) Process, $A^2/O$ Process, Step Feed Process, Bardenpho Process, Sequencing Batch Reactor (SBR) Process, Rotating Biological Contactor (RBC) Process and Oxidation Ditch. Respect to the phosphorus removal the report indicates that chemical precipitation is preferred over biological removal. Most of these processes will be described in the following chapters.

## CASE STUDY FOR SLAUGHTERHOUSES AND MEAT PROCESSING INDUSTRIES

The European Union and the United States of America impose the discharge limits for a certain industrial sector based on the attained concentrations achieved by the best available techniques. In this section, a comparative of these techniques and limits is performed for the slaughterhouse and meat processing sector. This industrial sector is characterized by the consumption of high amounts of water in the different processes; consequently, a high volume of wastewater is produced containing mainly solids, organic matter and nutrients (N and P). The concentration of nutrients in these effluents is important due to the nitrogen and phosphorus released from the breakdown of meat proteins. The information discussed below was obtained for the EU from the "Reference Document on Best Available Techniques in the Slaughterhouses and Animal By-products Industries" (SA BREF, 2005); and for the US from the "Technical Development Document for the Final Effluent Limitations Guidelines and Standards for the Meat and Poultry Products Point Source Category (40 CFR 432)" (EPA-821-R-04-011, 2004).

### European Union

The EU document recommends that, previous to the treatment, the minimization of the wastewater contamination generated in slaughterhouses needs to be performed by good practices. For example, the minimization of water use in each operational process, or the collection of by- wastes (e.g. skin, hair, etc.) as close to the source as possible, to prevent their contact with water. Furthermore, a good management of the selection and use of cleaning chemicals is essential to ensure that they do not kill the microorganisms in the biological processes of the WWTP.

Regarding the treatment of the effluents generated in this type of industries the EU indicates the necessity to perform the complete treatment in the same installation or a pre-treatment if the wastewater is discharged to an urban facility. However, some European countries, such as Denmark, Finland, Sweden and Norway take advantage of the organic matter content in this wastewater to be used for denitrification at municipal WWTPs, so only limited treatment is undertaken at the slaughterhouses.

The technologies that the EU recommends to treat the effluents generated in this industry are classified in Table 6 according to the compound removed. Regarding the BATs to remove nitrogen and phosphorous the document indicates the combination of the carbon oxidation, nitrification and dentrification steps in a single process. The use of a sequencing batch reactor (SBR), that allows alternate aerobic and anoxic periods during the operating cycle, is also suggested in the document to achieve a combination of carbon oxidation, nitrogen reduction and phosphorus removal. The document also indicates that the phosphorus can be removed by coagulant addition or biologically without coagulant addition.

The effluent discharge limits are imposed by the local authority for each industrial installation and can differ between them. These limits normally include suspended solids (SS), chemical oxygen demand (COD), biological oxygen demand (BOD), pH, ammonia nitrogen, total nitrogen, total phosphorus, free or emulsified grease, anionic synthetic detergents, daily flow and maximum weekly flow. They may also include sulphide concentrations, due to the potential for anaerobic decomposition of sludge. Discharge permits for slaughterhouses that carry out hide salting, may also include limits for chloride concentrations. The typical discharge levels and the emissions levels associated with the use of BATs for minimising wastewater emissions from slaughterhouses and animal by-products installations are summarized in Table 7.

*Table 6. Summary of technologies for treating slaughterhouse wastewater emissions*

| | TSS | Organics | Oils/fats/grease | Nitrates/ ammonia | Phosphorus |
|---|---|---|---|---|---|
| **Primary treatment** | | | | | |
| Mechanical screening | X | X | | | |
| Fat separation | X | X | X | X | X |
| Equalisation/balance tanks | | | | | |
| Dissolved air flotation<br>Dispersion flotation<br>Mechanical flotation | X<br>X<br>X | X | X | | |
| Coagulation/flocculation/precipitation | | X | X | X | X |
| Sedimentation/filtration/flotation | | X | X | | |
| **Secondary treatment** | | | | | |
| Anaerobic treatment, followed by anoxic step | | X | | | |
| Activated sludge/aeration lagoons | X | X | | X | X |
| Extended aeration | | X | | X | |
| Nitrification/denitrification | | | | X | |
| **Tertiary treatment** | | | | | |
| Filtration/coagulation/precipitation | | | | X | X |

(from SA BREF, 2005)

*Table 7. Values of the typical discharge level and the emission level associated with the BAT for the slaughterhouse sector*

| Pollutant | Typical Discharge Level (mg/L) | Emission Level Associated with BAT (mg/L) |
|---|---|---|
| COD | | 25-125 |
| BOD | <10-75 | 10-40 |
| SS | <30-60 | 5-60 |
| Total nitrogen | ≤15-65 | 15-40 |
| Ammonia | 10 | |
| Phosphate | 2 | 2-5 |

(from SA BREF, 2005)

## The United States of America

Similarly than in Europe, the U.S. document for meat and poultry industries effluents recommends the minimization in the origin of the load and flow of wastewater. To achieve this objective, the document indicates the necessity to monitor the wastewater generated in each process of the industry, to quantify the contribution of the different streams. One of the hot identified processes is the cleaning and sanitation activity, recommending the use of "dry cleaning" for solids materials removal instead of the common practice of using water as a "broom."

In the document the EPA selected a total of 54 pollutant compounds (24 classical and biological, 22 metals, and 6 pesticide) of concern in the wastewater generated by meat and poultry industries, to evaluate their possible regulation. The pollutants were considered based on their use or generation in the industry and the availability of an EPA-approved analytical method to analyse them. From all these pollutants the EPA finally selected for regulation pH, BOD, SS, hexane extractable materials (oil and grease), fecal coliforms, ammonia as nitrogen, and total nitrogen (total Kjeldahl nitrogen plus nitrite plus nitrate nitrogen). The document explains that although total phosphorus, orthophosphate and dissolved phosphorus are present in wastewater, these pollutants were not included in the final regulation because of the high cost of the treatment technology, unacceptable for some facilities (possible closure). The technologies for phosphorous removal were evaluated in the document (chemical-physical treatment), but they did not achieve a level of phosphorus reduction that justified the technology cost, and therefore, EPA did not consider those options economically achievable.

The document evaluates different technology options and describes with detail which option can be applied in function of the type of process (meat or poultry), the size of plant (small or high), the direct discharge or the discharge to the municipal facility, etc. The technology options evaluated by the EPA for the meat and poultry products industries are summarized in Table 8. The biological treatment is considered for the removal of BOD by an aerobic biological process.

The revision of the information regarding the effluents of meat and poultry industries as well as the removal efficiency of the different technologies available leads the EPA to establish the limitations for the effluents from the meat and poultry industries showed in Table 9.

*Table 8. Technology options considered by EPA for meat and poultry products facilities*

| Technology Option | Description |
|---|---|
| 1 | Biological treatment for BOD with aerobic process + limited nitrification + disinfection |
| 2 | Biological treatment for BOD with aerobic process + complete nitrification + disinfection |
| 3 | Option 2 + phosphorus removal |
| 4 | Option 2 + partial denitrification |
| 5 | Option 2 + partial denitrification + phosphorus removal |
| 6 | Option 2 + complete denitrification |
| 7 | Option 2 + complete denitrification + phosphorus removal |
| 8 | Option 2 +complete denitrification + chemical phosphorus removal + filtration |

(from EPA-821-R-04-011, 2004)

*Table 9. Limits established by the EPA for the effluents from meat and poultry products industry*

| | Pollutant | Daily Maximum Limitation (mg/L) | Monthly Average Limitation (mg/L) |
|---|---|---|---|
| Poultry | Ammonia (as N) | 8.0 | 4.0 |
| | $BOD_5$ | 26 | 16 |
| | TSS | 30 | 20 |
| | Oil & grease | 14 | 8 |
| | Total nitrogen | 147 | 103 |
| Meat | Ammonia (as N) | 8.0 | 4.0 |
| | Total nitrogen | 194 | 134 |

(from EPA-821-R-04-011, 2004)

## OTHER INITIATIVES

Around the world, the problem of nutrient pollution not only has been faced by means legislation development but also by environmental programs and initiatives developed by different national and international institutions worried about this problem. For example, the Convention for the Protection of the Marine Environment of the North-East Atlantic (OSPAR), the Helsinki Convention on the Protection of the Marine Environment of the Baltic Sea Area (HELCOM), the Convention on the Protection of the Black Sea (Bucharest Convention), the Convention for the Protection of the Mediterranean Sea (Barcelona Convention), etc. In the case of the OSPAR Convention, one of the fixed objectives is to reduce the nitrogen and phosphorus inputs by 50% with respect to the input levels corresponding to the year 1985. While in the HELCOM (2014), the targets are 13 and 41% reductions relative to 1997–2003 inputs for N and P, respectively.

An alternative worldwide approach is that of the World Health Organization (WHO), which claimed its worry about the nutrient pollution. It established the limit of 50 mg/L as a water quality standard for nitrate concentration, based on epidemiological evidence for methaemoglobinaemia in infants. This value has taken as a reference and is often used by programs aimed at controlling eutrophication.

Another example is the United Nations Environment Programme (UNEP) which is the leading environmental authority. It has focused on protecting coastal and marine waters from excessive, unregulated or illegal discharges of untreated wastewater. The UNEP developed some programs together with different institutions, like the WHO or the UNESCO, to tightening institutions and developing instruments to face environmental problems. The Strategic Action Plan on Wastewater is an example of this. Also, it hosts the Global Program of Action for the Protection of the Marine Environment from Land-based Activities (GPA), which published a global program where they identify the objectives to reduce nutrient pollution. Moreover, some activities are shown, including national, international and regional actions, policies and measures for the different states to minimize this problem.

The UNEP has a specific platform about nutrient pollution called Global Partnership on Nutrient Management (GPNM), which elaborates technical reports to provide information about the best practices to minimize nutrient pollution. The GPNM published an executive summary in collaboration with the International Nitrogen Initiative (Sutton et al., 2013) to provide a global overview of the nutrient pollution situation and to establish solutions to this challenge. They include actions in different fields, including agriculture, industry, transport, and waste or recycling. This was one of the key Messages for the United Nations Conference on Sustainable Development Rio+20 (2012). At this conference, proposals to the environmental protection were made by governments, private sectors, NGOs and other groups. Another report called "Foundations for Sustainable Nutrient Management" (2010) also sends the same message.

Besides these global platforms, the UNEP has specific programs of the different world regions which also take care of the nutrient pollution problem. One example is the Caribbean Environment Program (CEP), which also publishes technical reports and organizes conventions to assess the Caribbean countries, as the Cartagena Convention. Another example is the Mediterranean Action Plan (MAP), which started with the Barcelona Convention (1975). It also develops protocols to protect the marine environment and minimize marine pollution, including nutrient emissions.

The World Resources Institute (WRI) also takes care of the nutrient problem. It is a global research organization which involves more than 50 countries and proposes ideas to protect the natural resources. One of its publications is focused on the problem of eutrophication and proposes actions to face this. Monitoring programs, tightening institutions or creating fiscal and economic incentives are some of their suggestions.

Non-Governmental Organizations (NGO), like WWF or Greenpeace, have also been focusing a part of their work on making aware of nutrient emission problems and proposing solutions. WWF published reports about the eutrophication problems in the Baltic Sea, like "EU-trophication- linking tax money to eutrophication of the Baltic Sea" (2007) or "No more EU-trophication! How CAP reform can save the Baltic Sea" (2008). In the first one, they identify the different sources of nutrients and propose some solutions to the eutrophication problem, like the tightening of the national and international legislation, the increase in the regulations and standards for farming or the development of good agricultural practices. The second one is focused on the EU's Common Agricultural Policy (CAP), which is considered a cause of the eutrophication problem of the Baltic Sea as it promotes the intensification of the agricultural activities and thus the nutrient emissions. They propose a new Common Environment and Rural Policy (CERP), which aims the combat of eutrophication, to substitute the CAP by 2020.

Moreover, they elaborated a publication about nitrogen as a pollutant ("Nitrogen. Too much of a vital resource", 2015), in which they explain how the anthropogenic activities provoked an alteration of the nitrogen cycle. Their solutions are focused on the improvement of the efficiency of nitrogen use in agriculture and the food chain

Greenpeace has also published some reports about the problematic of nutrients pollution. One example is the technical report "Phosphorous in agriculture: Problems and Solutions". Their suggestions involve the reduction of phosphorus losses, mostly in agricultural activities, and the increase of phosphorous recovery and reuse. Also, they made a report about the impact of agricultural fertilizers in rivers, lakes and oceans (Tirado, 2008). The elimination of nutrient pollution and their main anthropogenic sources, the restoration of the vegetation of the affected zones to maximize the natural nutrient retention and the creation of initiatives to make aware of this problem are the different fields included in their proposals.

## CONCLUSION

The limits for nitrogen and phosphorus discharge to water bodies are based on environmental criteria; however when they are applied to industrial activities certain "exceptions" to these limits are observed. Despite the different legal frameworks around the world to set the limits for nutrients discharges into water bodies, the base for the requirements imposed on industrial activities is similar comparing the European Union and the United States of America. In both cases, the limits that the industries must fulfill concerning pollutants discharges are based on the technologies available to reduce them. Furthermore, a distinction is performed between the direct discharge and the discharge to a municipal WWTP. In general, the recommended techniques are for nitrogen removal based on the biological nitrification/denitrification process and for phosphorus removal the most appropriate is usually the chemical precipitation. The last years the action of different social partners (scientific community, NGOs, environmental programs, etc.) is increasing the pressure for the establishment of nutrients limits criteria more protective for the environment.

## ACKNOWLEDGMENT

This work was funded by the Spanish Government and FEDER through GRANDSEA (CTM2014-55397-JIN), FISHPOL (CTQ2014-55021-R) and REDNOVEDAR (CTQ2014-51693 REDC) projects, and by the Chilean Government through FONDECYT (1150285). The authors from the USC belong to CRETUS Strategic Partnership (AGRUP2015/02) and the Galician Competitive Research Group (GRC 2013-032), financed by the Galician Government.

## REFERENCES

Barthe, P., Chaugny, M., Roudier, S., & Delgado, L. (2015). *REF BREF. Best Available Techniques (BAT) Reference Document for the Refining of Mineral Oil and Gas*. Retrieved from The European IPPC Bureau Joint Research Centre website: http://eippcb.jrc.ec.europa.eu/reference/BREF/REF_BREF_2015.pdf

Black, M., Canova, M., Rydin, S., Scalet, B. M., Roudier, S., & Delgado, L. (2013). *Best Available Techniques (BAT) Reference Document for theTanning of Hides and Skins*. Retrieved from The European IPPC Bureau Joint Research Centre website: http://eippcb.jrc.ec.europa.eu/reference/BREF/TAN_Adopted552013.pdf

*Clean Water Act of 1972, 33 U.S.C. § 1251 et seq.* (2002). Retrieved from http://epw.senate.gov/water.pdf

Commission Directive 98/15/EC of 27 February 1998 amending Council Directive 91/271/EEC with respect to certain requirements established in Annex I thereof (1998) OJ L67/29.

Council Directive 76/160/EEC of 8 December 1975 concerning the quality of bathing water (1976) OJ L31/1.

Council Directive 76/464/EEC of 4 May 1976 on pollution caused by certain dangerous substances discharged into the aquatic environment of the Community (1976) OJ L129/23.

Council Directive 91/271/EEC of 21 May 1991 concerning urban waste-water treatment (1991) OJ L135/40.

Council Directive 91/676/EEC of 12 December 1991 concerning the protection of waters against pollution caused by nitrates from agricultural sources (1991) OJ L375/1.

Directive 2000/60/EC of the European Parliament and of the Council of 23 October 2000 establishing a framework for Community action in the field of water policy (2000) OJ L327/1.

Directive 2006/118/EC of the European Parliament and of the Council of 12 December 2006 on the protection of groundwater against pollution and deterioration (Daughter to 2000/60/EC) (2006) OJ L372/19.

Directive 2006/11/EC of the European Parliament and of the Council of 15 february 2006 on pollution caused by certain dangerous substances discharged into the aquatic environment of the community (2006) OJ L64/52.

Directive 2006/7/EC of the European Parliament and of the Council of 15 February 2006 concerning the management of bathing water quality and repealing Directive 76/160/EEC (2006) OJ L64/37.

Directive 2008/56/EC of the European Parliament and of the Council of 17 June 2008 establishing a framework for community action in the field of marine environmental policy (Marine Strategy Framework Directive) (2008) OJ L164/19.

Directive 2008/98/EC of the European Parliament and of the Council of 19 November 2008 on waste and repealing certain Directives (2008) OJ L312/3.

Directive 2010/75/EU of the European Parliament and of the Council of 24 November 2010 on industrial emissions (integrated pollution prevention and control) (2010) OJ L334/17.

Environmental Protection Agency. (2016). *State Development of Numeric Criteria for Nitrogen and Phosphorus Pollution*. Retrieved from https://www.epa.gov/nutrient-policy-data/state-development-numeric-criteria-nitrogen-and-phosphorus-pollution

EPA-820-F-15-096. (2015). *A compilation of cost data associated with the impacts and control of nutrient pollution*. Retrieved from https://www.epa.gov/sites/production/files/2015-04/documents/nutrient-economics-report-2015.pdf

EPA-821-R-04-011. (2004). *Technical Development Document for the Final Effluent Limitations Guidelines and Standards for the Meat and Poultry Products Point Source Category (40 CFR 432)*. Retrieved January 15, 2016, from http://nepis.epa.gov/Exe/ZyPDF.cgi/P1001576.PDF?Dockey=P1001576.PDF

EPA-823-R-07-002. (2007). *Bilogical Nutrient Removal Processes and Costs*. Retrieved from https://www.epa.gov/sites/production/files/documents/criteria_nutrient_bioremoval.pdf

European Commission. (2003). *CWW BREF. Integrated Pollution Prevention and Control, Reference Document on Best Available Techniques in Common Waste Water and Waste Gas Treatment / Management Systems in the Chemical Sector*. Retrieved from The European IPPC Bureau Joint Research Centre website: http://eippcb.jrc.ec.europa.eu/reference/BREF/cww_bref_0203.pdf

European Environment Agency. (2008). *Water quality and pollution by nutriets*. Retrieved from http://www.eea.europa.eu/themes/water/status-and-monitoring/state-of-surface-waters/water-quality-and-pollution-by-nutrients

European Environment Agency. (2016a). *European Pollutant Release and Transfer Register (E-PRTR)*. Retrieved January 15, 2016, from http://prtr.ec.europa.eu/#/home

European Environment Agency. (2016b). *Emission intensity of manufacturing industries in Europe*. Retrieved January 15, 2016, from http://www.eea.europa.eu/data-and-maps/indicators/emission-intensity-of-manufacturing-industries-1/assessment

Figueras, M. J., Polo, F., Inza, I., & Guarro, J. (1997). Past, present and future perspectives of the EU Bathing Water Directive. *Marine Pollution Bulletin*, *34*(3), 148–156. doi:10.1016/S0025-326X(97)00107-0

Karabulut, A., Bouraoui, F., Grizzetti, B., Bidoglio, G., & Pistocchi, A. (2014). *Managing Nitrogen and Phosphorus Loads to Water Bodies: Characterisation and Solutions. Towards Macro-Regional Integrated Nutrient Management*. Retrieved from The European IPPC Bureau Joint Research Centre website: http://publications.jrc.ec.europa.eu/repository/bitstream/JRC91624/lb-na-26822-en-n.pdf

Kourti, I., Corden, C., Grebot, B., Green, C., Cherrier, V., Pooley, M., . . . Claero, J. (2015). *Service contract for assessing the potential emission reductions deliverated by BAT conclusions adopted under the directive on industrial emissions (EID)*. Retrieved from https://circabc.europa.eu/sd/a/44aaf4c4-d716-4f02-91ab-a526b07ee6b7/Final%20report_20150501.pdf

Nitrates Action Programme. (2015). *Nitrates Action Programme and Phosphorus regulations 2015-2018*. Department of Agriculture and Rural Development. Retrieved from: https://www.daera-ni.gov.uk/sites/default/files/publications/doe/water-policy-nitrates-action-programme-201518-2015.PDF

Oenema, O., Bleeker, A., Braathen, N. A., Budňáková, M., Bull, K., Čermák, P., . . . Winiwarter, W. (2011). Nitrogen in current European policies. In The European Nitrogen Assessment. Cambridge, UK: Cambridge University Press. doi:10.1017/CBO9780511976988.007

Remus, R., Aguado, M. A., Roudier, S., & Delgado, L. (2013). *IS BREF. Best Available Techniques (BAT) Reference Document for Iron and Steel Production*. Retrieved from The European IPPC Bureau Joint Research Centre website: http://eippcb.jrc.ec.europa.eu/reference/BREF/IS_Adopted_03_2012.pdf

SA BREF. (2005). *Reference Document on Best Available Techniques in the Slaughterhouses and Animal By-products Industries*. Retrieved January 15, 2016, from http://eippcb.jrc.ec.europa.eu/reference/BREF/sa_bref_0505.pdf

Suhr, M., Klein, G., Kourti, I., Rodrigo, M., Giner, G., Roudier, S., & Delgado, L. (2015). *PP BREF. Best Available Techniques (BAT) Reference Document for the Production of Pulp, Paper and Board.* Retrieved from The European IPPC Bureau Joint Research Centre website: http://eippcb.jrc.ec.europa.eu/reference/BREF/PP_revised_BREF_2015.pdf

Sutton, M. A., Bleeker, A., Howard, C. M., Bekunda, M., Grizzett, B., de Vries, W., . . . Zhang, F. S. (2013). Our Nutrient World: The challenge to produce more food and energy with less pollution. In *Global Overview on Nutrient Management*. Centre for Ecology and Hydrology, Edinburgh on behalf of the Global Partnership on Nutrient Management and the International Nitrogen Initiative. Retrieved from: http://www.unep.org/gpa/documents/publications/ONW.pdf

Tirado, R. (2008). *Dead zones. How agricultural fertilizers kill our rivers, lakes and oceans.* Retrieved from Greenpeace webpage: http://www.greenpeace.org/canada/Global/canada/report/2008/7/dead-zones.pdf

van Eekert, M., Verdoes, N., & de Buisonjé, F. (2012). *Explorative research on innovative nitrogen recovery*. Amersfoort: STOWA. Retrieved from: http://www.stowa.nl/Upload/publicaties/STOWA%20 2012%2051_LR.pdf

## KEY TERMS AND DEFINITIONS

**BAT:** Best Available Techniques.
**BAT-AEL:** Best Available Technology-Associated Emission Levels.
**BREF:** Best Available Techniques Reference Document.
**CWA:** Clean Water Act.
**EEA:** European Environmental Agency.
**ELV:** Emission value limits.
**EPA:** Environmental Protection Agency.
**E-PRTR:** European Pollutant Release and Transfer Register.
**EU:** European Union.
**NPDES:** National Pollutant Discharge Elimination System.
**NVZ:** Nitrate Vulnerable Zones.
**P.E.:** Population equivalent.
**POTW:** Publicly owned treatment works.
**RBMP:** River Basin Management Plans.
**TMDL:** Total Maximum Daily Load.
**U.S.:** United States.
**UNEP:** United Nations Environment Programme.
**UWWTP:** Urban Wastewater Treatment Plant.

# Chapter 3
# Systems Based on Physical-Chemical Processes:
## Nutrient Recovery for Cycle Closure

**August Bonmatí-Blasi**
*IRTA, GIRO Joint Research Unit IRTA-UPC, Spain*

**Míriam Cerrillo-Moreno**
*IRTA, GIRO Joint Research Unit IRTA-UPC, Spain*

**Victor Riau-Arenas**
*IRTA, GIRO Joint Research Unit IRTA-UPC, Spain*

## ABSTRACT

*High strength waste streams, namely rejected water from a wastewater treatment plant, livestock slurry, and agro-food wastewater, are characterized by its high organic matter and nutrient content which favours processes aiming to recover energy and nutrients, instead of removing them. In this regard physical-chemical processes are suitable technologies to attain these objectives. Among others, stripping coupled with absorption, struvite precipitation, membrane separation, and vacuum evaporation, are all physical-chemical processes aiming to concentrate nutrients in a stream that can later be reused as fertilizer. In this chapter the main physical-chemical processes will be defined and described in terms of the objective of each process technique, their theoretical fundamentals, environmental effects (air, water and soil emissions, resource depletion), technical indicators (efficiencies, energy consumption, etc.), and by-product characteristics.*

## INTRODUCTION

An increasing food demand and the sharpening of agricultural detrimental effects on the environment raise urgent technological challenges. The inadequate nutrient supply, mainly nitrogen (N) and phosphorous (P), that characterizes agricultural practices in most developing countries, lead to soil quality depletion, water bodies and atmosphere pollution (Vitousek et al., 2009; Galloway et al., 2003).

DOI: 10.4018/978-1-5225-1037-6.ch003

On the other side, up to about 2-3% of the global energy consumption is ascribed to energy intensive fertilizer production processes (EU-BAT, 2006), with nitrogen fertilizers accounting for about 70% of it (Ramírez and Worrel, 2006). Almost all nitrogen fertilizers are synthesized from ammonia and around 80% of the world production capacity, estimated to be 109 Tg in 2003, is used for this purpose (EU-BAT, 2006). Direct energy requirements for the production of one Mg (tonne) of ammonia are between 40-45 GJ (Ahlgren et al., 2008) with an average cost of 100$ (Erisman et al., 2008). Moreover, fossil fuel use, mainly natural gas, is reported to represent 70-90% of these costs, thus exposing fertilizers and, indirectly, food sectors to international energy price fluctuations and instabilities (EIA, 2009).

Phosphorous, in its turn, is a mining resource with a very limited predicted lifetime; it is said that within the next 100 years it will be completed depleted (Zhang et al., 2015). This fact makes P a geo-strategic resource for most countries. Moreover, P recovery technologies and the subsequent marketing of phosphorous-rich solid products could have important positive environmental implications, as a replacement of inorganic-based P in the fertilizer industry would generate a global warming potential offset via reductions in climate emissions from P mining and production.

Management and technological solutions leading to an increase in nutrient use, and recycling efficiencies within agro-ecosystems, are receiving a renewed interest for their significant potential contribution in reducing fertilizer inputs and nutrient losses. Nevertheless, a development towards bio-base and circular economies is dependent on a more cost-effective nutrient recycling technologies from waste streams.

In this sense, both the quality and marketability of those end and by-products are relevant issues which could contribute to this economic balance. By-products from processing wastewater technologies (namely struvite, ammonia water, concentrates, etc.) are relatively new on the market, and legislations applicable to them are not always clear. Some products are not yet included and market operators are in doubt about their classification and requirements. Furthermore, due to associated logistic challenges, they have a considerably lower market price, even if also linked to market prices for mineral fertilisers (Foged et al., 2011). Besides this, end users are not always sure about the fertilization characteristics and harmlessness of these products; in this sense a lot of information is still missing.

In this chapter the main physical-chemical processes (Table 1) will be defined and described in terms of objectives of the technique, theoretical fundamentals, environmental effects (air, water and soil emissions, resources depletion, etc.), technical indicators (efficiencies, energy consumptions, etc.), side stream and by-products characteristics and destiny/use, etc. A detailed scheme of each technology and specificities for each waste stream application will also be presented.

## MECHANICAL SEPARATION TECHNOLOGIES

Mechanical separation can be an effective technique to remove solids from a (semi) liquid stream obtaining a liquid fraction containing soluble components such as mineral nitrogen (N), potassium (K) and other ions, and a solid nutrient-rich fraction containing the majority of the organic matter, including a significant proportion of phosphorus (P) (Jorgensen and Stoumann, 2009). Decanting by gravity or natural settling is usually used to easily obtain these two fractions. However, the long times and large spaces required, makes this technology not always feasible. A quicker separation can be obtained using mechanical separators which are able to reach, at the same time, high removal efficiencies and a thick stream with high dry matter and most of the nutrients.

*Table 1. Main physicochemical process technologies to recover nutrients from wastewater*

| | | |
|---|---|---|
| Physical base technologies | Mechanical separation technologies | • Sieves<br>• Screw pressing<br>• Filter belt<br>• Decanter centrifuge |
| | Membrane filtration | • Nanofiltration<br>• Ultrafiltration<br>• Reverse osmosis<br>• Other emerging technologies |
| | Evaporation and Drying | |
| Chemical base technologies | MAP (struvite) precipitation | |
| Physical-Chemical base technologies | Electrocoagulation | |
| | Stripping and Absorption | |

## Theoretical Fundamentals and Process Description

During the separation process, the distribution of nutrients between the two phases varies widely reflecting the solubility of the different compounds. For example, organic nitrogen is concentrated in the solid phase, while ammonia nitrogen remains at a similar concentration in the liquid phase. Phosphorous compounds, in their turn, are largely removed into the solid fraction, though pH can have a large effect.

Separation efficiencies of these mechanical separators may vary widely, due to the difference in the efficiency of each separator, and because separation efficiencies are affected by the variable physico-chemical composition of the streams (Moller et al., 2002)

Most separation processes are either set up to produce concentrated sludge or clarified effluents but rarely both in a single stage operation. Ensuring a clear supernatant is often at the expense of a diluted thick phase; conversely, achieving a concentrate filter cake compromises the clarification process. If both are required, then a multistage process is needed (Burton and Turner, 2003).

The performance of a separation process is often quoted as the simple ratio of the mass of a given component in the solid or concentrated phase divided by the presence of that same component in the feed stream (Equation 1). However, it should be considered that any separation process is also a method of dividing a stream. In this regard, if the apparent removal of a given component is of 25%, but the concentrate stream represents 25% of the original volume, this equates to a non-effective separation (Burton, 2007). Martinez et al. (1995) aiming to incorporate this issue, defined efficiency in more meaningful terms (Equation 2);

$$\eta = \left(S \times X_s / X_f \times F\right) \times 100 \tag{1}$$

$$\eta = S/F\left(X_s / X_f - 1\right) \times 100 \tag{2}$$

Where η is efficiency, $S$ is the concentrate solid stream (m$^3$/h), $F$ is the feed stream (m$^3$/h), $X_s$ is the concentration of a given component in $S$ (kg/m$^3$), and $X_f$ is the concentration of the same component in $F$ (kg/m$^3$).

Similarly, Moller et al. (2000) proposed a reduced separation efficiency index; where an index value of 0 indicates that the nutrients are distributed equally between the solid fraction and the liquid fraction, and a value of 1 indicates that all the nutrients are concentrated in the solid fraction.

However, the simple ratio of the mass balance is the efficiency index most commonly used.

## Usual Applications, Equipments, and Environmental Aspects

Mechanical separators are usually applied for the processing of streams with high un-dissolved solid content, namely livestock manure, sewage sludge whether raw or digested, agro-food wastewater, etc., and it is almost always a necessary first step if a further recovery technology is to be later applied.

The total solid content of a liquid stream is divided, in general terms, between those solids dissolved and those present as suspended matter. Beyond this, a useful division of the suspended matter can be made to outline the performance of the main groups of separating equipment available (Figure 1). The selection of one or another depends not only on the composition of the feed but on the final aim of the required processing technology (e.g. maximize nutrient recovery in the dry solid fraction, obtaining a liquid fraction without suspended solids, etc.).

There are different mechanical separators, the more common ones are described below (Flotats et al., 2011):

- **Sieves:** It consists of a screen of a specific pore size which allows solid particles smaller in size than the openings to pass through. The liquid flows through the screen and is drained off. There is a compromise between sieve size, separation performance, and risk of clogging. Sieve separators may be static or vibrant. Indeed, sieve clogging is one of the most common problems of static

*Figure 1. Common mechanical separator equipments (F: feed, L: Liquid phase, S: Solid phase) (Adapted from Moller et al., 2000, Burton and Turner, 2003)*

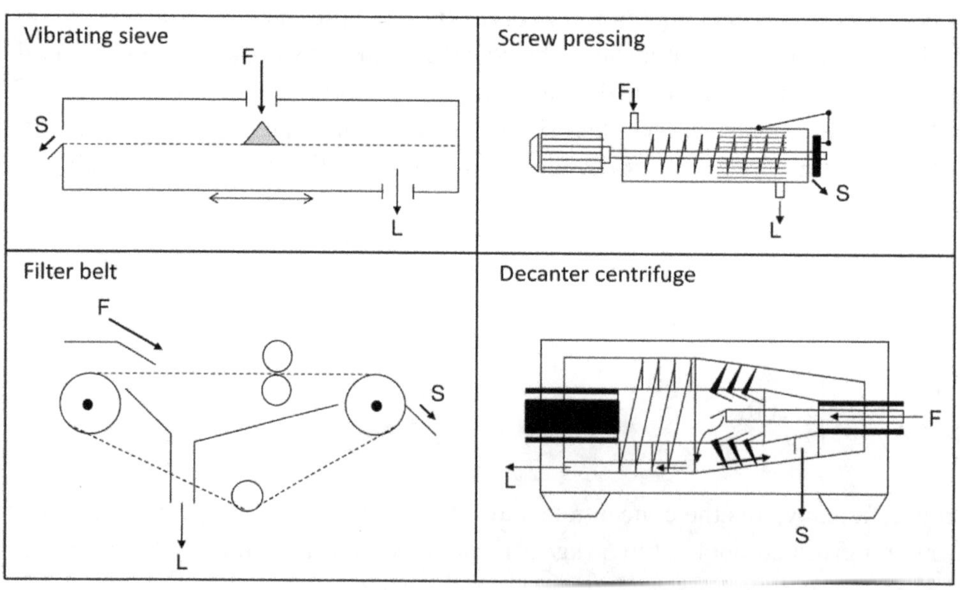

screens. Such risk is diminished in vibrant sieves due to vibration. If the flow is too high, a large amount of water can remain in the solid fraction.

- **Screw Pressing:** This is a process consisting in the application of pressure in order to separate the suspended and dissolved solids contained in a (semi) liquid, by filtration. The material to be separated enters into a cylindrical 0.5-1 mm screen by means of a screw. The liquid passes through the screen and is collected in a container surrounding the screen. At the end of the axle, the dry matter-rich fraction is pressed against a plate. The filter cake is compressed during pressure filtration, ensuring that the screw press produces a solid fraction with high dry matter content.
- **Filter Belt:** The filter belt separator also applies pressure to achieve separation. In this case the band filter is constantly turning on rollers to keep the material moving and to increase pressure on the material, whereby the liquid part will pass through the filter. The filter cake is continuously removed as the belt rotates, so that the loading and solid-fraction unloading areas are being constantly cleaned and changed over.
- **Decanter Centrifuge:** In decanter centrifuges, a centrifugal force is generated to cause the separation of solids from the liquid. There are two types of decanter centrifuges: vertical and horizontal. A horizontal decanter centrifuge uses a closed cylinder with a continuous turning motion (3000-4000 rpm). The centrifugal force separates solids and liquids at the wall, into an inner layer with a high dry matter concentration and an outer layer consisting of a liquid containing a suspension of colloids, organic components and salts. The solid and liquid phases are transported to either end of the centrifuge rotating the entire centrifuge at high speed and simultaneously rotating the conveyor at a speed slightly differing from the speed of the bowl (outer conical shell) (Flotats et al., 2011). The vertical decanter is designed and built primarily for high-temperature and/or pressurized operation. The vertical orientation lends itself to better and more reliable pressurization, with only one end to be provided with rotational seals. The amount of engineering required for pressurization and sealing makes the vertical decanters more expensive than the non-pressurised and open horizontal decanter. The availability of inexpensive vertical decanters is very limited, such that in some circumstances one has to select horizontal model even though a vertical one might have been preferred (Records and Sutherland, 2001).

Regarding possible environmental issues, some systems involve a high exposition of slurry to atmosphere, and thus, entail a risk of gaseous emissions in the form of volatile organic compounds (VOC) and/or $NH_3$, and odour problems. Nevertheless, this technology cannot be considered to have large environmental impacts.

## Nutrient Recovery by Mechanical Separation Technologies

As mentioned, nutrient recovery efficiency in the thick fraction is dependent on the characteristics of the equipment, the composition of the liquid stream and the use of additives such as coagulant-flocculants or polyelectrolytes. The separation efficiency for removal of dry matter and phosphorus (P) is ranked as follows: centrifugation> sedimentation> non-pressurized filtration> pressurized filtration. In general, total separation of N and $NH_4^+$ follows the same pattern, but separation efficiencies are lower than for dry matter and P (Hjorth, et al., 2009).

In Table 2, a summary of the efficiencies of the different devices applied on livestock slurry is shown.

*Table 2. Efficiency of different mechanical separators with livestock manure (adapted from Campos et al. 2004)*

| | Efficiency (%)[1] | | | |
|---|---|---|---|---|
| **Mechanical Separator** | **Dry Matter** | **COD** | **Nitrogen** | **Phosphorus** |
| Static sieve | 0-40 | 0-40 | 0-10 | 0-10 |
| Vibrant sieve | 0-40 | 0-50 | 0-10 | 0-30 |
| Filter belt | 10-60 | 30-60 | 10-40 | 20-80 |
| Screw pressing | 20-60 | 30-60 | 10-40 | 30-80 |
| Decanter centrifuge | 20-60 | 30-60 | 10-40 | 40-80 |

[1]% of the component removed from the feed stream

As can be observed, the range of efficiencies is quite wide, corresponding to a great variability in composition of this kind of streams and the adjustment of the different equipment.

## By-Products Characteristics and Economics Aspects

The characteristics of solid and liquid fractions resulting from mechanical separators mainly depend on the separation technique and it can be assumed as well, that the availability of nutrients for the plants can be also affected (Fangueiro et al., 2015). Furthermore, the inclusion of a previous process (e.g. anaerobic digestion or acidification) will also have a great influence over nutrient speciation, removal efficiencies and the characteristics of the two streams obtained (Moller, et al., 2000; Fangueiro et al., 2009).

The two fractions obtained with different organic matter, nutrient and ion composition should be further processed and/or used as fertilizers if an appropriate management is applied. The liquid phase is usually processed to further recover nutrients or remove them, e.g. reject water from sludge dehydrating is usually returned to the head of the WWTP; on the contrary, the liquid fraction of livestock manure is often used directly as fertilizer. The solid fraction can be used directly as fertilizer; but before usage, it is preferable to stabilize the organic matter, for example by means of composting.

Investment and operational costs of these technologies vary a lot depending on the equipment and the use of additives or not. Investment costs range between 10.000 € for sieves and 150.000 € for a decanter centrifuge. Operational costs are most influenced by the use of additives. In this case, it moves between 1.5 €/t and 3 €/t when using polyelectrolyte additives (Flotats et al., 2011).

# MEMBRANE FILTRATION

Membrane technology is a physical process for the separation of material mixtures in which a membrane acts as a filter, and the separated substances are neither thermally nor chemically nor biologically modified. This technology has grown in recent years due to growing concerns about water quality and pollution trends in relation to more complex global challenges such as rapid urbanization, increasing water demands in the domestic and industrial sectors, and problems associated with surplus nutrients in areas with high livestock density. In the last case, it could be partly solved by the use of membrane technology to concentrate manure nutrients in small volumes that could be exported as fertilizers to other agricultural regions. Membrane filtration offers concrete benefits in the production processes, such as the significant reduction in equipment size, efficiency boosting, energy savings, capital costs reduction, environmental impacts minimization, safety increase, and remote control and automation usage.

## Theoretical Fundamentals and Process Description

Membranes are typically classified depending on their pore size, applied pressure, molecular weight cut-off, and permeability (Table 3). The commonly accepted classification in the field of water and wastewater treatment includes microfiltration (MF), ultrafiltration (UF), nanofiltration (NF), and reverse osmosis (RO). Depending on wastewater composition and characteristics, as well as operational requirements, different materials are used for membranes. Membrane materials can be organic (e.g. cellulose or polymer membranes) or inorganic (e.g. ceramic or metal membranes). Compared to organic membranes, the advantages of inorganic ones are a high resistance against heat and chemicals with, correspondingly, a high regeneration capacity, as well as a long service life. Disadvantages are, above all, the higher investment due to the membrane material, and more expensive module constructions (Sadr and Saroj, 2015). Figure 2 shows a schematic diagram of the reverse osmosis process.

From a structural point of view, they can be symmetric (isotropic) or asymmetric (anisotropic). Symmetric membranes have a uniform composition and structure throughout; such membranes can be either porous or dense. Asymmetric membranes, on the other hand, consist of a number of layers, each with different structure and permeability properties. A typical asymmetric membrane has a relatively dense thin surface layer supported on an open, much thicker, microporous substrate. The surface layer performs the separation and is the principal flow barrier through the membrane. The open support layer provides mechanical strength. Inorganic membranes can be either isotropic or anisotropic (Pinnekamp

*Table 3. Membrane processes classification and operational conditions*

| Model | Membrane Process | Pore Size (μm) | Size of the Particles, Colloids, or Molecules to be Separated | Driving Force (Pressure Difference, Bar) | Application |
|---|---|---|---|---|---|
| Pore-flow | Microfiltration | 0.1-5 | Solids > 0.1 μm | 0.1-3 | Separation of solid matter from suspensions |
| Pore-flow | Ultrafiltration | 0.005-0.1 | 200,000 – 20,000 D* | 0.5-10 | Separation of macromolecules or colloids, disinfection |
| Solution-diffusion | Nanofiltration | 0.001-0.01 | 20,000 – 200 D* | 2-40 | Separation of dissolved organic molecules and polyvalent inorganic ions |
| Solution-diffusion | Reverse Osmosis | 0.0001-0.001 | < 200 D* | 5-70<br>In special cases up to 120 bar | Separation of organic molecules and all of ions |

*Dalton, numerically equivalent to the molecular weight in g/mol
(Modified from Pinnekamp and Friedrich, 2006)

*Figure 2. Schematic diagram of the reverse osmosis process*

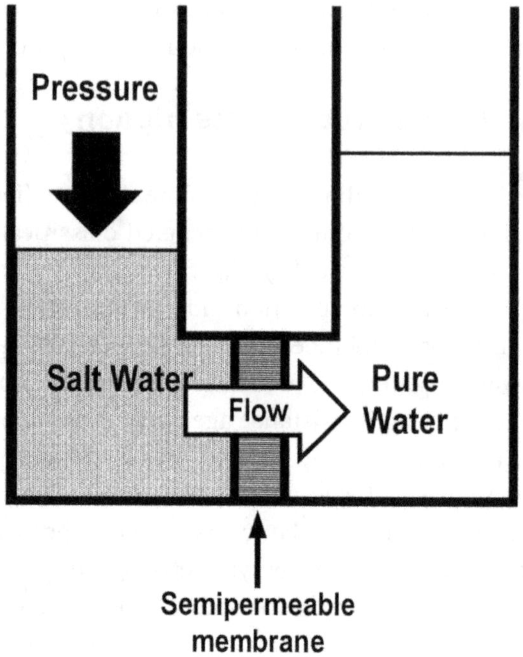


and Friedrich, 2006). The most important property of membranes is their ability to control the permeation rate of different species based on two main models (Table 3):

1. The solution-diffusion model, in which permeants dissolve in the membrane material and then diffuse through the membrane down a concentration gradient. The permeants are separated because of the differences in solubility of the materials in the membrane and the differences in the rates at which the materials diffuse through the membrane (NF, RO); and
2. The pore-flow model, in which permeants are transported by a pressure-driven convective flow through tiny pores. Separation occurs because one of the permeants is excluded (filtered) from some of the pores in the membrane, through which other permeants move on (MF, UF) (Baker, 2004).

Coarser membranes are associated with MF and the most selective membranes are associated with RO processes. Generally, suspended solids (including large microorganisms like protozoa and bacteria) are easily separated by MF whereas UF is used to remove particles down to size of around 20 nm. Moreover, NF is able to retain smaller organics and multivalent ions and RO is capable to remove all dissolved species (Table 3).

Often, membrane technologies are employed in a combination of physical, chemical, and biological processes, resulting in an efficient process configuration for a particular scenario of wastewater or waste treatment. For example, MBR technology combines activated sludge processes, or an anaerobic biodegradation process with MF, UF or RO.

## Usual Applications and Environmental Aspects

Although membrane technology was initially used almost exclusively for the desalination of sea and brackish water in arid zones, it has been tested and lately also applied for the separation of valuable materials from small water volumes and other waste streams: in biotechnology, in pharmaceutical and chemistry industries, in the metal-working industry, in the food and beverage industry, and in high strength wastewater such as livestock slurries. Specific applications of membrane processes for wastewater treatment are summarized in Table 4.

Due to its small footprint, flexible design and automated operation, membranes are a suitable option in developing countries for decentralized wastewater treatment and the extraction of valuable resources such as nutrients and energy.

## Nutrient Recovery by Membrane Processes

High-rejection membrane processes such as NF and RO have demonstrated huge potential in wastewater nutrient recovery. RO has been applied for urine concentration in a source-separation toilet system, achieving a concentration factor of five and high concentrations of ammonium, phosphate and potassium (Maurer et al., 2006). NF separation also exhibited a medium to high rejection of a range of nutrients, such as urea (Pronk et al., 2006), ammonium, phosphate and potassium (Niewersch et al., 2014, Schütte

*Table 4. Membrane processes application in municipal and industrial wastewater treatments*

| Membrane | Municipal Wastewater Treatment | Industrial Wastewater Treatment |
|---|---|---|
| **Micro and ultrafiltration** | • Separation of activated sludge and water<br>• Disinfection<br>• Pre-treatment prior to a reverse osmosis plant<br>• Phosphate removal after precipitation | • Waste water recycling and reuse as process water for different purposes<br>• Treatment of landfill leachate combined with a biological stage<br>• Recovery of water-based paint from spray booth effluents by concentration<br>• E.g. in the metal-working industry:<br>- Prolongation of the service life of electro-dipcoat bathes<br>- Concentration of water-oil emulsions<br>- Recycling of degreasing baths |
| **Nanofiltration** | For further treatment after biological methods for wastewater reuse (few cases) | • Relief of ions exchangers downstream reverse osmosis units<br>• Removal of color in the wastewater of the textile and the pulp and paper industry<br>• Demineralization of wastewater containing surfactants |
| **Reverse Osmosis** | For further treatment after biological methods for wastewater reuse (few cases) | • Concentration of drainage water from mines containing $CaSO_4$<br>• Dewatering of flushing water from photo laboratories for silver recovery<br>• Treatment of waste water from textiles dyeing (cotton and polyester dyeing)<br>• Concentration of cellulose washing water<br>• Recovery of phosphoric acid<br>• Treatment of wastewater from bleacheries<br>• Treatment of landfill leachate<br>• Treatment of livestock slurries |

(Modified from Pinnekamp and Friedrich, 2006)

et al., 2015). Häyrynen et al. (2009) studied the concentration of ammonium and nitrate in mine water by RO and NF membrane processes achieving retentions up to 96% for nitrate and 91% for ammonium. Retention of ammonium by NF membrane was 56.2% in this case.

For manure nutrient recovery, ammonia and potassium retention requires NF or RO, which have demonstrated a very good performance, whereas MF and UF membranes basically act as very efficient solid-liquid separators that can isolate nutrients associated with particles such as phosphorus (Masse et al., 2007). Konieczny et al. (2011) recovered water from pig slurry utilizing an integrated system consisting of centrifugation followed by a two-step UF followed by NF. However, the amount of ammonium in the produced water was too high, resulting in regenerated water suitable for irrigation. Guo and Jin (2014) investigated anaerobically digested cattle manure treatments with RO pre-treated by UF. It was observed that ammonia was rejected with a rate of 77.1% at the first pass and only 54.2% at the second pass, most probably due to the pH increase in the RO feed. However, the concentration of $NH_4^+$ in the final permeate was around 63 mg/L from the initial 836 mg/L. Rejection of other species such as K, Na and Cl was observed to be as high as 90%. Beyond laboratory experiences, RO has been applied in several pilot experiences at manure treatment plants to separate the liquid manure fraction into water that can be discharged and a concentrated salt solution, the so-called mineral concentrate (Velthof, 2011).

Fouling of the membranes by deposition of salts and the growth of microorganisms is one of the main problems when using reverse osmosis of the liquid fraction of slurry. Fouling of NF and RO membranes impairs membrane performance and shortens the membrane lifetime, thereby restraining productivity in nutrient recovery. Therefore, in reverse osmosis solids and organic matter should be previously removed as much as possible from the liquid fraction. This includes techniques such as UF, dissolved air flotation, low pressure membrane filtration and using cloth or paper filters, resulting in process flowcharts as the one showed in Figure 3. Thus, although the use of reverse osmosis for nutrient concentration and manure treatment is technically feasible with the proper pre-treatment, the steep maintenance and operation cost is the main hindrance to the application of this technology in the livestock manure sector.

*Figure 3. Treatment scheme of a manure treatment plant located in The Netherlands*
*(Modified from Velthof, 2011)*

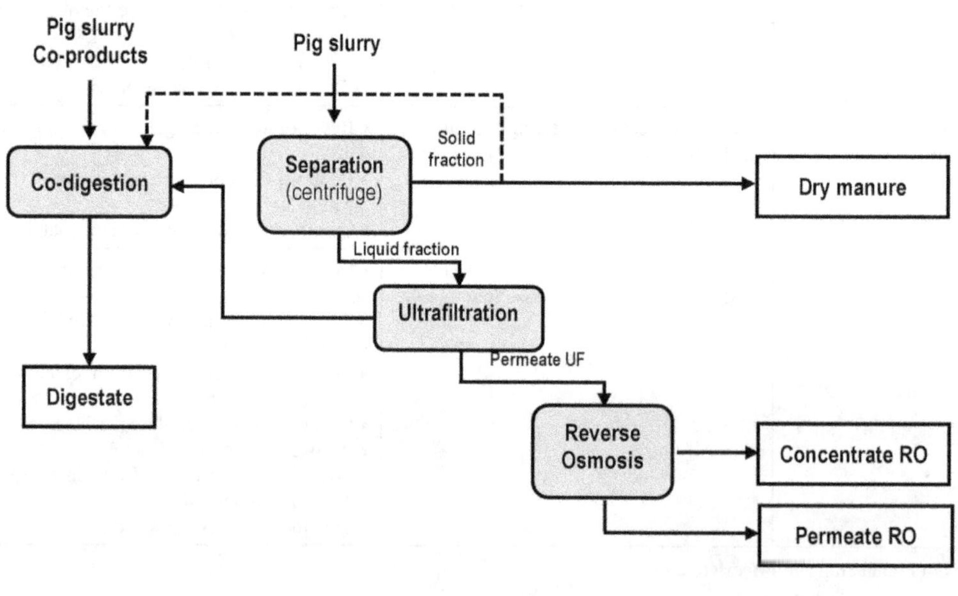

## By-Products Characteristics and Economics Aspects

The typology and characteristics of end and by-products obtained in a processing plant including a membrane process depends on the process flowchart but, in general, there are three effluents: solid fraction, concentrate and permeate.

The composition of the solid fraction is largely dependent on the separation technique, which leads to differences in the moisture content and, thus, to differences in nutrient contents of the solid fraction. This fraction contains most of the suspended solids, the organic matter and phosphorous. The levels of ammonium and potassium are largely determined by the levels of these nutrients in the untreated slurry. The composition of concentrates presents large differences depending on the processing plant, but in general, it can be said that they have a high content of N and K and low concentrations of P, as compared with the untreated slurry. The low content of organics could be classified as a mineral fertilizer. Finally, the permeate characteristics enable its direct discharge.

Based on the study carried out by Velthof (2011), the economic viability of plants treating animal slurry and other high strength streams with reverse osmosis heavily depends on the kind of slurry separator, mineral fertilizer prices, the price received from the providers of slurry, the prices of end products and of competitive products from slurry and fertilizers.

For example, the cost of the reverse osmosis installation unit is on average 7 to 8 € per ton of slurry treated. This applies to a lifespan of 10 years. The variability of the end products (low or high water content, much or little solid fraction) significantly affects profitability. Other factors affecting profitability are the further treatment of the solid fraction and the digestion of slurry. When the price paid by the providers of slurry is around 15 € per ton, or higher, plants are profitable. As a general practice in Europe, the costs of transport to the plant, weighing and sampling of the slurry are paid by the slurry supplier.

## Emerging Technologies for Nutrient Recovery by Membrane-Based Processes

Forward osmosis, membrane distillation and electrodialysis are three emerging membrane-based processes considered suitable to overcome the challenges in nutrient recovery, which could potentially represent a paradigm shift in nutrient management:

### Forward Osmosis (FO)

FO is a membrane operation developed from a natural process occurring in plants and human bodies. In FO, a semipermeable membrane is placed between two solutions of different concentrations: a concentrated draw solution and a more diluted feed solution. The driving force of the process is the chemical potential difference across the membrane which forces the water to flow from the low to high concentrated compartment, i.e. instead of hydraulic pressure; FO employs an osmotic pressure difference to drive the permeation of water across the membrane. In comparison with pressure-driven RO membrane filtration, FO has demonstrated a lower fouling propensity and a higher fouling reversibility (Blandin et al., 2016); also, a high rejection of phosphate and ammonium from activated sludge and wastewater by FO membranes has been reported (Huang et al., 2015; Xie et al., 2016). Moreover, FO has the potential to concentrate phosphate, ammonium, and magnesium in the source solution to facilitate struvite recovery due to the lack of hydraulic pressure and the occurrence of a reverse draw solute flux. These unique mass transfer properties of FO impel nutrient recovery from various waste streams such as urine, sewage, and sludge (Xie et al., 2016). Figure 4 shows a schematic diagram of the forward osmosis process.

*Figure 4. Schematic diagram of the forward osmosis process*

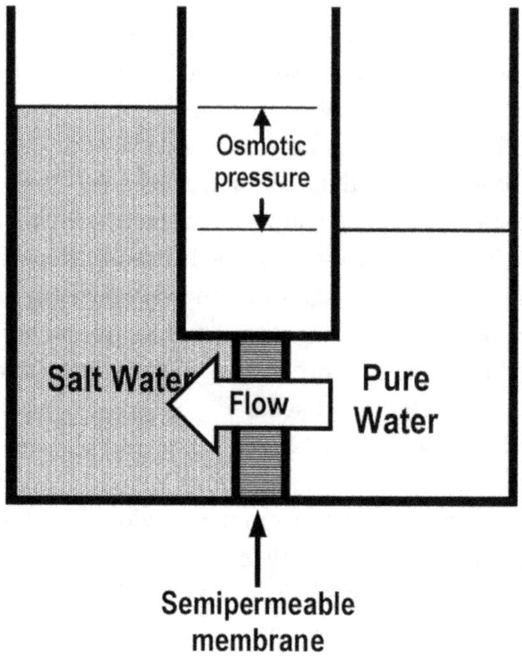


## Membrane Distillation (MD)

MD using gas-permeable membranes (GPM) is another prominent technique, which is not yet completely commercialized. Its transport mechanism is based on the volatility and vapour pressure of the compounds to be separated. These components can be concentrated either in the feed stream or the permeate streams. For example, non-volatile inorganic nutrient ions, such as potassium and phosphate, can be concentrated in the feed stream to facilitate subsequent nutrient precipitation (Quist-Jensen et al., 2015). Ammonia recovery can be another important application of a GPM process in wastewater nutrient recovery. Dissolved ammonia ($NH_3$) can cross through the micro-porous hydrophobic membrane and concentrate in an absorption solution on the other side of the membrane. The membrane is submerged in the liquid and the $NH_3$ is removed from the liquid before it escapes into the air; $NH_3$ permeates through the hydrophobic membrane reaching the acidic solution located on the other side of the membrane. Once in the acidic solution, $NH_3$ combines with free protons to form non-volatile $NH_4^+$ ions that are converted into a valuable ammonium salt fertilizer $(NH_4)_2SO_4$, (Garcia-González and Vanotti, 2015). These MD processes achieved more than 96% in ammonia recovery in the form of aqueous solution.

MD offers a number of advantages over alternative pressure-driven processes such as reverse osmosis. As the process is driven by temperature gradients, low-grade waste heat can be used, and expensive high-pressure pumps are not required. Membrane fluxes are comparable to reverse osmosis fluxes, so membrane areas are not excessive. Finally, the process is still effective with slightly reduced fluxes even with very concentrated solutions. This is an advantage over reverse osmosis, in which the feed solution osmotic pressure places a practical limit on the concentration of a salt in the feed solution to be processed (Baker, 2004).

Besides, gas-permeable membrane processes can be combined with other treatment technologies such as anaerobic digestion to improve their performance. The inhibition of methanogens by high $NH_3$ concentrations in manure severely inhibits the production of biogas. Hence, the use of gas-permeable membranes to recover $NH_3$ without damaging the carbonaceous material will improve the process.

One of the most innovative applications of gas-permeable membranes is its use coupled to bio-electrochemical systems (BES) with the objective of recovering ammonia from the anolyte without using energy-intensive stripping of the catholyte. $NH_4^+$ migration across a cation exchange membrane to maintain the charge balance, leads to a pH increase in the catholyte (pH >10) (Sotres et al., 2014), whence $NH_3$ can be recovered in an acidic solution using a hydrophobic membrane (Cheng et al., 2013).

### Electrodialysis (ED)

ED is an electro-membrane process, in which ions are transported through ion exchange membranes from one solution to another under the influence of a potential gradient. The electrical charges on the ions allow them to be driven through the membranes. Charging a voltage between two end electrodes generates the potential field required. As the ion exchange membrane has the ability to selectively transport ions with a positive or negative charge and to reject ions of the opposite charge, the useful concentration, removal or separation of electrolytes can be achieved. Therefore, it is possible that the excess nitrogen and phosphorus, which are found as ions in nutrient-rich streams such as swine liquid fraction, can be removed by this technique (Fukumoto and Haga, 2004). The ion separation is attained by ion-exchange membranes that include cation-selective, anion-selective, and bipolar membranes.

The mechanism of an ED process provides a selective tool for wastewater nutrient recovery. ED processes selectively partition phosphate from nutrient-rich streams containing various ions as a concentrated phosphate solution, achieving a concentration factor of up to 7. Similar selective phosphate enrichments by ED processes were also observed in urine nutrient recovery, resulting in a purified phosphate concentrate (Pronk et al., 2006). Phosphate selectivity in an ED process can be further enhanced by either adjusting the feed stream to the alkaline pH range or increasing current density. Better performance was expected based on the ED separation mechanism, where multivalent phosphate migrates more slowly than monovalent ions under the current field (Xie et al., 2016).

## EVAPORATION AND DRYING

The aim of evaporation and drying is to vaporize (applying heat) most of the water from a solution obtaining a concentrate that contains a desired product (in this case nutrients and organic matter). Apart from obtaining a concentrate with a lower water content and higher nutrient concentration than the original solution, another objective should be to obtain a purified condensate (water) that could be later reused (Bonmatí et al., 2003).

### Theoretical Fundamentals and Process Description

Evaporation could be performed at atmospheric pressure or under vacuum conditions. Vacuum evaporation is more common as the lower temperature for the process and the closed vessel used allow for a better emission control and an expected less energy consumption. When the pressure inside the evaporator unit

*Figure 5. Schematic diagram of the vacuum evaporation process (a) and drying process (b)*
*(adapted from Flotats et al., 2011)*

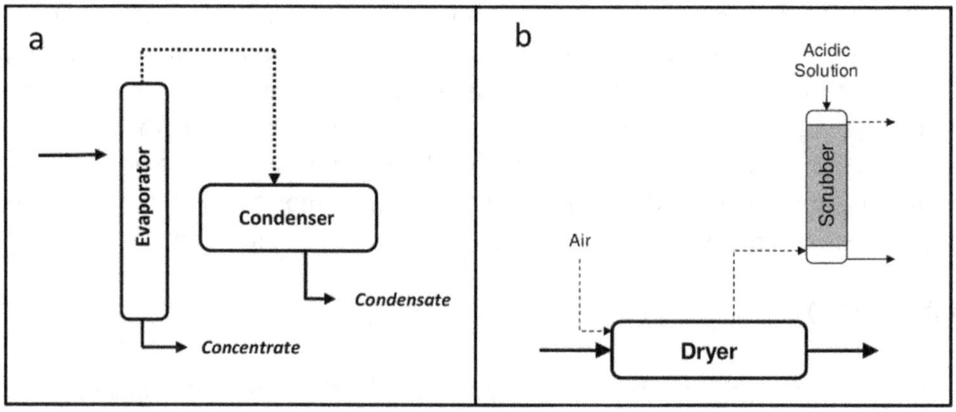


is reduced below the vapour pressure of the liquid, the liquid evaporates and the consequent concentration of the treated substrate occurs; evaporated water, then, is condensed by cooling and collected in the condensate tank (Figure 5a) (Chiumenti et al., 2013).

One limitation of evaporators is that they only work with fluids. Grüter *et al.* (1990) defined three different zones of sludge consistency in a dryer: a wet zone where the sludge is free-flowing, a sticky zone where the sludge is pasty, and a granular zone where the sludge is crumbly. This constitutes a limit to the final concentration: slurry cannot be converted into a completely dry product. The limit is about 20–25% TS, depending on the corrosion and viscosity of the concentrate. Prior removal of suspended solids is also necessary. Should this not be done, concentration would be limited as the particles form a matrix and the concentrate ceases to be a flowing liquid (Have and Chiappini, 1993). A further drying process is necessary to increase the dry matter concentration.

Drying works under the same principle, but in this case, the inflow is the solid fraction of a centrifuged slurry/sludge or the concentrate of an evaporation process. Most of the water (>85%) is removed applying heat, the concentrate reaching dry matter values >90%. In order to avoid the sludge sticky zone, dryers work in the granular zone (>60-65% of TS) where the concentrate is crumbly.

The dried product is often pelletized to increase its added value as fertilizer, pushing it through the holes of a matrix of the desired shape/size. Compression and temperature, as well as the composition of the raw material are the major factors affecting the process.

## General Applications and Environmental Aspects

This technology is mainly implemented in the field of the treatment of residuals from industrial production, in some cases with the aim of recovering chemicals, like, for example, in the galvanic industry. Vacuum evaporation has been applied for years also for the treatment of landfill leachate. In the agro-industry sector, this technology has been tested on the dairy industry, for the recovery of proteins from serum, in the olive oil industry, for the treatment of wastewater from mills and olive husks. Other tests have been performed on residues from the production of wood pulp and paper (Chiumenti et al., 2013). Regarding livestock manures, evaporation has been applied for processing pig slurry and the digestate from the co-digestion of biogas plants (Flotats et al., 2009). Although this process is applied usually at a large scale, there are some proposals for implanting it at a farm scale (Levasseur, 2004).

The increase in temperature causes the emission of several volatile compounds that can be present in the slurry leading to serious atmospheric pollution. Vacuum evaporation minimizes this effect since it occurs in closed systems, at lower temperature, and the exhausted gases can be easily treated. Considering its effects on air, water and soils, the process theoretically does not show air emissions, since the evaporated flow is recovered as a condensate. Thus, theoretically, this is a '0 gas emissions' process, including odours.

Adversely, drying is not usually operated in vacuum conditions and hence, all gaseous emissions from the dryer must be recovered (by filtration or scrubbing) to avoid ammonia ($NH_3$), organic volatiles (VOC) and odour emissions (Figure 5b).

## Nutrient Recovery by Evaporation

To ensure the recovery of nitrogen in the concentrate flow steam, and guarantee its absence in the recovered condensates, the ammonia ($NH_3$)-ammonium ($NH_4^+$) equilibrium must be modified, by means of pH control (usually adding a strong acid). If pH inside the evaporator is maintained under pH 5.5 it is guaranteed that ammonium will be recovered in the concentrate. Adversely, at this pH values, other volatile organics such as volatile fatty acids (VFA) are present in its un-ionized form (volatile) and can be easily transferred to the condensate, with manifest implications (organic contamination) in the condensate and post-treatment requirements.

Theoretical studies performed by Marks *et al.* (1994) showed that at pH=4, 1% of ammonia and 75% of acetic acid are transferred to the condensate and that, at pH=10, 93% of ammonia and 1% of acetic acid are transferred. This demonstrated that if a liquid has a high ammonia nitrogen and VFA content, it is not possible to prevent the volatilization of both components by adjusting its pH to a given value. A two-stage evaporation process or a previous treatment is necessary to achieve this. Furthermore, the non-ionized volatile organic compounds (VOCs) cannot be fixed by modifying pH. A previous treatment (i.e. stripping or anaerobic digestion) to remove VOCs is necessary in order to prevent their emissions during evaporation (Bonmatí, 2001).

For pH regulation, sulphuric acid or other strong acid should be added (Bonmatí et al., 2003). The acid consumption will depend on the slurry alkalinity. Reagents do not need to be reagent-grade. Consequently, in some applications $H_2SO_4$ sub-products of low purity have been used.

As it has been mentioned, the processed stream requires a previous acidification step to avoid ammonia emissions. The use of a strong acid has different implications:

- The choice of constructive materials (resistance to high temperature and low pH) which requires high quality materials (e.g. stainless steel).
- Risk of accidents and the requirement of Health and Safety training courses for the workers.
- Introduction of S forms (if acidification is performed with $H_2SO_4$) which could later appear (at low concentrations) in the condensates.

## By-Products Characteristics and Economics Aspects

Two by-products are obtained in this process, the concentrate and the condensate. If the process is operated correctly, the concentrate will contain most of the nutrients. Phosphorous and potassium will remain in the concentrate, and, as mentioned, if pH is set below 5.5 the majority of nitrogen will also be found in the concentrate. Likewise, if heavy metals are present in the raw manure, they will be found in the concentrate stream and, depending of their concentration, could compromise the use of the concentrate as a fertilizer.

On the other hand, condensates may be free of nutrients but can contain remarkable amounts of organic matter, as the low pH process will favour the volatilization of VOC which will later be transferred to the condensate. A previous process to remove VOC, such as stripping or anaerobic digestion, is a possible way to obtain a cleaner condensate (Bonmatí and Flotats, 2003).

In terms of energy consumption, 21 kWh/m$^3$ of electric energy and between 107 and 353 kWh/m$^3$ of heat have been obtained in a pilot experience at a farm scale unit treating 0.5 m$^3$/hour (Flotats et al., 2011). On the other hand, the increasing number of evaporation steps could significantly decrease energy consumption. For this reason a full scale evaporation system usually consists in at least two steps. Energy consumption in single-effect evaporators is very high and makes up for most of the cost of the evaporation system. Each evaporation step added reduces the energy consumption by 33%; although investment costs increase.

However, with clearly different evaporator designs, the process could be attractive for slurries/manures with high dry matter content, especially over 30% (e.g. poultry) due to the smaller volume of water to be removed and the higher yields of dry product (Burton and Turner, 2003). In this regard, the use of the residual energy from a CHP engine or other processes could be a way to make this process feasible.

## MAGNESIUM AMMONIUM PHOSPHATE (STRUVITE) PRECIPITATION

Typically, P in wastewater is removed biologically, using P- accumulating bacteria, or chemically, through precipitation by adding metal salts. Chemical precipitation introduces high costs and yields a large amount of chemical sludge to dispose of. Another option to recover nutrients is to create the physical-chemical conditions in the bulk solution (pH, ion strength, etc.) to favour the formation of different salts with low solubility products, and its subsequent precipitation to recover them. Among others, magnesium ammonium phosphate (MAP), usually called struvite, is the most common salt enabling the recovery of phosphorous and nitrogen from wastewaters.

### Theoretical Fundamentals of Map Formation

Struvite is a white crystalline substance composed of nitrogen (ammonia, $NH_4^+$), phosphorus (phosphate, $PO_4^{3-}$) and magnesium ($Mg^{2+}$) in equimolar relation ($MgNH_4PO_4{\cdot}6H_2O$), also called MAP (Magnesium Ammonium Phosphate), which is considered as a slow releasing and valuable fertilizer, as it is sparingly soluble under neutral and alkaline conditions but readily soluble in acid (Yetilmezsoy & Sapci-Zengin, 2009).

Struvite precipitation is produced in alkaline conditions when the concentration of $Mg^{2+}$, $NH_4^+$ and $PO_4^{3-}$ exceeds the solubility product, according to the following reaction (3):

$$Mg^{2+} + NH_4^+ + H_2PO_4^- + 6\,H_2O \rightarrow MgNH_4PO_4 \cdot 6\,H_2O + 2H^+ \quad (3)$$

Struvite crystal development occurs in two chemical phases: nucleation and crystal growth. In the nucleation phase, ions are arranged in the characteristic pattern of a crystalline solid, forming foundation sites for deposition of crystallizing particles which then grow into detectable crystals (Ohlinger et al., 1999).

A combination of physical and chemical parameters controls the complex mechanism of struvite precipitation. One of the main factors is pH, as it changes the concentration of free ions available for reaction. When pH increases, $Mg^{2+}$ and $NH_4^+$ concentrations decrease, as the first one complexes with hydroxides, and the second one increases its volatilization in the form of ammonia ($NH_3$). On the other hand, $PO_4^{3-}$ concentrations increase as the pH increases. pH is also involved in controlling struvite solubility, being minimal with pH values between 9 and 10.7 (Doyle and Parsons, 2002). So, pH is usually increased to favour struvite precipitation, generally using NaOH or $Mg(OH)_2$. Alternatively, $CO_2$ stripping can be used to reduce the use of alkali (Cerrillo et al., 2015) for, when $CO_2$ is liberated from solution, the change in carbonate speciation leads to a pH increase.

The molar ratio of N, Mg, and P can limit the precipitation process if one of the components is lacking in the wastewater, as can be the case of magnesium in wastewaters from anthropogenic, livestock or municipal leachate, where it may be necessary to be added in order to optimize the struvite crystallization process. Low contents of soluble phosphate in the wastewater may require a prior solubilization process and, in some cases, the solids content of the influent can slow down the struvite precipitation process (Tao et al., 2016).

The reactor design (agitation and sedimentation properties) influences the size of the obtained crystals and the easiness to recover them, as fine struvite crystals are difficult to settle or separate. The size of the aggregates influence notably in the economic value of the product. The typical struvite precipitation reactor consists in a mixed crystallization reactor with two zones. The lower zone is the reaction zone, and the upper section is the sedimentation zone, which prevents the formed particles to leave the reactor (Figure 6). The reactor could also include an aeration system to favour $CO_2$ stripping and pH increase, and prevent mechanical mixing (Cerrillo et al., 2015). The process temperature needs to be taken into account as it influences the solubility of the salts: low or ambient temperatures are preferable as maximum solubility of struvite is attained at 50 °C (Doyle and Parsons, 2002).

Finally, the ionic strength of competitive cations, such as $Ca^{2+}$ and $Na^+$ that can form other salts, especially with phosphates, as newberyite ($MgHPO_4 \cdot 3H_2O$), bobierrite ($Mg_3(PO_4)_2 \cdot 8H_2O$) or hydroxiapatite ($Ca_5(PO_4)_3OH$) can hinder struvite obtainment (Cerrillo et al., 2015). Even depending on the relative concentrations of $Ca^{2+}$ and $Na^+$, struvite formation may be inhibited (Doyle & Parsons, 2002; Moerman et al., 2009).

The analysis of the thermodynamics equilibrium and kinetics reaction can help to predict and explain the formation of struvite in different conditions (Crutchik and Garrido, 2016). However, industrial wastewater can be a complex matrix containing various compounds which might interfere with struvite precipitation, its quality as a fertiliser, and the length and shape of the crystals, as impurities block active

*Figure 6. Lab scale struvite precipitation reactor (left), and pilot plant at SAT Sant Mer Biogas Plant (Vilademuls, Girona) (right)*

growth sites by adsorption to crystal surfaces and inhibit crystal growth (Le Corre et al., 2009), leading to a possible necessary detailed study for each wastewater. As an example, Figure 7 shows struvite crystals obtained with the liquid fraction of pig slurry and the crystals formed with synthetic water applying the same operational conditions.

## Usual Applications and Environmental Aspects

Struvite precipitation technology has been studied and applied to different wastewaters, as food industry wastewater (Moerman et al., 2009), dairy manure (Uludag-Demirer et al., 2008), poultry manure (Yetilmezsoy and Sapci-Zengin, 2009), digested pig and cow slurry (Huang et al., 2011; Cerrillo et al.,

*Figure 7. Stuvite crystals obtained from pig slurry (left), and synthetic wastewater (right) (Bonmatí, A. ©)*

2015) or even human urine (Liu et al., 2013). Most of these studies show that high amounts of $NH_4^+$ and $PO_4^{3-}$ can be removed when adding the limiting reactive to wastewater (usually $Mg^{2+}$ and $PO_4^{3-}$), and working at an optimal pH. Reported $NH_4^+$ and $PO_4^{3-}$ recovery efficiencies vary depending on the substrate and the operational conditions, but are usually above 95% (Kataki et al., 2016).

Apart from the benefits obtained when recovering P and N from wastewater, some environmental issues can be attributed to these systems if they are not operated correctly. Highly mixed reactors, or those equipped with air flow systems, can induce the removal of $NH_3$ from the solution, as the relatively high operational pH and agitation creates a removal pathway for dissolved ammonia to volatilize from the solution. Moreover, the air flow dilutes the concentration of the gas-phase $NH_3$ equilibrium and increases the driving force for dissolving $NH_3$ to separate the gaseous phase (Yetilmezsoy and Sapci-Zengin, 2009).

## By-Products Characteristics and Economics Aspects

The positive value of struvite obtained from calf manure (Schuiling & Andrade, 1999), municipal wastewater (González-Ponce et al., 2009) or anaerobically digested pig slurry (Cerrillo et al., 2015), among others, has already been demonstrated. The advantages of struvite as a solid fertilizer relative to the direct land application of manure as a liquid fertilizer include ease of application, low transportation costs, reduced bulk storage, and the avoidance of odour and pathogen contamination issues (Tao et al., 2016). Moreover, struvite recovered from wastewater reduces the necessity of using chemical fertilizers obtained from phosphate rocks, a non-renewable geological reserve. Struvite dissolves slowly in the environment to release nutrients at a rate suitable for crop uptake. Its low solubility prevents struvite from contributing to typical fertilization problems such as crop burn and generation of nutrient-rich runoff (Nkoa, 2014).

Struvite production costs have been established between 4 and 15 \$/kg $N\text{-}NH_4^+$, depending on the ammonia concentration in the influent (Türker & Çelen, 2007). As the economic feasibility of struvite is heavily influenced by reagent costs, several low cost magnesium sources as by-products of industrial processes, such as magnesite pyrolysate (Huang et al., 2011), or other magnesium rich renewable sources such as sea water and brine (Liu et al., 2013), have been proposed to replace high grade magnesium. The investment for the precipitation reactor can be of around 0.12 $€/m^3_{REACTOR}$, according to Jaffer et al. (2002).

There are few crystallization plants at full scale, as there is still a lack of knowledge of struvite properties as a fertilizer by end users; and its quality, market value, difficulties in the crystallization process, pH control and reagent costs may be a hindrance to widespread struvite precipitation technology. However, some commercial full scale systems are already in the market, for example: PEARL, NuReSys, PHOSPAQ, AirPrex, etc. Commercial struvite recovery units handling municipal wastewater sludge are in operation in countries like Japan, Canada, England, USA, Australia, Sweden, The Netherlands and Italy (Kataki et al., 2016). Struvite sale price has been reported to be approximately 198-1885 \$/t in Australia and Japan (Le Corre et al., 2009), mainly depending on its purity.

# ELECTROCOAGULATION

The objective of electro coagulation (EC) is to destabilize suspended, emulsified or dissolved particles within an aqueous media, by applying an electric current. Electricity destabilizes those particles favouring the aggregation of colloidal particles just like chemical reagents do in a conventional coagulation-flocculation process.

*Figure 8. Scheme of an electrocoagulation reactor*

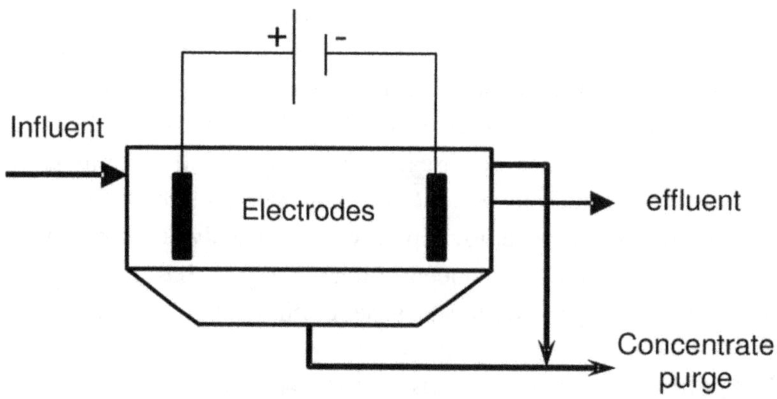


## Theoretical Fundamentals and Process Description

EC is an electrolytic process that has been already used for the treatment of various liquid wastes and consisting of dissolution of sacrificial anodes (iron or aluminium) upon application of a current between two electrodes. This provokes the dissolution of the electrode to yield metal ions ($Fe^{2+}$, $Fe^{3+}$ or $Al^{3+}$) at the anode, which are involved in the neutralization of the negative charges on the colloids of the effluent. The cathodic reaction involves the formation of hydrogen gas and hydroxide ions which favour the formation of flocks; these flocks are then easily separated by decantation or filtration, depending on their density.

EC offers some advantages over chemical dosing: it has simple equipment requirements and can be readily automated; it reduces the chemical cost using cheaper materials; gas bubbling provides for a gentle mixing which promotes coagulation and helps form bigger flocks; and gas bubbling carries some particles up to the surface of the liquid in a kind of flotation, which may then be easily separated. The main disadvantages of this method are lack of a systematic approach to EC reactor design and operation, the need to replace electrodes at regular intervals, high electricity costs and anode passivation (Shalaby et al., 2014; Lin et al., 2015).

The EC process occurs in three steps. In the first step, a coagulant is formed due to anode oxidation. In the second step, pollutants are destabilized. In the last step, destabilized matters are aggregated. The most common electrode materials for electrocoagulation are aluminium and iron. These are cheap, readily available, and have proven to be effective (Gharibi et al., 2010).

If the effluent has a high enough electrical conductivity, the electrochemical process can generate an electrical field and/or produce a field 'in situ' by sacrificial electrodes, without adding additional ionic species The process alters the surface charge of the effluent and enhances P and solids settling.

When aluminium is used as electrode material, the reactions are as follows:

Cathode: $3H_2O + 3e^- \rightarrow 3/2\ H_{2(g)} + 3OH^-$ (4)

Anode: $Al \rightarrow Al^{3+} + 3e^-$ (5)

Solution: $Al^{3+}(aq) + 3H_2O \rightarrow Al(OH)_3 + 3H^+(aq)$ (6)

When iron is used as electrode material, the reactions are:

Cathode: $3H_2O + 3e^- \rightarrow 3/2\ H_{2(g)} + 3OH^-$ (7)

Anode: $4Fe(s) \rightarrow 4Fe^{2+}{}_{(aq)} + 8e^-$ (8)

Solution (with dissolved $O_2$): $4Fe^{2+}{}_{(aq)} + 10H_2O_{(l)} + O_{2(g)} \rightarrow 4Fe(OH)_3 + 8H^+{}_{(aq)}$ (9)

Overall reaction: $4Fe_{(s)} + 10\ H_2O_{(l)} + O_{2(g)} \rightarrow Fe(OH)_{3(s)} + 4H_{2(g)}$ (10)

The release of hydroxyls and metal ions, and the formation of metal hydroxides cause the precipitation of different dissolved ions. As an example, phosphates precipitation reactions are as follows.

$H_3PO_4 + 3OH^- \rightarrow 3H_2O + PO_4^{3-}$ (11)

$Al^{3+} + PO_4^{3-} \rightarrow AlPO_4$ (12)

$Al + H_3PO_4 \rightarrow AlPO_4 + 3/2H_2$ (13)

In the solution, $H_3PO_4$ is neutralized by the hydroxyl ions released at the cathode (Equation 11). So, for pH lower than 6.5, aluminium phosphate is formed before aluminium hydroxide; consequently, the $PO_4^{3-}$ ions obtained react with $Al^{3+}$ ions produced at the anode (Equation 5). This results in the precipitation of aluminium phosphate (Equation 11). The overall reaction is best summarized by Equation 13 (Tchamango et al., 2010).

## Usual Applications and Environmental Aspects

EC has been a promising technology aiming to destabilize and remove fine particles from water and wastewaters since the beginning of the 20th century. Presently, it is considered as a prospectively competitive technology for the removal of pollutants from supply water, urban wastewater, and also for the treatment of industrial effluents such as those generated in the agro-food industry, metalworking and textile industries (Lacasa et al., 2013). EC has been successfully used to remove different kinds of pollutants such as solids, nitrogen, organic compounds, heavy metals, dyes, hardness, chemical oxygen demand (COD), and important anions such as nitrate, fluoride and phosphate (Shalaby et al., 2014). Furthermore, it has demonstrated its superior performance in treating effluents containing suspended solids, oil and grease, and other similar substances (Zhao et al., 2010). Thus, EC is also effective at removing phosphorus and solids from digestates without any chemical addition.

Regarding its environmental side effects, the performance of some systems entails a high exposition of manure/slurry to atmosphere (high stirred systems), and thus, risk of gaseous emissions (COV) and odour problems could arise. Also during flotation (aeration) a high proportion of ammonia can be discharged into the air. It is therefore necessary to treat or collect the exhaust air coming from the flotation equipment.

## Nutrient Recovery by Electrocoagulation

As mentioned, EC has proven to be a good option, in terms of efficiency, for the removal of several components such as total solids, COD, N and P. Tchamango et al. (2010) obtained N and P removals of 81% and 89%, respectively. Zhao et al. (2010) achieved P removals between 70-90%, whereas Lekuona (2004) reported higher P (>99%) but lower N (>60%) removals with electrocoagulation combined with a band filter, as can be seen in Table 5. The efficiency of the electrocoagulation process depends on many operational parameters such as the initial pH, initial pollutant concentration, temperature, nature of electrolyte and current density (Shalaby et al., 2014).

Besides, EC has been also successfully applied for removing nitrate $NO_3^-$ from different liquid streams such as wastewater, groundwater (Lacasa et al., 2011), surface water (Koparal et al., 2002) and drinking water (Vasudevan et al., 2010), among others.

## By-Products, Characteristics, and Economics Aspects

Settled solids resulting from EC have high nitrogen contents, phosphorus and other minerals, which can be used as fertilizer. Although the solids separated from the EC unit still contains considerable amount of water, they can be dried in the open air or by waste heat from the electric generator which is generally set up near the anaerobic digester. However, if the original influent contained heavy metals or other pollutants, they will end up in this solid fraction and its usefulness as a fertilizer could be compromised.

In terms of energy consumption, 20-22 kW·h per $m^3$ of pig slurry liquid fraction with about 1.1-1.3% TS has been reported (Lekuona, 2004). Additionally, EC represents an improvement in separation efficiency compared to other techniques (by electric supply) but sacrifice electrodes, consisting in iron plats, should be regularly replaced due to the release of $Fe^{2+}$ in the slurry.

As commented before, simple equipment is required for EC and it can be readily automated. Moreover, EC reduces chemical costs using cheaper materials. Therefore, the cost investment for this technology is mainly related to the cost of electrodes. The chemical and electrical costs come to 20¢$/m^3$ and 6¢$/m^3$, respectively (Zhao et al., 2010) when settling is used as dewatering strategy and a sacrificial electrode as chemical input.

*Table 5. Efficiency of EC for various components*

| | Concentration in Slurry (mg/l) | Concentration in Effluent (mg/l) | Removal (%) |
|---|---|---|---|
| Total Suspended Solids (TSS) | 20,000-30,000 | 100-200 | >99 |
| Chemical Oxygen Demand (COD) | 25,000-35,000 | 1,500-2,500 | >90 |
| Total N | 2,000-3,000 | 600-800 | >60 |
| Total P | 200-350 | <2 | >99 |

(Lekuona, 2004)

## AMMONIA AIR STRIPPING AND ABSORPTION

Ammonia air stripping is a physical-chemical process aiming to transfer volatile ammonia (strongly dependent on pH and temperature) from a liquid phase to a gas phase by means of liquid-gas (air or steam) intimate contact (usually in counter current), and its subsequent recovery in an acidic solution as ammonium salt, or by condensation. The process is usually performed in columns with packing material to favour the contact between liquid and gas (Frear et al., 2011).

### Theoretical Fundamentals and Process Description

The amount of ammonia that can be stripped from a liquid waste or absorbed in the acidic solution is, to a great extent, dependent on two thermodynamic equilibria:

- Ammonia dissociation equilibrium in the liquid: $NH_{3\,(l)} \leftrightarrow NH_{3\,(g)}$
- And, ammonia gas/liquid equilibrium: $NH_4^+ \leftrightarrow NH_3 + H^+$

In the case of a diluted ammonia solution, Henry's Law can be applied to represent the equilibrium between phases (Equation 14).

$$y = \frac{K_H}{P} \cdot x \qquad (14)$$

Where: $y$ is the concentration of ammonia in the gas (molar fraction), $x$ is the ammonia concentration in the liquid (molar fraction), $P$ is the total pressure in the system (bar) and $K_H$ is Henry's constant (bar).

Like all equilibrium constants, Henry's constant is strongly influenced by temperature. The temperature dependency of $K_H$ can be expressed according to Dasgupta and Dong (1986). Ammonia equilibrium in aqueous solution is pH and temperature dependant (Equation 15).

$$[NH_3] = \frac{[NH_3 + NH_4^+]}{1 + \frac{[H^+]}{K_a}} = \frac{[NH_3 + NH_4^+]}{1 + 10^{pK_a - pH}} \qquad (15)$$

Where: $[NH_3]$ is the free-ammonia concentration, $[NH_3 + NH_4^+]$ is the total ammonia concentration, $[H^+]$ is the hydrogen ion concentration, and $pK_a$ is the acid ionisation constant for ammonia which can be expressed as function of temperature (T) by the following Equation (16) obtained by polynomial regression of data from Lide (1993); the higher the pH and temperature, the higher the free ammonia fraction.

$$pK_a = 4 \cdot 10^{-8} \cdot T^3 + 9 \cdot 10^{-5} \cdot T^2 - 0.0356 \cdot T + 10.072 \qquad (16)$$

As seen in Figure 9, if ammonia air stripping is performed at room temperature, a high pH is required to move the ammonia equilibrium towards free ammonia. However, with an increase in temperature a lower pH is required. pH is usually set between 9 and 10 by means of base addition or previous $CO_2$ stripping. On the other hand, ammonia absorption is clearly favoured when operating at pH values below 5, with almost no temperature effect.

*Figure 9. Free ammonia fraction as a function of pH at different temperature levels*
*(data from Perry, 1992)*

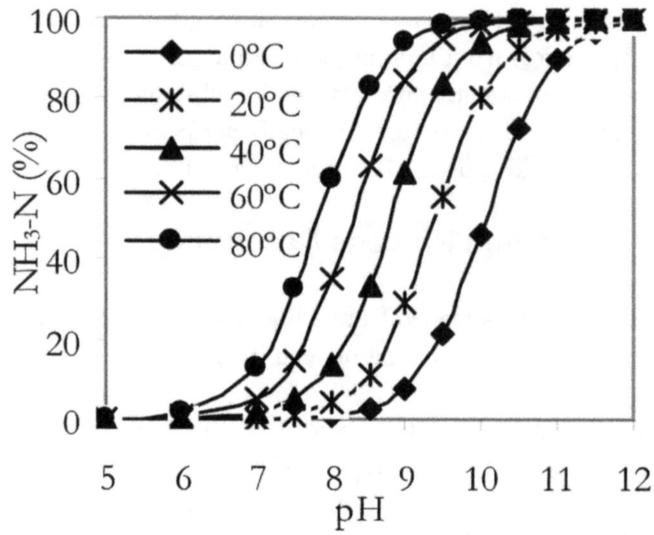


Figure 10 shows a schematic diagram of the stripping-absorption technology. The process is usually performed in vertical columns where the liquid phase (ammonia riche stream) is introduced in the upper part while the gaseous phase enters in counter current from the bottom. To enhance liquid/gas contact the columns are usually filled with packing material (specifically shaped pieces of inert material). Stripped ammonia is recovered either by absorption in a second column, with a counter current acid solution, or by vapour condensation, obtaining $NH_4OH$ or ammonia salts.

Up to 95% of ammonia reduction can be achieved under optimal conditions and almost complete ammonia recovery by absorption in acid solutions is possible with only a little acid stoichiometric ex-

*Figure 10. Schematic diagram of the stripping-absorption process*

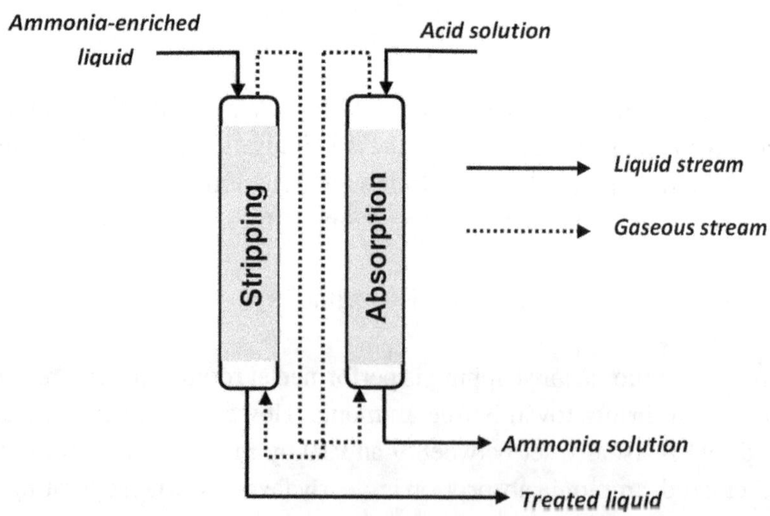

cess (1.1:2 $H_2SO_4$:$NH_3$). With respect to the reagents used, NaOH, $Ca(OH)_2$ or other bases are added to increase pH (if $CO_2$ stripping is not enough), whereas $H_2SO_4$, $HNO_3$ or $H_3PO_4$ solutions are commonly used to absorb the $NH_3$ from the gas phase.

## Usual Applications and Environmental Aspects

Ammonia stripping has already been successfully applied in anaerobic digestion (AD) supernatants from municipal wastewater treatment plants, landfill leachate, and industrial wastewater at a commercial scale. In the livestock sector, the process has been applied, both as pre- and post-treatment for anaerobic digestion, to pig slurries, cattle and fermented chicken manures, with a wide range of reported removal efficiencies. The use of digested slurry showed clear advantages compared with raw slurry, achieving efficiencies above 90% even without pH adjustment (Bonmatí & Flotats, 2003, Laureni et al., 2013, Jiang et al., 2014). However, not many cases can be found when it comes to farm-based AD systems. In the Lombardy Region (Italy), there are some facilities that have been implemented as part of a treatment chain mainly as post-treatment of the anaerobic digestion supernatant (Moscatelli & Fabbri, 2008) (Figure 11).

## Nutrient Recovery by Stripping-Absorption

The removal efficiency of a stripping process depends on a number of operational parameters. As mentioned, the main parameters are pH and temperature, as they have a direct effect over ammonia speciation ($NH_4^+$/$NH_3$), but other parameters such as the characteristics of the inflow stream should be also taken into account. In this regard, a distinct relationship between the inflow characteristics and removal efficiency has not yet been clearly established, but the content of organic matter is likely to be the most relevant parameter (Laureni et al., 2013).

*Figure 11. Ammonia stripping plant located in an agricultural biogas plant in Brescia, Italy - CRPA, 2015*

Slurries with an initial low organic matter (< 10 $g_{COD}$ $L^{-1}$) content led to higher N removal efficiencies (> 90%) when stripping was performed at 50 °C and pH was not modified. In contrast, nitrogen removal efficiency was limited to about 50% when organic matter concentrations were high, above 27 $g_{COD}$ $L^{-1}$ (Laureni et al., 2013). Similarly, Bonmatí and Flotats (2003) compared fresh and digested pig slurries in terms of stripping efficiencies at high temperature (80 °C). The authors reported higher N removals with anaerobically digested slurry (90%) than with the fresh substrate (63%). This behaviour could be explained by the capacity of the organic matter present to bind cations (e.g. $NH_4^+$) with the consequent reduction of the strippable ammonia available (Laureni et al., 2013).

The combination of ammonia stripping with anaerobic digestion processes presents other advantages besides diminishing the organic matter content. The biological process itself increases the ammonia content; part of the organic nitrogen is hydrolyzed to ammonia; and it slightly increases pH reducing alkali requirements. Furthermore, as stripping is usually performed at high temperatures, the biogas produced during anaerobic digestion can partially or totally provide for the necessary heat for stripping at high temperature.

## By-Products Characteristics and Economic Aspects

Stripping/absorption processes resulted in a liquid ammonia solution or solid ammonia salts, produced after a crystallization process. Both, products could be used directly as fertilizers or sold for other industrial applications (WWT of paper industry). However, the products recovered should have a marketable quality, in terms of nutrient concentration and organic matter contamination.

Organic matter contamination of the recovered ammonium solution, in terms of TOC and VOC concentration, strongly depends on the initial organic matter content of the treated slurry and, to some extent, on the operational pH. The higher the organic matter content, the higher the contamination; and oppositely, the higher the pH, the lower the contamination (Laureni et al., 2013).

Investment costs could be set between 0.25 and 0.5 M€ for treating 10-15 $m^3$/h, and operational costs amount to 0.66€/$m^3$ for NaOH and 0.21 €/$m^3$ for $H_2SO_4$ as well as 2.5-4.5 €/kg of stripped nitrogen (only for the stripping column). At least an equivalent range of values should be considered for the absorption step (Flotats et al., 2011). The by-product obtained (ammonia salts) can be used as raw material for the production of liquid fertilizers. An ammonium-sulfate solution with low organic contamination and a nitrogen concentration of 6% w/w, can be sold at a price of about 0.35 € $kg^{-1}$ N (Laureni et al., 2012).

## CONCLUSION

In the current juncture of increasing raw materials and energy sources scarcity, organic waste (e.g. livestock manure and other industrial wastewaters) arises as an attractive resource which, if processed properly, can yield renewable energy and be a source of added-value products (e.g. nutrient fertilizers) while, in turn, reducing the environmental impacts associated to its management.

The development and implementation of cost effective processing technologies aiming to recover valuable products as nutrients, is crucial to redirect the development towards a circular economy. In this chapter, the main process technologies already in operation, as well as the new technological developments have been reviewed, showing that these technologies are ready but have different hindering issues which should be overcome before it can spread worldwide. Operational costs, the quality of the recovered products and the need to open the market for this kind of products are the main limiting factors. Nevertheless, these recovering technologies present good potential to be widely implemented in a near future.

## REFERENCES

Ahlgren, S., Baky, A., Bernesson, S., Nordberg, A., Norén, O., & Hansson, P. A. (2008). Ammonium nitrate fertiliser production based on biomass –Environmental effects from a life cycle perspective. *Bioresource Technology*, *99*(17), 8034–8041. doi:10.1016/j.biortech.2008.03.041 PMID:18440225

Baker, R. W. (2004). *Membrane Technology and Applications*. Wiley. doi:10.1002/0470020393

Blandin, G., Vervoort, H., Le-Clech, P., & Verliefde, A. R. D. (2016). Fouling and cleaning of high permeability forward osmosis membranes. *J Wat Proc Eng*, *9*, 161–169. doi:10.1016/j.jwpe.2015.12.007

Bonmatí, A. (2001). *Thermal energy uses for the improvement of pis slurry anaerobic digestion ant the recovery of useful products*. Catalonia, Spain: University of Lleida.

Bonmatí, A., Campos, E., & Flotats, X. (2003). Concentration of pig slurry by evaporation: Anaerobic digestion as the key process. *Water Science and Technology*, *48*(4), 189–194. PMID:14531439

Bonmatí, A., & Flotats, X. (2003a). Air stripping of ammonia from pig slurry: Characterisation and feasibility as a pre- or post- treatment to mesophilic anaerobic digestion. *Waste Management (New York, N.Y.)*, *53*(3), 261–272. doi:10.1016/S0956-053X(02)00144-7 PMID:12737968

Bonmatí, A., & Flotats, X. (2003b). Pig slurry concentration by vacuum evaporation: Influence of previous mesophilic anaerobic digestion process. *J Air Waste Manage*, *53*(1), 21–31. doi:10.1080/10473289.2003.10466112 PMID:12568250

Burton, C. H. (2007). The potential contribution of separation technologies to the management of livestock manure. *Livestock Science*, *112*(3), 208–216. doi:10.1016/j.livsci.2007.09.004

Burton, C. H., & Turner, C. (2003). *Manure management. Treatment strategies for sustainable agriculture* (2nd ed.). Silsoe: Bedfors UK.

Campos, E., Illa, J., Magrí, A., Palatsi, J., Solé-Mauri, F., & Flotats, X. (2004). *Guía de los tratamientos de deyecciones ganaderas*. Agencia de Residuos de Cataluña. Retrieved from http://www.arc-cat.net/es/altres/purins/guia/pdf/guia_dejeccions.pdf

Cerrillo, M., Palatsi, J., Comas, J., Vicens, J., & Bonmatí, A. (2015). Struvite precipitation as a technology to be integrated in a manure anaerobic digestion treatment plant – removal efficiency, crystal characterization and agricultural assessment. *Journal of Chemical Technology and Biotechnology (Oxford, Oxfordshire)*, *90*(6), 1135–1143. doi:10.1002/jctb.4459

Cheng, K. Y., Kaksonen, A. H., & Cord-Ruwisch, R. (2013). Ammonia recycling enables sustainable operation of bioelectrochemical systems. *Bioresource Technology*, *143*, 25–31. doi:10.1016/j.biortech.2013.05.108 PMID:23774293

Chiumenti, A., da Borso, F., Chiumenti, R., Teri, F., & Segantin, P. (2013). Treatment of digestate from a co-digestion biogas plant by means of vacuum evaporation: Tests for process optimization and environmental sustainability. *Waste Management (New York, N.Y.)*, *33*(6), 1339–1344. doi:10.1016/j.wasman.2013.02.023 PMID:23562449

Crutchik, D., & Garrido, J. M. (2016). Kinetics of the reversible reaction of struvite crystallisation. *Chemosphere*, *154*, 567–572. doi:10.1016/j.chemosphere.2016.03.134 PMID:27085317

Dasgupta, P. K., & Dong, S. (1986). Solubility of ammonia in liquid water and generation of trace levels of standard gaseous ammonia. *Atmospheric Environment*, *20*(3), 565–570. doi:10.1016/0004-6981(86)90099-5

Doyle, J., & Parsons, S. (2002). Struvite formation, control and recovery. *Water Research*, *36*(16), 3925–3940. doi:10.1016/S0043-1354(02)00126-4 PMID:12405401

Erisman, J. W., Sutton, M. A., Galloway, J. N., Klimont, Z., & Winiwarter, W. (2008). How a century of ammonia synthesis changed the world. *Nature Geoscience*, *1*(10), 636–639. doi:10.1038/ngeo325

EU-BAT. (2006). *Reference document on best available techniques for the manufacture of large volume inorganic chemicals – ammonia, acids and fertilizers*. Sevilla, Spain: Integrated Pollution Prevention and Control. European Comissions.

Fangueiro, D., Coutinhob, J., Borges, L., & Vasconcelosa, E. (2015). Recovery efficiency of nitrogen from liquid and solid fractions of pig slurry obtained using different separation technologies. *Journal of Plant Nutrition and Soil Science*, *178*(2), 229–236. doi:10.1002/jpln.201400261

Fangueiro, D., Ribeiroa, H., Vasconcelosa, E., Coutinhob, J., & Cabrala, F. (2009). Treatment by acidification followed by solid-liquid separation affects slurry and slurry fraction scomposition and their potential of N mineralization. *Bioresource Technology*, *100*(20), 4914–4917. doi:10.1016/j.biortech.2009.04.032 PMID:19464881

Flotats, X., Bonmatí, A., Fernandez, B., & Magrí, A. (2009). Manure treatment technologies: On-farm versus centralized strategies NE of Spain as case study. *Bioresource Technology*, *100*(22), 5516–5526. doi:10.1016/j.biortech.2008.12.050 PMID:19231173

Flotats, X., Foged, H. L., Bonmati, A., Palatsi, J., Magri, A., & Schelde, K. M. (2011). *Manure Processing technologies*. Technical Report No. II concerning "Manure Processing Activities in Europe" to the European Commission, Directorate-General Environment. Project reference: ENV.B.1/ETU/2010/0007. Retrieved from http://hdl.handle.net/2117/18944

Foged, H. L., Flotats, X., Bonmati, A., Palatsi, J., & Magri, A. (2011). End and by-products from livestock manure processing - general types, chemical composition, fertilising quality and feasibility for marketing. Technical Report No. III concerning "Manure Processing Activities in Europe" to the European Commission. Directorate-General Environment. Retrieved from http://hdl.handle.net/2117/18945

Frear, C., Liao, W., Ewing, T., & Chen, S. (2011). Evaluation of co-digestion at a commercial dairy anaerobic digester. *Clean-Soil, Air. Water (Basel)*, *39*, 697–704.

Fukumoto, Y., & Haga, K. (2004). Advanced treatment of swine wastewater by electrodialysis with a tubular ion exchange membrane. *Animal Science Journal*, *75*(5), 479–485. doi:10.1111/j.1740-0929.2004.00216.x

Galloway, J. N., Aber, J. D., Erisman, J. W., Seitzinger, S. P., Howarth, R. W., Cowling, E. B., & Cosby, B. J. (2003). The nitrogen cascade. *Bioscience*, *53*(4), 341–356. doi:10.1641/0006-3568(2003)053[0341:TNC]2.0.CO;2

Garcia-González, M. C., & Vanotti, M. B. (2015). Recovery of ammonia from swine manure using gas-permeable membranes: Effect of waste strength and pH. *Waste Management (New York, N.Y.)*, *38*, 455–461. doi:10.1016/j.wasman.2015.01.021 PMID:25687948

Gharibi, H., Mahvi, A. H., Chehrazi, M., Sheikhi, R., & Hosseini, R. (2010). Phosphorous removal from wastewater effluent using electro-coagulation by aluminum and iron plates. *Analytical and Bioanalytical Chemistry*, *2*, 165–177.

González-Ponce, R., López-de-Sá, E. G., & Plaza, C. (2009). Lettuce Response to Phosphorus Fertilization with Struvite Recovered from Municipal Wastewater. *HortScience*, *44*, 426–430.

Grüter, H., Matter, M., Oehlmann, K. H., & Hicks, M. D. (1990). Drying of sewage sludge- and important step in waste disposal. *Water Science and Technology*, *22*(12), 57–63.

Guo, X., & Jin, X. (2014). Purification of UF-treated anaerobically digested manure wastewater by two-pass reverse osmosis. *Desalination and Water Treatment*, *52*(16-18), 3027–3034. doi:10.1080/19443994.2013.797671

Have, P. J. W., & Chiappini, U. (1993). *Processing of manure surpluses.* European Conference. Environment Agriculture Stock Farming in Europe, Montova, Italy.

Häyrynen, K., Pongrácz, E., Väisänen, V., Pap, N., Mänttäri, M., Langwaldt, J., & Keiski, R. L. (2009). Concentration of ammonium and nitrate from mine water by reverse osmosis and nanofiltration. *Desalination*, *240*(1-3), 280–289. doi:10.1016/j.desal.2008.02.027

Hjorth, M., Christensen, K. V., Christensen, M. L., & Sommer, S. G. (2010). Solid-liquid separation of animal slurry in theory and practice. A review. *Agron. Sust. Devel*, *30*(1), 153–180. doi:10.1051/agro/2009010

Huang, H., Xu, C., & Zhang, W. (2011). Removal of nutrients from piggery wastewater using struvite precipitation and pyrogenation technology. *Bioresource Technology*, *102*(3), 2523–2528. doi:10.1016/j.biortech.2010.11.054 PMID:21146405

Huang, L. Y., Lee, D. J., & Lai, J. Y. (2015). Forward osmosis membrane bioreactor for wastewater treatment with phosphorus recovery. *Bioresource Technology*, *198*, 418–423. doi:10.1016/j.biortech.2015.09.045 PMID:26409853

Jaffer, Y., Clark, T. A., Pearce, P., & Parsons, S. A. (2002). Potential phosphorus recovery by struvite formation. *Water Research*, *36*(7), 1834–1842. doi:10.1016/S0043-1354(01)00391-8 PMID:12044083

Jiang, A., Zhang, T., Zhao, Q.-B., Li, X., Chen, S., & Frear, C. (2014). Evaluation of an integrated ammonia stripping, recovery, and biogas scrubbing system for use with anaerobically digested dairy manure. *Biosystems Engineering*, *119*, 117–126. doi:10.1016/j.biosystemseng.2013.10.008

Jorgensen, K., & Stoumann, J. L. (2009). Chemical and biochemical variation in animal manure solids separated using different commercial separation technologies. *Bioresource Technology*, *100*(12), 3088–3096. doi:10.1016/j.biortech.2009.01.065 PMID:19272772

Kataki, S., West, H., Clarke, M., & Baruah, D. C. (2016). *Phosphorus recovery as struvite from farm, municipal and industrial waste: Feedstock suitability, methods and pre-treatments*. Waste Manage. doi:10.1016/j.wasman.2016.01.003

Konieczny, K., Kwiecińska, A., & Gworek, B. (2011). The recovery of water from slurry produced in high density livestock farming with the use of membrane processes. *Separation and Purification Technology*, *80*(3), 490–498. doi:10.1016/j.seppur.2011.06.002

Koparal, A. S., & Öğütveren, Ü. B. (2002). Removal of nitrate from water by electroreduction and electrocoagulation. *Journal of Hazardous Materials*, *89*(1), 83–94. doi:10.1016/S0304-3894(01)00301-6 PMID:11734348

Lacasa, E., Cañizares, P., Sáez, C., Fernández, F. J., & Rodrigo, M. A. (2011). Removal of nitrates from groundwater by electrocoagulation. *Chemical Engineering Journal*, *171*(3), 1012–1017. doi:10.1016/j.cej.2011.04.053

Lacasa, E., Cañizares, P., Sáez, C., Martínez, F., & Rodrigo, M. A. (2013). Modelling and cost evaluation of electro-coagulation processes for the removal of anions from water. *Separation and Purification Technology*, *107*, 219–227. doi:10.1016/j.seppur.2013.01.035

Laureni, M., Palatsi, J., Llovera, M., & Bonmatí, A. (2013). Influence of pig slurry characteristics on ammonia stripping efficiencies and quality of the recovered ammonium-sulfate solution. *Journal of Chemical Technology and Biotechnology (Oxford, Oxfordshire)*, *88*(9), 1654–1662. doi:10.1002/jctb.4016

Le Corre, K. S., Valsami-Jones, E., Hobbs, P., & Parsons, A. (2009). Phosphorus recovery from wastewater by struvite crystallization: A review. *Environmental Science & Technology*, *39*(6), 433–477. doi:10.1080/10643380701640573

Lekuona, A. (2004). Planta de tratamiento de purines de Egiluze. RETEMA. *Revista Técnica de Medio Ambiente.*, *103*, 20–24.

Levasseur, P. (2004). *Traitement des effluents procins. Guide Paractique des Procédés*. Institut Technique du Porc.

Lide, D. (Ed.). (1993). *CRC Handbook of Chemistry and Physics. A Ready-reference Book of Chemical and Physical Data*. Boca Raton, USA: CRC Press.

Lin, H., Gan, J., Rajendran, A., Rodrigues Reis, C. E., & Hu, B. (2015). Phosphorus Removal and Recovery from Digestate after Biogas Production. In Biofuels - Status and Perspective (pp. 517-546). Academic Press.

Liu, B., Giannis, A., Zhang, J., Chang, V. W. C., & Wang, J. Y. (2013). Characterization of induced struvite formation from source-separated urine using seawater and brine as magnesium sources. *Chemosphere*, *93*(11), 2738–2747. doi:10.1016/j.chemosphere.2013.09.025 PMID:24134888

Marks, A. L., Luthy, R. G., & Diweker, U. M. (1994). Semi-continuous evaporation model for leachate treatment process evaluation. *Environment and Progress*, *13*(4), 278–289. doi:10.1002/ep.670130419

Martinez, J., Burton, C. H., Sneath, R. W., & Farrent, J. W. (1995). A study of the potential contribution of sedimentation to aerobic treatment processes for pig slurry. *J. Agric. Ng. Res*, *61*, 8–96.

Masse, L., Massé, D. I., & Pellerinb, Y. (2007). The use of membranes for the treatment of manure: A critical literature review. *Biosystems Engineering*, *98*(4), 371–380. doi:10.1016/j.biosystemseng.2007.09.003

Maurer, M., Pronk, W., & Larsen, T. A. (2006). Treatment processes for source-separated urine. *Water Research*, *40*(17), 3151–3166. doi:10.1016/j.watres.2006.07.012 PMID:16949123

Moerman, W., Carball, M., Vandekerckhove, A., Derycke, D., & Verstraete, W. (2009). Phosphate removal in agro-industry: Pilot- and full-scale operational considerations of struvite crystallization. *Water Research*, *43*(7), 1887–1892. doi:10.1016/j.watres.2009.02.007 PMID:19264340

Møller, H. B., Lund, I., & Sommer, S. G. (2000). Solid-liquid separation of livestock slurry: Efficiency and cost. *Bioresource Technology*, *74*(3), 223–229. doi:10.1016/S0960-8524(00)00016-X

Møller, H. B., Sommer, S. G., & Ahring, B. K. (2002). Separation efficiency and particle size distribution in relation to manure type and storage conditions. *Bioresource Technology*, *85*(2), 189–196. doi:10.1016/S0960-8524(02)00047-0 PMID:12227544

Moscatelli, G., & Fabbri, C. (2008). Strippaggio, tecnica efficient per l'abbattimento dell'azoto. *L'informatore Agrario*, *18*, 45–50.

Niewersch, C., Battaglia Bloch, A. L., Yüce, S., Melin, T., & Wessling, M. (2014). Nanofiltration for the recovery of phosphorus: Development of a mass transport model. *Desalination*, *346*, 70–78. doi:10.1016/j.desal.2014.05.011

Nkoa, R. (2014). Agricultural benefits and environmental risks of soil fertilization with anaerobic digestates: A review. *Agron Sustain Dev*, *34*(2), 473–492. doi:10.1007/s13593-013-0196-z

Ohlinger, K. N., Young, T. M., & Schroeder, E. D. (1999). Kinetics effects on preferential struvite accumulation in wastewater. *Journal of Environmental Engineering*, *125*, 730–737. doi:10.1061/(ASCE)0733-9372(1999)125:8(730)

Perry, R. H. (1992). *Perry's Chemical Engineers' Handbook* (6th ed.). McGraw Hill.

Pinnekamp, J., & Friedrich, H. (2006). *Membrane technology for waste water treatment. Municipal water and waste management* (Vol. 2). FiW-Verlag.

Pronk, W., Palmquist, H., Biebow, M., & Boller, M. (2006). Nanofiltration for the separation of pharmaceuticals from nutrients in source-separated urine. *Water Research*, *40*(7), 1405–1412. doi:10.1016/j.watres.2006.01.038 PMID:16530802

Quist-Jensen, C. A., Macedonio, F., & Drioli, E. (2015). Membrane technology for water production in agriculture: Desalination and wastewater reuse. *Desalination*, *364*, 17–32. doi:10.1016/j.desal.2015.03.001

Ramírez, C. A., & Worrell, E. (2006). Feeding fossil fuels to the soil. An analysis of energy embedded and technological learning in the fertilizer industry. *Resources, Conservation and Recycling*, *46*, 75–93.

Records, A., & Sutherland, K. (2001). *Decanter Centrifuge Handbook*. Amsterdam: Elsevier.

Sadr, S. M., & Saroj, D. P. (2015). *Membrane technologies for municipal wastewater treatment. In Advances in Membrane Technologies for Water Treatment: Materials, Processes and Applications* (pp. 443–463). Woodhead Publishing. doi:10.1016/B978-1-78242-121-4.00014-9

Schuiling, R. D., & Andrade, A. (1999). Recovery of struvite from calf manure. *Environmental Technology*, *20*(7), 765–768. doi:10.1080/09593332008616872

Schütte, T., Niewersch, C., Wintgens, T., & Yüce, S. (2015). Phosphorus recovery from sewage sludge by nanofiltration in diafiltration mode. *Journal of Membrane Science*, *480*, 74–82.

Shalaby, A., Nassef, E., Mubark, A., & Hussein, M. (2014). Phosphate removal from wastewater by electrocoagulation using aluminium electrodes. *American Journal of Environmental Engineering and Science*, *1*(5), 90–98.

Sotres, A., Cerrillo, M., Viñas, M., & Bonmatí, A. (2015). Nitrogen recovery from pig slurry in a two-chambered bioelectrochemical system.[PubMed]. *Bioresource Technology*, *194*, 373–382.

Tao, W., Fattah, K. P., & Huchzermeier, M. P. (2016). Struvite recovery from anaerobically digested dairy manure: A review of application potential and hindrances. *Journal of Environmental Management*, *169*, 46–57. doi:10.1016/j.jenvman.2015.12.006 PMID:26720329

Tchamango, S., Nanseu-Njiki, C. P., Ngameni, E., Hadjiev, D., & Darchen, A. (2010). Treatment of dairy effluents by electrocoagulation using aluminium electrodes. *The Science of the Total Environment*, *408*(4), 947–952. doi:10.1016/j.scitotenv.2009.10.026 PMID:19900696

Türker, M., & Çelen, I. (2007). Removal of ammonia as struvite from anaerobic digester effluents and recycling of magnesium and phosphate. *BioresourceTechnol*, *98*(8), 1529–1534. doi:10.1016/j.biortech.2006.06.026 PMID:17084077

Uludag-Demirer, S., Demirer, G. N., Frear, C., & Chen, S. (2008). Anaerobic digestion of dairy manure with enhanced ammonia removal. *Journal of Environmental Management*, *86*(1), 193–200. doi:10.1016/j.jenvman.2006.12.002 PMID:17257738

Vasudevan, S., Epron, F., Lakshmi, J., Ravichandran, S., Mohan, S., & Sozhan, G. (2010). Removal of $NO^{3-}$ from Drinking Water by Electrocoagulation – An Alternate Approach. *Clean – Soil, Air. Water (Basel)*, *38*, 225–229.

Velthof, G. L. (2011). Synthesis of the research within the framework of the Mineral Concentrates Pilot. Wageningen, Alterra, Alterra report 2224.

Vitousek, P. M., Naylor, R., Crews, T., David, M. B., Drinkwater, L. E., Holland, E., & Zhang, F. S. et al. (2009). Nutrient Imbalances in Agricultural Development. *Science*, *324*(5934), 1519–1520. doi:10.1126/science.1170261 PMID:19541981

Xie, M., Shon, H. K., Gray, S. R., & Elimelech, M. (2016). Membrane-based processes for wastewater nutrient recovery: Technology, challenges, and future direction. *Water Research*, *89*, 210–221. doi:10.1016/j.watres.2015.11.045 PMID:26674549

Yetilmezsoy, K., & Sapci-Zengin, Z. (2009). Recovery of ammonium nitrogen from the effluent of UASB treating poultry manure wastewater by MAP precipitation as a slow release fertilizer. *Journal of Hazardous Materials*, *166*(1), 260–269. doi:10.1016/j.jhazmat.2008.11.025 PMID:19097699

Zhang, Y., Desmidt, E., Van Looveren, A., Pinoy, L., Meesschaert, B., & Van der Bruggen, B. (2013). Phosphate separation and recovery from wastewater by novel electrodialysis. *Environmental Science & Technology*, *47*(11), 5888–5895. doi:10.1021/es4004476 PMID:23651001

Zhang, Y., Zhang, W., & Pan, B. (2015). Struvite-based phosphorus recovery from the concentrated bioeffluent by using HFO nanocomposite adsorption: Effect of solution chemistry. *Chemosphere*, *141*, 227–234. doi:10.1016/j.chemosphere.2015.07.023 PMID:26246192

Zhao, Q., Zhang, T., Frear, C., Bowers, K., Harrison, J., & Chen, S. (2010). *Phosphorous Recovery Technology in Conjunction with Dairy Anaerobic Digestion. Climate Friendly Farming: Research Report 2010 – 001*. CSANR.

# Chapter 4
# Chemical and Biological Processes for Nutrients Removal and Recovery

**Dafne Crutchik Pedemonte**
*University of Verona, Italy*

**Nicola Frison**
*University of Verona, Italy*

**Carlota Tayà**
*University of Vic-Central University of Catalonia, Spain*

**Sergio Ponsa**
*University of Vic-Central University of Catalonia, Spain*

**Francesco Fatone**
*University of Verona, Italy*

## ABSTRACT

*This chapter gives an overview on the main technologies for nutrient removal from industrial wastewater by focusing on principles and operational parameters of real applications. A plethora of technologies can achieve the nutrients removal from wastewater depending mainly on their concentration and forms; however, biological nitrification and denitrification and chemical precipitation are the most common processes used today to remove nitrogen and phosphorus, respectively. Stripping, adsorption and membrane based processes for nutrients recovery can be economically viable only when nitrogen concentration is higher than 1.5-2 gN/L. On the other hand, phosphorus recovery should always be pursued and struvite crystalization is the most common option that should be evaluated together with biological phosphorus accumulation in sludge or plants for the following post-processing and valorization.*

## INTRODUCTION

Nitrogen and phosphorus are essential elements of life, however, the continuous release of excess nitrogen and phosphorus into the environment during naturally occurring processes (e.g. weathering, leaching, erosion) and anthropogenic activities (e.g. surface run-off from agricultural and livestock activities, wastewater treatment plants, etc.), which may lead to eutrophication (i.e. algae bloom and oxygen depletion

DOI: 10.4018/978-1-5225-1037-6.ch004

in aqueous systems) in natural water systems. The eutrophication problem is expected to increase due to the increasing population, agricultural intensification and industrialization. Nevertheless, eutrophication can be reversed by reducing the nutrients load to water bodies. One such solution is to decrease the load of nutrients discharged from wastewater treatment plants. Thus, efficient and reliable nutrients removal and/or recovery technologies are required. Nitrogen and phosphorus removal/recovery from wastewater can be achieved by biological or physical-chemical processes. The selection of nutrients removal/recovery technologies is determined by the nitrogen and phosphorus concentrations, and consequently the cost-effectiveness of the process. The nutrients removal/recovery technologies can be classified into three concentrations ranges (Mulder, 2003):

- Diluted wastewater with phosphorus concentration <40 mg P/L and ammonium concentration <0.1 g N/L. In this nutrients concentration range, biological nutrients removal is the preferable process based on cost-effectiveness. In addition, chemical phosphorus precipitation can be an option for phosphorus removal from diluted wastewater.
- Concentrated wastewater with phosphorus concentration >60 mg P/L and ammonium concentration in the range 0.1-5 g N/L. Chemical phosphorus precipitation processes are identified as interesting alternatives for the simultaneous recovery of phosphorus and nitrogen as an agricultural fertilizer (struvite) or phosphorus recovery as a raw material for the phosphorus industry (calcium phosphates). For nitrogen, biological treatment processes are also to be preferred.
- Concentrated wastewater with ammonium concentration >5 g N/L. In this range, the physical-chemical process for ammonium removal can be technically and economically feasible (e.g. ammonia stripping, selective ion exchange and breakpoint chlorination).

Thus, this chapter summarizes the methods and technologies commonly used for nitrogen and phosphorus removal/recovery from wastewater. While the processes for nutrients removal, such as biological nutrients removal or chemical phosphorus precipitation from diluted wastewater, have been widely adopted in nutrient removal plants, strategies for its recovery are still being studied. One such technology is the phosphorus recovery by struvite crystallization from wastewater streams with high phosphorus concentration. Moreover, ammonium removal from concentrated wastewater by ammonia stripping can be used because of its simple operation and high efficiency.

## NITROGEN AND PHOSPHORUS IN INDUSTRIAL WASTEWATER

The characteristics of industrial wastewater vary essentially with the type of industry and the type of industrial process used (Table 1). Agro-industrial effluents, like piggery effluents or fishery discharges, slaughterhouses and meat processing, food processing, frequently combine high concentration both in ammonium, phosphorus and organic matter. For many industrial wastewaters, especially in chemical and petrochemical industry, reduction of effluent total nitrogen is difficult because of batch production operations and the refractory nature of many complex organic nitrogen compounds in process chemicals.

*Table 1. Industrial wastewater composition in terms of nitrogen and phosphorous content (mg/L)*

| Type of Wastewater | Total Nitrogen | Total Phosphorus |
|---|---|---|
| Slaughterhouse poultry | 10-190 ($NH_4^+$-N) | - |
| Distillery | 1200 ($NH_4^+$-N) | - |
| Brewery | 25-35 (TKN) | 20-30 |
| Vegetable processing | - | 1-30 |
| Seafood processing | 25-405 | 5-95 |
| Dairy product | 60-310 | 1-40 |
| Personal care and pharmaceutical products | 15-130 | 3-30 |
| Meat processing | 90-700 (TKN) | 12-120 |

TKN: Total Kjeldahl Nitrogen
(Wun-Jern, 2006)

## REMOVAL VS RECOVERY: FEASIBILITY AND SUSTAINABILITY

There are different technologies for nutrients recovery from nitrogenous industrial wastewater. Besides the solid/liquid separation technologies are often involved to separate the organic and mineral forms of nitrogen, there are two main groups of technologies aiming the removal and recovery of nitrogen, respectively:

- Physical-chemical processes which separate ammonia and concentrate it in other flows, either solid, liquid or gaseous, that can be further processed to recover ammonium salts;
- Biological processes which convert ammonia into nitrogen gas that returns to the atmosphere.

Although the nutrients recovery always appears the desirable option, its sustainability for removal or recovery at large-scale applications should be evaluated as well, both from the technical, environmental, administrative/regulatory, and finally, economic aspects. However, the latter is closely linked to the market for recovered materials and the ability to compete with traditional chemical fertilizers. Maurer et al. (2003) compared the energy consumption of different technologies for N recovery from source separated urine and calculated that:

1. N recovery by air stripping followed by acid condensation and $(NH_4)_2SO_4$ production needs 90 MJ/kg N;
2. Biological removal with nitrification/denitrification with methanol as substrate needs 109 MJ/kg N;
3. Biological removal with the completely autotrophic process (partial nitritation + anaerobic ammonia oxidation) needs only 19 MJ/kg N;
4. The Haber-Bosch process to produce ammonia from $N_2$ needs 45 MJ/kg N.
5. It follows that the combination of the conventional biological removal process with the Haber-Bosch process (154 MJ/kg N) results less convenient than the N recovery with air stripping (90 MJ/kgN), but completely autotrophic processes plus the Haber-Bosch process could be more energetically convenient (64 MJ/kg N).

The conventional technologies for the removal of phosphorus from wastewater are chemical phosphate precipitation by adding metal salts or biological phosphorus removal processes. However, in these technologies, the phosphorus is not recovered as a sustainable product, but rather like a waste sludge that needs to be treated or disposed of in landfills. An important requisite for phosphorus recovery technologies is generating a final product with high quality, high purity and free from pollutants. Thus, the aim of phosphorus recovery technologies must be to obtain a product that has a potential to be used as fertilizer or as raw material for phosphorus industry. Thus, the technologies of nutrients recovery should not only be technically feasible, but they should also be cost-effective. The economic viability is, therefore, crucial to decide on the implementation of a nutrients recovery technology.

## Market and Opportunities of Recovered Nutrients

During the last decades, the steadily growing global demand for grain crops has generated higher crop prices and higher needs of fertilizers (N and P based) with the consequent rise in the prices (Figure 1). The consumption of fertilizer in 2011 was estimated at 176.8 million tonnes. With a successive growth of 1.9 percent per year, it is expected to reach 194.1 million tonnes by the end of 2016 (FAO, 2012).

Nitrogen is recovered from concentrated industrial wastewater often as ammonium sulfate which has potential applications as fertilizer. Ammonium sulfate account for about 4.7% of the world nitrogen fertilizer market (www.ihs.com, Chemical Economics Handbook Ammonium Sulfate). There are no serious environmental concerns involved with the use of ammonium sulfate as a fertilizer material. The global ammonium sulfate market size was estimated at USD 2.86 billion in 2014, and the forecast is for a further increase in the next seven years (Grand View Research, 2016). Although the total nitrogen fertilizer market grew substantially between 1987 and 2013 and it is expected further growth for 2018, the bulk of the world demand for nitrogen fertilizer is being filled by urea, another solid fertilizer material with a higher nitrogen content.

*Figure 1. Price for urea and super-phosphate procurement during the years from 1960-2013*
*Source: USDA, 2013*

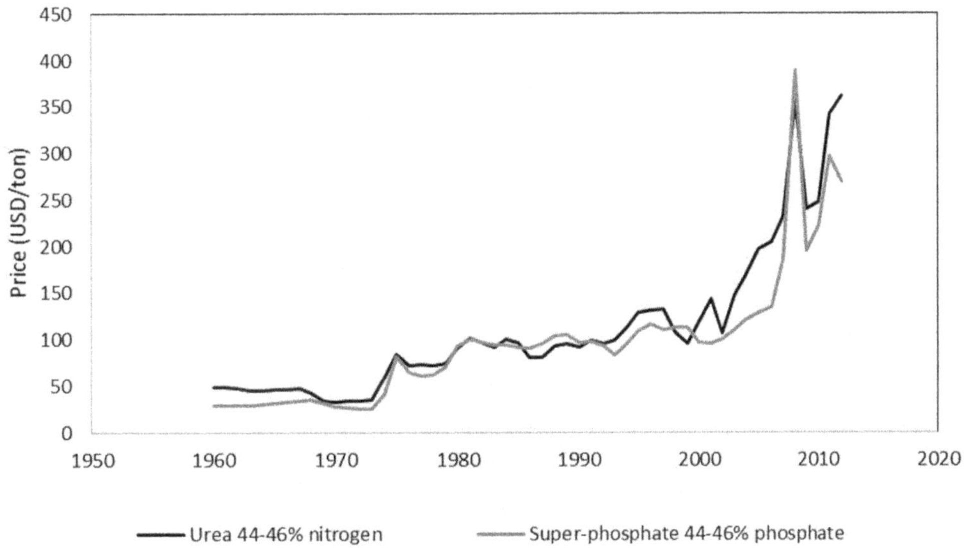

Struvite is considered as a potentially marketable phosphorus product from wastewater streams. Some potential applications of struvite are: secondary phosphate ore (as raw material), source of phosphate in other industrial processes or slow-release fertilizer (Schuiling & Andrade, 1999). The possible use of struvite as a raw material is associated with the production of elementary phosphorus or P-based fertilizer. Another use could be as a raw material in magnesium phosphate cement of building materials (p.e. fire resistant board material).

The most promising use of struvite is as a slow release fertilizer. Struvite can have low rates of leach and, thus, to slowly release nutrients (El Diwani, Rafie, Ibiari, & El-Aila, 2007; Le Corre, Valsami-Jones, Hobbs, & Parsons, 2005; de-Bashan & Bashan, 2004). The fertilizer properties of struvite have been studied by numerous researches. These studies are in agreement that the struvite is an excellent fertilizer for crops, ornamentals, vegetables, flower boards, turf grass, tree seedlings and garden grass (Rahman, Liu, Kwag, & Ra, 2011; Uysal, Yilmazel, & Demirer, 2010; González-Ponce & García López, 2007).

## FUNDAMENTALS

### Biological Nutrient Removal: Principles

#### Biological Nitrogen Removal

Biological nitrogen removal in activated sludge (AS) systems is based on two main separated but complementary processes called nitrification and denitrification (Figure 2).

*Figure 2. Biological transformations in the N cycle. Blue and green arrows represent the conventional nitrification and heterotrophic denitrification process, respectively. The anammox process is presented in red*

*(Adapted from Tayà, 2013)*

*Source: Tayà, 2013*

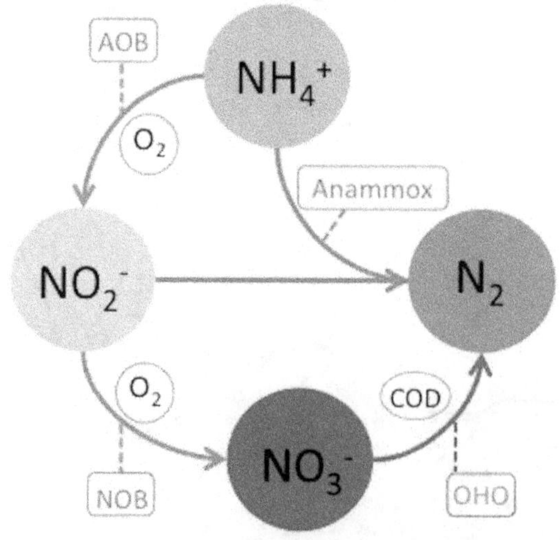

## Nitrification

Nitrification in AS systems is defined as the biological oxidation of the ammonium present in wastewater to its most oxidized form (nitrate). In AS systems, this process is performed by two different nitrifying bacteria communities, and thus, nitrification is considered as a two-step process. Ammonia Oxidizing Bacteria (AOB) firstly oxidize ammonium to nitrite (nitritation, Equation 1) and Nitrite Oxidizing Bacteria (NOB) perform the nitrate formation (nitratation, Equation 2). The nitrification process has the following stoichiometric equations (Wiesmann, 1994):

$$NH_4^+ + \frac{3}{2}O_2 \rightarrow NO_2^- + H_2O + 2H^+ \, AOB \, (\textit{Nitrosonomas spp., Nitrosospira spp.,...}) \quad (1)$$

$$NO_2 + \frac{1}{2}O_2 \rightarrow NO_3 \, NOB \, (\textit{Nitrobacter spp., Nitrospira spp.,...}) \quad (2)$$

If the synthesis of new bacteria is considered (assuming that the empirical formulation of bacterial cells is $C_5H_7NO_2$), the above equations are modified as follows (Haug & McCarty, 1972):

$$55\ NH_4^+ + 76\ O_2 + 5\ CO_2 \rightarrow 54\ NO_2^- + 52\ H_2O + 109\ H^+ + C_5H_7NO_2 \, \textit{Nitritation} \quad (3)$$

$$400NO_2^- + 195O_2 + 2H_2O + NH_4^+ + 5CO_2 \rightarrow 400NO_3^- + H^+ + C_5H_7NO_2 \, \textit{Nitratation} \quad (4)$$

It is important to take into account the affectations or inhibitions in the nitrification process. Temperature, pH (Metcalf & Eddy, 2003), free ammonia (FA) and free nitrous acid (FNA) concentration (Anthonisen, Loehr, Prakasam & Srinath, 1976; Kim, Lee, & Kellet, 2006; Torà, Lafuente, Baeza & Carrera, 2010) or the lack of inorganic carbon presence (Wett & Rauch, 2003; Guisasola, Petzet, Baeza, Carrera, & Lafuente, 2007; Torà *et al.*, 2010; Ganigué, Volcke, Puig, Balaguer & Colprim 2012) are one of the most deeply studied parameter that could affect this process.

Current knowledge on nitrification process has attributed to the nitritation step the majority of nitrous oxide ($N_2O$) and nitric oxide (NO) emissions detected in WWTP (Kim, Mishurov & Kiely 2010; Wunderlin, Mohn, Joss, Emmenegger & Siegrist 2012). Among them, $N_2O$ has received increasing attention due to the detrimental effects that it can cause as greenhouse gases (GHG) and as a toxic to living organisms (Guerrero, 2014).

### *Denitrification*

The following step inside de nitrogen removal is the denitrification. Denitrification is the reduction of nitrate to nitrogen gas ($N_2$), which is removed from the system achieving the wastewater nitrogen removal. This process is composed of different steps (Equation 5) starting with nitrate reduction to nitrite, followed by nitrite reduction to nitric oxide and nitrous oxide and finally to nitrogen gas (Gujer, Henze, Takahashi & van Loosdrecht 1999).

$$NO_3^- \rightarrow NO_2^- \rightarrow NO \rightarrow N_2O \rightarrow N_2 \quad (5)$$

The stoichiometry of total denitrification to $N_2$ when using ethanol as carbon source is showed in Equation 6 (Wiesmann, 1994):

$$12NO_3^- + 5C_2H_5OH^- \rightarrow 6N_2 + 10CO_2 + 9H_2O + 12OH^- \quad (6)$$

The synthesis of new bacteria follows the stoichiometry presented in equation 7 (Liu, Qin & Yang, 2007):

$$97NO_3^- + 50C_2H_5OH \rightarrow 46N_2 + 75CO_2 + 84H_2O + 97OH^- + 5C_5H_7NO_2 \quad (7)$$

Conventional denitrification in AS (i.e. heterotrophic denitrification) is mainly conducted by one microbial population called ordinary heterotrophic organisms (OHO). Under anoxic conditions, where the electron acceptor is nitrite or nitrate, OHO can use them for organic matter (organic electron donor) oxidation to $CO_2$ and $H_2O$. Although the amount of organic carbon present in urban wastewaters is commonly enough for total nitrogen denitrification, some industrial influents have low COD/N ratios limiting this process. Among other possible solutions, a biodegradable external carbon source can be used for these cases. As an example, in the north of Europe the addition of methanol is a widespread practice to enhance the denitrification capacity of urban WWTP (Purtschert, Siegrist & Gujer, 1996), however treating industrial wastewater an external carbon source has to be added to complete the denitrification and therefore the BNR.

For some years, it is being studied the denitrification process as a contributor of $N_2O$ and NO emissions because they are intermediate compounds of nitrate or nitrite reduction to nitrogen gas (Equation 5). Although low emissions are related to denitrification in comparison with nitrification for normal WWTP operation, some specific conditions have been focused as possible factors to favour $N_2O$ and NO accumulation under anoxic conditions. Carbon source limitations, low pH or high nitrite concentrations are some examples (Thörn & Sörenson, 1996; Itokawa, Hanaki, & Matsuo, 2007). At present, there is still a lack of studies about understanding the influence of environmental factors and operational configurations on $N_2O$ and NO emissions, being then current hot research topics (Guerrero, 2014).

## Advanced Nitrogen Removal Processes

Novel advanced N removal processes have been developed in the recent years mainly focused on reducing organic carbon necessities for denitrification step. Most of them are based on nitrification and denitrification via nitrite pathway (i.e. shortcut biological N removal). This process has the following stoichiometry (Henze, van Loosdrecht, Ekama, & Brdjanovic, 2008):

$$NH_4^+ + \frac{3}{2}O_2 \rightarrow NO_2 + H_2O + 2H^+ \textit{ Nitritation (AOB)} \quad (8)$$

$$8\ NO_2^- + 3\ CH_3COOH \rightarrow 4\ N_2 + 6\ CO_2 + 2\ H_2O + 8OH^- \textit{ Denitritation (OHO)} \quad (9)$$

The first step consists in a partial nitrification where ammonium is oxidized to nitrite (nitritation) but not to nitrate. Consequently, total NOB washout is mandatory to ensure nitratation suppression. Several methods have been proposed so far for selecting AOB against NOB, which are based on decreasing growth rate of the NOB over AOB under some specific operational conditions.

For complete biological N removal through nitrite, a second step for nitrite reduction, autotrophically or heterotrophically, has to be included. Comparing with conventional nitrification/denitrification process, the combination of partial nitritation (Equation 8) and heterotrophic denitritation (Equations 9) has some advantages since it results in a decrease of 25% in oxygen requirements, around 30-40% less of carbon source requirements for denitrification step and a reduction of 40% in biomass production (Turk & Mavinic, 1987; Van Hulle et al., 2010). In addition, the anoxic volume can be reduced because denitrification rates from nitrite are around 1.5-2 times faster than denitrification from nitrate (Peng & Zhu, 2006; Aslan & Dahab, 2008).

Regarding autotrophic denitrification, the anaerobic ammonium oxidation (anammox) is the most innovative nitrogen removal process. Under anoxic conditions, anammox microorganisms oxidize ammonium to nitrogen gas by using nitrite as an electron acceptor. Hence, differently to heterotrophic denitritation, for anammox process only 50% of the ammonium entering the system has to be previously oxidized to nitrite. As can be observed in anammox stoichiometry (Equation 10) no carbon source is needed during this process, and around a 10% of the nitrogen treated is converted to nitrate (Strous, Heijnen, Kuenen & Jetten, 1998).

$$NH_4^+ + 1.32NO_2^- + 0.0066HCO_3^- + 0.13H^+ \rightarrow 1.02N_2 + 0.26NO_3^- + 2.03H_2O + 0.0066CH_2O_{0.5}N_{0.15} \quad (10)$$

This process presents different advantages (Liu et al., 2008):

1. Overall energy savings around 40-50% related to low oxygen requirements,
2. There is no need for carbon source supply and
3. The amount of sludge produced is lower due to low biomass substrate yield of anammox bacteria.

On the contrary, the slow growth rate of such microorganisms, the inhibitory effect of oxygen presence and the complexity to find a proper inoculum enriched in anammox results in long-term start-up periods. In spite of this, several treatment processes that combine nitritation and anammox have been already developed: SHARON/Anammox (two reactors system; Hwang, Min, Choi & Yun, 2005), CANON (complete autotrophic N removal over nitrite; Vázquez-Padin et al., 2009), OLAND (oxygen-limited autotrophic nitrification and denitrification; Vlaeminck et al., 2010) or NAS (floc-based partial nitritation and anammox process; Desloover et al., 2011).

Some other autotrophic denitrification processes have also been studied during last decade not based on anammox bacteria. In these cases, autotrophic denitrification is carried out by using hydrogen ($H_2$) as electron donor (Lee & Rittman, 2002; Rezania, Oleszkiewicz & Cicek, 2007) or sulfur compounds such as $H_2S$, S, $S_2O_3^{-2}$ or $SO_3^{-2}$ (Batchelor & Lawrence, 1978; Sierra-Alvarez et al., 2007; Fajardo, Mosquera-Corral, Campos & Méndez 2012; Mora, Lopez, Gamisans & Gabriel 2014). As also occurred with anammox microorganism, the organic carbon source is neither needed during denitrification. This process presents the same advantages as anammox systems over heterotrophic denitrification. However, the high alkalinity required during the process and the sulfate production due to the oxidation of sulfur compounds constitute the main drawbacks.

## Biological Phosphorus Removal: Enhanced Biological Phosphorus Removal (EBPR) Process

EBPR aims at removing and recovering phosphorus using biological processes and thus, without the addition of chemicals, which increases the costs and produces a chemical undesired sludge. This process is based on the enrichment of activated sludge with polyphosphate accumulating organisms (PAO) under alternating anaerobic/aerobic or anaerobic/anoxic conditions. Under anaerobic conditions, PAO can uptake organic substrates (preferably volatile fatty acids, VFA) and store them as a polyhydroxyalkanoates (PHA). This storage requires energy provided by hydrolysis of poly-phosphates (poly-P) resulting in P-release into solution. The reduction power is provided mainly by the catabolism of stored glycogen. The release of phosphate produced under anaerobic conditions is compensated by its aerobic P-uptake. In the aerobic stage, PHA is used as a carbon and energy sources for growth, and the replenishment of internal glycogen and poly-P pools (Mino, van Loosdrecht & Heijnen, 1998). A schematic representation of the PAO metabolism most widely accepted is presented in Figure 3.

The two most common VFA present in municipal wastewater are acetate and propionate. The metabolism of acetate and propionate by PAO has been amply studied (Oehmen, Zeng, Yuan & Keller., 2005; Pijuan, 2004; Hood & Randall, 2001; Lemos, Serafim, Santos, Reis & Santos, 2003; Lu, Oehmen, Virdis & Yuan, 2006; Chen, Randall & McCue 2004; Thomas, Wright, Blackall, Urbain & Keller, 2003). Depending on the carbon source used, PHA polymers are produced with a different composition. Pijuan et al. (2009) studied the effect of carbon sources on the type and quantity of PHA synthesis obtained with different EBPR microbial communities enriched with acetate or propionate. It was shown that the same carbon source produced different PHA composition in each community. The PHA composition was also strongly influenced by the carbon source. Acetate produced PHB-PHV copolymer with a much higher amount of PHB (70%), propionate produced PHV (53%) or PH2MV (49%) copolymer, while other non-identified PHA monomers were also detected with butyrate (Tayà, 2013).

*Figure 3. Schematic representation of PAO metabolism*
*(Tayà, 2013)*
*Source: Tayà, 2013*

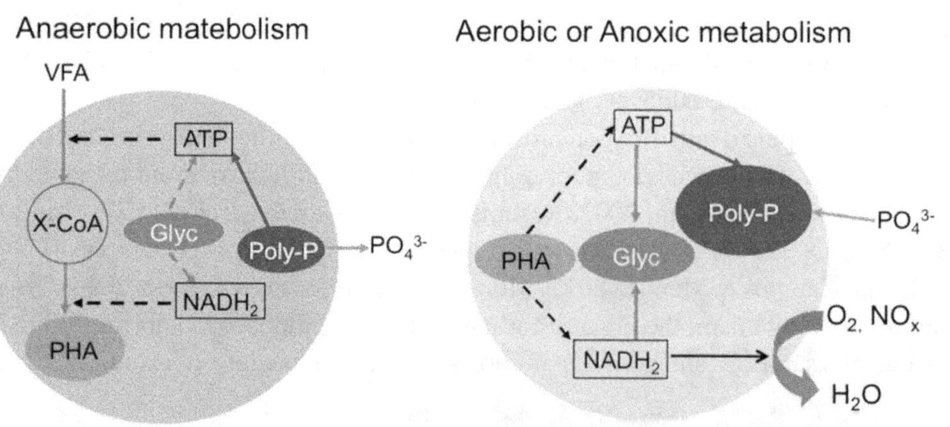

Through microbial techniques, *Acinetobacter* was first proposed to be the primary organism responsible for P removal in EBPR (Fuhs & Chen, 1975), and was long believed to be the sole PAO present in EBPR plants. However, the use of fluorescence *in situ* hybridization (FISH), 16S rRNA-based clone libraries or denaturing gradient gel electrophoresis (DGGE), showed that a high diversity of phylogenetic groups are present in lab- and full-scale EBPR sludge (Seviour, Mino & Onuki, 2003; Hesselmann, Werlen, Hahn, van der Meer & Zehnder 1999; Crocetti et al., 2000; Liu et al., 2001; Ngunyen, Le, Hansen, Nielsen, & Nielsen 2011).

However, one of the large known failures on EBPR systems is the presence of other microorganisms able to survive under anaerobic/aerobic conditions and compete with PAO for donor electron (carbon source). These microorganisms are able to uptake the carbon source under anaerobic conditions without any P activity (Cech & Hartman, 1990, 1993; Crocetti, Banfield, Keller, Bond & Blackall, 2002; Liu, Mino, Nakamura, & Matsuo, 1994; Nielsen, et al., 1999; Satoh, Mino & Matsuo, 1994; Wong, Tan, Ng, & Liu, 2004). The term GAO (glycogen accumulating organisms) was proposed by Mino, Liu, Kurisu, & Matsuo, (1995), and is defined as the phenotype of organisms that store glycogen aerobically and consume it anaerobically as their primary source of energy for taking up carbon sources and storing them as PHA. These microbial competitions in the EBPR systems is influenced by several factors, such as the carbon source, pH, temperature and aeration (Filipe, Daigger, & Grady, 2001; Brdjanovic et al.,1998a,b; Oehmen, Zeng, Yuan & Keller, 2005; Oehmen, Saunders, Vives, Yuan, & Keller, 2006a: Oehmen et al., 2006b; Lopez-Vazquez et al., 2008; 2009; Lu, Oehmen, Virdis, & Yuan, 2006).

## Main Schemes Applied for Conventional and Advanced BNR for Industrial WWTP

Treatment of wastewater by means of biological processes has been widely implemented for classical urban and industrial wastewater. Moreover, there are many biological nutrient removal systems to deal with nutrient removal such as: Phoredox (A/O) (Figure 4), $A^2/O$ (Figure 5), Five-stage Bardenpho (Figure 6), UCT (University of Cape Town) (Figure 7), modified UCT (Figure 8) and VIP (Virginia Initiative Plant) (Figure 9) (Metcalf & Eddy, 2003).

*Figure 4. Phoredox or A/O process*

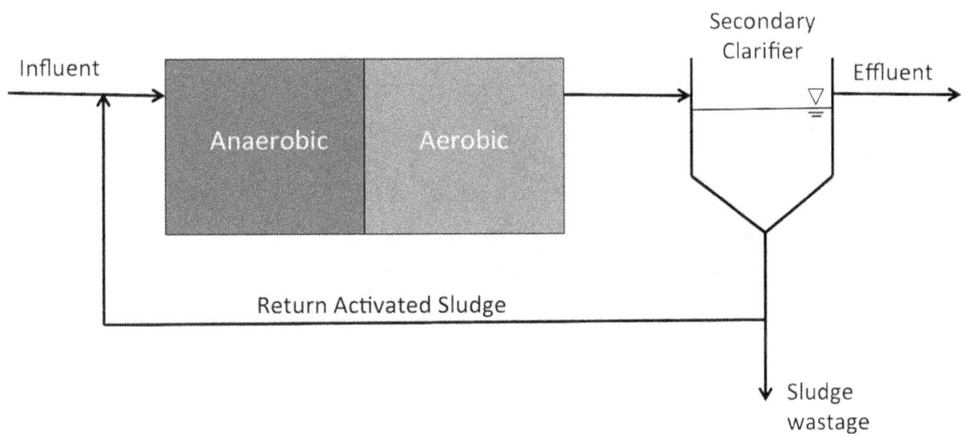

*Figure 5. A2/O process*

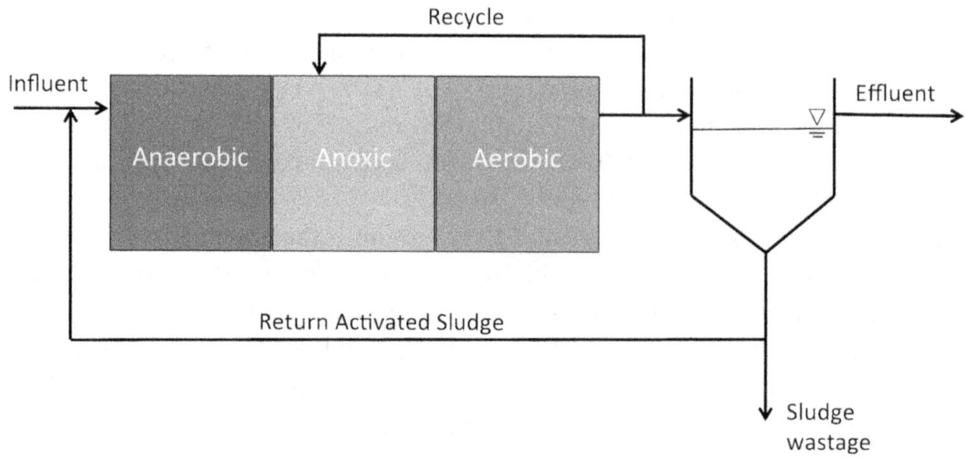


*Figure 6. Modified Bardenpho (5-stage) process*

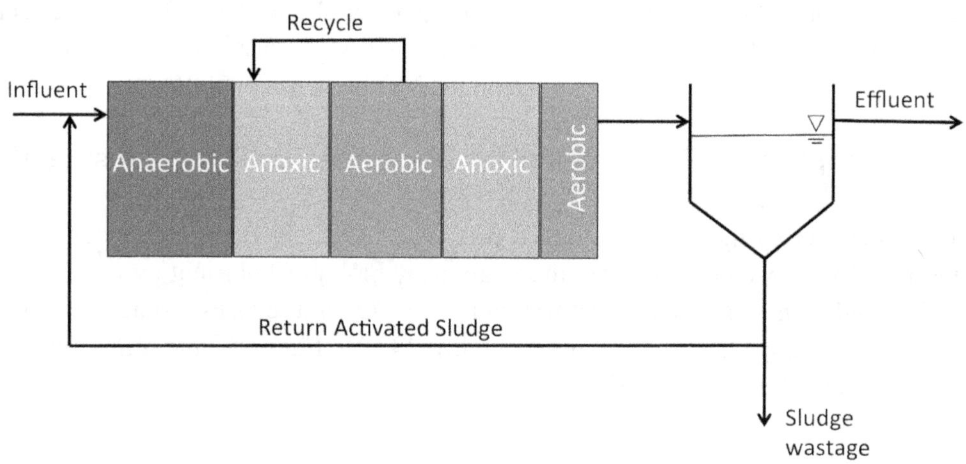


*Figure 7. University of Cape Town (UCT) process*

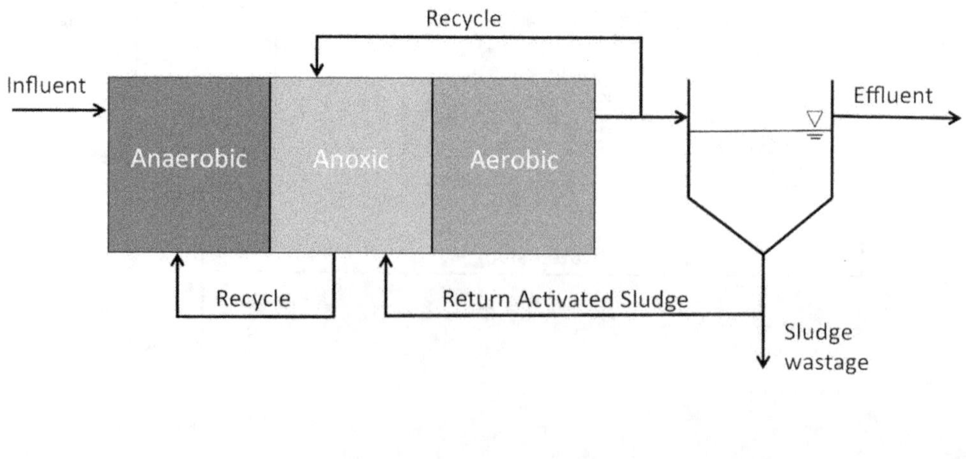

*Figure 8. Modified UCT process*

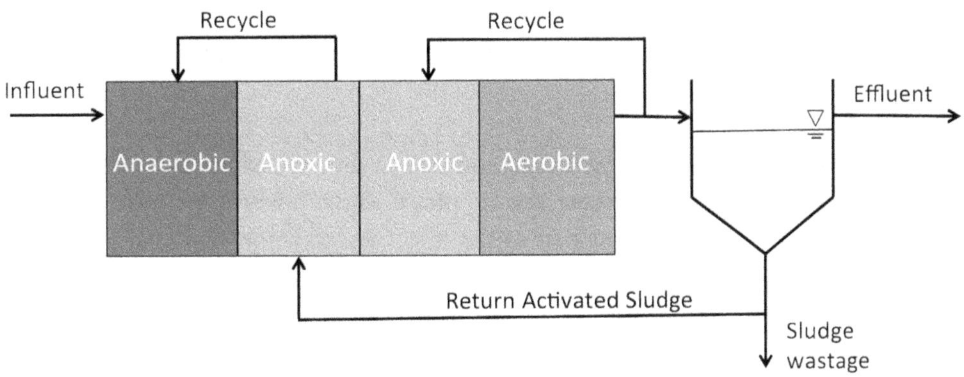


*Figure 9. Virginia Initiative Plant (VIP) process*

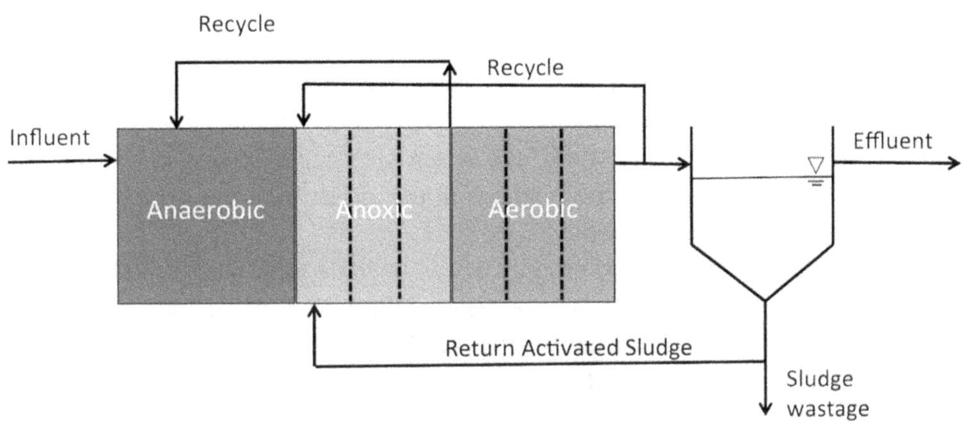


Figure 4 describes the A/O process, which is a patented version of Phoredox by *Air Products and Chemicals, Inc.* The main difference is the use of multiple-staged anaerobic and aerobic reactors. This process is used for combined carbon oxidation and phosphorus removal. In this process there is no nitrification, and the anaerobic retention time is 30 min to 1 h to provide the selective conditions for the biological phosphorus removal (Tayá, 2013).

$A^2/O$ process (Figure 5) is patented by *Air Products and Chemicals, Inc.* It is a modified A/O process to provide an anoxic zone for denitrification. The operation is relatively simple compared to other processes. With this process it is achieved simultaneous nitrogen and phosphorus removal. In the anoxic zone there is a low oxygen concentration, but nitrite and nitrate can be used as an electron acceptor from the recycle of the aerobic zone. However, it requires high COD/P ratios.

On the other hand, Barnard (1975) first achieved phosphorus removal in a mainstream process later called Bardenpho process. Figure 6 shows a 5-stage Bardenpho process, which provides anaerobic, anoxic, aerobic stages for phosphorus, nitrogen and carbon removal. A second anoxic stage is provided for additional denitrification using nitrate produced in the aerobic stage as the electron acceptor and the endogenous organic carbon as the electron donor. The final aerobic stage is used to strip residual nitrogen gas from solution to prevent rising problems and to minimize the P-release in the final clarifier (Metcalf & Eddy, 2003).

The UCT process (Figure 7) was developed at the University of Cape Town and is similar to the $A^2/O$ process. There are two differences:

1. The return activated sludge is recycled to the anoxic reactor instead of the anaerobic stage, and
2. There is an additional recycle from the anoxic phase to the anaerobic phase. This configuration decreases the amount of nitrate recycled to the anaerobic reactor, increasing the organic matter availability for EBPR.

The modified UCT process (Figure 8) splits the anoxic zone into two stages. The nitrate rich recycle from the aerobic zone is recycled to the head of the second anoxic stage. The nitrate contained in the return activated sludge is recycled to the first anoxic stage where it is denitrified. Next, the denitrified returned activated sludge is recycled from the end of the first anoxic stage to the head of the anaerobic stage and mixed with the incoming wastewater

The VIP process (Figure 9) (from Virginia Initiative Plant in Norfolk, US) is similar to $A^2/O$ and UCT processes except for the methods used for recycling. The return of activated sludge is discharged to the inlet of the anoxic zone along with nitrified recycle from the aerobic zone. The mixed liquor from the anoxic zone is returned to the head end of the anaerobic zone. The organic matter in the process influent is stabilized through anaerobic mechanisms in the anaerobic zone, which reduces the oxygen requirements of the process (Metcalf & Eddy, 2003).

The Sequencing Batch Reactor (SBR) (Figure 10) is a single tank where the biological processes and settler occur under distinguished temporal sequence. The SBR is often applied for the treatment of both domestic and industrial wastewater because allowed high removal of the pollutant and flexibility of the operations. The principle of the SBR is the operation in cycles which involves the same pattern of

*Figure 10. Sequencing Batch Reactor process*

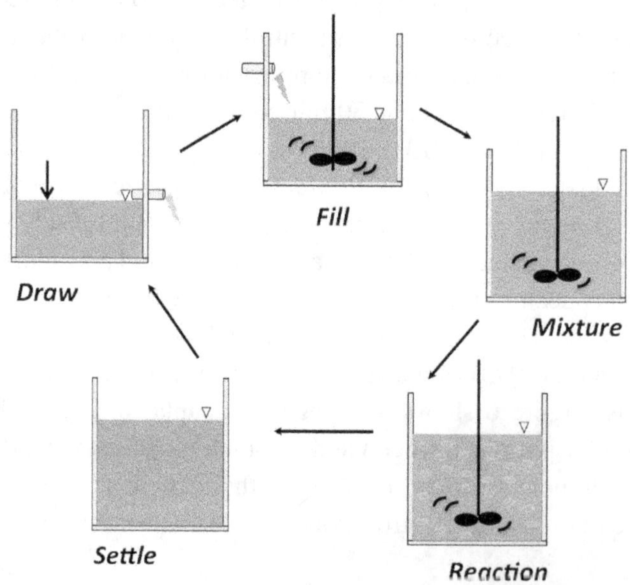

phases, which typically are: 1) Filling; 2) Batch reaction wise (aerobic and/or anoxic and/or anaerobic); 3) Settling; 4) Discharging; 5) Idle. For continuous-flow applications, at least two SBR tanks must be provided so that one tank receives flow while other completes the treatment cycle (Metcalf & Eddy, 2003).

## Nitrogen Stripping, Adsorption, and Vacuum Distillation, Processes: Principles

### Stripping of Ammonia

Ammonia removal from high nitrogenous wastewaters can be accomplished by air or steam stripping. Ammonia stripping is a well-known application to concentrate nitrogen in concentrated wastewater treatment (e.g. landfill leachate, supernatants of anaerobic digestion processes, specific flows of the petrochemical industry), while very few applications in sewage treatment are reported in the literature (Capodaglio, Hlavínek, & Raboni, 2015).

The stripping of ammonia requires a high pH (around 10 to 12), which is usually achieved by addition of base in significant amounts:

$$NH_4^+ + NaOH \rightarrow NH_3 + H_2O + Na^+ \quad (11)$$

$$NH_4^+ + Ca(OH)_2 + HCO_3^- \rightarrow NH_3 + CaCO_3 + 2H_2O \quad (12)$$

The equilibrium is governed by pH and temperature, according to the Equation 13 describing the reaction's ionization constant ($K_{eq}$).

$$K_{eq} = \frac{[NH_4^+][OH^-]}{[NH_3]} = 1.8 \cdot 10^{-5} \text{ at } 25\ ^\circ\text{C} \quad (13)$$

The stripping factor (S) describes the balance between gas and water at the upper end of the stripping column (Morales et al., 2013). The minimum required amount of air ($Q_{Air}$) in relation to the supernatant ($Q_{Liquid}$) is in accordance with Equation (14) (Siegrist, Laureni, & Udert, 2013) to be at least $1/H_{NH3}$ where $H_{NH3}$ is the $NH_3$ Henry's constant.

$$S = \frac{Q_{AIR}}{Q_{LIQ}} \geq \frac{1}{H_{NH_3}} \quad (14)$$

There are two ways to strip the ammonia out the water (Figure 11):

a. Air stripping: air is bubbled through the wastewater and takes up the gaseous ammonia. This gas has to be washed out in a second reactor under acidic conditions.

$$2NH_3 + H_2SO_4 \rightarrow (NH_4)_2SO_4 \quad (15)$$

*Figure 11. Typical flow diagrams of the stripping processes with air (a) and steam (b)*
*(Adapted from: Raboni, Torretta, Viotti, & Urbini, 2013)*
*Source: Adapted from Raboni et al., 2013*

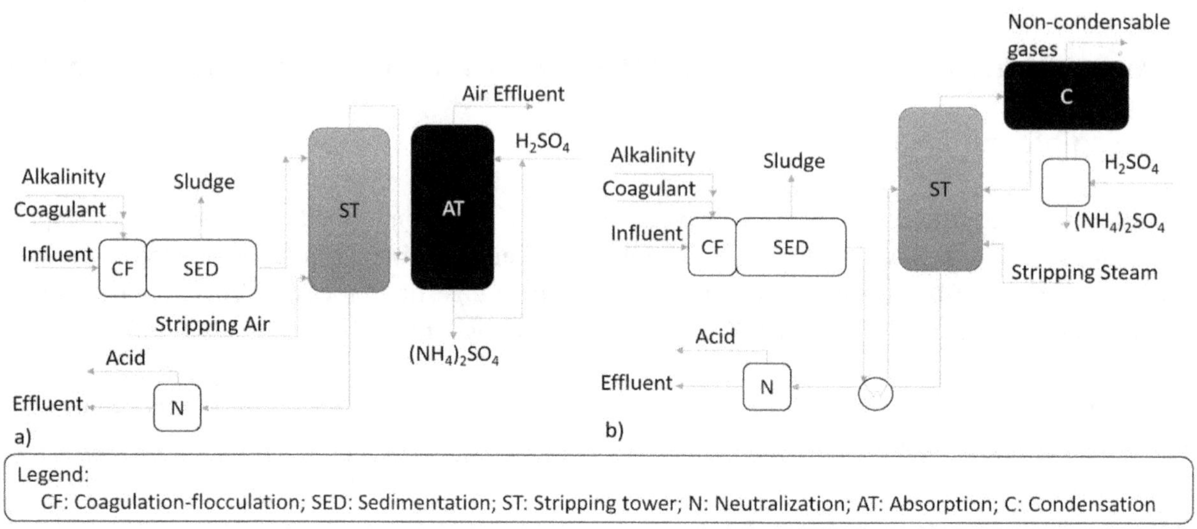


b. Steam stripping: steam is blown through the water and after condensation a concentrated ammonia solution is produced.

This process requires 2.4 kg/kg N calcium oxide for increasing the pH above 9 and 7.0 kg sulphuric acid each kg of ammonia nitrogen absorbed (Purtschert, Siegrist, & Gujer, 1996). The electricity requirement for the aeration is an estimated 26.3 MJ/kg N and the following energy data are used for the calculations (Patyk & Reinhard, 1997).

The stripping process is not applicable for wastewater of low strength. Certain wastewater, like from manure, from the production of PU-foams or digester supernatant in municipal WWTP, can be treated economically.

Air stripping is often used in the treatment of municipal solid waste (MSW), landfill leachate (or groundwater polluted by the same leachate). An example of the process is represented in Figure 12. The system consists of: 1) coagulation-flocculation at pH > 11; 2) water heating at 38°C and ammonia air stripping; 3) ammonia recovery by absorption with a sulfuric acid solution.

Raboni et al. (2013) obtained a reduction of ammonia from 199 mg/L (as N) down to 22.5 mg/L (as N). The plant showed severe scaling issues, with an increase in weight of the single filling components (Pall rings) up to seven times their original weight. The progressive scaling produced an 18% loss of ammonia removal efficiency, caused by the reduction of both the packing interfacial surface and the air/water ratio due to head loss increase. Periodical (every six months) chemical washing of the stripping towers with a 30% HCl solution was carried out to avoid scaling build-up. Closed-loop air recirculation allowed the water temperature inside the column to be maintained in the range of 25-38°C, even during winter, but caused a gradual decrease in the solution's pH within the tower. This was due to the acidic characteristics of the stripping air coming from the ammonia absorption tower. Subsequent experiences proved the actual possibility of achieving high ammonia removal efficiencies (greater than 95%) by operating the stripping towers at 38°C with liquid rates below 4.5 $m^3/(m^2 \cdot h)$ and air/water ratios above 1500 $Nm^3/m^3$ (Viotti & Gavasci, 2015).

*Figure 12. Diagram of a physical-chemical process for leachate treatment, with wastewater pre-heating and a closed-loop air stripping process*
*(Adapted from: Raboni et al. 2013)*
*Source: Adapted from Raboni et al., 2013*

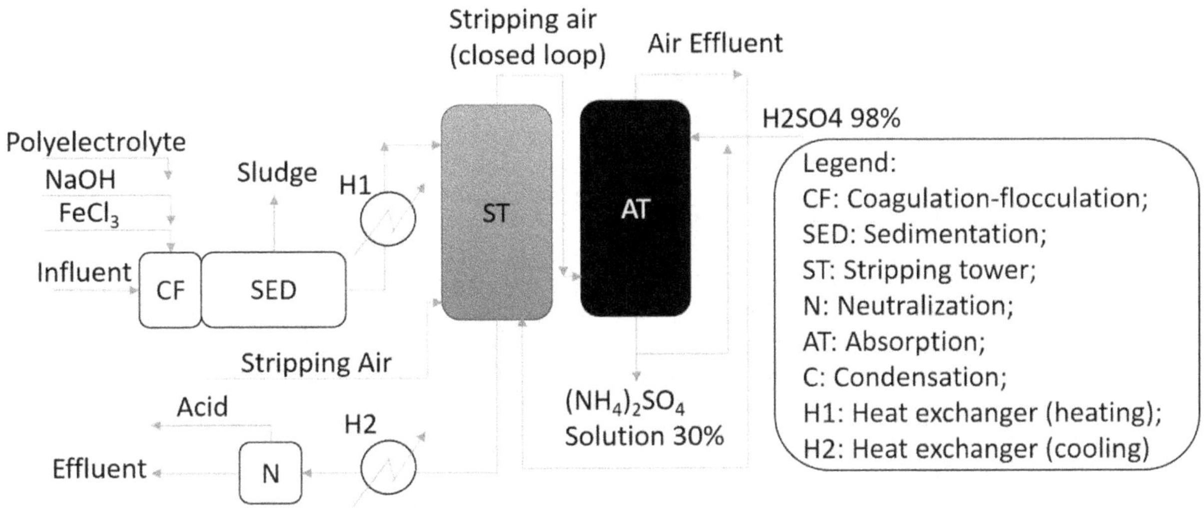


## Vacuum Distillation

In the case of highly concentrated streams vacuum distillation is a viable solution for ammonia recovery (EPA, 2013).

Vacuum flash distillation processes (Figure 13) have been applied to ammonia recovery from centrate of manure digested sludge, using as a heat source the biogas generated by the anaerobic digestion (Orentlicher, Brown, & Hall, 2009). The ammonia can be removed from approximately 1000 mg $NH_3$–N/L to less than 100 mg $NH_3$–N/L operating at a pH value of 9.5 and a temperature of 60°C under vacuum (Orentlichter et al., 2009). Other combinations of temperature and pH achieve similar high free ammonia fractions, allowing an effective stripping process. The resulting ammonia-rich distillate is a marketable product or it can be converted into an ammonium sulfate solution. If desired, such a solution can be further concentrated, using multiple vacuum distillation stages.

## Adsorption

Zeolites and other mineral like clay or tuff have ion exchange properties and are used to adsorb ammonium when is present at low concentration. A common approach is to pump wastewater through columns packed with zeolite. Ammonia concentrations of 25 to 50 mg $NH_4$-N/L can be lowered to levels around 1 mg $NH_4$-N/L (zeolites: Booker, Priestley, & Fraser, 1996; Nishimura, Somiya, Tsuno, & Iwabu, 1996; Nguyen & Tanner, 1998; Komarowski & You, 1997; Cooney, Booker, Shallcross, & Stevens, 1999; Lahav & Green, 2000; tuff: Ibrahim, 2001; clay: Gisvold, Odegaard, & Follesdal, 2000).

Buday (1994) used cation exchange resins to reuse ammonia from wastewater of an ammonium nitrate fertiliser plant. The exhausted resins were regenerated with nitric acid, gaining an ammonium nitrate solution that is recycled into the fertiliser production. Chen et al. (1999) successfully tested the use of

*Figure 13. Diagram of a typical flash distillation process for ammonia recovery*

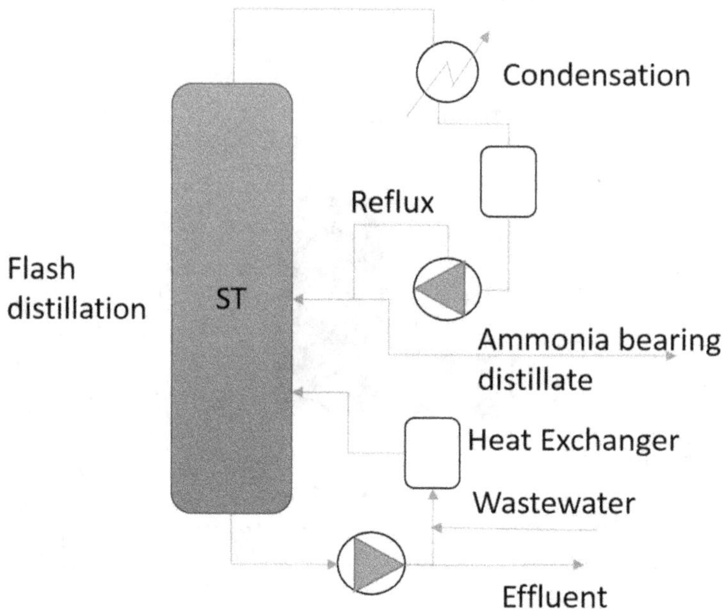


different cellulose acetate fibres for the adsorption and recovery of proteins. Kioussis et al. (2000) developed polymer hydrogels that remove nitrate, nitrite and phosphate from aquaculture wastewater. They claim to achieve following qualities: 10 mg N/L for nitrate and 0.08 mg N/L for nitrite. Gel regeneration is accomplished by flushing 1 M sodium hydroxide solution.

## Chemical Phosphorus Precipitation and Crystallization: Principles

### Chemical Phosphorus Precipitation Processes

The chemical phosphorus precipitation by addition of metal salts is widely used due to its simple operation and implementation. However, the disadvantages of chemical precipitation are high sludge production and costs of chemicals. The sludge produced during chemical phosphorus precipitation cannot be applied to agriculture lands because it has a low bioavailability (Morse, Brett, Guy, & Lester, 1998). Chemical phosphorus precipitation process is widely used to remove phosphorus from diluted wastewater streams. Generally, chemical phosphorus precipitation process is based on the addition of metal salts to wastewater, producing an insoluble metal phosphate precipitate. This precipitate is separated from the liquid phase using conventional separation technologies such as sludge filters or centrifuges. The phosphate precipitation efficiency by metal salts addition is highly dependent on the solids separation process. Even though phosphorus removal by adding metal salts is usually used, it is not appealing for recovering phosphorus because the obtained precipitates are difficult to manage, and so they are often disposed to landfills. This is because the aluminium in aluminium phosphates may be toxic to many plants and some soil organisms. Meanwhile, phosphorus in iron phosphates is usually considered to be unavailable for plants, therefore reducing its suitability for replacing phosphate fertilisers (Rittman, Mayer, Westerhoff, & Edwards, 2011). Iron phosphates are no significant industrial outlets for recovered ion phosphates because its recycle is very expensive at present. Regardless, phosphorus in aluminium phosphates can be used as raw material for phosphorus production by using the electrothermal process of Thermphos (Schipper et al., 2001).

The most important parameter for the chemical phosphate precipitation process is the dose of metal salts added. The amount of metal salts used influences the performance and the economy of the precipitation process. The metal ions should be added in excess relative to phosphate concentration to achieve a low concentration of phosphate in the effluent (Rybicki, 1997). The amount of metal salt added cannot be determined only by the concentration of phosphate, and it should be determined for each case. It also depends on the wastewater characteristics, operational pH, method of chemical addition, flocculation method, chemical addition feeding point(s) and time after chemical addition, and it is usually determined by on-site testing (EPA, 2010). It is possible to calculate only an approximate dose of metal salts for removing phosphorus. Several side reactions can take place alongside the precipitation of phosphate during the phosphate removal process. The EPA Design Manual (1987) recommends a 20% excess of aluminium ions to reduce phosphate by around 80%, while the 100% excess is suggested to obtain a phosphate removal of 95%.

## Phosphorus Crystallization Processes

Phosphorus crystallization processes can be considered separately than phosphorus precipitation by adding metal salts because the chemistry involved in both technologies is quite different. Like in phosphorus precipitation processes, phosphorus crystallization allows transferring soluble phosphorus from wastewater to solid phase. However, phosphorus crystallization process is only used for removing phosphorus from wastewater but also generates a final product that has a potential to be used as fertilizer or as raw material for phosphorus industry.

### *Calcium Phosphate Precipitation*

Phosphorus can be recovered from wastewater by crystallization processes either using lime to produce calcium phosphates, or adding magnesium salts to produce magnesium ammonium phosphate (struvite). Calcium phosphates are less valuable than struvite as a multiple source of nutrient and fertilizer. However, calcium phosphate can be recovered and recycled into phosphate and fertilizer industries. Calcium phosphate is a suitable form of recovered phosphorus because they are similar to phosphate rocks. Different calcium phosphate species may precipitate from wastewater. The nature of calcium phosphate is influenced by the wastewater characteristics and the operating pH. Hydroxyapatite ($Ca_5(PO_4)_3OH$) is thermodynamically the most stable calcium phosphate species, but hydroxyapatite does not precipitate readily. Amorphous calcium phosphate ($Ca_3(PO_4)_2 \cdot n\ H_2O$), dicalcium phosphate dihydrate or brushite ($CaHPO_4 \cdot 2\ H_2O$), dicalcium phosphate anhydrous or monenite ($CaHPO_4$) and octacalcium phosphate ($Ca_8H_2(PO_4)_6 \cdot 5\ H_2O$) can be precursors of hydroxyapatite, depending on the operational conditions.

As mentioned above, lime is usually used to recover phosphorus from wastewater. This calcium compound can be added either as quicklime ($CaO$) or hydrated lime ($Ca(OH)_2$). Lime is an effective phosphate removal agent, but it produces large volumes of sludge. This chemical has been widely used to remove phosphorus from wastewater, but its use has been declining due large amounts of lime are required to obtain low levels of phosphorus (Metcalf & Eddy, 2007; EPA, 1987). Thus, phosphorus recovery by adding lime can produce large volumes and high chemicals costs.

Usually, the dose of lime added to the precipitation system is higher than the amount required to remove phosphate from wastewater. Lime reacts with phosphate to form insoluble precipitates of calcium phosphates, and it can also react with the alkalinity of wastewater to produce calcium carbonates (EPA, 2010). The operating pH must be fixed at high values to obtain a high phosphate removal efficiency (i.e. pH values higher than 10), and the amount of lime added must be approximately 1.5 times the total alkalinity (expressed as mg $CaCO_3$/L) (Rybicki, 1997). Thus, the amount of produced precipitates will be influenced by the level of bicarbonate alkalinity in wastewater.

## Struvite Crystallization

The crystallization of struvite is one of the most recommended technologies to phosphorus recovery from concentrated phosphorus wastewater, such as effluents from anaerobic digesters or sludge dewatering process (Desmidt et al., 2015). The reason is that the crystallization of struvite is not only used for recovering phosphorus from wastewater but also generates a product with a commercial value as a fertilizer.

Struvite is a magnesium ammonium phosphate salt ($MgNH_4PO_4{\cdot}6H_2O$) that usually crystallizes in a reversible reaction and produces crystals with a $Mg^{2+}$:$NH_4^+$:$PO_4^{3-}$ molar ratio of 1:1:1. Struvite can be used as a slow release fertilizer, either directly or after processing by fertilizer industry (El Diwani et al., 2007; de-Bashan & Bashan, 2004). The formation of struvite is conditioned by a combination of factors, including the presence of seed crystals, thermodynamics and mass transfer between solid and liquid phases, and reaction kinetics. At the same time, these factors are influenced by several physical-chemical parameters such as pH, the chemical composition of the wastewater, mixing energy and temperature (Doyle & Parsons, 2002). The most important parameters that must be controlled are: pH, temperature and presence of other ions.

The pH is one of the key factors in struvite crystallization process. Theoretically, there should be a direct relation between the efficiency of phosphorus recovery as struvite and pH. This is because the ionic equilibriums involved in the struvite crystallization are very pH dependent, which in turn could affect the struvite formation (Andrade & Schuiling, 2001). A slight variation in the pH can significantly affect the crystallization of struvite because it varies the distribution of the struvite constituents, affecting their availability for struvite formation. The pH range in which struvite may crystallize is between 7.0 and 11.5, but the suitable pH range is from 7.5 to 9.0 (Rahman et al., 2014). The solubility of struvite is also conditioned by the pH of the solution. It is highly soluble at acidic pH and highly insoluble at alkaline pH (Doyle & Parsons, 2002).

Temperature also influences the struvite crystallization process, struvite characteristics and its thermal stability. Temperature affects the struvite solubility in a way that its solubility increases with the temperature increase, and thus the formation of struvite is more difficult to occur at high temperatures (Le Corre, Valsami-Jones, Hobbs, & Parsons, 2009). So, struvite solubility product is inversely proportional to the variation of temperature (Ben Moussa, Tlili, Batis, & Ben Amor, 2011; Aage, Andersen, Blom, & Jense, 1997).

The presence of other ions in solution, also known as foreign ions, affects the performance of struvite crystallization process because these foreign ions can interfere with struvite crystallization by competing for magnesium and phosphate to form other precipitates species. So, the efficiency of struvite crystallization is affected by the presence of foreign ions, such as calcium and/or carbonate, and also the formation of calcium and/or carbonates precipitates affects the purity of the produced struvite. In this regard, the struvite crystallization is more affected by the presence of calcium at low phosphate concentrations (<40 mg $PO_4^{3-} \cdot L^{-1}$) (Desmidt et al., 2013). Crutchik & Garrido (2011) indicated that a high ammonium concentration could be highly beneficial to the reaction of struvite crystallization, so to obtain a relatively pure solid phase in terms of struvite purity. Thus, it could be possible to obtain a high purity regarding struvite, instead of precipitates of calcium phosphates in high ammonium wastewater.

Moreover, the economic viability of struvite crystallization is highly influenced by the availability of magnesium in the raw wastewater, the cost of alkali used or the cost magnesium source used. This is because wastewater usually contains a low amount of magnesium compared with phosphate and nitrogen and it is necessary to supply a magnesium source to promote the crystallization of struvite. In the literature, the use of high grade magnesium sources, such as $MgCl_2$, $MgSO_4$, MgO or $Mg(OH)_2$, has been reported (Rahaman et al., 2014; Hutnik et al., 2013; Jaffer, Clark, Pearce, & Parsons, 2002; Nelson, Mikkelsen, & Hesterberg, 2003). Nevertheless, the utilization of these high grade and expensive magnesium sources can lead to high operating costs compromising the viability of the process. In this sense, the use of low cost magnesium sources, such as industrial grade MgO or $Mg(OH)_2$, seawater or bittern, can significantly reduce the cost of struvite crystallization. MgO and $Mg(OH)_2$ can be used in a dual function, as magnesium and alkali sources. Unlike for NaOH, the use of MgO and $Mg(OH)_2$ does not produce drastic pH changes. Meanwhile, seawater, brine and bittern can be added to promote struvite crystallization for those urban or industrial WWTP located near the seashore (Liu, Giannis, Zhang, Chang, & Wang, 2013; Señoráns, Barros, García, & Garrido, 2011; Lee, Weon, Lee, & Koopman, 2003; Matsumiya, Yamasita, & Nawamura, 2000).

## FULL SCALE APPLICATIONS

### Biological Nutrients Removal by SBR from Industry

The SBR system has demonstrated high removal efficiency, and thus, it is an optimal process for the treatment of nitrogenous industrial wastewater, such as landfill, textile effluent, tannery, meat product processing, toxic and slaughterhouse wastewaters, because of its simplicity in operation and economic aspects. Table 2 shows the typical application of SBR with various kinds of wastewaters. Among them, the leachate showed a very low $BOD_5$/COD ratio, a high concentration of COD, $NH_3$-N, heavy metals, and other toxic compounds which decrease the activity of the biomass and the treatment capacity of the SBR as a consequence (Uygur & Kargi, 2004). Typically, the treatment of leachate for nutrients removal using the SBR is reported in the literature (Guo, Abbas, Chen, Liu, & Chen, 2010; Spagni, & Marsili-Libelli, 2009; Klimiuk & Kulikowska,2006). Although leachate may have a very low $BOD_5$/COD ratio (0.09 to 0.3), with the addition of an external carbon source high nutrients removal efficiency can be achieved.

The variability in the operating conditions of the SBR is mainly related to the type of the wastewater fed and the quality of the final effluent. Table 3 reports the operating conditions for the treatment applied to leachate.

*Table 2. Type of industrial wastewater treated using Sequencing Batch Reactor*

| Type of Wastewater | Pollutants/Removal | Removal | Reference |
|---|---|---|---|
| Landfill leachate | COD | <83 | Timur & Ozturk (1999) |
| Dairy wastewater | COD<br>Total Nitrogen | 80.2<br>38.3 | Li & Zhang (2002) |
| Landfill leachate | COD<br>$NH_4$-N<br>Phosphate | 75<br>44<br>44 | Uygur & Kargi (2004) |
| Tannery wastewater | COD<br>$NH_4$-N | 80-82<br>83-99 | Ganesh, Balaji, & Ramanujam (2006) |
| Slaughterhouse wastewater | COD<br>Total nitrification | 96<br>96 | Li et al. (2008) |
| Landfill leachate | Nitrification<br>Nitrogen<br>COD | 98<br>95<br>20-30 | Spagni & Marsili-Labelli (2009) |
| Meat products processing wastewater | COD<br>$BOD_5$<br>$NH_4$-N | 99<br>98<br>71 | Rodriguez, Pino, & Penuela (2011) |
| Piggery wastewater | COD<br>TN<br>TP | 98<br>98<br>98 | Tilche et al. (2001) |

*Table 3. Industrial wastewater (leachate) treated using Sequencing Batch Reactor*

| N° | Reference | HRT (d) | SRT (d) | MLSS (mg/L) | OLR (kg COD/m³d) |
|---|---|---|---|---|---|
| 1 | Timur & Oztuk (1999) | 1.5-10 | 9-22 | - | - |
| 2 | Kennedy & Lentz (2000) | 0.5-1.0 | - | - | 0.4-9.4 |
| 3 | Neczaj, Okoniewska, & Kacprzak (2005) | - | 10 | 4000 | - |
| 4 | Klimiunk & Kulikowsk (2006) | 2-12 | 17-51 | - | - |
| 5 | Leatinen, Luonsi, & Vilen (2006) | - | 10-40 | - | - |
| 6 | Bu et al. (2010) | - | - | 4400 | 1.7 |

HRT: hydraulic retention time; SRT: solids retention time; MLSS: mixed liquor suspended solids; OLR: organic loading rate.

## Nitrogen Recovery from Industry and Agrowaste

### Nitrogen and Phosphorus Recovery from Anaerobic Effluent in Glattbrugg, Switzerland

A full-scale stripping reactor including a $CO_2$ pre-stripper was located in the Kloten/Opfikon Wastewater Treatment Plant (60,000 population equivalent, Glattbrugg, Switzerland). It was the first of these reactors installed in Switzerland (Böhler & Liebi, 2012a). The ammonia stripping process was set in operation for the first time in fall 2010. Some of the characteristics of the stripping reactor were summarized in Table 4, while the scheme of the overall process is shown in Figure 14. The stripping columns (ST) use special UV-stabilized Fluorpolymere packing with a honeycomb shape (CF 12F, Hewitech GmbH &

Co, Ochtrup, Germany). This material has 12 mm of channel opening, a specific surface of 300 $m^2/m^3$ and a cross-flow channel structure. The ammonium rich stream is fed to the top of the $CO_2$-pre stripper where the air is fluxed in counter flow. The $CO_2$ stripping increases the pH of the liquid due to the displacement of carbonic acid system equilibrium (Cohen & Kirchmann, 2004).

To avoid a $CO_2$ accumulation in the column the $CO_2$ rich gas was discharged after this unit, and fresh air was provided. The pH of the liquid in the ammonia stripper was increased using sodium hydroxide up to a fixed value to favor the equilibrium displacement towards the $NH_3$, as showed in Equation (16):

$$NH_4^+ + OH^- \leftrightarrow NH_{3(aq)} + H_2O \tag{16}$$

*Table 4. Dimensions and operational condition of the stripping system*

| Parameter | Value | Units |
|---|---|---|
| Column height ($CO_2$ and Ammonia) | 10 | M |
| Packing bed height | 6 | M |
| Diameter | 0.98 | M |
| Liquid inflow | 5-12 | $m^3/h$ |
| Air flow to ammonia stripper | 1000-6250 | $Nm^3/h$ |
| Gas flow to the $CO_2$ stripper | 50-200 | $Nm^3/h$ |
| Average $NH_4$-N concentration inlet | 1000 | mg/L |
| Average $NH_4$-N concentration effluent | 30 | mg/L |
| Maximum removal rate | 99 | % |

*Figure 14. Flow-chart of the co-treatment of sludge liquid and treated urine*
*(Adapted from: Morales, Boehler, Buettner, Liebi, & Siegrist, 2013)*
*Source: Adapted from Morales et al., 2013*

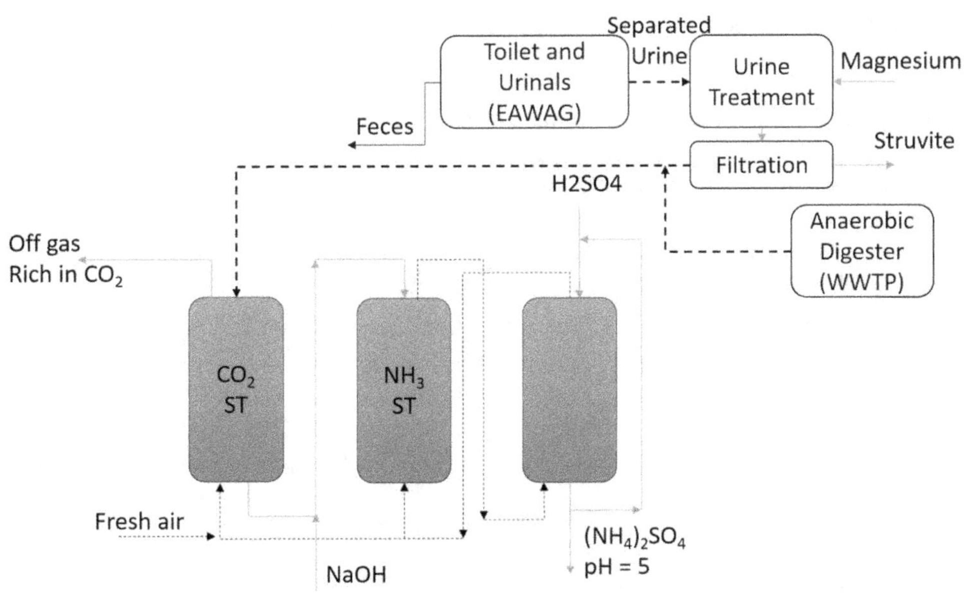

This liquid was fed by the top of the stripper reactor, where the ammonia solved in the liquid phase is stripped to the gas phase. The flow rate of the urine maintained up to around 5.1 $m^3/h$, while the operation $Q_{Air}/Q_{liq}$ ratio used in the air stripping unit was in the range of 700–750 $Nm^3/m^3$ at 60 °C, because higher air to liquid ratios resulted in higher electricity consumption (Böhler, Büttner, Morales, Liebi, Schachtler, & Siegrist, 2012b). Then, the ammonia rich gas was washed in the sorber unit using a sulfuric acid solution ($H_2SO_4$ at 32% solution, 1.16 kg/L, Thommen-Furler AG, Büren, Switzerland) and ammonia sulfate was recovered (Equation 15) obtaining a final density solution of 1.2 kg/L.

Acid cleaning of the stripping reactor is executed daily using hydrochloric acid at pH 0–2 (HCl at 35%, Thommen-Furler AG, Büren, Switzerland) with an automated pH control. Each kg of N removed need at least 3.5 kg of $H_2SO_4$ and will produce 4.7 kg of $(NH_4)_2SO_4$. In a cost study for a full-scale stripping plant with a $CO_2$ pre-stripper, the estimated costs for treatment were around 3–4.5 €/kg N recovered (Zuleeg et al., 2012).

## Cold Stripping and Water Recovery from Liquid Manure in Remedello (Brescia, Italy)

The N-Free® plants is a patented process installed in the North of Italy, which is considered a Nitrate Vulnerable Zone by the Nitrate Directive (91/676/EEC).

The N-Free® process (Fiolini e Savani s.r.l., Brescia, Italy) is characterized by a series of physical/chemical treatments which allow separation and concentration of the liquid fraction of the digestate and a final stripping for the production of ammonium sulfate (Figure 15). The N-Free® unit operated in batch mode and treated up to 100 $m^3$/day of raw anaerobic digestate (about 14 $m^3$ each batch cycle). The overall process scheme is composed by the following steps:

1. Rough solid/liquid separation achieved by a screw press (SC) separator (Doda, Mantova-Italy).
2. Conditioning (with polyamide flocculant) of the liquid fraction from the screw press and effective solid/liquid separation by decanter centrifuge (DC) (MAMMOTH 570/3, Pieralisi, Jesi-Italy).
3. Ultrafiltration (UF) of the liquid effluent from the decatenter by a 40 kDa grafted polyacrylonitrile membranes (PAN) (Orelis, France), operating at 3.5 to 4.5 bar of exercise pressure.
4. Treating of the UF permeate in a two-stage reverse osmosis unit (Dow, USA).

*Figure 15. Simplified scheme of N-Free® plant; (1) Screw press separation, (2) Centrifuge separation, (3) Ultrafiltration-UF, (4) Reverse Osmosis – RO, (5) Zeolites refining, (6) Cold ammonia stripping*
*(Adapted from: Ledda, Schievano, Salati, & Adani, 2013)*
*Source: Adapted from Ledda et al., 2013*

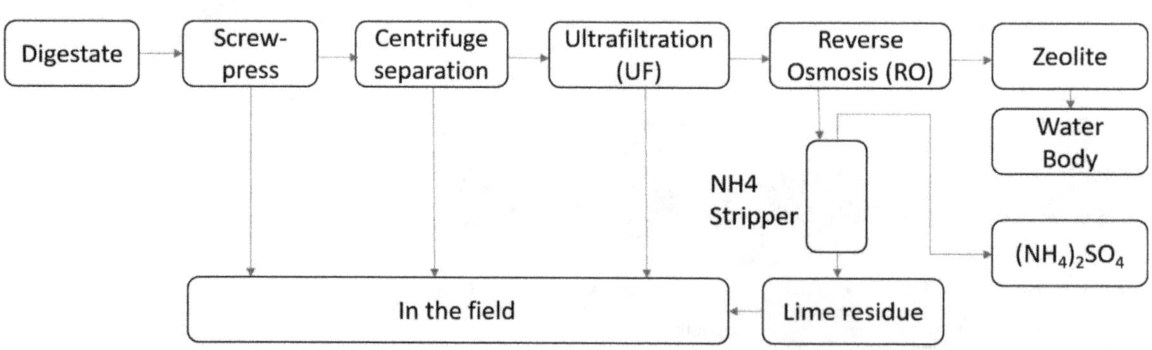

5. Refining in zeolites bed of the final reverse osmosis permeate before the final discharge in the water body.
6. Cold stripping of the reverse osmosis (RO) concentrate by air blowing recirculation and through addition of 15-18 kg/m$^3$ sludge of lime which allows pH raising up to 12-12.5. At these conditions, the equilibrium $NH_4/NH_3$ is completely shifted to $NH_3$ which is easily stripped as gaseous ammonia by the controlled air flow. Air plus ammonia is then scrubbed with sulfuric acid producing liquid ammonium sulfate (8% on a wet weight basis as N).

Finally, the cold stripping resulted in RO ammonia reduction of 78% from the anaerobic digestate, a significant reduction of the exportable volumes, leading a more sustainable livestock farming.

Electrical consumptions and use of reagents during the operation led to a total cost of 2 €/m$^3$ of digestate, while up to 4.2 €/m$^3$ considering personnel cost for assistant and depreciation charge. However, the income could be related to the savings regarding less transportation volume required (about 1 €/m$^3$ of digestate) and incomes related with ammonium sulfate trade (about 0.5 €m$^3$ of ammonia sulfate at 6-8% as N).

## Chemical Phosphorus Precipitation from Industry

The DHV Crystalactor© was developed by DHV Consulting Engineers (The Netherlands) to remove calcium from drinking water. Later, this technology was used to remove other compounds, such as phosphate, from wastewater streams. This technology utilizes a cylindrical fluidized bed reactor to recover phosphorus by phosphate precipitation in which sand is used as seed material. Phosphorus can be recovered either by calcium phosphate precipitation or by struvite crystallization, which takes place over these sand grains (Giesen, 1999; Eggers, Dirkzwager, & Van der Honing, 1991). For calcium phosphates precipitation, $Ca(OH)_2$ solution is added to the precipitation system and the dosage is controlled by the measurement of operational pH. Efficient calcium phosphate crystallization requires an operational pH around 9. Most of phosphate recovery technologies are based on the same crystallization chemistry of the DHV Crystalactor© technology, but with some variation (e.g. in seeding materials).

## Struvite Recovery from Industry

The full-scale facilities of struvite crystallization have the potential to produce a commercial product, which could be used in combination with fertilizers from mined phosphate. Struvite can be produced using different crystallization technologies. The technology chosen will likely be a determining factor in the quality of the produced struvite. The particle size of struvite may vary with the operating conditions and reactor configuration used. According to particle size, struvite can be divided into two categories: fine crystals and pellets. In order to obtain a product suitable for application to agricultural lands, struvite pellets must be produced. Struvite pellets are much easier to handle and to transport than fine crystals. Thus, the quality product determines the user-friendliness of struvite, and consequently it can influence the cost-effectiveness of struvite crystallization process. Up to date, the main technologies to recover phosphate by struvite crystallization at full-scale are DHV Crystalactor©, Unitika Phosnix© process, Ostara PEARL© process, Phospaq™ process, NuReSys© process and Multiform Harvest™ (Table 5). Currently, several full-scale facilities based on the aforementioned technologies have been developed in Europe. These facilities tend to be concentrated in certain countries: Belgium, The Netherlands, Germany

*Table 5. Phosphorus recovery technologies at full-scale (in operation)*

| Technology Name | Crystalactor© | Ostara Pearl® | Multiform Harvest™ | NuReSys© | Phosphaq™ |
|---|---|---|---|---|---|
| Type of reactor | FBR | FBR | FBR | CSTR | CSTR with diffused air |
| Name of product recovered | Struvite, CaP, MgP | Crystal Green® | Struvite | BioStru® | Struvite |
| Efficiency of recovery (%) | 10-40% $NH_3$-N<br>85-95% P (struvite)<br>>90% P (CaP) | 10-40% $NH_3$-N<br>80-90% P | 10-40% $NH_3$-N<br>80-90% P | 5-20% N<br>>85% P | 10-40% $NH_3$-N<br>80% P |
| N° of full scale installations | 4 | 8 | 2 | 7 | 3 |

and United Kingdom, and the struvite crystallization technologies implemented in these countries are: Ostara PEARL©, NuReSys© and Phospaq™.

The first applications of struvite crystallization at full-scale were implemented in Japan. The Unitika Phosnix© process is based on the struvite crystallization in a fluidized bed reactor (FBR), in which the crystallization of struvite is promoted by adding sodium hydroxide and magnesium hydroxide as alkali and magnesium sources, respectively. Additionally, struvite seed crystals are added to facilitate the phosphorus recovery by struvite crystallization (Ueno & Fuji, 2001). Another technology developed is the Ostara PEARL© process, which is based on the phosphorus crystallization using a two-phase fluidized bed reactor. This system is composed by a column comprising some sections. The diameter of the column increases stepwise between the sections. This technology uses magnesium chloride when required, and sodium hydroxide as magnesium and alkali sources to facilitate the struvite crystallization (Koch, Mavinic, Yonemitsu, & Britton, 2009). Like in the Unitika Phosnix© process, Ostara PEARL© process uses recycled struvite as seed material. The Phospaq™ process recovers phosphorus by struvite crystallization in a continuous stirred tank reactor (CSTR) aerated. In this technology struvite crystallization is facilitated by the increase of pH via stripping of carbon dioxide ($CO_2$) and the addition of magnesium oxide as magnesium source (Abma, Driessen, Haarhuis, & van Loosdrecht, 2010). This technology removes biological degradable chemical oxygen demand (COD), phosphate and ammonium in the same reactor. Aeration provides optimal conditions for struvite crystallization and the oxygen for the biological conversion of COD. The NuReSys© process is based on two CSTRs: a stripping reactor and a stirred crystallization reactor, in which magnesium chloride and sodium hydroxide are added to promote struvite crystallization (Moerman, Carballa, Vandekerckhove, Derycked, & Verstraete, 2009). Finally, Multiform Harvest™ uses a cone-shaped fluidize-bed crystallizer for struvite crystallization. Sodium hydroxide and magnesium chloride are added to produce struvite crystals from a stream with high concentrations of nitrogen and phosphorous (Bowers & Westerman, 2005).

## CONCLUSION

This chapter presents and discusses the main options for nutrient management in industrial wastewater treatment, including the parameters to optimize the processes and the different technology readiness levels. Phosphorus recovery should always be pursued in order to recycle a critical raw material. Struvite recovery is a robust technology which optimization is mainly related to the reactor configuration and magnesium source. However, other possible technologies for P recovery, including enhanced biological P removal or novel ion exchange resins should be promoted especially for low concentrated streams. On the other hand, nitrogen recovery should be evaluated case by case on the basis of its economic and environmental sustainability. Ammonia stripping and recovery as ammonium sulfate is the most traditional and reliable system, but it could be limited by the legislation restriction. If the biological process can be applied, nitrogen deammonification and its recovery via Haber-Bosh process are energetically favorable. In the case of biological processes, the via-nitrite pathway represents an advance over conventional activated sludge systems, and the best carbon source may enable the optimization of denitritation with the concurrent anoxic phosphorus uptake.

## ACKNOWLEDGMENT

Carlota Tayà is grateful for the TECNIOSPRING grant received from the government of Catalonia and ACCIÓ (TECSPR14-1-0033). The authors Carlota Tayà and Sergio Ponsá are members of BETA Tech. Center (TECNIO Network) and BETA Reseach Group (2014 SGR 1286).

## REFERENCES

91/676/EEC. EEC. (1991). *Nitrates Directive. European Union.*

Aage, H. K., Andersen, B. L., Blom, A., & Jense, I. (1997). The solubility of struvite. *Journal of Radioanalytical and Nuclear Chemistry*, *223*(1-2), 213–215. doi:10.1007/BF02223387

Abma, W. R., Driessen, W., Haarhuis, R., & van Loosdrecht, M. C. M. (2010). Upgrading of sewage treatment plant by sustainable and cost-effective separate treatment of industrial wastewater. *Water Science and Technology*, *61*(7), 1715–1722. doi:10.2166/wst.2010.977 PMID:20371929

Andrade, A., & Schuiling, R. D. (2001). The chemistry of struvite crystallization. *Mineralogical Journal (Ukraine)*, *23*, 37–46.

Anthonisen, A. C., Loehr, R. C., Prakasam, T. B. S., & Srinath, E. G. (1976). Inhibitions of nitrification by ammonia and nitrous-acid. *Journal - Water Pollution Control Federation*, *48*(5), 835–852. PMID:948105

Aslan, S., & Dahab, M. (2008). Nitritation and denitritation of ammonium-rich wastewater using fluidized-bed biofilm reactors. *Journal of Hazardous Materials*, *156*(1-3), 56–63. doi:10.1016/j.jhazmat.2007.11.112 PMID:18206304

Batchelor, B., & Lawrence, A. W. (1978). A kinetic model for autotrophic denitrification using elemental sulphur. *Water Research*, *12*(12), 1075–1084. doi:10.1016/0043-1354(78)90053-2

Ben Moussa, S., Tlili, M. M., Batis, N., & Ben Amor, M. (2011). Influence of temperature on struvite precipitation by CO2-deagassing method. *Crystal Research and Technology*, *46*(3), 255–260. doi:10.1002/crat.201000571

Böhler, M., Büttner, S., Morales, N., Liebi, C., Schachtler, M., & Siegrist, H. (2012b). *Recovery of nutrients from ammonia rich sludge liquids and urine for the production of fertilizer by full scale air stripping or membrane stripping*. Paper presented at the IWA World Congress on Water, Climate and Energy, Dublin, Ireland.

Böhler, M., & Liebi, C. (2012a). Reduzierter NaOH-Verbrauch bei der Ammoniak-Strippung durch den Einsatz einer vorgeschalteten CO2-Strippung—Betriebserfahrungen auf der ARA Kloten-Opfikon (In German). Reduction of base consumption in the free ammonia stripping process by means of a pre treatment by CO2-stripping—Operational experiences at WWTP Kloten/Opfikon. In *Proceedings of 8th Conference on Nitrogen Rejection Load—State of the Art*. Atemis GmbH.

Booker, N. A., Priestley, A. J., & Fraser, I. H. (1999). Struvite formation in wastewater treatments plants: Opportunities for nutrient recovery. *Environmental Technology*, *20*(7), 777–782. doi:10.1080/09593332008616874

Bouropoulos, N. Ch., & Koutsoukos, P. G. (2000). Spontaneous precipitation of struvite from aqueous solutions. *Journal of Crystal Growth*, *213*(3-4), 381–388. doi:10.1016/S0022-0248(00)00351-1

Bowers, K. E., & Westerman, P. W. (2005). Performance of cone-shaped fluidized bed struvite crystallizers in removing phosphorus from wastewater. *Transactions of the ASAE. American Society of Agricultural Engineers*, *48*(3), 1227–1234. doi:10.13031/2013.18523

Bu, L., Wang, K., Zhao, Q. L., Wei, L. L., Zhang, J., & Yang, J. C. (2010). Characterization of dissolved organic matter during landfill leachate treatment by sequencing batch reactor, aeration corrosive cell-Fenton, and granular activated carbon in series. *Journal of Hazardous Materials*, *179*(1-3), 1096–1105. doi:10.1016/j.jhazmat.2010.03.118 PMID:20430522

Buday, M. (1994). Reutilization of Ammonia from wastewater using cation-exchange resins. *Water Science and Technology*, *30*, 111–119.

Capodaglio, A.G., Hlavínek, P., & Raboni, M. (2015). Physico-chemical technology for nitrogen removal from wastewater: a review. *Ambiente & Agua-An interdisciplinary Journal of Applied Science*, 1980-1993.

Chen, L. A., Carbonell, R. G., & Serad, G. A. (1999). Recovery of proteins and other biological compounds using fibrous materials: I. Adsorption by salt addition. *Journal of Chemical Technology and Biotechnology (Oxford, Oxfordshire)*, *74*(8), 733–739. doi:10.1002/(SICI)1097-4660(199908)74:8<733::AID-JCTB111>3.0.CO;2-4

Chen, Y., Randall, A. A., & McCue, T. (2004). The efficiency of enhanced biological phosphorus removal from real wastewater affected by different ratios of acetic to propionic acid. *Water Research*, *38*(1), 27–36. doi:10.1016/j.watres.2003.08.025 PMID:14630100

Cohen, Y., & Kirchmann, H. (2004). Increasing the pH of wastewater to high levels with different gases CO2 stripping. *Water, Air, and Soil Pollution*, *159*(1), 265–275. doi:10.1023/B:WATE.0000049185.69759.b0

Cooney, E. L., Booker, N. A., Shallcross, D. C., & Stevens, G. W. (1999). Ammonia removal from wastewater using natural Australian zeolite. I. Characterization of the zeolite. *Separation Science and Technology*, *34*(12), 2307–2327. doi:10.1081/SS-100100774

Crutchik, D., & Garrido, J. M. (2011). Struvite crystallization versus amorphous magnesium and calcium phosphate precipitation during the treatment of a saline industrial wastewater. *Water Science and Technology*, *64*(12), 2460–2467. doi:10.2166/wst.2011.836 PMID:22170842

de-Bashan, L. E., & Bashan, Y. (2004). Recent advances in removing phosphorus from wastewater and its future use as fertilizer (1997-2003). *Water Research*, *38*(19), 4222–4246. doi:10.1016/j.watres.2004.07.014 PMID:15491670

Desloover, J., De Clippeleir, H., Boeckx, P., Du Laing, G., Colsen, J., Verstraete, W., & Vlaeminck, S. E. (2011). Floc-based sequential partial nitritation and anammox at full scale with contrasting N2O emissions. *Water Research*, *45*(9), 2811–2821. doi:10.1016/j.watres.2011.02.028 PMID:21440280

Desmidt, E., Ghyselbrecht, K., Zhang, Y., Pinoy, L., van der Bruggen, B., Verstraete, W., & Meesschaert, B. et al. (2015). Global phosphorus scarcity and full-scale P-recovery techniques: A review. *Critical Reviews in Environmental Science and Technology*, *45*(4), 336–384. doi:10.1080/10643389.2013.866531

Dockhorn, T. (2009). About the economy of phosphorus recovery. In *Proceedings of the International Conference on Nutrient Recovery from Wastewater Streams*. Vancouver, Canada: IWA Publishing.

Doyle, J. D., & Parsons, S. A. (2002). Struvite formation, control and recovery. *Water Research*, *36*(16), 3925–3940. doi:10.1016/S0043-1354(02)00126-4 PMID:12405401

Eggers, E., Dirkzwager, A. H., & Van der Honing, H. (1991). Full-scale experiences with phosphate crystallization in a crystalactor. *Water Science and Technology*, *23*, 819–824.

El Diwani, G., Rafie, S., Ibiari, N., & El-Aila, H. (2007). Recovery of ammonia nitrogen from industrial wastewater treatment as struvite slow releasing fertilizer. *Desalination*, *214*(1-3), 200–214. doi:10.1016/j.desal.2006.08.019

EPA. (1987). *Process design manual for phosphorus removal. US EPA Proceedings 625/1-87/001.* Washington, DC: Environmental Protection Agency.

EPA. (2010). *Nutrient control design manual. US EPA Proceedings 600/R-10/100*. Washington, DC: Environmental Protection Agency.

Fajardo, C., Mosquera-Corral, A., Campos, J. L., & Méndez, R. (2012). Autotrophic denitrification with sulphide in a sequencing batch reactor. *Journal of Environmental Management*, *113*, 552–556. doi:10.1016/j.jenvman.2012.03.018 PMID:22704254

FAO. (2011, June). *Current world fertilizer trends and outlook to2015*. Retrieved from http://www.fao.org/3/a-av252e.pdf

Ganesh, R., Balaji, G., & Ramanujam, R. A. (2006). Biodegradation of tannery wastewater using sequencing batch reactor– Respirometric assessment. *Bioresource Technology*, *97*(15), 1815–1821. doi:10.1016/j.biortech.2005.09.003 PMID:16230010

Ganigué, R., Volcke, E. I. P., Puig, S., Balaguer, M. S., & Colprim, J. (2012). Impact of influent characteristics on a partial nitritation SBR treating high nitrogen load wastewater. *Bioresource Technology*, *111*, 62–69. doi:10.1016/j.biortech.2012.01.183 PMID:22405759

Giesen, A. (1999). Crystallization process enables environmental friendly phosphate removal at low costs. *Environmental Technology*, *20*(7), 769–775. doi:10.1080/09593332008616873

Gisvold, B., Odegaard, H., & Follesdal, M. (2000). Enhanced clay aggregate filters. *Water Science and Technology*, *41*, 409–416.

González-Ponce, R., & García López, D. M. E. (2007). Evaluation of struvite as fertilizer: A comparison with traditional P sources. *Agrochimica*, *51*, 301–308.

Grand View Research. (2016). *Ammonium Sulfate Market Analysis By Product (Solid Ammonium Sulfate, Liquid Ammonium Sulfate), By Application (Fertilizers, Feed Additives, Industrial Use) And Segment Forecasts To 2022*. Grand View Research.

Guerrero, J. (2013). *Improving EBPR stability in WWTPs aiming at simultaneous carbon and nutrient removal: from modeling studies to experimental validation.* (PhD Thesis). Universitat Autònoma de Barcelona, Spain.

Guisasola, A., Petzet, S., Baeza, J. A., Carrera, J., & Lafuente, J. (2007). Inorganic carbon limitations on nitrification: Experimental assessment and modelling. *Water Research*, *41*(2), 277–286. doi:10.1016/j.watres.2006.10.030 PMID:17169397

Gujer, W., Henze, M., Takahashi, M., & van Loosdrecht, M. C. M. (1999). Activated Sludge Model No. 3. *Water Science and Technology*, *29*(1), 183–193. doi:10.1016/S0273-1223(98)00785-9

Guo, J. S., Abbas, A. A., Chen, Y. P., Liu, Z. P., & Chen, P. (2010). Treatment of landfill leachate using a combined stripping, Fenton, SBR, and coagulation process. *Journal of Hazardous Materials*, *178*(1-3), 699–705. doi:10.1016/j.jhazmat.2010.01.144 PMID:20188464

Haug, R. T., & McCarty, P. L. (1972). Nitrification with submerged filters. *Journal - Water Pollution Control Federation*, *44*, 2086.

Henze, M., van Loosdrecht, M., Ekama, G., & Brdjanovic, D. (2008). *Biological Wastewater Treatment*. London, UK: IWA Publishing.

Hood, C. R., & Randall, A. A. (2001). A biochemical hypothesis explaining the response of enhanced biological phosphorus removal biomass to organic substrates. *Water Research*, *35*(11), 2758–2766. doi:10.1016/S0043-1354(00)00573-X PMID:11456176

Hutnik, N., Kozik, A., Mazienczuk, A., Piotrowski, K., Wierzbowska, B., & Matynia, A. (2013). Phosphates (V) recovery from phosphorus mineral fertilizers industry wastewater by continuous struvite reaction crystallization process. *Water Research*, *47*(11), 3635–3643. doi:10.1016/j.watres.2013.04.026 PMID:23726699

Hwang, I. S., Min, K. S., Choi, E., & Yun, Z. (2005). Nitrogen removal from piggery waste used combined SHARON and ANAMMOX process. *Water Science and Technology*, *52*(10-11), 487–494. PMID:16459825

Ibrahim, K. M. (2001). Evaluation of Jordanian Faujasite tuff by comparison with other natural and synthetic zeolites. *Environmental Geology*, *40*(4-5), 440–445. doi:10.1007/s002540000235

Itokawa, H., Hanaki, K., & Matsuo, T. (2007). Nitrous oxide production in high-loading biological nitrogen removal process under low cod/n ratio condition. *Water Research*, *35*(3), 657–664. doi:10.1016/S0043-1354(00)00309-2 PMID:11228962

Jaffer, Y., Clark, T. A., Pearce, P., & Parsons, S. A. (2002). Potential phosphorus recovery by struvite formation. *Water Research*, *36*(7), 1834–1842. doi:10.1016/S0043-1354(01)00391-8 PMID:12044083

Kennedy, K. J., & Lentz, E. M. (2000). Treatment of landfill leachate using sequencing batch and continuous flow upflow anaerobic sludge blanket (UASB) reactors. *Water Research*, *34*(14), 3640–3656. doi:10.1016/S0043-1354(00)00114-7

Kim, D. G., Mishurov, M., & Kiely, G. (2010). Effect of increased N use and dry periods on N2O emission from a fertilized grassland. *Nutrient Cycling in Agroecosystems*, *88*(3), 397–410. doi:10.1007/s10705-010-9365-5

Kim, D. J., Lee, D. I., & Kellet, J. (2006). Effect of temperature and free ammonia on nitrification and nitrite accumulation in landfill leachate and analusis of itr nitrifying bacterial community by FISH. *Bioresource Technology*, *97*(3), 459–468. doi:10.1016/j.biortech.2005.03.032 PMID:15927463

Kioussis, D. R., Wheaton, F. W., & Kofinas, P. (2000). Reactive nitrogen and phosphorus removal from aquaculture wastewater effluent using polymer hydrogels. *Aquacultural Engineering*, *23*(4), 315–332. doi:10.1016/S0144-8609(00)00058-3

Klimiuk, E., & Kulikowska, D. (2006). The influence of hydraulic retention time and sludge age on the kinetics of nitrogen removal from leachate in SBR. *Polish Journal of Environmental Studies*, *15*(2), 283–289.

Koch, F. A., Mavinic, S., Yonemitsu, N., & Britton, A. T. (2009). *Fluidized bed wastewater treatment*. United States patent US 7622047 B2.

Komarowski, S., & Yu, Q. (1997). Ammonium ion removal from wastewater using Australian natural zeolite: Batch equilibrium and kinetic studies. *Environmental Technology*, *18*(11), 1085–1097. doi:10.1080/09593331808616628

Lahav, O., & Green, M. (2009). Ammonium removal from primary and secondary effluents using a bioregenerated ion-exchange process. *Water Science and Technology*, *42*, 179–185.

Laitinen, N., Luonsi, A., & Vilen, J. (2006). Landfill leachate treatment with sequencing batch reactor and membrane bioreactor. *Desalination*, *191*(1-3), 86–91. doi:10.1016/j.desal.2005.08.012

Le Corre, K. S., Valsami-Jones, E., Hobbs, P., & Parsons, S. A. (2005). Impact of calcium on struvite crystal size, shape and purity. *Journal of Crystal Growth*, *283*(3-4), 514–522. doi:10.1016/j.jcrysgro.2005.06.012

Le Corre, K. S., Valsami-Jones, E., Hobbs, P., & Parsons, S. A. (2009). Phosphorus recovery from wastewater by struvite crystallization: A review. *Critical Reviews in Environmental Science and Technology*, *39*(6), 433–477. doi:10.1080/10643380701640573

Ledda, C., Schievano, A., Salati, S., & Adani, F. (2013). Nitrogen and water recovery from animal slurries by a new integrated ultrafiltration, reverse osmosis and cold stripping process: A case study. *Water Research*, *47*(16), 6157–6166. doi:10.1016/j.watres.2013.07.037 PMID:23948560

Lee, K. C., & Rittmann, B. E. (2002). Applying a novel autoydrogenotrophic hollow-fiber membrane biofilm reactor for denitrification of drinking water. *Water Research*, *36*(8), 2040–2052. doi:10.1016/S0043-1354(01)00425-0 PMID:12092579

Lee, S. I., Weon, S. Y., Lee, C. W., & Koopman, B. (2003). Removal of nitrogen and phosphate from wastewater by addition of bittern. *Chemosphere*, *51*(4), 265–271. doi:10.1016/S0045-6535(02)00807-X PMID:12604078

Lemos, P. C., Serafim, L. S., Santos, M. M., Reis, M. A. M., & Santos, H. (2003). Metabolic pathways for propionate utilisation by phosphorus accumulating organisms in activated sludge:13 C labelling and in vivo nuclear magnetic resonance. *Applied and Environmental Microbiology*, *69*(1), 241–251. doi:10.1128/AEM.69.1.241-251.2003 PMID:12514001

Li, J. P., Healy, M. G., Zhan, X., Norton, D., & Rodgers, M. (2008). Effect of aeration rate on nutrient removal from slaughterhouse wastewater in intermittently aerated sequencing batch reactors. *Water, Air, and Soil Pollution*, *192*(1-4), 251–261. doi:10.1007/s11270-008-9652-9

Li, X., & Zhang, R. (2002). Aerobic treatment of dairy wastewater with sequencing batch reactor systems. *Bioprocess and Biosystems Engineering*, *25*(2), 103–109. doi:10.1007/s00449-002-0286-9 PMID:14505010

Liu, B., Giannis, A., Zhang, J., Chang, V., & Wang, J.-Y. (2013). Characterization of induced struvite formation from source-separated urine using seawater and brine as magnesium sources. *Chemosphere*, *93*(11), 2738–2747. doi:10.1016/j.chemosphere.2013.09.025 PMID:24134888

Liu, S. T., Yang, F. L., Xue, Y., Gong, Z., Chen, H. H., Wang, T., & Su, Z. C. (2008). Evaluation of oxygen adaptation and identification of functional bacteria composition for anammox consortium in non-woven biological rotating contactor. *Bioresource Technology*, *99*(17), 8273–8279. doi:10.1016/j.biortech.2008.03.006 PMID:18439820

Liu, Y., Qin, L., & Yang, S. F. (2007). *Microbial granulation technology for nutrient removal wastewater*. New York: Nova Science Publisher.

Lu, H., Oehmen, A., Virdis, B., & Yuan, Z. (2006). Obtaining highly enriched cultures of Candidatus Accumulibacter phosphates through alternating carbon sources. *Water Research*, *40*(20), 3838–3848. doi:10.1016/j.watres.2006.09.004 PMID:17070894

Matsumiya, Y., Yamasita, T., & Nawamura, Y. (2000). Phosphorus removal from sidestreams by crystallization of magnesium-ammonium-phosphate using seawater. *Journal of the Chartered Institution of Water and Environmental Management*, *14*(4), 291–296. doi:10.1111/j.1747-6593.2000.tb00263.x

Maurer, M., Schwegler, P., & Larsen, T. A. (2003). Nutrient in urine: Energetic aspects of removal and recovery. *Water Science and Technology*, *48*(1), 37–46. PMID:12926619

Metcalf, E., & Eddy, H. (2003). *Wastewater engineering: Treatment, disposal and reuse*. New York: Ed. Mc Graw-Hill.

Metcalf, E., & Eddy, H. (2007). *Wastewater treatment*. New York: Ed. Mc Graw-Hill.

Mino, T., van Loosdrecht, M. C. M., & Heijnen, J. J. (1998). Microbiology and biochemistry of enhanced biological phosphate removal process. *Water Research*, *32*(11), 3193–3207. doi:10.1016/S0043-1354(98)00129-8

Moerman, W., Carballa, M., Vandekerckhove, A., Derycked, D., & Verstraete, W. (2009). Phosphate removal in agro-industry: Pilot- and full-scale operational considerations of struvite crystallization. *Water Research*, *43*(7), 1887–1892. doi:10.1016/j.watres.2009.02.007 PMID:19264340

Mora, M., Lopez, R. L., Gamisans, X., & Gabriel, D. (2014). Coupling respirometry ans titrimetry for the characterization of the biological activity of SO-NR consortium. *Chemical Engineering Journal*, *251*, 111–115. doi:10.1016/j.cej.2014.04.024

Morales, N., Boehler, M. A., Buettner, S., Liebi, C., & Siegrist, H. (2013). Recovery of N and P from Urine by Struvite Precipitation Followed by Combined Stripping with Digester Sludge Liquid at Full Scale. *Water (Basel)*, *5*(3), 1262–1278. doi:10.3390/w5031262

Morse, G. K., Brett, S. W., Guy, J. A., & Lester, J. N. (1998). Review: Phosphorus removal and recovery technologies. *The Science of the Total Environment*, *212*(1), 69–81. doi:10.1016/S0048-9697(97)00332-X

Mulder, A. (2003). The quest for sustainable nitrogen removal technologies. *Water Science and Technology*, *48*, 67–75. PMID:12926622

Neczaj, E., Okoniewska, E., & Kacprzak, M. (2005). Treatment of landfill leachate by sequencing batch reactor. *Desalination*, *185*(1-3), 357–362. doi:10.1016/j.desal.2005.04.044

Nelson, N. O., Mikkelsen, R. L., & Hesterberg, D. L. (2003). Struvite precipitation in anaerobic swine lagoon liquid: effect of pH and Mg:P ratio and determination of rate constant. *Bioresource Technology*, *89*(3), 229–236. doi:10.1016/S0960-8524(03)00076-2 PMID:12798112

Nguyen, M. L., & Tabber, C. C. (1998). Ammonium removal from wastewater using natural new Zealand zeolites. *New Zealand Journal of Agricultural Research*, *41*(3), 427–446. doi:10.1080/00288233.1998.9513328

Nishimura, F., Somiya, I., Tsuno, H., & Iwabu, H. (1996). Development of a combined bac and Bz reactor for removal of nitrogen in wastewater from sludge drying process. *Water Science and Technology*, *34*(1-2), 145–151. doi:10.1016/0273-1223(96)00504-5

Oehmen, A., Zeng, R. J., Yuan, Z., & Keller, J. (2005). Anaerobic metabolism of propionate by polyphosphate-accumulating organisms in enhanced biological phosphorus removal systems. *Biotechnology and Bioengineering*, *91*(1), 43–53. doi:10.1002/bit.20480 PMID:15880463

Orentlicher, M. (2012). Overview of nitrogen removal technologies and application/use of associated end products. In *Proceedings of Got Manure Tradeshow and Conference*. New York: Cornell University.

Orentlicher, M., Brown, S., & Hall, C. (2009). Centrate ammonia reduction with ARP®: pilot data with synthetic and actual wastewaters. In *Proceedings of WEF/WERF Nutrient Removal Specialty Conference*. New York: Water Environment Federation.

Patyk, A., & Reinhardt, G. A. (1997). *Düngemittel – Energie- und Stoffstrombilanzen*. Vieweg. doi:10.1007/978-3-663-08014-5

Peng, Y., & Zhu, G. (2006). Biological nitrogen removal with nitrification and denitrification via nitrite pathway. *Applied Microbiology and Biotechnology*, *73*(1), 15–26. doi:10.1007/s00253-006-0534-z PMID:17028876

Pijuan, M. (2004). *Effect of different carbon sources and continous aeration on EBPR process*. (PhD Thesis). Universitat Autònoma de Barcelona, Spain.

Pijuan, M., Casas, C., & Baeza, J. A. (2009). Polyhydroxyalkanoate synthesis using different carbon sources by two enhanced biological phosphorus removal microbial communities. *Process Biochemistry*, *44*(1), 97–105. doi:10.1016/j.procbio.2008.09.017

Purtschert, I., Siegrist, H., & Gujer, W. (1996). Enhanced denitrification with methanol at WWTP Zürich-Werdhölzli. *Water Science and Technology*, *33*(12), 117–126. doi:10.1016/0273-1223(96)00465-9

Raboni, M., Torretta, V., Viotti, P., & Urbini, G. (2013). Experimental plant for the chemico-physical treatment of groundwater polluted by MSW leachate, with ammonia recovery. *Revista Ambiente & Agua*, *8*(3), 22–32.

Rahaman, M., Mavinic, D. S., Meikleham, A., & Ellis, N. (2014). Modeling Phosphorus Removal and Recovery from Anaerobic Digester Supernatant through Struvite Crystallization in a Fluidized Bed Reactor. *Water Research*, *51*, 1–10. doi:10.1016/j.watres.2013.11.048 PMID:24384559

Rahman, M., Liu, Y. H., Kwag, J.-H., & Ra, C. S. (2011). Recovery of struvite from animal wastewater and its nutrient leaching loss in soil. *Journal of Hazardous Materials*, *186*(2-3), 2026–2030. doi:10.1016/j.jhazmat.2010.12.103 PMID:21236571

Rahman, M., Salleh, M. A. M., Rashid, U., Ahsan, A., Hossain, M. M., & Ra, C. S. (2014). Production of slow release crystal fertilizer from wastewaters through struvite crystallization - A review. *Arabian Journal of Chemistry*, *7*(1), 139–155. doi:10.1016/j.arabjc.2013.10.007

Rezania, B., Oleszkiewicz, J. A., & Cicek, N. (2007). Hidrogen-dependent denitrification of water in an anaerobic submerged membrane bioreactor coupled with a novel hydrogen delivery system. *Water Research*, *41*(5), 1074–1080. doi:10.1016/j.watres.2006.11.016 PMID:17207833

Rittman, B. E., Mayer, B., Westerhoff, P., & Edwards, M. (2011). Capturing the lost phosphorus[Review]. *Chemosphere*, *84*(6), 846–853. doi:10.1016/j.chemosphere.2011.02.001 PMID:21377188

Rodriguez, D. C., Pino, N., & Penuela, G. (2011). Monitoring the removal of nitrogen by applying a nitrification–denitrification process in a sequencing batch reactor (SBR). *Bioresource Technology*, *102*(3), 2316–2321. doi:10.1016/j.biortech.2010.10.082 PMID:21075619

Rybicki, S. (1997). *Phosphorus removal from wastewater: A Literature Review*. Advanced Wastewater Treatment. Report No 1. Stockholm, Sweden: Swedish Reports.

Schipper, W. J., Klapwijk, A., Potjer, B., Rulkens, W. H., Temmink, B. G., Kiestra, F. D. G., & Lijmbach, A. C. M. (2001). Phosphate recycling in the phosphorus industry. *Environmental Technology*, *22*(11), 1337–1345. doi:10.1080/09593330.2001.9619173 PMID:11804355

Schuiling, R. D., & Andrade, A. (1999). Recovery of struvite from calf manure. *Environmental Technology*, *20*(7), 765–768. doi:10.1080/09593332008616872

Señoráns, L., Barros, S., García, G., & Garrido, J. M. (2011). Phosphorus recovery by struvite crystallization in a wastewater from a frozen fish factory. *Tecnología del Agua*, *335*, 26–35. (in Spanish)

Siegrist, H., Laureni, M., & Udert, K. M. (2013). Transfer into the gas phase: Ammonia stripping. In T.A. Larsen, K.M. Udert, & J. Lienert (Eds.), Source Separation and Decentralization for Wastewater Management (pp. 337-350). London, UK: IWA.

Sierra-Alvarez, R., Beristan-Cardoso, R., Salazer, M., Gómez, J., Razo-Flores, E., & Field, J. A. (2007). Chemolithotrophic denitrification with elemental sulphur for groundwater treatment. *Water Research*, *41*(6), 1253–1262. doi:10.1016/j.watres.2006.12.039 PMID:17296214

Spagni, A., & Marsili-Libelli, S. (2009). Nitrogen removal via nitrite in a sequencing batch reactor treating sanitary landfill leachate. *Bioresource Technology*, *100*(2), 609–614. doi:10.1016/j.biortech.2008.06.064 PMID:18707876

Strous, M., Heijnen, J. J., Kuenen, J. G., & Jetten, M. S. M. (1998). The sequencing batch reactor as a powerful tool for the study of slowly growing anaerobic ammonium-oxidizing microorganisms. *Applied Microbiology and Biotechnology*, *50*(5), 589–596. doi:10.1007/s002530051340

Sukalyan, S., Tabish, N., & Jeffrey, J. (2015). Nitrogen and Phosphorus Recovery from Wastewater. *Water Pollution*, *1*(3), 155–166.

Tayà, C. (2013). *Facing current EBPR bottlenecks in view of full-scale implementation*. (PhD. Thesis). Universitat Autònoma de Barcelona, Spain.

Tchobanoglous, G., Burton, F. L., & Stensel, H. D. (2003). *Wastewater engineering: Treatment and reuse*. Boston: Ed. Mc Graw-Hill.

Thomas, M., Wright, P., Blackall, L., Urbain, V., & Keller, J. (2003). Optimisation of Nossa BNR plant to improve prformance and reduced operating costs. *Water Science and Technology*, *47*(12), 141–148. PMID:12926681

Thörn, M., & Sörensson, F. (1996). Variation of nitrous oxide formation in the denitrification basin in a wastewater treatment plant with nitrogen removal. *Water Research*, *30*(6), 1543–1547. doi:10.1016/0043-1354(95)00327-4

Tilche, A., Bortone, G., Malaspina, F., Piccinini, S., & Stante, L. (2001). Biological nutrient removal in a full-scale SBR treating piggery wastewater: Results and modelling. *Water Science and Technology*, *43*(3), 363–371. PMID:11381929

Timur, H., & Ozturk, I. (1999). Anaerobic sequencing batch reactor treatment of landfill leachate. *Water Research*, *33*(15), 3225–3230. doi:10.1016/S0043-1354(99)00048-2

Torà, J. A., Lafuente, J., Baeza, J. A., & Carrera, J. (2010). Combined effect of inorganic carbon limitation and inhibition by free ammonia and free nitrous acid on ammonia oxdazing bacteria. *Bioresource Technology*, *101*(15), 6051–6058. doi:10.1016/j.biortech.2010.03.005 PMID:20356734

Turk, O., & Mavinic, D. S. (1987). Benefits of using selective inhibition to remove nitrogen from highly nitrogenous wastes. *Environmental Technology Letters*, *8*(9), 419–426. doi:10.1080/09593338709384500

Ueno, Y., & Fujii, M. (2001). Three years experience of operating and selling recovered struvite from full-scale plant. *Environmental Technology*, *22*(11), 1373–1381. doi:10.1080/09593332208618196 PMID:11804359

*USDA*. (2013). Retrieved from: www.ers.usda.gov/Data/FertilizerUse/

Uygur, A., & Kargi, F. (2004). Biological nutrient removal from pre-treated landfill leachate in a sequencing batch reactor. *Journal of Environmental Management*, *71*(1), 9–14. doi:10.1016/j.jenvman.2004.01.002 PMID:15084355

Uysal, A., Yilmazel, Y. D., & Demirer, G. N. (2010). The determination of fertilizer quality of the formed struvite from effluent of a sewage sludge anaerobic digester. *Journal of Hazardous Materials*, *181*(1-3), 248–254. doi:10.1016/j.jhazmat.2010.05.004 PMID:20605682

Van Hulle, S. W. H., Vandeweyer, H. J. P., Meesschaert, B. D., Vanrollenghem, P. A., Dejans, P., & Dumoulin, A. (2010). Engineering aspects and practical application of autotrophic nitrogen removal from nitrogen rich streams. *Chemical Engineering Journal*, *162*(1), 1–20. doi:10.1016/j.cej.2010.05.037

Vázquez-Padín, J. R., Pozo, M. J., Jarpa, M., Figueroa, M., Franco, A., Mosquera-Corral, A., & Méndez, R. et al. (2009). Treatment of anaerobic sludge digester effluents by the CANON process in an air pulsing SBR. *Journal of Hazardous Materials*, *166*(1), 336–341. doi:10.1016/j.jhazmat.2008.11.055 PMID:19117673

Viotti, P., & Gavasci, R. (2015). Scaling of ammonia stripping towers in the treatment of groundwater polluted by the leachate of municipal solid waste landfill: Study of the causes and the effects on the stripping performance. *Revista Ambiente & Agua*, *10*(2), 240–252.

Vlaeminck, S. E., Terada, A., Smets, B. F., De Clippeleir, H., Schaubroeck, T., Bolca, S., & Verstraete, W. et al. (2010). Aggregate size and architecture determine microbial activity balance for one-stage partial nitritation and Anammox. *Applied and Environmental Microbiology*, *76*(3), 900–909. doi:10.1128/AEM.02337-09 PMID:19948857

Wett, B., & Rauch, W. (2003). The role of inorganic carbon limitation in biological nitrogen removal of extremely ammonia concentrated wastewater. *Water Research*, *37*(5), 1100–1110. doi:10.1016/S0043-1354(02)00440-2 PMID:12553985

Wiesmann, U. (1994). Biological nitrogen removal from wastewater. *Advances in Biochemical Engineering/Biotechnology*, *51*, 113–154. doi:10.1007/BFb0008736 PMID:8165950

Wun-Jern, N. (2006). *Industrial Wastewater Treatment*. London, UK: Imperial College Press.

Wunderlin, P., Mohn, J., Joss, A., Emmenegger, L., & Siegrist, H. (2012). Mechanisms of N2O production in biological wastewater treatment under nitrifying and denitrifying conditions. *Water Research*, *46*(4), 1027–1103. doi:10.1016/j.watres.2011.11.080 PMID:22227243

Zuleeg, S., Wüthrich, F., Egli, C., Hürlimann, M., Grömping, M., Böhler, M., & Siegrist, H. (2012). ARA Altenrhein—Rücklaufbehandlung Variantenvergleich. ARA Altenrhein/Ingenieurbüro AG. (in German)

## KEY TERMS AND DEFINITIONS

**Aerobic Process:** Biological process in which, microorganisms depend on oxygen for their metabolic activities.

**Anaerobic Process:** Biological process that occur in the absence of any form of oxygen available for metabolic activities of microorganisms.

**Anoxic Process:** Biological process, that occur in the absence of molecular gaseous oxygen, but other forms of oxygen can be present and they are used for metabolic activities of microorganisms.

**Biodegradable:** A substance that can be broken down by biological processes (e.g. by bacteria, fungi, and other microorganisms) under aerobic, anoxic and/or anaerobic conditions.

**Biological Nutrient Removal (BNR) Processes:** An adapted process of the conventional activated sludge process where anaerobic and anoxic zones are included to biological reactor to create the conditions in which microorganisms are able to remove carbon, nitrogen and phosphorus.

**Chemical Precipitation:** A chemical reaction, in which two or more compounds react to form a precipitate, resulting in the formation of a separable solid substance from a solution.

**Nutrient:** Any element that is essential for the growth of animals, plants and humans. Usually nitrogen and phosphorus in wastewater streams.

**Nutrient Recovery:** Recovery of nutrients from wastewater streams through physico-chemical processes (e.g. precipitation) or other biological technologies.

**Struvite:** A magnesium ammonium phosphate salt ($MgNH_4PO_4 \bullet 6H_2O$) that usually precipitates from wastewater with high concentrations of phosphorus. This precipitate is a crystalline substance that may be used in agriculture as slow release fertilizer.

# Chapter 5
# Fundamentals of the Biological Processes for Nitrogen Removal

**Francisco J. Cervantes**
*IPICYT, Mexico*

**Flor Cuervo-López**
*Universidad Autónoma Metropolitana-Iztapalapa, Mexico*

**Jorge Gómez Hernández**
*Universidad Autónoma Metropolitana-Iztapalapa, Mexico*

## ABSTRACT

*In this chapter the fundamental aspects involved in biological treatment processes applied for the removal of nitrogen from wastewaters are described. A comprehensive review of the literature is provided including kinetic, microbiological and biochemical aspects of nitrification, denitrification and anammox, as well as key operational parameters affecting the processes. This information is relevant for designing wastewater treatment processes applied for the removal of nitrogen from wastewaters. The information is also essential for predicting and controlling the performance of these treatment processes.*

## INTRODUCTION

Excess nitrogen discharged from industrial wastewaters represents a threat to water reservoirs as it promotes several environmental problems. In addition, nitrogen pollution could induce adverse effects on human health. Besides the nitrogen pollution linked to intensive agriculture and combustion of fossil fuels, an important source of this contamination comes from industries demanding different nitrogenous compounds. For instance, $NH_3$ is used as a raw material to create multiple products, such as nylon, plastics, resins, glues, melamine, animal/fish/shrimp feed supplements, and explosives. A recent global assessment has pointed out three major environmental problems caused by nitrogen pollution in aquatic ecosystems:

1. It can drastically decrease the pH of freshwater ecosystems without much acid-neutralizing capacity, leading to acidification of these water bodies;

DOI: 10.4018/978-1-5225-1037-6.ch005

2. It can stimulate or enhance the development, maintenance and proliferation of primary producers, resulting in eutrophication of aquatic ecosystems; and
3. It can also impair the ability of aquatic animals to survive, grow and reproduce as a consequence of direct toxicity of nitrogenous compounds (Cervantes 2009).

Additionally, nitrogen pollution can have important economic effects in different sectors. For instance, human sickness and death, resulting directly (*e.g.* ingested nitrate and nitrite from polluted drinking water) or indirectly (*e.g.* aerosol exposure to algal toxins, consumption of contaminated seafood causing poisoning syndromes) from nitrogen pollution, can have high economical expenses because of lost wages and work days, and because of medical treatment and investigation (Camargo & Alonso, 2006). Many countries have strictly regulated the discharge of nitrogen-rich wastewaters into aquatic ecosystems to prevent these environmental, economic and public health concerns.

Biological processes represent sustainable technologies to remove nitrogen from industrial wastewaters. This chapter describes the fundamental aspects of biological processes involved in nitrogen removal. Kinetic, microbiological and biochemical aspects will be described as well as key operational parameters affecting the processes. The chapter will firstly describe aspects related to nitrification, followed by denitrification and anaerobic ammonium oxidation (anammox).

## NITRIFICATION

Ammonia removal from wastewaters can be achieved by physicochemical or biological processes. The first alternative is fast and efficient, but may not be sustainable, as the pollutant is usually transferred from one place to another, or because the treated water is not attractive due to increased saline concentration, bad odor, or extreme end pH values. Most typical physicochemical methods are stripping, electrochemical removal, precipitation or a combination of them (Bonmati & Floats, 2003; Chen, 2004). On the other hand, biological ammonia removal is usually carried out in two steps, one aerobic and the other anoxic. The former is known as nitrification, and the latter is referred to as denitrification.

Nitrification was firstly described by Winogradski (1890), who isolated and characterized *Nitrosomonas* and *Nitrobacter* species from soil. From the microbiological and physiological points of view, nitrification has deeply been described by Prosser (1989) and Francis et al. (2007). Nitrification is a key process intensively interacting with the carbon, nitrogen, and sulfur cycles. Nitrification is a catabolic pathway where ammonia is oxidized to nitrate. It is known as a respiratory process, which yields energy for microbial growth. It is important to note that the produced energy throughout the catabolism is not always used for microbial growth, as other cell functions, such as maintenance, may consume it, or it may be dissipated to the environment as heat. Under controlled conditions, ammonia can completely be oxidized to nitrate, but the produced energy not necessarily be utilized for biosynthetic reactions. In this case, the energy produced is dissipated and, therefore, nitrification turns into an energetically dissimilative process where biomass production is negligible (Texier & Gomez, 2002, 2007). This fact is relevance in wastewater treatment systems, which minimize sludge generation while achieving nitrogen removal.

## Microbiology

Nitrification occurs in two steps; ammonia is firstly oxidized to nitrite, which is subsequently oxidized to nitrate. These processes are usually carried out by two different microbial groups, ammonia and nitrite oxidizing, which are distributed among four subphyla, within of the domain *of Proteobacteria*. Recent studies have reported that within the domain Archaea, there are strains that can oxidize ammonia to nitrite (Hatzenpichler, 2012). From this domain, diverse studies of the phylum Thaumarchaeota have been made and isolated the species *Nitrosopumilus maritimus, Nitrososphaera viennensis*. Due to the relevance of nitrifying microorganisms from Archaea, in the nitrogen cycle and the environment, at present this domain is intensely being studied. Several nitrifying microorganisms have an ammonia monooxygenase enzyme (AMO), which oxidizes ammonia to nitrite (Treusch et al., 2005). However, recent results suggest that these microorganisms only oxidize ammonia to hydroxylamine, and a different enzyme performs the oxidation of hydroxylamine to nitrite (Vajralaa et al., 2013; Banning et al., 2015). Results have shown that the two domains show differences in metabolism, genetics and morphology. Moreover, the Proteobacteria domain seems to be in higher abundance than Archaea. Therefore, Proteobacteria are possibly predominant organisms in wastewater treatment plants, since experimental evidence has suggested that these bacteria seem to be less sensitive to inhibition by ammonia and organic matter (Verhamme et al., 2011).

Nitrifying Proteobacteria are Gram negative, obligated aerobic, lithoautotrophic microorganisms. *Nitrosomonas* and *Nitrobacter* are the most studied genera, but they are not the only ones who carry out the oxidation of ammonia to nitrate. Table 1 presents nitrifying microorganisms most frequently mentioned and further details can be found in Koops and Pommerening-Röser (2001). The morphology of the nitrifying microorganisms is quite diverse, including spherical, spiral and flagellated forms. All these microorganisms have ammonia as nitrogen and energy source and $CO_2$ as carbon source. Nitrifying organisms are known as obligated lithoautotrophs and strict aerobes. However, there is also experimental evidence, particularly in microbial consortia, showing that nitrifying organisms can tolerate moderate concentrations of organic compounds. The tolerance level seems to be linked to biomass concentration, to the chemical structure of organic matter and its concentration (Texier & Gomez, 2002; Zepeda et al., 2007).

*Table 1. Location of some ammonia-oxidizing and nitrite-oxidizing microorganisms*

| | Genera | Habitat |
|---|---|---|
| Ammonia-oxidizing | *Nitrosomonas* | Soils and many sources of water |
| | *Nitrosolobus* | Soils mainly |
| | *Nitrosococcus* | Soils and different sources of water |
| | *Nitrospira* | Soils and sweet water mainly |
| Nitrite-oxidizing | *Nitrobacter* | Soils and salt and sweet water |
| | *Nitrospira* | Salt waters |
| | *Nitrococcus* | Marine waters |

Recently, *Nitrospira* species were demonstrated to have the capacity to completely oxidize ammonia to nitrate. Evidence indicated that these strains encode all the enzymes necessary for ammonia oxidation via nitrite to nitrate in their genomes, and indeed completely oxidize ammonium to nitrate to conserve energy (Van Kessel et al., 2015). This achievement opens new possibilities for performing nitrification with one strain, but further research is still necessary.

## Biochemical Aspects

### Ammonia Oxidation

To understand the poor microbial growth of nitrifying microorganisms, thermodynamic aspects must be considered. Gibbs free energy change ($\Delta G°'$) is a useful variable because it predicts the spontaneity of reactions, besides the potential work to perform. Nitrification is a redox process including two reactions:

Ammonia oxidation (Equation 1)

$$NH_3 + 1.5\ O_2 \rightarrow NO_2^- + H^+ + H_2O \qquad \Delta G^{\circ\prime} - 198.\ 5\ KJ\ /\ reaction \qquad (1)$$

Nitrite oxidation (Equation 2)

$$NO_2^- + ½\ O_2 \rightarrow\ NO_3^- \qquad \Delta G^{\circ\prime} - 59.5 KJ\ /\ reaction \qquad (2)$$

As compared to glucose oxidation, ammonia oxidation to nitrate has a $\Delta G°'$ very low. Adenosine triphosphate (ATP) production is therefore limited. Thus, nitrifying biomass production will always be scant. Equation 2 suggests that nitrite oxidizers will grow less than ammonium oxidizers.

A second approach in order to explain the microbial growth and its kinetic aspects is related to the biochemistry of nitrification, including the enzymes involved. Ammonia oxidation to nitrite is performed by two enzymatic reactions (Equation 3 and 4):

$$NH_4^+ + \ 0.5O_2 \rightarrow NH_2OH \ + \ H^+ \qquad (3)$$

$$NH_2OH \ + O_2 \rightarrow NO_2^- + \ H^+ + \ H_2O \qquad (4)$$

Ammonia oxidase enzyme (AMO) is the first enzyme of the nitrification pathway, which is a metalloenzyme that oxidizes ammonia to hydroxylamine. The enzyme is catalytically diverse since, besides ammonia, also oxidizes aromatic compounds of low solubility in water. It has been seen that hydrocarbons may also cause competitive and noncompetitive inhibition (Keener & Arp, 1994). AMO activities can be detected in the cytoplasm and at the membrane, but catalytic activities may correspond to different enzymes as they have different biochemical properties (Gilch et al., 2009). This enzyme also shows activity under anaerobic conditions with nitrite as terminal electron acceptor (see anammox section). AMO is strongly sensitive to inhibitors, such as acetylene, allythiourea, nitrite and ammonia (Stein & Arp, 1998).

There are variables that have a high level of control on the enzyme activity, as the pH value due to modification of the active site of the enzyme. Likewise, this variable also determines the predominant species between ammonia and ammonium. Both chemical species are different substrates for AMO enzyme. The enzymatic rate of oxidation of ammonia and ammonium into nitrite can be determined by using the Michaelis-Menten equation. It has experimentally been observed that the Km (affinity constant for the enzyme) value for ammonia is lower than that for ammonium (Schmidt & Bock, 1998). The Km value for ammonia in the pH range of 7-8 is around 21 μM and the maximum specific rate is close to 7.86 mg $NH_4^+$-N/mg volatile suspended solids (VSS)-d while, for ammonium, it is around 175 times higher (Schimidt & Bock, 1998). Thus, pH is a key factor on the oxidation of ammonia to nitrite.

The next step in nitrification is the oxidation of hydroxylamine to nitrite by hydroxylamine oxidoreductase (HAO) enzyme, which is also a homotrimeric metalloenzyme located in the periplasmic space. The complete homotrimeric structure is necessary for enzymatic activity. HAO is an essential enzyme for energy generation through ammonia oxidation (Cardevall et al., 2013). HAO catalyzes the oxidation of hydroxylamine to nitrite in all ammonia-oxidizing bacteria and four electrons are released, which are split between AMO and the respiratory chain. Two electrons are transferred back to AMO enzyme. The main studies have been in the *Nitrosomonas* genus. It has been found that HAO also has nitric oxide as an end product, but only when $O_2$ is at low concentrations. The Km value for this enzyme is higher than that for AMO. According to Frijlink et al., (1992), the Km value of HAO is 180 μM with low variation in the pH range of 7-8. Nonetheless, remarkable changes in maximum specific rate have been observed at this interval (1.06 and 4.47 mg N/mg VSS-d). The high Km constant reflects that this is the limiting step for nitrite formation and therefore, the bottleneck for the nitrification process. Nonetheless, this enzyme is less sensitive to inhibition than AMO. It is necessary to make clear that rate should be pondered with care since these are response variables that change as a function of independent variables.

### Nitrite Oxidation

Nitrite oxidation to nitrate is carried out by nitrite oxidase (NO), a metalloenzyme that is quite sensitive to cyanide, allylthiourea and nitrite. NO is also strongly inhibited by organic compounds, just like AMO (Ginestet et al., 1998).

The enzyme activity is strongly linked to the pH, and the influence of this variable also determines the availability of the substrate. At the pH of 7.5, the Km is higher for ammonium (the protonated species), than for ammonia (free ammonia). It has been observed that the Km for NO is around 3.6 mM at pH 7.6 (Park & Bae, 2009), which is higher than the Km of AMO. In consequence, the reaction rate for nitrite formation might be considered as the bottleneck of nitrification. Likewise, it has also been stated that when ammonia is in excess, it is observed the inhibition of NO (Silva et al. 2011). Nonetheless, oxidation of nitrite to nitrate is even more sensitive to inhibition by nitrite (in fact, nitrous acid), so that the levels of this should be below the threshold of inhibition. The free nitrous acid concentration increases when the pH is below the neutral value. The nitrous acid, even at very low concentration (the inhibition constant, Ki, is around 0.1 mg $L^{-1}$), affects the energy production rate (Vadivelu, 2007). Thus, it is imperative to avoid accumulation of free nitrous acid; otherwise, the rate of nitrification may significantly decrease.

## Kinetics of Axenic and Mixed Nitrifying Cultures

Taking into account the thermodynamic ($\Delta G^{\circ\prime}$) features of the nitrification process, the microbial biomass yield for any nitrifying population will always be lower than 0.1 mg biomass/mg N. Nitrification is a foremost dissimilative process with poor energy yield (Equations 1 and 2). Nitrifying biomass production is inexorably low. This characteristic may be correlated to kinetic aspects. Thus, if nitrification rate were based on the concentration of biomass, it would be a very slow process. Nitrification would then be an inappropriate process for wastewater treatment. However, it must be kept in mind that the respiratory rate is independent on the microbial growth rate, but this is dependent on the respiratory process. Moreover, there is evidence that the environmental conditions for microbial growth might be different from those of respiration (Kim et al., 2006). Thus, nitrification must be focused on the respiratory point of view, and not so much on microbial growth. Nevertheless, more attention has been paid to optimize microbial growth, rather than the respiration process. In fact, it has been seen that the respiration rate is higher than the microbial growth. Regarding this, using synthetic wastewater and a stabilized nitrifying sludge at pH 7, an almost complete conversion of ammonium to nitrate was achieved, with a yield (Y, response variable that indicates the pathway predominant) close to 1. Thus, the process was considered as exclusively dissimilative (Silva et al., 2011). Therefore, nitrification can be considered for nitrogen removal from wastewaters, in spite of the scarce microbial growth. However, if a faster nitrification is required, a bioreactor with high cell density can be used (Lazarova et al., 1998).

It is worth to note that conventional activated sludge processes are usually considered comparable to nitrification, but both processes are different. The former is mainly biosynthetic and nitrification is a catabolic process. Likewise, nitrification is a lithoautotrophic process with little tolerance to the presence of organic matter. The enzymes (AMO and NO) may be inhibited causing poor nitrification performance (Keener & Arp, 1994; Texier & Gómez, 2007; Zepeda et al., 2007). On the other hand, activated sludge systems support high levels of organic matter (chemical oxygen demand, COD). Thus, it is clear that the environmental conditions to perform both processes are different. According to experimental evidence, C/N ratio, oxygen concentration, pH control and COD level might be the main key parameters to consider in wastewater treatment for simultaneous removal of COD and nitrogen.

Michaelis-Menten equation is used to describe enzymatic processes. However, there are many works where this equation is also applied to the kinetics of microbial growth, instead of the Monod model. Although the profiles tendencies in both processes may be similar, the meaning of both equations is not the same. In this sense, the Ks constant may be considered as the sum of each Km of all metabolic pathways. According to Cuervo-López et al. (2009) the metabolism of microbial cultures can be stated by Equation 5, where $q_s$ is the specific consumption rate of substrate, a constant of first order which is dependent on the environmental culture conditions; $\mu$ is the specific growth rate; $q_p$ is the specific formation rate of product; $Y_{x/Scx}$ and $Y_{p/Scp}$ are respectively, the yields for biomass and products respect to the consumed substrate. Therefore, Equation 5 summarizes the substrate utilization rate in terms of biosynthetic process ($\mu/Y_{X/Scx}$) and the respiratory or catabolic process ($q_p/Y_{P/Scp}$).

$$q_s = \frac{\mu}{Y_{x/Scx}} + \frac{q_p}{Y_{p/Scp}} \tag{5}$$

At well-defined nitrifying conditions, microbial growth is negligible due to the low energy production and/or environmental conditions (low (C/N ratio). Therefore, the first term of the above equation might also be considered negligible. Thus, the equation can be reduced to Equation 6:

$$q_s = \frac{q_p}{Y_{p/_{Scp}}} \tag{6}$$

In terms the Monod equation, the expression to determine de specific ammonia consumption rate is expressed by Equation 7:

$$q_s = \frac{q_{\max}\left[S\right]}{Y_p\left(K_s + \left[S\right]\right)} \tag{7}$$

Equation 7 may be used to calculate the affinity constant of certain microbial culture for the substrate (Ks) and the maximum respiratory rate ($q_{max}$). Instead of microbial growth rate, it was determined the substrate consumption and product formation (Gonzalez-Blanco et al., 2013). Table 2 summarizes specific nitrifying rates achieved by pure and mixed cultures.

Many attempts have been made in order to develop reactors with high conversion rates selectively achieving partial nitrification (nitritation: conversion of ammonia to nitrite) in order to connect this process with anammox, which ultimately saves operational costs. The main parameters considered to accomplish nitritation are oxygen concentration, biomass concentration, solid retention time, and temperature (He et al. 2015).

*Table 2. Specific rates values of nitrifying axenic and mixed sludge in batch and continuous culture*

| Ammonia Oxidizing | Batch | Continuous | Reference |
|---|---|---|---|
| Nitrosomonas europaea | 0.088 | 0.063 | Skinner & Walker (1961) |
| N. europaea | 0.036 | 0.064 | Helder & de Vries (1983) |
| N. Europaea | 0.017 | 0.039 | Keen & Proseer (1987) |
| N. Oceanus | 0.014 | ---- | Glover (1985) |
| Sludge | 0.040 | ---- | Antoniou et al. (1990) |
| Nitrosomonas sp | 0.019 | ---- | Beccari et al. (1979) |
| Nitrobacter sp | 0.023 | ---- | Blackburne et al. (2007) |
| **Nitrite Oxidizing** | | | |
| Nitrobacter sp | 0.039 | 0.033 | Gay & Corman (1984) |
| Nitrococcus mobilis | 0.033 | ---- | Glover (1985) |
| Nitrobacter sp | 0.043 | 0.035 | Kenn & Prosser (1987) |
| Sludge | ---- | 0.060 | Poduska & Andrews (1975) |
| Sludge | 0.510 | ---- | Texier & Gómez (2004) |
| Nitrobacter | | | Blackburne et al. (2007) |

* (mg $NH_4$+-N/mg VSS·d)

## DENITRIFICATION

The biological dissimilative transformation of nitrate ($NO_3^-$) or nitrite ($NO_2^-$) into molecular nitrogen under anoxic conditions with energy conservation is known as denitrification (Zumft, 1997). Coupled to nitrification, denitrification is up to now the most suitable system used for nitrogen removal from wastewater because of its low operational costs, its respiratory process stability and its high removal efficiencies (Lin et al., 2009).

### Denitrification Processes

In order to perform this dissimilative process, microorganisms require concomitant oxidation of an electron donor as energy source. Organotrophic denitrification occurs when organic sources are used as electron donors whereas autotrophic or chemolithotrophic denitrification exists when inorganic compounds as energy source are oxidized coupled to the reduction of N oxides. According to ΔG°' values, exergonic conditions of organotrophic and autotrophic processes are illustrated in Table 3. Regarding organotro-

*Table 3. Some ΔG°' values of organotrophic and autotrophic denitrification*

| Compound | Equation | ΔG°' ($kJ/NO_3^-$) |
|---|---|---|
| | Organotrophic denitrification | |
| Acetic acid | $CH_3COOH + 1.6NO_3^- \rightarrow 2CO_2 + 0.8N_2 + 1.6OH^- + 1.2H_2O$ | -527 |
| Ethanol | $C_2H_6O + 2.4NO_3^- + 0.4H^+ \rightarrow 1.2N_2 + 2HCO_3^- + 2.2H_2O$ | -513 |
| Lactic acid | $C_3H_6O_3 + 2.4NO_3^- \rightarrow 1.2N_2 + 3HCO_3^- + 1.2H_2O + 0.6H^+$ | -525 |
| Glucose | $C_6H_{12}O_6 + 4.8NO_3^- \rightarrow 2.4N_2 + 6HCO_3^- + 1.2H^+ + 2.4H_2O$ | -560 |
| Phenol | $C_6H_6O + 5.6NO_3^- + 0.2H_2O \rightarrow 2.8N_2 + 6HCO_3^- + 0.4H^+$ | -503 |
| *p*-Cresol | $C_7H_8O + 6.8NO_3^- \rightarrow 3.4N_2 + 7HCO_3^- + 0.2H^+ + 0.4H_2O$ | -504 |
| Benzene | $C_6H_6 + 6NO_3^- \rightarrow 3N_2 + 6HCO_3^-$ | -496 |
| Toluene | $C_7H_8 + 7.2NO_3^- + 0.2H^+ \rightarrow 3.6N_2 + 7HCO_3^- + 0.6H_2O$ | -489 |
| Xylene | $C_8H_{10} + 8.4NO_3^- + 0.4H^+ \rightarrow 4.2N_2 + 8HCO_3^- + 1.2H_2O$ | -492 |
| | **Sulfur Autotrophic Denitrification** | |
| Sulfide | $S^{-2} + 1.6NO_3^- + 1.6H^+ \rightarrow SO_4^{-2} + 0.8N_2 + 0.8H_2O$ | -465 |
| Partial sulfide | $S^{-2} + NO_3^- + 2H^+ \rightarrow S^0 + NO_2^- + H_2O$ | -130 |
| Partial sulfide | $S^{-2} + 4NO_3^- \rightarrow SO_4^{-2} + 4NO_2^-$ | -126 |
| Elemental sulphure | $S^0 + 1.2NO_3^- + 0.4H_2O \rightarrow SO_4^{-2} + 0.6N_2 + 0.8H^+$ | -456 |
| Tiosulphate | $S_2O_3^{-2} + 1.6NO_3^- + 0.2H_2O \rightarrow 2SO_4^{-2} + 0.8N_2 + 0.4H^+$ | -479 |
| | **Hydrogen Autotrophic Denitrification** | |
| Partial | $H_2 + NO_3^- \rightarrow NO_2^- + H_2O$ | -163 |
| Partial | $1.5H_2 + NO_2^- + H^+ \rightarrow 0.5N_2 + 2H_2O$ | -397 |
| Total | $2.5H_2 + NO_3^- + H^+ \rightarrow 0.5N_2 + 3H_2O$ | -560 |
| | **Methane Autotrophic Denitrification** | |
| Nitrate | $5CH_4 + 8NO_3^- + 8H^+ \rightarrow 4N_2 + 5CO_2 + 14H_2O$ | -96 |
| Nitrite | $3CH_4 + 8NO_2^- + 8H^+ \rightarrow 4N_2 + 3CO_2 + 10H_2O$ | -116 |

phic processes, lower ΔG°' values are obtained than autotrophic denitrification where hydrogenotrophic denitrification appeared to be the lowest endergonic process; therefore, lower biomass production is expected to obtain under these conditions:

1. **Organotrophic Process** *:* This process requires the presence of organic compounds as electron source for energy generation. Organic matter is also used for sustaining microbial growth. Methanol has been the most common organic substrate used for conducting denitrification (Lu et al., 2014). Nevertheless many other organics have been used for accomplishing denitrification, such as glucose, acetate, lactate, sucrose, ethanol and phenol (Martínez-Gutiérrez et al., 2012). Likewise, several recalcitrant compounds as benzene, toluene and *p*-xylene (all of them known as BTX), *p*-cresol, and 2-chlorophenol (Peña-Calva et al., 2004; Martínez-Hernández et al., 2009; Cervantes et al., 2009) and some emergency organic compounds (EOCs) such as antibiotics and personal care compounds have been successfully removed by organotrophic denitrification (Suarez et al., 2010). Thus, organotrophic denitrification treatment could be successfully applied for simultaneously elimination of carbon and nitrogen compounds present in municipal or industrial wastewaters. However, type of electron source and the C/N ratio play an important role in determining the removal efficiency of the denitrifying process (Cuervo-López et al., 2009; Xu et al., 2015).
2. **Autotrophic Process:** When no organic matter is available, several autotrophic processes can be conducted, such as hydrogen autotrophic denitrification, sulfur autotrophic denitrification and methane autotrophic denitrification, when hydrogen ($H_2$), sulfur compounds and methane are respectively used as electron donors (Modin et al., 2007; Karanasios et al., 2010; He et al., 2015). Carbon dioxide is the carbon source used in these processes for microbial growth, but methane can be used as carbon and energy source (Martínez et al., 2008). Nevertheless, slow denitrification rates, sulfate production in effluents and alkalinity consumption among other factors, are the main shortcomings of autotrophic denitrification for the complete replacement of organotrophic processes (Xu et al., 2015).
    a. **Hydrogen Autotrophic Denitrification**: Reduction of $NO_3^-$ to $N_2$ with $H_2$ results in hydrogen autotrophic denitrification. This process has been used for removing selectively $NO_3^-$ or $NO_2^-$ from drinking water where the presence of organic compounds is very low (Vasiliadou et al., 2006; Karanasios et al., 2010). The relatively low cost of $H_2$, its clean origin, high energy efficiency, low biomass yield and no by-products generation, have converted it in an excellent electron source for carrying out nitrate removal (Karanasios et al., 2010; Wang et al., 2015). Nevertheless, the explosive atmosphere that could be created in treatment plants, besides the low denitrifying rates obtained, are the main constraints (Vasiliadou et al., 2006).
    b. **Sulfur Autotrophic Denitrification**: When sulfurous compounds, such as elemental sulfur ($S^0$), sulfide ($S^{2-}$), thiosulfate ($S_2O_3^{2-}$) or sulfite ($SO_3^{2-}$) are used as energy source for the reduction of nitrate to $N_2$, sulfur autotrophic denitrification occurs. Accumulation of sulfurous compounds in aquatic environments can cause several problems, such as toxicity, bad odors, materials corrosion and high levels of chemical oxygen demand (Fajardo et al., 2014). Therefore, sulfur autotrophic denitrification process draws great attention due to its ecological significance, as it can be used for the simultaneous elimination of sulfurous and nitrogenous compounds from different water environments (Shao et al., 2010). In fact, the process can be applied for eliminating low concentrations of nitrate and sulfide as those found in drinking water, ground water and landfill leachate (Tandukar et al., 2009), but also it can be used for treating high-

salinity wastewaters (Koening & Liu 2004), petrochemical effluents (Vaiopoulou et al., 2005) or fermentation wastewaters (Can-Dogan et al., 2010) where higher concentrations of these compounds are present. Nevertheless, for achieving complete sulfurous compounds oxidation, a multistep sulfur oxidation is required consequently various intermediate inorganic sulfur compounds might be produced. In fact, oxidation of $S^{2-}$ by autotrophic denitrification process can lead to the formation of $S^0$ or sulfate ($SO_4^{2-}$), depending on the established N/S ratio and the environmental conditions (Beristain et al., 2006; Can-Dogan et al., 2010).

c. **Methane Autotrophic Denitrification**: Methane has been considered a suitable electron source for carrying out denitrification of drinking water, wastewater or leachates, where the presence of organic compounds is poor (Modin et al., 2007). These authors refers that methane oxidation linked to denitrification can occur either by aerobic or anaerobic processes, respectively known as aerobic methane oxidation coupled to denitrification (AME-D) or anaerobic methane oxidation coupled to denitrification (ANME-D). Anaerobic oxidation is considered a safer process since the risk of explosion of methane and oxygen is avoided. The potential use of methane as electron source for sustaining reduction of nitrate to $N_2$ has been reported since 1970s by some authors as Sollo et al. (1976) and Mason (1977) in flasks, activated carbon fluidized reactors or biodisc units, respectively supplied with methane. However, denitrification linked to methane oxidation was not effectively demonstrated in these works as production of $N_2$ was not assessed, endogenous decay of organic matter could be the main removal process instead of methane oxidation and because of the denitrifying rates determined were very low. Later on, Islas-Lima et al. (2004) and Raghoebarsin et al. (2006) respectively reported anoxic denitrifying sludge and archaea-bacteria consortium, capable of carrying out anaerobic denitrification with methane. Recently, Haroon et al. (2013) demonstrated anaerobic methane oxidation coupled to nitrate reduction and anammox process, whereas He et al. (2015) reported anaerobic methane oxidation linked to nitrite reduction in anaerobic sequencing batch reactors, where the main problem detected was the bioreactor startup. Nevertheless, production of $N_2$ was not assessed in this work.

3. **Mixotrophic Process** : As a result of industrial activities, mixtures of nitrate, organic compounds and sulfide are often present in effluents, such as those generated from petrochemical industries. Combination of organotrophic and autotrophic denitrification processes might result in a suitable option for the simultaneous removal of these pollutants. Therefore, researchers have focused their studies on the performance of mixotrophic denitrifying processes, where organic and inorganic compounds serve as energy source for the nitrate reduction. Removal of nitrate and sulfide in a mixture with acetate (Xu et al., 2015), methanol (Zhao et al., 2012), phenol (Beristain Cardoso et al., 2009) or p-cresol (Meza et al., 2008; Cervantes et al., 2009) among other organic compounds, have been conducted using microbial consortia.

## Microbiology

Denitrifying microorganisms are ubiquitously distributed in nature, where their remarkable characteristic is their facultative anoxic respiration. Denitrifying organisms must have anoxic respiration utilizing nitrate or nitrite as electron acceptors and reduce them to $N_2$. The presence of the nitrite oxide reductase enzyme is also mandatory. Nevertheless, some bacteria are only capable of reducing nitrate to nitrite or $N_2O$, thus, accumulation of nitrogen intermediates could be an important constraint, e.g. greenhouse gases generation, and should be avoided during wastewater treatment.

Different taxonomic groups are involved in denitrification. Many genera can use organic sources as *Bacillus, Halomonas, Pseudomonas* (González-Domenech et al., 2008), *Acidovorax, Alcaligenes, Azospirillum, Flavobacterium* (Calderer et al., 2014), *Wolinella, Paracoccus, Rhizobium, Agrobacteria* (Tiedje, 1988), whereas some genus as *Rhodobacter and Rhodopseudomonas* are phototrophic (Zumft, 1997). Most of organotrophic denitrifying microorganisms are able to grow at maximum rate in mesophilic range at neutral pH in soil, sediments and lakes; nevertheless, some of them have been isolated from marine systems, sulfur springs and wastewater treatment plants (Knowles, 1982).

Autotrophic denitrifiers belong to α-, β-, γ- and ε-proteobacteria group where so far, 10 genus have been identified as *Paracoccus, Thiobacillus, Thermotrix, Thioalkalivibrio, Thiohalomonas, Thiohalophilus, Thioalkalispira, Thiohalorhabdus, Thiomicrospira* and *Sulfurimonas*. According to their optimal growth pH (6.5-8.6) and temperature (close to 30 °C), most autotrophic microorganisms are neutrophilic and mesophilic (Kellerman & Griebler, 2009). Some of them are capable of carrying out autotrophic denitrification up to 65-75 °C, such as *Thermotrix thiopara*, a thermophilic β-proteobacteria isolated from sulfur spring. *Sulfurimonas denitrificans*-related species were identified as autotrophic denitrifiers in the transition zones of between oxide and anoxic/sulfidic water layers in the sea basin (Labrenz et al., 2005). *Thiobacillus denitrificans* (Kelly and Wood 2000) and *Sulfurimonas lithotrophicum* (Mills et al. 2004) β and ε-proteobacteria species have been isolated from marine sediments. Five extremophiles species belonging to γ-proteobacteria have been isolated from soda lakes and hypersaline lakes: *Thioalkalivibrio denitrificans, Thiohalomonas denitrificans,Thiohalorhabdus denitrificans, Thioalkalivibrio thiocyanodenitroficans* and *Thiohalophilus thiocyanoxidans*. So far, two species belonging to ε-proteobacteria have been isolated and cultured from terrestrial habitats such as oil field brines and oil-contaminated groundwater: *Thiomicrospira* CVO (Gevertz et al., 2000) and *Sulfuricurvum kujiense* (Kodama & Watanabe, 2004).

Most microorganisms reported as hydrogen-oxidizing denitrifiers also belong to *Proteobacteria* class. *Thauera, Hydrogenophaga, Rhodocyclus, Pseudomonas, Acinetobacter, Aeromonas, Shewanella, Acidovorax* and *Paracoccus* are some of the genera frequently isolated from diverse $H_2$-dependent denitrification reactors (Karanasios et al., 2010). Nevertheless, considering that hydrogenotrophic denitrification environment is highly selective, limited bacteria species have been used in pure cultures, such as *Ralstonia eutropha* (formerly classified as *Alcaligenes eutrophus*) and *Rhodocyclus sp.* (Lee & Rittmann 2000). Recently, by means of pyrosequencing technology and 16S rRNA gene characterization analyses indicated that *Proteiniclasticum* genus as well as *Gemmata, Planctomyces* and *Hyphomicrobium* were found to be dominant in a continuous hydrogenotrophic denitrification reactor (Wang et al., 2015). The optimum pH value for conducting hydrogenotrophic denitrification is around 7.0-7.6.

Mixotrophic denitrification is generally carried out with consortia where facultative autotrophic bacteria such as *Thiobacillus, Pseudomonas* or *Paracoccus*, which are able to conduct sulfur oxidation in presence of nitrate or nitrite to $N_2$, are commonly present in addition with organotrophic denitrifiers. Nevertheless, it has been reported that pure cultures of *Pseudomonas sp.* C27 can perform both autotrophic and organotrophic denitrification in mixotrophic cultures amended with sulfur, nitrate and acetate (Chen et al., 2013; Guo et al., 2014). It has been reported that under mixotrophic conditions where methanol, ethanol and sulfur were the electron donors, the dominant organisms belong to the genera *Methyloversatilis, Thauera, Geobacter, Azospira, Longilinea, Hypomicrobium Thiobacillus* and

*Sulfurimonas* (Zhang et al., 2015). Authors suggest that the last two genera were mainly related with the sulfur consumption (autotrophic denitrification) whereas the rest played an important role in the organotrophic denitrification. Recently, Xu et al. (2015b) have reported that at mixotrophic conditions, sequences similar to *Thauera, Thiobacillus, Chromatiales* and *Xanthomonadaceae* were predominant in a sequential batch reactor supplemented with sulfide, nitrate and acetate. Mixotrophic denitrification culture conditions have been commonly set at pH values ranging between 8.5 to 9.5 and temperature values between 25-35 °C (Zhao et al., 2012; Chen et al., 2013; Zhang et al., 2015).

It is generally accepted that microorganisms involved in anaerobic methane oxidation coupled to denitrification are consortia containing anaerobic methanotrophic archaea, which produce acetate, methanol or formate and bacteria able to oxidize these compounds as electron sources coupled to nitrate or nitrite reduction (Raghoebarsing et al., 2006). The former authors reported in a reactor conducting ANME-D, microbial populations of acetate and methanol degrading denitrifying bacteria related to *Mesorhizobium plurifarium, Stenotrophomonas sp.* and *Hypomicrobium sp.* in addition to a methanotrophic strain related to *Methylocystis sp.* So far, there is no evidence that ANME-D is completely carried out by a single microorganism (Modin et al., 2007) as there is no known methanotrophic organism that can carry out complete denitrification (Knowles 2005), this is, the oxidation of methane coupled to a complete reduction of nitrate to $N_2$. Recently, it has been shown the simultaneous concurrence of anaerobic oxidation of methane, nitrate reduction and anammox conducted by a recently described methanotrophic bacteria (*Candidatus "Methanoperedens nitroreducens"*) able to consume methane and reduce nitrate to nitrite, which in conjunction with an anammox bacteria related to *Candidatus Kuenenia stuttgartiensis,* converted the nitrite and ammonium into $N_2$ by the anammox process (Haroon et al., 2013). ANME-D has been commonly assessed at temperature values ranging from 22 to 30 °C and pH values between 7-7.5.

## Biochemical Aspects

According to Mariotti (1986), denitrification can be described by Equation 8:

$$2\ NO_3^- + 10\ e- \ + 12H^+ \rightarrow N_2 + 6H_2O \qquad \Delta G^{o\,\prime} = -560.3\ KJ\ /\ NO_3 \tag{8}$$

Irrespectively whether organic or autotrophic process is conducted, denitrification has been described as a modular organization in which every biochemical reaction is catalyzed by specific reductase enzymes (Cuervo-López et al., 2009). The reactions take place when environmental conditions become anaerobic (Berks et al., 1995). Four endergonic enzymatic reactions take place in the cell where reduced ubiquinone ($UQH_2$); ubiquinone (UQ); reduced cytochrome($c^{2+}$); oxidized cytochrome ($c^{3+}$) have been identified (Lalucat et al., 2006).

At first, nitrate is reduced to nitrite by nitrate reductase *(Nar).* The reaction can take place in the cell membrane and periplasmic space. A subsequent reduction of nitrite to nitric oxide is carried out by one of two nitrite reductase (*Nir*), *CuNir* or the cytochrome $cd_1$, both located at the periplasmic space. Afterwards, in the cell membrane, nitric oxide is reduced to nitrous oxide by the enzyme nitric oxide reductase (*Nor).* Finally, nitrous oxide is reduced to $N_2$ by the enzyme nitrous oxide reductase (*Nos)* which is located at the periplasmic space. All of these denitrifying reactions and their thermodynamic data are listed below (Table 4).

*Table 4. Denitrifying reactions*

| Reaction | Enzyme Involved | $\Delta G^{o\prime}$ (KJ/$NO_X$) |
|---|---|---|
| $NO_3^- + UQH_2 \rightarrow NO_2^- + UQ + H_2O$ | Nitrate reductase *(Nar)* | -163.2 |
| a) $NO_2^- + Cu^{1+} + 2H^+ \rightarrow NO + H_2O + Cu^{2+}$<br>or<br>b) $NO_2^- + c^{2+} + 2H^+ \rightarrow NO + H_2O + c^{3+}$ | Nitrite reductase (*Nir*)<br>Cu*Nir*<br>Or<br>cytochrome $cd_1$ | -73.2 |
| $2NO + 2c^{2+} + 2H^+ \rightarrow N_2O + H_2O + 2c^{3+}$ | Nitric oxide reductase (*Nor*) | -153.2 |
| $N_2O + 2c^{2+} + 2H^+ \rightarrow N_2 + H_2O + 2c^{3+}$ | Nitrous oxide reductase (*Nos*) | -341.4 |

Several environmental factors, such as oxygen concentration, presence of nitrogen oxides, pH, temperature, type and concentration of electron source, C/N ratio and S/N ratio affect denitrification.

## Kinetics of Denitrifying Cultures

According to Equation 5, in the particular case of denitrification processes, it is desirable to obtain dissimilative processes where small values of $\mu$ and high substrate conversions to $N_2$ and bicarbonate can be achieved, thus, minimal efforts for sludge post-treatment should be required. It is noteworthy that differences in specific consumption rates ($q_s$) of denitrifying processes are expected to occur when using distinct electron donors, as $q_S$ values depend on the concentration and type of substrate. In fact, according to Xu et al., (2015), lower consumption rates have been reported for autotrophic denitrification as compared to organotrophic process.

For kinetic purposes and considering that denitrification is a respiratory process where reduction of nitrate is simultaneously linked to electron donor oxidation, the specific substrate consumption rate ($q_S$) can be expressed for nitrate or electron donor. Likewise, specific production rate can be described in terms of $N_2$ or $HCO_3^-$ (Texier et al., 2013).

### Growth Rates

There are studies in the literature indicating the feasibility of using $H_2$, sulfur and methane as electron and carbon source for sustaining growth of denitrifying bacteria.

Several kinetic data of different denitrification processes are illustrated in Table 5. Monod model is generally used for determining $\mu_{max}$ and $k_S$ values; nevertheless, Rezania et al., (2005) have successfully assumed zero order kinetics model for nitrate ($q_{NO3-}$) and $H_2$ ($q_{H2}$) consumption rates, so, $q_{NO3-}$ or $q_{H2}$ values only depended on biomass concentration and maximum $q_{NO3-}$ or $q_{H2}$. Very small ks values ranging from 0.16-0.18 mg N/L and 0.0009 to 0.0066 mg $H_2$/L for nitrate and hydrogen are respectively found in the literature (Smith et al., 1994) confirming this assumption. A microbial consortium consisting of two microorganisms, a bacterium and an archaeon was enriched over 16 months for anaerobic denitrification with methane (Raghoebarsing et al., 2006). It was determined in batch cultures an apparent affinity constant for methane close to 0.0096 mg $CH_4$/L, which is higher than those reported for $H_2$ in hydrogenotrophic denitrification. In spite of no growth rates were determined for this culture, authors reported that its growth was very slow. Trautwein et al. (2008) have reported the ability of the strain *Aromatoleum aromaticum* for growing with toluene as carbon and electron source. The authors

*Table 5. Kinetic parameters for some organo-, auto- or mixo-trophic denitrification processes*

| Parameters | References | | | |
|---|---|---|---|---|
| | Vasiliadou et al., (2006) | Raghoebarsing et al. (2006) | Fajardo et al. (2014) | González Blanco et al. (2013) |
| Electron donor | $H_2$ | $CH_4$ | $S^{2-}$ and $NO_2$ | *p*-cresol |
| Inoculum-Culture condition | Mixed culture-batch | Bacteria-archaea consortium-batch | Denitrifying sludge | Denitrifying sludge |
| μmax (1/h) | 0.0485 | - | qmax 176 (mg $N_2$/ gVSS-d | qmax 227 (mg $N_2$/ gVSS-d |
| $k_S$ (mg $NO_3^-$/L) | 28.63 | 0.0096 mg $CH_4$/L | 10.7 mg $NO_2^-$-N/L | 20.32 mg *p*-cresol-C/ L |
| $Y_{X/S}$ (mg biomass/mg $NO_3^-$-N | 0.4207 | - | - | - |
| $K_i$ | - | - | 34.7 mg $NO_2^-$-N/L | 35.75 mg *p*-cresol-C/L |

reported a growth rate of 0.012 $h^{-1}$ at a toluene concentration of 6.4 mg/L. Nevertheless, growth rate was significantly decreased to 0.003 $h^{-1}$ as toluene concentration was increased to 68.1 mg/L, indicating an inhibitory effect of toluene.

## Denitrifying Rates

As it has been previously pointed out, denitrification processes can be discussed either in terms of substrate consumption rates for nitrate or nitrite consumption ($q_{NO2-}$) or production rates ($N_2$, ($q_{N2}$) or $HCO_3^-$ ($q_{HCO3-}$) generation). Nevertheless, most of the data available in literature only refer to substrate consumption rates so; it is not possible to discern whether the process is more linked to assimilation or denitrification. Generally, this is the case for autotrophic denitrification.

Several kinetic data based on denitrifying cultures show differences in $q_{NO3-}$ depending on the electron donor, nitrate concentration, inoculum size, use of axenic cultures or consortia, among others as Tables 6 and 7 illustrate. Differences in units for describing $q_{NO3-}$ values are also noted as in some cases no biomass concentration is given by the authors.

Studies conducted on hydrogen autotrophic denitrification have resulted in small denitrifying rates (Table 6). Some works have pointed out that increases in nitrate concentration, temperature and pH results in higher $q_{NO3-}$ (An et al., 2010; Wang et al., 2015). Comparable $q_{NO3-}$ values (g $NO_3^-$-N/g VSS·d) at similar operating conditions (25 °C, pH 6.8-8 and 0.5-6.61 g biomass/L) for hydrogenotrophic denitrification (0.38-0.74), organotrophic process with methanol (0.43-0.61) and sulfur autotrophic process (0.14-0.19) have been pointed out by Rezania et al. (2005). These results suggest the feasibility of hydrogenotrophic process for treating underground water where low nitrate concentration is present. Nevertheless, accumulation of nitrite might be an important drawback necessary to avoid for the successfully application of hydrogenotrophic denitrification in the treatment of wastewaters with higher nitrate content as those generated by some industries.

Autotrophic denitrification with methane has been demonstrated in batch cultures by Islas-Lima et al. (2004) and Raghoebarsing et al. (2006) using denitrifying sludge and a consortium formed by bacteria and archaea, respectively, as these authors have assessed $N_2$ formation and methane depletion during cultures. The former authors reported that the ratio between nitrate depletion and nitrogen gas production was close to one, thus $q_{N2}$ values are similar to those calculated for $q_{NO3-}$. In spite of Raghoebarsing

*Table 6. Kinetic data for hydrogen and methane autotrophic denitrification processes*

| Initial $NO_3^-$ conc. ($mg\ NO_3^-$-N/L) | Type of inoculum and conc. (gVSS/L) | T (°C) | pH | $q_{NO3-}$ | $q_{N2}$ | Reference |
|---|---|---|---|---|---|---|
| Hydrogen denitrification | | | | | | |
| 25 | Activated sludge, batch culture, 0.5 | 25 ± 1<br>12 ± 1 | 7.5 - 9.5<br>7.5 - 9.5 | 0.38-0.74<br>0.21-0.28<br>g $NO_3^-$-N/g VSS·d | ND*<br>ND* | Rezania et al., (2005) |
| 31 - 201 | Activated sludge, batch culture, 0.33 | 30± 1 | 6.4 - 7 | 0.07<br>g $NO_3^-$-N/L·d | ND* | Vasiliadou et al., (2006) |
| 23 - 155 | Mixed batch culture | 35± 1 | 7 | 0.91<br>g $NO_3^-$-N/L·d | ND* | Wang et al., (2015) |
| 20 - 80 | Axenic batch culture, *Alcaligenes eutrophus* | 30± 1 | 8.5 - 9 | 0.19-0.82 1/d (calculated as first order kinetic constant) | ND* | An et al., (2010) |
| **Methane Denitrification** | | | | | | |
| 50 | Bacteria-archaea consortium batch, 0.1 | 30 | 7 | ND* | 6.16<br>mg $N_2$/g protein·h | Raghoebarsing et al. (2006) |
| 35 | Denitrifying sludge, batch,6.5, | 30 | 7 | 0.25<br>mg $NO_3^-$-N/g VSS·d | 0.25<br>mg $N_2$-N/g VSS·d | Islas-Lima et al., (2004) |
| 223 | Enriched culture of *Methanoperedens nitroreducens and Kuenenia stuttgartiensis,* pulse-feed reactor, 1.1 | 22± 2 | 7-7.5 | 219<br>mg $NO_3^-$-N/d | 187<br>mg $N_2$-N/d | Haroon et al. (2013) |
| 7 mg N-$NO_2^-$/L | SBR reactor, enriched methanotrophs-organotrophic denitrifiers | 30 | ND* | 2.96<br>mg $NO_2^-$-N/g VSS·d<br>0.84 mg $CH_4$-C/g VSS·d | ND* | He et al. (2015) |

*ND, not determined

et al. (2006) reported higher $q_{N2}$ values than those determined by Islas-Lima et al. (2004), these values remained much lower than those reported for mixotrophic or organotrophic denitrification processes. Haroon et al. (2013) demonstrated anaerobic methane oxidation coupled to nitrate reduction, where the nitrite formed in presence of ammonium is further reduced to $N_2$ by means of anammox process. Based on data obtained by isotope labeling batch test, authors demonstrated that all the nitrate and methane supplemented to the reactor was oxidized to $CO_2$ and $N_2$. Recently, He et al. (2015b) determined in a sequencing batch reactor (SBR) the existence of nitrite reduction coupled to methane oxidation. Authors reported a 200-fold increase in nitrite removal rate after 609 days of SBR operation (12.22 mg N/L·d). Batch assays indicated that on day 609, 17 and 83% of total $q_{NO2-}$ (2.69 mg $NO_2^-$-N/g VSS·d) can be respectively attributed to nitrite-dependent anaerobic methane oxidation and organotrophic denitrification. Nevertheless, in this work is not possible to determine whether nitrite elimination is associated to assimilation or denitrification processes as $q_{N2}$ were not determined. In spite of the feasibility for conducting denitrification coupled to methane oxidation has been shown, more research has to be done in order to apply this process to wastewater treatment.

*Table 7. Kinetic data for autotrophic/mixotrophic sulfur denitrification and organotrophic denitrification processes*

| Type of inoculum and biomass concentration (g/L) | Culture conditions | Initial substrate concentration (mg/L) | Final products | Denitrifying rates | Reference |
|---|---|---|---|---|---|
| Enriched mixed consortium. Batch. | Autotrophic | N: 104; $S^{2-}$: 180 | $N_2$, $NO_3^-$, $NO_2^-$, $S^o$ and $S_2O_3^{2-}$ | 15.12 mg $NO_3^-$-N/L·d<br>84.48 mg $S^{-2}$-S/L·d | Lee & Wong (2014) |
| Enriched autotrophic mixed culture. Batch. 1.0 | Autotrophic<br>T: 35 °C | $NO_3^-$-N: 78; $S^{2-}$:150<br>$NO_2^-$-N: 29; $S^{2-}$:25<br>$NH_4^+$-N: 10-100 | $N_2$<br>$N_2$, $NO_2^-$, $N_2O$ | 159 mg $N_2$/g VSS·d<br>78.11 mg $N_2$/g VSS·d | Fajardo et al. (2014) |
| Enriched mixed consortium. Batch. | Mixotrophic | N: 104; $S^{2-}$: 180;<br>Acetate-C:160 | $N_2$, $NO_X$, $S^o$ and $S_2O_3^{2-}$ | 32.28 mg $NO_3^-$-N/L·d | Lee & Wong (2014) |
| Enriched mixed consortium. Membrane fluidized bed continuous reactor | Mixotrophic | N: 25-80<br>Methanol-C: 12 - 54<br>Ethanol-C: 12 - 52 | $N_2$, $S^o$ | 1.44 g $NO_3^-$-N/L·d<br>3.84 g $NO_3^-$-N/L·d | Zhang et al. (2015) |
| Anaerobic sludge, 1.7. SBR. | Mixotrophic | N: 108 - 194<br>$S^{2-}$: 200 - 533<br>Acetate-C: 105 - 153 | $S^0$ or $SO_4$ | 0.153 g $NO_3^-$-N/L·d<br>1.55 - 1.69 g S/L·d<br>1.74 - 1.99 g Ac-C/L·d | Xu et al. (2015b) |
| Denitrifying sludge. Batch. 1. | Mixotrophic | N: 64-120; $S^{2-}$: 25-100;<br>*p*-Cresol-C: 44 | $N_2$, $NO_2^-$, $S^o$ | 344-308 mg $NO_3^-$-N/g VSS·d<br>166-248 mg S/g VSS·d<br>288-111 mg C/g VSS·d | Cervantes et al. (2009) |
| Denitrifying sludge. Batch. 1. | Organotrophic | *p*-cresol-C: 44;<br>$NO_3^-$-N: 50 | $N_2$, $NO_2^-$ | 200 mg $NO_3^-$-N/g VSS·d<br>180 mg C/g VSS·d | Cervantes et al. (2009) |
| Denitrifying sludge. Batch. 1.1 | Organotrophic | Acetate-C: 95<br>$NO_3^-$-N: 75<br>Toluene-C: 70<br>$NO_3^-$-N: 70<br>*m*-xylene-C: 70<br>$NO_3^-$-N: 75 | $N_2$, $HCO_3^-$<br>$N_2$, $HCO_3^-$<br>$N_2$, $NO_2$, $N_2O$, $HCO_3$ | 0.29 mg $NO_3^-$/mg VSS·d<br>0.36 mg Acetate-C / mg VSS·d<br>0.025 mg Toluene-C/ mg VSS·d<br>0.0038 mg *m*-xylene-C/mg VSS·d | Peña-Calva et al. (2004) |
| Denitrifying sludge. Batch. 2. | Organotrophic | *p*-cresol-C: 40<br>$NO_2^-$-N: 83 | $N_2$, $HCO_3^-$ | 169 mg $N_2$/g VSS·d | González-Blanco et al. (2013) |

Fajardo et al. (2014) conducted studies on sulfide autotrophic denitrification assessing the effect of different concentrations (mg/L) of sulfide (25-300), nitrate ($NO_3^-$-N: 13-156), nitrite ($NO_2^-$-N: 29-232) and ammonium ($NH_4^+$-N: 10-100) on denitrifying rates. No effect of ammonium was observed on the denitrification rate, whereas with nitrate as electron acceptor the highest denitrifying rate was obtained at 35°C, 150 mg $S^{2-}$/L and 78 mg $NO_3^-$-N/L. The maximum $q_{N2}$ was obtained at 35 mg $NO_2^-$-N/L and the kinetic behavior was adequately represented by the Haldane model. Therefore, sulfur denitrification can be successfully applied for eliminating sulfur and nitrate concentrations up to 150 and 80 mg $NO_3^-$-N/L, respectively.

Denitrifying sulfide removal has also been evaluated by means of autotrophic and mixotrophic sulfide denitrification in batch assays by Lee & Wong (2014). Authors reported $q_{NO3-}$ and sulfide oxidation rate ($q_{S2-}$) of 15.12 mg N/L·d and 84.48 mg S/L·d. In spite of complete sulfide depletion occurred, its oxidation was incomplete as a mixture of $N_2$, $S^0$ and $S_2O_3^{2-}$ was registered after 60 h of culture. Mixotrophic denitrification of acetate, sulfide and nitrate was completed in 40 h resulting in $q_{NO3-}$ 2.4 times higher than the one obtained in autotrophic denitrification. Likewise, the end products were formed by $N_2$, $S^0$ and $S_2O_3^{2-}$, but data for $q_{N2}$ were not provided in any case. In spite of mixotrophic denitrification appeared to be a faster process, incomplete denitrification in both cases might be related to the C/N and S/N ratios as well as initial substrate concentrations, where nitrate should be the limiting substrate. Xu et al. (2015) assessed the performance of autotrophic, organotrophic and mixotrophic denitrification when 146 mg Acetate-C/L and 821 mg Sulfide-S/L at several S/C and S/N ratios were the electron donors for the reduction of nitrate. Complete elimination of the three compounds was achieved in the three cases, where $SO_4^{2-}$ was the product formed. Likewise nitrite accumulation was registered in all cases. The highest nitrite accumulation was obtained at autotrophic denitrification, followed by mixotrophic and finally at organotrophic process. Authors suggest that in mixotrophic conditions, the $NO_3^-$ to $NO_2^-$ and $NO_2^-$ to $N_2$ conversions are limited regarding to organotrophic or autotrophic denitrification. It is important to note that in mixotrophic denitrification, autotrophic and heterotrophic processes occurred at sequential reaction rates. Recently, mixotrophic denitrification where acetate, sulfide and acetate were simultaneously and completely consumed has been reported in a sequential batch reactor by Xu et al. (2015). In this work, sulfide was converted to $S^0$ or $SO_4^{2-}$ according to the defined S/N ratio. The oxidation kinetics of sulfide and acetate followed a zero-order model. According to the rate constant obtained, acetate consumption rate ($q_A$) was higher than the $q_{S2-}$ for sulfide, thus organotrophic denitrification step is faster than the autotrophic one. Likewise, authors concluded that $q_{NO3-}$ was determined by the sulfide/acetate ratio. A fluidized bed membrane reactor has been used for successfully achieving mixotrophic denitrification when ethanol or methanol plus sulfur particles were used as electron donors (Zhang et al., 2015). Better kinetic performance was registered with ethanol (3.84 g/L·d). According to mass balances, authors confirmed that carbon sources were completely consumed and in both cases the nitrate consumed was mainly converted to $N_2$ (89-96%) and a small fraction to biomass (6-8%), indicating a dissimilative denitrification process. Likewise, lower $NO_2^-$ and $N_2O$ production was observed in the ethanol-fed process, suggesting that ethanol-fed-mixotrophic denitrification using membrane fluidized reactors might be used for wastewater treatment process. Mixotrophic denitrification with different sulfide concentrations and phenolic compounds has been reported in batch assays by Cervantes et al. (2009), where *p*-cresol was used as electron source. Their results pointed out two kinetically rate limiting steps, nitrite reduction and $S^0$ oxidation, where nitrite reduction was the most evident rate-limiting step in the mixotrophic denitrification. Nevertheless, under these conditions, a promoting effect on $q_{NO3-}$ was determined by the simultaneous removal of *p*-cresol and sulfide as lower $q_{NO3-}$ was obtained in organotrophic denitrification.

Organotrophic denitrification has been extensively reported in the literature as different carbonaceous compounds have been successfully used as electron donors. One of the most commonly carbon compound used for evaluating kinetic behavior of denitrification is acetate, as the rates obtained at these conditions are generally used as reference in both batch and continuous denitrifying cultures. Regarding this, complete acetate and nitrate consumption and their conversion to $CO_2$ and $N_2$ has been respectively reported by Peña-Calva et al. (2004). Organotrophic denitrification with several concentrations of *p*-cresol (10-65 mg C/L) and nitrite as electron acceptors was evaluated in batch cultures (González-Blanco et al., 2013), where these compounds were completely consumed and converted to $HCO_3^-$ and $N_2$ respectively

According to a first order model, both $q_{NO2-}$ and $q_{N2}$ increased as initial *p*-cresol concentration increased, reaching a maximum $q_{N2}$ value at 40 mg C/L. Aromatic compounds such as toluene and xylene isomers represent some other electron donors which have been used for conducting organotrophic denitrification by Peña-Calva et al. (2004). These authors evaluated the denitrification process in presence of several concentrations of toluene (15-100 mg C/L) and *m*-xylene (15-70 mg C/L). Toluene was completely consumed and converted into $CO_2$ whereas up to 50% of *m*-xylene remained at the end of the cultures. Consumption rates of toluene ($q_T$) and *m*-xylene ($q_{mX}$) increased as their initial concentration also increased. According to the maximum rates obtained for each compound, the highest $q_{NO3}$ was achieved with acetate followed by toluene and finally for *m*-xylene. In fact, $q_T$ was almost 17 times lower than $q_A$ but almost 6 times higher than $q_{mX}$, which in turn, was almost 100 times lower than that obtained for acetate. As $N_2$ was the mainly product generated during toluene and *m*-xylene denitrification, authors suggest that denitrification is a suitable option for treating wastewater contaminated with these compounds.

In summary, according to the literature, the highest and lowest denitrifying rates have been obtained with organotrophic and methane autotrophic processes as follows:

Organotrophic> sulfur mixotrophic > sulfur autotrophic> hydrogen autotrophic> methane.

Kinetic data should be considered when designing autotrophic, mixotrophic or organotrophic denitrifying reactors to treat industrial wastewaters or underground water containing aromatic or aliphatic carbonaceous, sulfurous and nitrogenous contaminants.

## Inhibitory Factors

Autotrophic and organotrophic denitrification rates might be affected by several environmental factors, such as pH, temperature and type and concentration of electron source, among others. Some factors influencing the denitrifying rates and their main associated effects on the denitrifying processes are presented in Table 8.

It has been widely observed that denitrification can properly operate within pH values ranging from 6 to 9. Denitrification rates, particularly at nitrite and nitrous oxide reduction steps are strongly affected by pH values lower than 5. This situation has been associated to several facts (Cuervo-López et al., 2009) where $HNO_2$ formation or the dissociation of the *Nir-Nor* enzymatic complex seem to generate the most negative effects. In autotrophic denitrifying processes, particularly when hydrogen is the electron donor, decreases in $q_{NO3-}$ at acidic pH are strongly related to $CO_2$ stripping, whereas at pH values higher than 8.5, diminishes in $q_{NO3-}$ appeared to be associated to nitrite accumulation (Rezania et al., 2005). Diminishes up to 4 fold times in $q_{NO3-}$ and suppression of denitrifying process are also provoked as a result of nitrate concentrations higher than 80 mg $NO_3^-$-N/L as several researchers have reported (Vasiliadou et al., 2006; An et al., 2010; Wang et al., 2015). This effect has been associated to nitrite accumulation. Therefore, pH values, nitrite and nitrate concentrations are critical factors for determining consumption and production rates in autotrophic denitrification with hydrogen. Sulfide concentrations higher than 200 mg/L exerted an inhibitory effect on autotrophic denitrification as denitrifying rates diminished significantly (Beristain et al., 2006; Fajardo et al., 2014). Likewise, incomplete denitrification was determined at 300 mg $S^{2-}$/L as nitrite and nitrous oxide were present in significant amounts (Fajardo et al., 2014). These authors also reported the effect of different initial nitrite and sulfide concentrations as electron acceptor or donor, where according to Haldane model, nitrite concentration appeared to be the most inhibiting factor (inhibition coefficient, ki = 34.7 mg $NO_2^-$-N/L, Table 8) indicating that this nitrogenous compound exerted an inhibitory effect on nitrous oxide reduction step. Similarly to organotrophic and autotrophic denitrifica-

*Table 8. Factors influencing denitrifying rates in organotrophic or autotrophic processes*

| Inoculum/Denitrifying Process | Factor Evaluated | Effects | Reference |
|---|---|---|---|
| **Hydrogen Denitrification** | | | |
| Activated sludge- batch culture | pH>8.5<br>pH< 7 | Decrease in $q_{NO3-}$* due to $NO_2^-$ accumulation.<br>Decrease in $q_{NO3-}$* due to $CO_2$ stripping | Rezania et al. (2005) |
| Mixed culture- batch culture | $[NO_3^-]$ > 165 mg $NO_3$-N/L | Decrease in $q_{NO3-}$* and bacterial growth | Vasiliadou et al. (2006) |
| Mixed culture- batch culture | $[NO_3^-]$ > 105 mg $NO_3^-$-N/L, pH 7, 35°C, C/N = 30 | $q_{NO3-}$**= 0 due to $NO_2^-$ accumulation. | Wang et al. (2015) |
| Pure culture *Alcaligenes eutrophus*- batch culture | $[NO_3]$ > 80 mg $NO_3^-$-N/L | Decrease in $q_{NO3-}$*** | An et al. (2010) |
| Mixed culture-mixotrophic denitrification-batch culture | $[CH_3COONa]$ = 0.5 g/L<br>$[Na_2S]$ = 2 g/L | $NO_2^-$ accumulation.<br>Decrease in $q_{NO3-}$*** | Xu et al. (2015b) |
| Organotrophic denitrification<br>Mixotrophic | [*p*-cresol-C] = 45<br>$[S^{2-}]$ > 25 mg/L | $NO_2^-$ and $N_2O$ accumulation.<br>Decrease in $q_{NO2-}$+<br><br>Decrease in $q_{NO2-}$+ | Cervantes et al. (2009) |
| Sulfur Autotrophic denitrification | $[S^{2-}]$ > 200 mg/L<br>$[NO_2^-]$>35 mg/L | Decrease in $q_{N2}$++<br>$NO_2^-$ and $N_2O$ accumulation. | Fajardo et al. (2014) |
| Organotrophic denitrification | [*p*-cresol-C]>45 mg/L | Decrease in $q_{N2}$++<br>$NO_2^-$ accumulation | González-Blanco et al. (2013) |
| Organotrophic denitrification | [toluene-C]=100 mg/L<br>[*m*-xylene-C] = 45 mg/L | Decrease in $q_{N2}$++, $q_{NO3}^-$ and $q_T^x$<br>Decrease in $q_{NO2-}$ and $q_{N2O}$<br>Accumulation of carbon intermediaries | Peña-Calva et al. (2004) |

*$q_{NO3-}$: Nitrate removal rate (g $NO_3^-$-N/m³·d)
**$q_{NO3-}$: Nitrate removal rate (mg $NO_3^-$/L·h)
***$q_{NO3-}$: Nitrate removal rate (1/d, calculated as first order kinetic constant)
+$q_{NO2-}$: Nitrite removal rate (mg $NO_2^-$-N/VSS·d)
++$q_{N2}$: Denitrifying rate (mg $N_2$-N/VSS·d)
$q_T^x$: Toluene consumption rate (mg toluene-C/g VSS·d)

tion, nitrite accumulation seemed to be a crucial fact for inhibition in mixotrophic denitrification. Nitrite accumulation has been attributed to the lower nitrite reduction rate regarding to nitrate reduction rate. In organotrophic denitrification, nitrite accumulation has been reported with phenolic compounds such as *p*-cresol (Meza-Escalante et al., 2008; Cervantes et al., 2009). Sulfide also exerts inhibition on nitrite oxidation rate in both, autotrophic and mixotrophic denitrification (Cervantes et al., 2009). This effect has been partially associated to precipitation of trace elements which are essential for the activity of nitrite reductase enzymes (Berks et al., 1995). In mixotrophic denitrification assays where acetate and sulfide were the electron sources, diminishes in $q_{NO3-}$ were observed by Xu et al (2015). Authors determined that the presence of acetate and sulfide, respectively affected the activity of autotrophic and organotrophic denitrifiers, resulting in a decrease in $q_{NO3-}$ values, thus more research are needed to elucidate this effect. In organotrophic denitrification with nitrite as electron acceptor, concentrations of *p*-cresol higher than 45 mg C/L provoked inhibition in the process as $q_{N2}$ decreased (González-Blanco et al., 2013). In spite of most of the initial *p*-cresol was consumed, important quantities of nitrite and organic intermediates were detected at the end of the cultures, suggesting that *p*-cresol consumption was carried out by denitrification and fermentation. The inhibitory effect of toluene and *m*-xylene on denitrification has been evidenced in batch assays by Peña-Calva et al. (2004). A significant decrease in $q_T$, (mg of toluene-C/mg of VSS d) was determined at 100 mg of toluene-C/L, likewise, the metabolic mineralization of toluene diminished resulting in lower specific production rates of $HCO_3^-$ and $N_2$. A decrease in $q_{NO2-}$ and $q_{N2O}$ was observed in the assays supplemented with *m*-xylene, suggesting an inhibitory effect on nitrite and nitrous oxide reductase enzymes. In summary, nitrite and nitrous oxidation steps appeared to be the most sensitive in autotrophic, mixotrophic and organotrophic denitrification processes.

## ANAMMOX

Anaerobic ammonium oxidation (anammox) is a chemolithotrophic process in which microorganisms anaerobically oxidize ammonium into molecular nitrogen ($N_2$) using nitrite as terminal electron acceptor. Before anammox was discovered, the only recognized microbial process capable to oxidize ammonium was nitrification. The Austrian physicochemist, Engelbert Broda (1977), based on evolutionary and thermodynamic grounds, already predicted the existence of this microbial process within the nitrogen cycle. Oceanographers also reported unexplainable loss of ammonium from anoxic marine environments (Hamm and Thompson, 1941). Furthermore, anammox was also observed during the operation of a fluidized bioreactor treating effluents from a yeast factory in The Netherlands, when nitrate was supplied to prevent malodorous caused by sulfide (Mulder et al., 1995). This serendipity was further elucidated confirming that anammox was a biologically mediated process using nitrite (instead of the nitrate initially added) as electron acceptor (Van de Graaf et al., 1996).

An enormous interest was created on anammox since its discovery and intensive research has been performed in many groups around the world. After two decades of its discovery, it is known that anammox bacteria are widespread and have been detected in soil, groundwater, wastewater treatment plants, fresh water and marine sediments, lakes, estuaries, oxygen minimum zones and continental shelves in the oceans, polar regions, hot springs, deep-sea hydrothermal vents and hypersaline sites. Is has been estimated that about 50% of the annual fixed nitrogen loss could be attributed to anammox activity (Arrigo, 2005; Lam and Kuypers, 2011).

### Microbial Diversity

Successful enrichments of anammox bacteria have been developed in different treatment systems with a reasonable good retention capacity of biomass, which is crucial considering the extremely low growth rate of anammox bacteria (Kartal et al., 2013). Reactor configurations used for anammox enrichments include sequential batch reactors (SBR), upflow anaerobic sludge bed (UASB) reactors, rotating biological contactors, membrane bioreactor (MBR), as well as moving bed and packed bioreactors using distinct supporting materials (He et al., 2015; Hu et al., 2013; Lackner et al., 2014). Overall, ten anammox species have been described, including seven enrichments cultures available. None of these were obtained as classical pure cultures, thus all have the taxonomical status of "*Candidatus*", which are distributed in five different genera:

1. *Kuenenia*, represented by *Kuenenia stuttgartiensis* (Strous et al., 2006);
2. *Brocadia* (three species: *B. anammoxidans, B. fulgida,* and *B. sinica*) (Strous et al., 1999; Kartal et al., 2008; Oshiki et al., 2011);

3. *Anammoxoglobus*, represented by *A. propionicus* (Kartal et al., 2008);
4. *Jettenia* with one species reported (*J. asiatica;* Quan et al. 2008);
5. *Scalindua* including four species (*S. brodae, S. sorokinii, S. wagneri,* and *S. profunda*) (Schmid et al. 2003; Woebken et al. 2008; Van de Vosseberg et al. 2012).

Species from the first four genera have been enriched from sludge samples derived from wastewater treatment facilities. *Scalindua,* on the other hand, has been detected in marine environments, as well as freshwater ecosystems and wastewater treatment plants. Furthermore, over 2000 16S rRNA gene sequences affiliated with anammox bacteria have been deposited to the Genbank, reflecting an unappreciated variety of species. Microdiversity is especially large for *Scalindua* (Kartal et al. 2013).

Phylogenetic analysis revealed that all anammox enrichments belong to the phylum *Planctomycetes,* which form a super-phylum together with *Verrucomicrobia* and *Chlamydiae.* Anammox bacteria deeply branch as a monophyletic clade within the *Planctomycetes.* Their phylogenetic position together with a wide range of specific cellular, physiological and molecular characters endow anammox bacteria their own order *Brocadiales* (Kartal et al. 2013). Several molecular tools have been developed to detect anammox bacteria in environmental samples and their description can be found elsewhere (Li & Gu 2011; Kim et al. 2013).

## Particular Cell Features

As *Planctomycetes,* anammox bacteria have complicated cell plan comprising intracellular organelles. Indeed, anammox cell is composed of three compartments, each surrounded by a membrane bilayer (Van Niftrik and Jetten, 2012). The outermost membrane encloses both the cell and the outer compartment, the so-called paryphoplasm. The second membrane surrounds the riboplasm that harbors the nucleoid and ribosomes. Like in the cytoplasm of common prokaryotes, the transcription, translation, and household machinery are presumed to take place in this compartment (Kartal et al., 2013). The third, innermost, highly curved membrane bounds the central cell structure: the anammoxosome. Scientific reports show that this structure is fully closed and that it is inherent to the daughter cells during cell division (Van Niftrik and Jetten, 2012). The key enzymes responsible of catalyzing anammox (see below details) are hypothesized to reside in the anammoxosome.

Another specific feature of anammox bacteria is the nature of the cell membrane constituents. Their glycerolipids contain a combination of ester-linked (typical of Bacteria and Eukarya), fatty acids or ether-linked (typical of Archaea) long-chain alcohols. Particularly, membranes of anammox bacteria have saturated C17-C20 fatty acids and alcohols that are fused by *cis*-ring junctions to make ladder-like ("ladderane") cyclobutane and cyclohexane ring systems (Schmid et al., 2005).

## Biochemical Aspects

Based on genome assembly of the anammox bacterium *Kuenenia stuttgartiensis*, a set of three reactions (Equations 9-11) involving hydrazine ($N_2H_4$) and nitric oxide (NO) was proposed (Strous et al., 2006) to explain the overall anammox stoichiometry (Equation 12):

$$NO_2^- + 2H^+ + e^- \rightarrow NO + H_2O \qquad \left(E_0' = +0.38\ V\right) \tag{9}$$

$$NO + NH_4^+ + 3e^- + 2H^+ \rightarrow N_2H_4 + H_2O \qquad \left(E_0' = +0.06\ V\right) \tag{10}$$

$$N_2H_4 \rightarrow N_2 + 4H^+ + 4e- \qquad \left(E_0' = -0.75\ V\right) \tag{11}$$

$$NH_4^+ + NO_2^- \rightarrow N_2 + 2H_2O \qquad (\Delta G^{\circ\prime} = -357\ kJ\ mol^{-1}) \tag{12}$$

Deep understanding on the molecular mechanisms involved in anammox was described by Kartal et al. (2011). Highly enriched culture of *K. stuttgartiensis* (comprising more than 95% of microbial population) was used to elucidate these mechanisms based on transcriptomics and proteomics, as well as inhibitor and isotopic evidences. Activity of key purified enzymes of the anammox process was also documented. Transcriptomics and proteomics indicated that *K. stuttgartiensis* expressed cd1 nitrite:nitric oxide reductase with the potential ability to reduce nitrite to NO. This possibility was investigated by incubating *K. stuttgartiensis* with ammonium, nitrite (2mM each) and 100 mM NO scavenger PTIO (2-phenyl-4,4,5,5,-tetramethylimidazoline-1-oxyl-3-oxide). When PTIO was introduced at the start of the incubation or when it was added to active cells, anammox activity was inhibited. Further, the cells were incubated with ammonium and nitrite (2mM each) in the presence of DAF2-DA (10 mM) that reacts with NO to form a fluorescent product. Sampled *K. stuttgartiensis* cells displayed the characteristic green fluorescence indicating NO production. In control experiments without nitrite or with added PTIO, there was no detectable fluorescent signal (Kartal et al., 2011). Furthermore, when acetylene (15 mM) was added, the anammox reaction was inhibited. Acetylene inhibits aerobic ammonium oxidation by binding covalently to the ammonia monooxygenase, the ammonia-activating enzyme of aerobic ammonium oxidizers (Gilch et al., 2009). Apparently, it also interfered with the ammonium-activating step of anammox cells (Equation 10). Importantly, acetylene inhibition resulted in an immediate accumulation of NO; hydroxylamine accumulation was not observed, consistent with the role of NO as the direct precursor for $N_2H_4$ (Kartal et al., 2011).

The complementary reduction of NO and its simultaneous condensation with ammonium to produce $N_2H_4$ (Equation 10) was also confirmed. To investigate whether $N_2H_4$ could be produced directly from NO, cell suspensions were incubated with NO (0.1mM) and ammonium (2 mM). A transient accumulation of hydrazine (18 μM) was observed (Kartal et al., 2011), albeit at a much lower concentration (200–500 μM) for incubations with hydroxylamine (Kartal et al., 2008). This is consistent with Equations 9–11 because the major part of the produced $N_2H_4$ would be oxidized to $N_2$ as expected from the overall reaction and NO could be supplied at much lower concentrations (Equation 13).

$$NO + NH_4^+ \rightarrow \tfrac{1}{4}N_2H_4 + H_2O + \tfrac{3}{4}N_2 + H^+ \tag{13}$$

The anammox pathway is completed by the oxidation of $N_2H_4$ to $N_2$ (Equation 11). $N_2H_4$ is known as an alternative substrate for octahaem-hydroxylamine oxidoreductases (HAOs), which catalyze the conversion of hydroxylamine to nitrite in aerobic ammonium oxidizers (Hooper et al., 1997). It was confirmed that the *K. stuttgartiensis* genome encoded ten divergent paralogues of this enzyme (Kartal et al., 2011). Six expressed paralogues belonged to the 'type II' hydrazine/hydroxylamine oxidoreductases (HZO/HAO). Two related 'type II' HZO/HAO and one divergent octahaem cytochrome c were also detected at lower levels (4–15%) and one was not detected. Purified HZO/HAO-like proteins (kustc0694 and kustc1061) from *K. stuttgartiensis* appear to be closely related to two enzymes of unknown function isolated from an anammox enrichment culture KSU-1 (Shimamura et al., 2008). Both enzymes catalyzed the four-electron oxidation of $N_2H_4$ to $N_2$ with cytochrome c as the artificial electron acceptor. When they were incubated with $^{30}N_2H_4$ and cytochrome c, $^{30}N_2$ ($^{15}N^{15}N$) was produced stoichiometrically, in agreement with Equation 11. Interestingly, Kustc1061 also oxidized hydroxylamine to NO (rather than nitrite) with a higher rate (6 mmol/min-mg protein). In contrast, kustc0694 did not catalyze this reaction and hydroxylamine and NO were powerful inhibitors of $N_2H_4$ oxidation, suggesting kustc0694 was the dedicated hydrazine dehydrogenase (HDH) in *K. stuttgartiensis*. Furthermore, the inhibition of kustc0694 explained the transient accumulation of $N_2H_4$ in the presence of hydroxylamine or NO.

## Kinetic Aspects

Anammox is a chemolitotrophic process, which involves microorganisms having low growth rates and yields due to the low Gibbs free energy provided by the reaction. Anammox Gibbs free energy and its activation energy are -357 kJ/(mol $NH_4^+$) and 70 kJ/(mol $NH_4^+$), respectively, according to Vázquez-Padín et al., (2009).

Like most microbial processes, anammox can be described by a Monod equation (Vázquez-Padín et al., 2009) that indicates the dependency of the bacteria growing on a limiting substrate, including a bacteria decay coefficient (Equation 14):

$$\mu = \mu_{max} \frac{C_S}{K_S + C_S} - k_d \tag{14}$$

Where: $\mu$ = specific growth rate (1/d)
$\mu_{max}$ = maximum specific growth rate (1/d)
$K_s$ = substrate saturation constant (g/L)
$C_S$ = substrate concentration (g/L)
$k_d$ = decay constant (g/(g-d))

If ammonium is the limiting substrate, Equation 14 would be written as Equation 15:

$$\mu = \mu_{max} \frac{C_{NH_4}}{K_{NH_4} + C_{NH_4}} - k_d \tag{15}$$

Where:
$K_{NH4}$ = ammonium saturation constant (g/L)
$C_{NH4}$ = ammonium concentration (g/L)

If the limiting substrate is nitrite, the employed equation would be Equation 16:

$$\mu = \mu_{max} \frac{C_{NO_2}}{K_{NO_2} + C_{NO_2}} - k_d \qquad (16)$$

Where:
$K_{NO2}$ = nitrite saturation constant (g/L)
$C_{NO2}$ = nitrite concentration (g/L)

Typical kinetic parameters for an anammox enrichment culture, of the type *Candidatus* "Brocadia anammoxidans" are included in Table 9. The low growing rates of these microorganisms imply, on one hand, long start-up periods for anammox systems, and on the other hand, low amounts of sludge produced, solving the problem of the disposal of sludge in excess. These facts will imply the need of systems with a good biomass retention but also reduction of operational costs (Vázquez-Padín et al., 2009).

## Key Factors Affecting Anammox Processes

### Temperature

As any biological process, anammox follows the principle of Arrhenius equation, which implies that the highest the temperature of incubation, the highest metabolic activity is accomplished, until a threshold limit in which anammox cells are either physiologically altered or denaturalized. The activation energy of anammox bacteria is similar to AOB (63-72 kJ/mol) (Hendrickx et al., 2012). Anammox activity has been reported in a wide range of temperatures (6-43 °C), although the reaction rate sharply drops at temperatures lower than 15 °C or higher than 40 °C (Zhu et al., 2008). Despite high anammox activities have been reported at low temperatures, for example 2.28 g N/L-d at 16 °C (Ma et al., 2013), a dramatic drop in activity has been reported when temperatures lower than 15 °C have been applied (Guo et al., 2015; Laureni et al., 2015).

*Table 9. Kinetic parameters of Candidatus "Brocadia anammoxidans"*

| | |
|---|---|
| Anammox rate | 45 nmol $NH_4^+$/(min·mg protein) |
| Maximum growth rate | 0.0031/h |
| Duplication time | 10.6 d |
| Ks (ammonium) | 5 μM |
| Ks (nitrite) | < 5 μM |
| Biomass yield | 0.066 mol C/(mol $NH_4^+$) |
| Oxygen | Reversible inhibition |
| Gibbs free energy | -357 kJ/(mol $NH_4^+$) |
| Biomass yield | 0.07 mol C/mol C |

(Strous et al., 1998)

## Organic Matter

Traditionally, anammox have successfully been applied to wastewaters with a high nitrogen content, but limited amount of organic matter (Pynaert et al., 2004). Nevertheless, due to its lithoautotrophic nature and sensitivity, anammox has faced difficulties when treating effluents with considerable COD levels. For example, poor (e.g. 13-22%) ammonium removal could be achieved when applying the anammox concept in anaerobically digested poultry manure (Dong and Tollner, 2003). Previous studies, however, have shown that addition of ammonium as an alternative electron donor to denitrifying reactors avoid accumulation of intermediates (e.g. nitrite and $N_2O$) due to the coupling between anammox and denitrification (Cervantes et al., 1999; 2001). Since then, multiple studies have confirmed that the coupling between anammox and organotrophic denitrification can promote the simultaneous removal of nitrogen and organic matter from wastewaters (for updated information see the review by Kumar and Lin, 2010). The C/N ratio is a key parameter for allowing both respiratory processes simultaneously.

## Dissolved Oxygen (DO) Concentration

The reversibility of oxygen inhibition observed in anammox enrichments has important implications for the relevance of the anammox process. Certainly, to remove ammonium from wastewaters, anammox process has to be supplied with nitrite as an electron acceptor. Thus complete nitrogen removal can only be accomplished when part of the ammonium is firstly oxidized to nitrite in a preceding partial-nitrifying step. Under oxygen limited conditions, the supplied ammonium is partly oxidized to nitrite. The produced nitrite can then be converted to nitrogen gas with the remaining ammonium by anammox bacteria. Maintaining the proper concentration of DO in single reactors concomitantly conducting partial nitrification and anammox is crucial to obtain a sustainable, robust process. A recent study demonstrated that anammox bacteria can be alive in aerobic tanks of municipal wastewater treatment plants, even when DO exceeds 2 mg/L, which expands the range of DO concentrations at which anammox could be active (Wang et al., 2015).

## Substrate Concentration

Several anammox consortia have shown a high affinity for both ammonium and nitrite with typical affinity constants below 5 μM (Strous et al., 1999). The anammox process has not been inhibited by ammonium or by the by-product nitrate up to concentrations of at least 1 g N/L. Nevertheless, in the presence of 0.1 g $NO_2^-$-N/L, the process is strongly inhibited (Strous et al., 1999). Prolonged exposure of different anammox consortia to relatively high concentrations of nitrite (up to 0.02-0.4 g $NO_2^-$-N/L) collapsed the respiratory process causing production of ammonium (Cervantes et al., 1999; Strous et al., 1999). Supply of traces of either hydrazine or hydroxylamine to nitrite-inhibited anammox consortia fully restored the metabolic activity (Strous et al., 1999).

## FUTURE RESEARCH DIRECTIONS

Further research should be focused on the following important issues:

- Optimize nitrogen removing biological processes to minimize the emission of greenhouse gases during their application and understand the operational aspects contributing to these issues.
- Develop molecular techniques to monitor the performance of nitrogen removing processes.
- Develop new treatment concepts integrating the different biological processes involved in the nitrogen cycle to remove nitrogen from wastewater in a single bioreactor.
- Develop strategies to recover nitrogenous contaminants as valued by-products in wastewater treatment facilities.

## CONCLUSION

The present chapter described kinetic, microbiological and biochemical aspects of nitrification, denitrification and anammox, as well as key operational parameters affecting the processes. Great scientific and technological advances have been achieved to understand and optimize biological treatment processes to remove nitrogenous compounds from wastewaters. Nitrification, denitrification and anammox are now implemented at full-scale, but several challenges remain, such as the development of novel treatment concepts with the capacity to mitigate the emission of greenhouse gases (methane and nitrous oxide) and to recover nitrogenous compounds as valued by-products from industrial waste streams.

## REFERENCES

An, Y., Li, T., Jin, Z., Dong, M., & Li, Q. (2010). Nitrate degradation and kinetic analysis of the denitrification system composed of nanoparticles and hydrogenotrophic bacteria. *Desalination*, *252*(1-3), 71–74. doi:10.1016/j.desal.2009.10.023

Arrigo, K. R. (2005). Marine microorganisms and global nutrient cycles. *Nature*, *437*(7057), 349–355. doi:10.1038/nature04159 PMID:16163345

Banning N. C., Maccarone L. D., Fisk L. M., & Murphy D. V. (2015). Ammonia-oxidising bacteria not archaea dominate nitrification activity in semi-arid agricultural soil. *Scientific Reports*. doi: 10.1038/srep11146

Beristain, R., Sierra-Alvarez, R., Rowlette, P., Flores, E. R., Gomez, J., & Field, J. A. (2006). Sulfide oxidation under chemolithoauthotrophic denitrifying conditions. *Biotechnology and Bioengineering*, *95*(6), 1148–1157. doi:10.1002/bit.21084 PMID:16807929

Beristain-Cardoso, R., Texier, A. C., Sierra-Álvarez, R., Field, J., Razo-Flores, E., & Gomez, J. (2009). Effect of initial sulfide concentration on phenol oxidation under denitrifying conditions. *Chemosphere*, *74*(2), 200–205. doi:10.1016/j.chemosphere.2008.09.066 PMID:18990426

Brannan, D. K., & Caldwell, D. E. (1980). *Thermotrix thiopara*: Growth andmetabolism of a newly isolated thermophile capable of oxidizing sulfur and sulfur compounds. *Applied Microbiology and Biotechnology*, *88*, 1027–1042.

Broda, E. (1977). Two kinds of lithotrophs missing in nature. *Zeitschrift fur Allgemeine Mikrobiologie*, *17*(6), 491–493. doi:10.1002/jobm.3630170611 PMID:930125

Calderer, M., Marti, V., de Pablo, J., Guivernau, M., Prenafreta-Boldú, F. X., & Viñan, M. (2014). Effects of ehnanced denitrification on hydrodinamics and microbial community structure in a soil column system. *Chemosphere*, *111*, 112–119. doi:10.1016/j.chemosphere.2014.03.033 PMID:24997907

Camargo, J. A., & Alonso, A. (2006). Ecological and toxicological effects of inorganic nitrogen pollution in aquatic ecosystems: A global assessment. *Environment International*, *32*(6), 831–849. doi:10.1016/j.envint.2006.05.002 PMID:16781774

Can-Dogan, E., Turker, M., Dagasan, L., & Arslan, A. (2010). Sulfide removal from industrial wastewaters by lithotrophic denitrification using nitrate as an electron acceptor. *Water Science and Technology*, *62*(10), 2286–2293. doi:10.2166/wst.2010.545 PMID:21076214

Cervantes, F. J. (2009). *Environmental Technologies to Treat Nitrogen Pollution: Principles and Engineering*. IWA Publishing.

Cervantes, F. J., de la Rosa, D. A., & Gómez, J. (2001). Nitrogen removal from wastewaters at low C/N ratios with ammonium and acetate as electron donors. *Bioresource Technology*, *79*(2), 165–170. doi:10.1016/S0960-8524(01)00046-3 PMID:11480925

Cervantes, F. J., Meza-Escalante, E. R., Texier, A. C., & Gómez, J. (2009). Kinetic limitations during the simultaneous removal of p-cresol and sulfide in a denitrifying process. *Journal of Industrial Microbiology & Biotechnology*, *36*(11), 1417–1424. doi:10.1007/s10295-009-0628-6 PMID:19672641

Cervantes, F. J., Monroy, O., & Gómez, J. (1999). Influence of ammonium on the performance of a denitrifying culture under heterotrophic conditions. *Applied Biochemistry and Biotechnology*, *81*(1), 13–21. doi:10.1385/ABAB:81:1:13 PMID:15304781

Chen, C., Ho, K. L., Liu, F. C., Ho, M., Wang, A., Ren, N., & Lee, D. J. (2013). Autotrophic and heterotrophic denitrificataion by a newly isolated strain *Pseudomonas* sp. C27. *Bioresource Technology*, *145*, 351–356. doi:10.1016/j.biortech.2012.12.027 PMID:23434385

Chen, G. (2004). Electrochemical technologies in wastewater treatment. *Separation and Purification Technology*, *38*(1), 11–41. doi:10.1016/j.seppur.2003.10.006

Cuervo-López, F., Martínez Hernández, S., Texier, A. C., & Gómez, J. (2009). Denitrification for Wastewater Treatment. In F. J. Cervantes (Ed.), *Environmental Technologies to Treat Nitrogen Pollution: Principles and Engineering* (pp. 23–65). London: IWA Publication.

Cuervo-López, F. M., Martínez-Hernández, S., Texier, A. C., & Gómez, J. (2009). Denitrification for wastewater treatment. In F. Cervantes (Ed.), *Environmental Technologies to treat nitrogen pollution. Principles and Engineering* (pp. 41–65). International Water Association Publishing.

Dong, X., & Tollner, E. W. (2003). Evaluation of anammox and denitrification during anaerobic digestion of poultry manure. *Bioresource Technology*, *86*(2), 139–145. doi:10.1016/S0960-8524(02)00154-2 PMID:12653278

Fajardo, C., Mora, M., Fernández, I., Mosquera-Corral, A., Campos, J. L., & Méndez, R. (2014). Cross effect of temperatura, pH and free ammonia on autotrophic denitrification process with sulphide as electron donor. *Chemosphere*, *97*, 10–15. doi:10.1016/j.chemosphere.2013.10.028 PMID:24216266

Francis, C. A., Beman, J. M., & Kuypers, M. M. M. (2007). New processes and players in the nitrogen cycle: The microbial ecology of anaerobic and archaeal ammonia oxidation. *The ISME Journal*, *1*(1), 19–27. doi:10.1038/ismej.2007.8 PMID:18043610

Frijlink, M. J., Abee, T., Laanbroek, H. J., de Boer, W., & Konings, W. N. (1992). The bioenergetics of ammonia and hydroxylamine oxidation in *Nitrosomonas europaea* at acid and alkaline pH. *Archives of Microbiology*, *157*(2), 194–199. doi:10.1007/BF00245290

Gevertz, D., Telanj, A. J., Voordouw, G., & Jenneman, G. E. (2000). Isolation and characterization of strains CVO and FWKO B, two novel nitrate reducing, sulfide-oxidizing bacteria isolated from oil field brine. *Applied and Environmental Microbiology*, *66*(6), 2491–2501. doi:10.1128/AEM.66.6.2491-2501.2000 PMID:10831429

Gilch, S., Vogel, M., Lorenz, M. W., Meyer, O., & Schmidt, I. (2009). Interaction of the mechanism-based inactivator acetylene with ammonia monooxygenase of Nitrosomonas europaea. *Microbiology*, *155*(1), 279–284. doi:10.1099/mic.0.023721-0 PMID:19118368

Ginestet, P., Audic, J. M., Urbain, V., & Block, J. C. (1998). Estimation of nitrifying bacterial activities by measuring oxygen uptake in the presence of the metabolic inhibitors allylthiourea and azide. *Applied and Environmental Microbiology*, *64*, 2266–2268. PMID:9603846

González-Blanco, G., Cuervo-López, F., Cervantes, F. J., Beristain-Cardoso, R., & Gómez, J. (2013). Nitrite as oxidizing power for p-cresol removal using a denitrifying consortium: Kinetic study. *Journal of Chemical Technology and Biotechnology (Oxford, Oxfordshire)*, *88*(12), 2176–2180. doi:10.1002/jctb.4082

González-Domenech, C. M., Béjar, V., Martínez-Checa, F., & Quesada, E. (2008). *Halomonas nitroreducens* sp. nov., a novel nitrate- and nitrite-reducing species. *International Journal of Systematic and Evolutionary Microbiology*, *58*(4), 872–876. doi:10.1099/ijs.0.65415-0 PMID:18398185

Guo, H., Chen, C., Lee, D. J., Wang, A., Gao, D., & Ren, N. (2014). Couple carbon, sulfur and nitrogen cycles of mixotrophic growth of *Pseudomonas* sp. C27 under denitrifying sulfide removal conditions. *Bioresource Technology*, *171*, 120–126. doi:10.1016/j.biortech.2014.08.035 PMID:25189517

Guo, Q., Xing, B.-S., Li, P., Xu, J.-L., Yang, C.-C., & Jin, R.-C. (2015). Anaerobic ammonium oxidation (anammox) under realistic seasonal temperature variations: Characteristics of biogranules and process performance. *Bioresource Technology*, *192*, 765–773. doi:10.1016/j.biortech.2015.06.049 PMID:26111630

Haroon, M., Hu, S., Shi, Y., Imelfort, M., Keller, J., Hugenholtz, P., & Tyson, G. W. et al. (2013). Anaerobic oxidation of methane coupled to nitrate reduction in a novel archael lineage. *Nature*, *500*(7464), 567–572. doi:10.1038/nature12375 PMID:23892779

Hatzenpichler, R. (2012). Diversity, Physiology, and Niche Differentiation of Ammonia-Oxidizing Archaea. *Applied and Environmental Microbiology*, *78*(21), 7501–7510. doi:10.1128/AEM.01960-12 PMID:22923400

He, S., Niu, Q., Ma, H., Zhang, Y., & Li, Y. Y. (2015). The Treatment Performance and the Bacteria Preservation of Anammox: A Review. *Water, Air, and Soil Pollution*, *226*(5), 163. doi:10.1007/s11270-015-2394-6

He, Z., Wang, J., Zhang, X., Cai, C., Geng, S., Zheng, P., & Hu, B. et al. (2015). Nitrogen removal from wastewater by anaerobic methane-driven denitrification in a lab-scale reactor: Heterotrophic denitrifiers associated with denitrifying methanotrophs. *Applied Microbiology and Biotechnology*, *99*(24), 10853–10860. doi:10.1007/s00253-015-6939-9 PMID:26342737

Hendrickx, T. L. G., Wang, Y., Kampman, C., Zeeman, G., Temmink, H., & Buisman, C. J. N. (2012). Autotrophic nitrogen removal from low strength waste water at low temperature. *Water Research*, *46*(7), 2187–2193. doi:10.1016/j.watres.2012.01.037 PMID:22349000

Islas-Lima, S., Thalasso, F., & Gómez, J. (2004). Evidence of anoxic methane oxidation coupled to denitrification. *Water Research*, *38*(1), 13–16. doi:10.1016/j.watres.2003.08.024 PMID:14630098

Karanasios, K. A., Vasiliadou, I. A., Pavlou, S., & Vayenas, D. V. (2010). Hydrogenotorophic denitrification of potable water: A review. *Journal of Hazardous Materials*, *180*(1-3), 20–37. doi:10.1016/j.jhazmat.2010.04.090 PMID:20471745

Kartal, B., de Almeida, N. M., Maalcke, W. J., Op den Camp, H. J. M., Jetten, M. S. M., & Keltjens, J. T. (2013). How to make a living from anaerobic ammonium oxidation. *FEMS Microbiology Reviews*, *37*(3), 428–461. doi:10.1111/1574-6976.12014 PMID:23210799

Kartal, B., Maalcke, W. J., de Almeida, N. M., Cirpus, I., Gloerich, J., Geerts, W., & Strous, M. et al. (2011). Molecular mechanism of anaerobic ammonium oxidation. *Nature*, *479*(7371), 127–130. doi:10.1038/nature10453 PMID:21964329

Kartal,, B., van Niftrik,, L., Rattray,, J., van de Vossenberg,, J.L., & Schmid,, M.C., Sinninghe Damst, E. J., … Strous, M. (2008). Candidatus 'Brocadia fulgida': An autofluorescent anaerobic ammonium oxidizing bacterium. *FEMS Microbiology Ecology*, *63*, 46–55. doi:10.1111/j.1574-6941.2007.00408.x PMID:18081590

Keener, W. K., & Arp, D. J. (1994). Transformations of aromatic compounds by *Nitrosomonas europea*. *Applied and Environmental Microbiology*, *60*, 1914–1920. PMID:16349282

Kellerman, C., & Griebler, C. (2009). *Thiobacillus tiophillus*, sp. Nov., a chemolitoautotrophic, thiosulfate-oxidizing bacterium isolated from contaminated aquifer sediments. *International Journal of Systematic and Evolutionary Microbiology*, *59*(3), 583–588. doi:10.1099/ijs.0.002808-0 PMID:19244446

Kelly, D. P., & Wood, A. P. (2000). Reclasification of some species of *Thiobacillus* to the newly designated genera *Acidithiobacillus* gen. nov. *Halothiobacillus* gen. nov. and *Thermithiobacillus* gen. nov. *International Journal of Systematic and Evolutionary Microbiology*, *50*(2), 511–516. doi:10.1099/00207713-50-2-511 PMID:10758854

Kim, J., Lim, J., & Lee, C. (2013). Quantitative real-time PCR approaches for microbial community studies in wastewater treatment systems: Applications and considerations. *Biotechnology Advances*, *31*(8), 1358–1373. doi:10.1016/j.biotechadv.2013.05.010 PMID:23747590

Knowles, R. (1982). Denitrification. *Microbiological Reviews*, *46*(1), 43–70. PMID:7045624

Knowles, R. (2005). Denitrifiers associated with methanotrophs and their potential impact on the nitrogen cycle. *Ecological Engineering*, *24*(5), 441–446. doi:10.1016/j.ecoleng.2005.01.001

Kodama, Y., & Watanabe, K. (2004). *Sulfuricurvum kujiense* gen. nov. sp. nov., a facultative anaerobic chelithoautotrophic, sulfur oxidyinz bacterium isolated from an underground crude-oil storage cavity. *International Journal of Systematic and Evolutionary Microbiology*, *54*(6), 2297–2300. doi:10.1099/ijs.0.63243-0 PMID:15545474

Koening, A., & Liu, L. H. (2004). Autotrophic denitrification of high-salinity wastewater using elemental sulfur: Batch tests. *Water Environment Research*, *76*(1), 37–46. doi:10.2175/106143004X141564 PMID:15058463

Koops, H. P. & Pommerening-Röser, A. (2001). Distribution and ecophysiology of the nitrifying bacteria emphasizing cultured species. *FEMS Microbiology Ecology*, *37*, 1–9. doi:10.1111/j.1574-6941.2001.tb00847.x

Kumar, M., & Lin, J. G. (2010). Co-existence of anammox and denitrification for simultaneous nitrogen and carbon removal—Strategies and issues. *Journal of Hazardous Materials*, *178*(1-3), 1–9. doi:10.1016/j.jhazmat.2010.01.077 PMID:20138428

Kurt, M., Dunn, J. I., & Bourne, R. J. (1987). Biological denitrification of drinking water using autotrophic organisms with $H_2$ in a fluidized-bed biofilm reactor. *Biotechnology and Bioengineering*, *29*(4), 493–501. doi:10.1002/bit.260290414 PMID:18576478

Kuypers, M. M. M., Sliekers, A. O., Lavik, G., Schmid, M., Jørgensen, B. B., Kuenen, J. G., & Jetten, M. S. M. et al. (2003). Anaerobic ammonium oxidation by anammox bacteria in the Black Sea. *Nature*, *422*(6932), 608–611. doi:10.1038/nature01472 PMID:12686999

Labrenz, M., Jost, G., Pohl, C., Beckmann, S., Martens-Habbena, W., & Jurgens, K. (2005). Impact of different in-vitro electron donor/acceptor conditions on potential chemolithoautotrophic communities from marine pelagic redoxiclines. *Applied and Environmental Microbiology*, *71*(11), 6664–6672. doi:10.1128/AEM.71.11.6664-6672.2005 PMID:16269695

Lackner, S., Gilbert, E. M., Vlaeminck, S. E., Joss, A., Horn, H., & Van Loosdrecht, M. C. M. (2014). Full-scale partial nitritation/anammox experiences - an application survey. *Water Research*, *55*, 292–303. doi:10.1016/j.watres.2014.02.032 PMID:24631878

Lalucat, J., Bennasar, A., Bosch, R., García-Valdés, E., & Palleroni, N. J. (2006). Biology of *Pseudomonas stutzeri*. *Microbiology and Molecular Biology Reviews*, *70*(2), 510–547. doi:10.1128/MMBR.00047-05 PMID:16760312

Lam, P., & Kuypers, M. M. M. (2011). Microbial nitrogen cycling processes in oxygen minimum zones. *Annual Review of Marine Science*, *3*(1), 317–345. doi:10.1146/annurev-marine-120709-142814 PMID:21329208

Laureni, M., Weissbrodt, D. G., Szivak, I., Robin, O., Nielsen, J. L., Morgenroth, E., & Joss, A. (2015). Activity and growth of anammox biomass on aerobically pre-treated municipal wastewater. *Water Research*, *80*, 325–336. doi:10.1016/j.watres.2015.04.026 PMID:26024830

Li, M., & Gu, J. D. (2011). Advances in methods for detection of anaerobic ammonium oxidizing (anammox) bacteria. *Applied Microbiology and Biotechnology*, *90*(4), 1241–1252. doi:10.1007/s00253-011-3230-6 PMID:21476137

Ma, B., Wang, S., Cao, S., Miao, Y., Jia, F., Du, R., & Peng, Y. (2016). Biological nitrogen removal from sewage via anammox: Recent advances. *Bioresource Technology*, *200*, 981–990. doi:10.1016/j.biortech.2015.10.074 PMID:26586538

Martínez-Gutiérrez, E., González-Márquez, H., Martínez-Hernández, S., Texier, A.-C., Cuervo-López, F. M., & Gómez, J. (2012). Effect of phenol and acetate addition on 2-chlorophenol consumption by a denitrifying sludge. *Environmental Technology*, *33*(12), 1375–1382. doi:10.1080/09593330.2011.627882 PMID:22856312

Martínez-Hernández, S., Olguín, E. J., Gómez, J., & Cuervo-López, F. M. (2009). Acetate enhances the specific consumption rate of toluene under denitrifying conditions. *Archives of Environmental Contamination and Toxicology*, *57*(4), 679–687. doi:10.1007/s00244-009-9321-z PMID:19387525

Meza-Escalante, E. R., Anne-Claire, T., Cuervo-López, F., Gómez, J., & Cervantes, F. (2008). Inhibition of sulfide on the simultaneous removal of nitrate and p-cresol by a denitrifying sludge. *Journal of Chemical Technology and Biotechnology (Oxford, Oxfordshire)*, *83*(3), 372–377. doi:10.1002/jctb.1823

Mills, H. J., Martínez, R. J., Story, S., & Sobecky, P. A. (2004). Identification of members of of the metabolically active microbial populations associated with *Beggiatoa* species mat communities from Gulf of Mexico cold-deep sediments. *Applied and Environmental Microbiology*, *7*(9), 5447–5458. doi:10.1128/AEM.70.9.5447-5458.2004 PMID:15345432

Modin, O., Fukushi, K., & Yamamoto, K. (2007). Denitrification with methane as external carbon source. *Water Research*, *41*(12), 2726–2738. doi:10.1016/j.watres.2007.02.053 PMID:17433401

Mulder, A., Vandegraaf, A. A., Robertson, L. A., & Kuenen, J. G. (1995). Anaerobic ammonium oxidation discovered in a denitrifying fluidized-bed reactor. *FEMS Microbiology Ecology*, *16*(3), 177–183. doi:10.1111/j.1574-6941.1995.tb00281.x

Park, S., & Bae, W. (2009). Modeling kinetics of ammonium oxidation and nitrite oxidation under simultaneous inhibition by free ammonia and free nitrous acid. *Process Biochemistry*, *44*(6), 631–640. doi:10.1016/j.procbio.2009.02.002

Peña-Calva, A., Olmos, D. A., Viniegra, G. G., Cuervo, L. F., & Gómez, J. (2004). Denitrification in presence of benzene, toluene, and *m*-xylene. *Applied Biochemistry and Biotechnology*, *119*(3), 195–208. doi:10.1007/s12010-004-0002-0 PMID:15591614

Prosser, J. I. (1989). Autotrophic nitrification in bacteria. *Advances in Microbial Physiology*, *30*, 125–181. doi:10.1016/S0065-2911(08)60112-5 PMID:2700538

Pynaert, K., Smets, B. F., Beheydt, D., & Verstraete, W. (2004). Start-up of autotrophic nitrogen removal reactors via sequential biocatalyst addition. *Environmental Science & Technology*, *38*(4), 1228–1235. doi:10.1021/es030081+ PMID:14998042

Raghoebarsing, A. A., Pol, A., van de Pas-Schoonen, K. T., Smolders, A. J. P., Ettwig, K. F., Rijpstra, W. I. C., & Strous, M. et al. (2006). A microbial consortium couples anaerobic methane oxidation to denitrification. *Nature*, *440*(7086), 918–921. doi:10.1038/nature04617 PMID:16612380

Rezania, B., Cicek, N., & Olezkiewics, J. A. (2005). Kintics of hydrogen-dependent denitrification under varying pH and temperature conditions. *Biotechnology and Bioengineering*, *92*(7), 900–906. doi:10.1002/bit.20664 PMID:16116656

Schmid, M. C., Maas, B., Dapena, A., van de Pas-Schoonen, K., van de Vossenberg, J., Kartal, B., & Strous, M. et al. (2005). Biomarkers for in situ detection of anaerobic ammonium-oxidizing (anammox) bacteria. *Applied and Environmental Microbiology*, *71*(4), 1677–1684. doi:10.1128/AEM.71.4.1677-1684.2005 PMID:15811989

Schmidt, I., & Bock, E. (1998). Anaerobic ammonia oxidation by cell-free extracts of *Nitrosomonas eutropha*. *Antonie van Leeuwenhoek*, *73*(3), 271–278. doi:10.1023/A:1001572121053 PMID:9801772

Shimamura, M., Nishiyama, T., Shinya, K., Kawahara, Y., Furukawa, K., & Fujii, T. (2008). Another multiheme protein, hydroxylamine oxidoreductase, abundantly produced in an anammox bacteriumbesides the hydrazine-oxidizing enzyme. *Journal of Bioscience and Bioengineering*, *105*(3), 243–248. doi:10.1263/jbb.105.243 PMID:18397776

Silva, C. D., Gómez, J., & Beristain-Cardoso, R. (2011). Simultaneous removal of 2-chlorophenol, phenol, p-creso and p-hydroxybenzaldehyde under nitrifying conditions: Kinetic study. *Bioresource Technology*, *102*(11), 6464–6468. doi:10.1016/j.biortech.2011.03.105 PMID:21504846

Silva, D. D., Cuervo-López, F. M., Gómez, J., & Texier, A.-C. (2011). Nitrite effect on ammonium and nitrite oxidizing processes in a nitrifying sludge. *World Journal of Microbiology & Biotechnology*, *27*(5), 1241–1245. doi:10.1007/s11274-010-0573-6

Smith, R. L., Ceazan, M. L., & Brooks, M. H. (1994). Autotrophic, hydrogen oxidizing, denitrifying bacteria in groundwater, potential agents for bioremediation of nitrate contamination. *Applied and Environmental Microbiology*, *60*(6), 949–955. PMID:16349284

Stein, L. Y., & Arp, D. J. (1998). Loss of ammonia monooxygenase activity in *Nitrosomonas europaea* upon exposure to nitrite. *Applied and Environmental Microbiology*, *64*, 4098–4102. PMID:9758853

Strous, M., Fuerst, J. A., Kramer, E. H., Logemann, S., Muyzer, G., van de Pas-Schoonen, K. T., & Jetten, M. S. M. et al. (1999). Missing lithotroph identified as new planctomycete. *Nature*, *400*(6743), 446–449. doi:10.1038/22749 PMID:10440372

Strous, M., Heijnen, J. J., Kuenen, J. G., & Jetten, M. S. M. (1998). The sequencing batch reactor as a powerful tool for the study of slowly growing anaerobic ammonium-oxidizing microorganisms. *Applied Microbiology and Biotechnology*, *50*(5), 589–596. doi:10.1007/s002530051340

Strous, M., Pelletier, E., Mangenot, S., Rattei, T., Lehner, A., Taylor, M. W., & Le Paslier, D. et al. (2006). Deciphering the evolution and metabolism of an anammox bacterium from a community genome. *Nature*, *440*(7085), 790–794. doi:10.1038/nature04647 PMID:16598256

Suarez, S., Lema, J. M., & Omil, F. (2010). Removal of pharmaceutical and personal care products (PPCPs) under nitrifying and denitrifying conditions. *Water Research*, *44*(10), 3214–3224. doi:10.1016/j.watres.2010.02.040 PMID:20338614

Tandukar, M., Pavlostathis, S. G., & Cervantes, F. J. (2009). Autotrophic denitrification for the removal of nitrogenous and sulphurous contaminants from wastewaters. In F. Cervantes (Ed.), *Environmental Technologies to treat nitrogen pollution. Principles and Engineering* (pp. 319–371). International Water Association Publishing.

Texier, A. C., & Gomez, J. (2002). Tolerance of nitrifying sludge to different concentrations of *p*-cresol. *Biotechnology Letters*, *40*(4), 321–332. doi:10.1023/A:1014057907953

Texier, A. C., & Gomez, J. (2007). Simultaneous nitrification and p-cresol oxidation in a nitrifying sequencing batch reactor. *Water Research*, *41*(2), 315–322. doi:10.1016/j.watres.2006.10.017 PMID:17126873

Texier, A. C., Gomez, J., & Cuervo-López, F. M. (2013). Inhibitory, toxic and structure effects of toluene on microbial consortia involved in wastewater treatment. In M. Palminteri (Ed.), *Toluene: Chemical Properties, Application and Toxicology* (pp. 93–124). Nova Science Publishers, Inc.

Tiedje, J. M. (1988). *Biology of anaerobic microorganisms* (A. J. B. Zehnder, Ed.). Wiley.

Trautwein, K., Kühner, S., Wöhlbrand, L., Halder, T., Kuchta, K., Steinbuchel, A., & Rabus, R. (2008). Solvent Stress Response of the Denitrifying Bacterium '' *Aromatoleum aromaticum*'' Strain EbN1. *Applied and Environmental Microbiology*, *74*(8), 2267–2274. doi:10.1128/AEM.02381-07 PMID:18263750

Treusch, A. H., Leininger, S., Kietzin, A., Schuster, S. C., Klenk, H. P., & Schleper, C. (2005). Novel genes for nitrite reductase and Amo-related proteins indicate a role of uncultivated mesophilic Crenarchaeota in nitrogen cycling. *Environmental Microbiology*, *7*(12), 1985–1995. doi:10.1111/j.1462-2920.2005.00906.x PMID:16309395

Vaiopoulou, E., Melidis, P., & Aivasidis, A. (2005). Sulfide removal in wastewater from petrochemical industries by autotrophic denitrification. *Water Research*, *39*(17), 4101–4109. doi:10.1016/j.watres.2005.07.022 PMID:16139866

Vajralaa, N., Martens-Habbenab, W., Sayavedra-Sotoa, L. A., Schauerc, A., Bottomleyd, P. J., Stahlb, D. A., & Arpa, D. J. (2013). Hydroxylamine as an intermediate in ammonia oxidation by globally abundant marine archaea. *Proceedings of the National Academy of Sciences of the United States of America*, *110*(3), 1006–1011. doi:10.1073/pnas.1214272110 PMID:23277575

Van de Graaf, A. A., de Bruijn, P., Robertson, L. A., Jetten, M. S. M., & Kuenen, J. G. (1996). Autotrophic growth of anaerobic ammonium-oxidizing micro-organisms in a fluidized bed reactor. *Microbiology*, *142*(8), 2187–2196. doi:10.1099/13500872-142-8-2187

Van de Vossenberg, J., Woebken, D., Maalcke, W. J., Wessels, H. J. C. T., Dutilh, B. E., Kartal, B., & Jetten, M. S. M. et al. (2013). The metagenome of the marine anammox bacterium 'Candidatus Scalindua profunda' illustrates the versatility of this globally important nitrogen cycle bacterium. *Environmental Microbiology*, *15*(5), 1275–1289. doi:10.1111/j.1462-2920.2012.02774.x PMID:22568606

Van Kessel, M. A. H. J., Speth, D. R., Albertsen, M., Per, H., Nielsen, P. H., Op den Camp, H. J. M., & Lücker, S. et al. (2015). Complete nitrification by a single microorganism. *Nature*, *528*, 555–559. PMID:26610025

Van Niftrik, L., & Jetten, M. S. M. (2012). Anaerobic ammonium oxidizing bacteria: Unique microorganisms with exceptional properties. *Microbiology and Molecular Biology Reviews*, *76*(3), 585–596. doi:10.1128/MMBR.05025-11 PMID:22933561

Vasiliadou, I. A., Pavlou, S., & Vallenas, D. V. (2006). A kinetic study of hydrogenotrophic denitrification. *Process Biochemistry*, *41*(6), 1401–1408. doi:10.1016/j.procbio.2006.02.002

Vázquez-Padín, J. R., Fernández, I., Mosquera-Corral, A., Campos, J. L., & Méndez, R. (2009). The anammox process. In F. J. Cervantes (Ed.), *Environmental technologies to treat nitrogen pollution* (pp. 67–91). London: IWA Publishing.

Verhamme, D. T., Prosser, J. I., & Nicol, G. N. (2011). Ammonia concentration determines differential growth of ammonia-oxidising archaea and bacteria in soil microcosms. *The ISME Journal*, *5*(6), 1067–1071. doi:10.1038/ismej.2010.191 PMID:21228892

Wang, H., He, Q., Chen, D., Wei, L., Zou, Z., Zhou, J., & Zhang, H. (2015). Microbial community in hydrogenotrophic denitrification reactor based on pyrosequencing. *Applied Microbiology and Biotechnology*, *99*(24), 10829–10837. doi:10.1007/s00253-015-6929-y PMID:26286515

Wang, S., Peng, Y., Ma, B., Wang, S., & Zhu, G. (2015). Anaerobic ammonium oxidation in traditional municipal wastewater treatment plants with low-strength ammonium loading: Widespread but overlooked. *Water Research*, *84*, 66–75. doi:10.1016/j.watres.2015.07.005 PMID:26210031

Winogradsky, S. (1890). Recherches sur les organismes de la nitrification. *Annales de l'Institut Pasteur*, *4*, 257–275.

Xu, G., Peng, J., Feng, C., Fang, F., Chen, S., Xu, Y., & Wang, X. (2015). Evaluation of simultaneous autotrophic and heterotrophic processes and bacterial community structure analysis. *Applied Microbiology and Biotechnology*, *99*(15), 6527–6536. doi:10.1007/s00253-015-6532-2 PMID:25825049

Zepeda, A., Texier, A. C., & Gómez, J. (2007). Batch nitrifying cultures in presence of mixtures of benzene, toluene, and *m*-xylene. *Environmental Technology*, *28*(3), 355–360. doi:10.1080/09593332808618796 PMID:17432387

Zhang, L., Zhang, C., Hu, C., Liu, H., Bai, Y., & Qu, J. (2015). Sulfur-based mixotrophic denitrification corresponding to different electron donors and microbial profiling in anoxic fluidized-bed membrane bioreactors. *Water Research*, *85*, 422–431. doi:10.1016/j.watres.2015.08.055 PMID:26364226

Zhao, Y., Zhang, B., Feng, C., Huang, F., Zhang, P., Zhang, Z., & Sugiura, N. et al. (2012). Behavior of autotrophic denitrification and heterotrophic denitrification in an intensified biofilm-electrode reactor for nitrate-contaminated drinking water treatment. *Bioresource Technology*, *107*, 159–165. doi:10.1016/j.biortech.2011.12.118 PMID:22244955

Zhu, G., Peng, Y., Li, B., Guo, J., Yang, Q., & Wang, S. (2008). Biological removal of nitrogen from wastewater. In D. Whitacre (Ed.), *Reviews of Environmental Contamination and Toxicology* (Vol. 192, pp. 159–195). New York: Springer. doi:10.1007/978-0-387-71724-1_5

Zumft, W. G. (1997). Cell biology and molecular basis of denitrification. *Microbiology and Molecular Biology Reviews*, *64*, 533–616. PMID:9409151

## KEY TERMS AND DEFINITIONS

**Ammonia Monoxygenase:** Enzyme that oxidizes ammonia to hydroxylamine.
**Anammox:** Anaerobic ammonium oxidation.
**Nitrification:** Microbial process to oxidize ammonia into nitrite or nitrate.
**Denitrification:** Microbial process to oxidize ammonia into nitrite or nitrate.
**Denitrification:** Microbial respiratory process in which nitrate serves as terminal electron acceptor.
**Hydroxylamine Oxidoreductase:** Enzyme that oxidize hydroxylamine to nitrite.
**Lithotrophic:** Microorganism that utilizes inorganic compounds as energy source.
**Organitrophic:** Microorganism that utilizes organic compounds as energy source.

# Chapter 6
# Autotrophic Denitrification Processes

**Marisol Belmonte**
*Pontificia Universidad Católica de Valparaíso-Universidad de Playa Ancha, Chile*

**Carmen Fajardo**
*Universidad Autónoma Metropolitana-Iztapalapa, Mexico*

**Javiera Belén Toledo-Alarcón**
*Institut National de la Recherche Agronomique, France*

**Daniel Valenzuela Heredia**
*University Adolfo Ibáñez, Chile*

**Lorena Jorquera**
*Pontificia Universidad Católica de Valparaíso, Chile*

**Ramón Méndez**
*University of Santiago de Compostela, Spain*

**Estela Tapia-Venegas**
*Pontificia Universidad Católica de Valparaíso, Chile*

**Gonzalo Ruiz-Filippi**
*Pontificia Universidad Católica de Valparaíso, Chile*

## ABSTRACT

*Effluents coming from anaerobic digesters are characterized by a COD/N ratio between 2 and 10, high ammonia $NH_4^+$ concentrations about 500 mg/L and a temperature range of 25-35 ºC. To remove nitrogen from these effluents biological processes as the autotrophic denitrification with sulfur compounds, hydrogen or methane can be applied. The main goal of this chapter is to describe and evaluate the use of these processes from an economic point of view. The methanotrophic denitrification is the cheapest alternative to remove nitrate from effluents with low COD/N ratios.*

## INTRODUCTION

Anaerobic digestion is the process most utilized for organic matter removal in industrial wastewater with high Chemical Organic Demand/Nitrogen (COD/N) ratio (> 10 g/g), such as coming from marine products industries, livestock industry, agri-food industry, aquaculture, among others.

This process achieves relative high removal efficiency (> 80%) of carbonaceous compounds of wastewater but has a low efficiency (< 20%) regarding the nitrogen removal.

DOI: 10.4018/978-1-5225-1037-6.ch006

Effluents coming from anaerobic digesters are characterized by a COD/N ratio between 2 and 10 g/g, high ammonium ($NH_4^+$) concentrations (about 500 mg/L) and a temperature range of 25-35 ºC. To remove nitrogen from these effluents, the biological processes based on the biogeochemical nitrogen cycle are the most commonly using. However, the application of this process depends on the COD/N ratio that is present in the wastewater (Campos *et al.*, 2010).

When the COD/N ratio of the wastewater is higher than 5, the combination of conventional nitrification (sequential $NH_4^+$ oxidation to nitrite ($NO_2^-$) and nitrate ($NO_3^-$)) and heterotrophic denitrification (nitrate or nitrite reduction to nitrogen gas ($N_2$)) processes are the most used as treatment (Ahn, 2006). While the COD/N ratio of the wastewater is lower than 5, the addition of an external carbon source is necessary to remove nitrogen by heterotrophic denitrification, being more expensive the treatment. Autotrophic denitrification is considered an advanced process to remove nitrogen in effluents with low COD/N ratio (Ahn, 2006) (Figure 1). This process is characterized by using inorganic compounds such as sulfur compounds, hydrogen ($H_2$) or methane ($CH_4$) (Equations 1, 2 and 3) as electron donors to remove nitrogen and inorganic carbon as a carbon source. Therefore, autotrophic denitrification could be an attractive and suitable alternative for the treatment of wastewater with low COD/N ratios as effluents coming from anaerobic digesters, in comparison with the heterotrophic denitrification.

$$5S^0 + 6NO_3^- + 2H_2O \rightarrow 3N_2 + 5SO_4^{2-} + 4H^+ \tag{1}$$

$$5H_2 + 2NO_3^- \rightarrow N_2 + 4H_2O + 2OH^- \tag{2}$$

$$5CH_4 + 8NO_3^- + 8H^+ \rightarrow 4N_2 + 5CO_2 + 14H_2O \tag{3}$$

*Figure 1. Partial cycle of nitrogen: nitrification and denitrification*

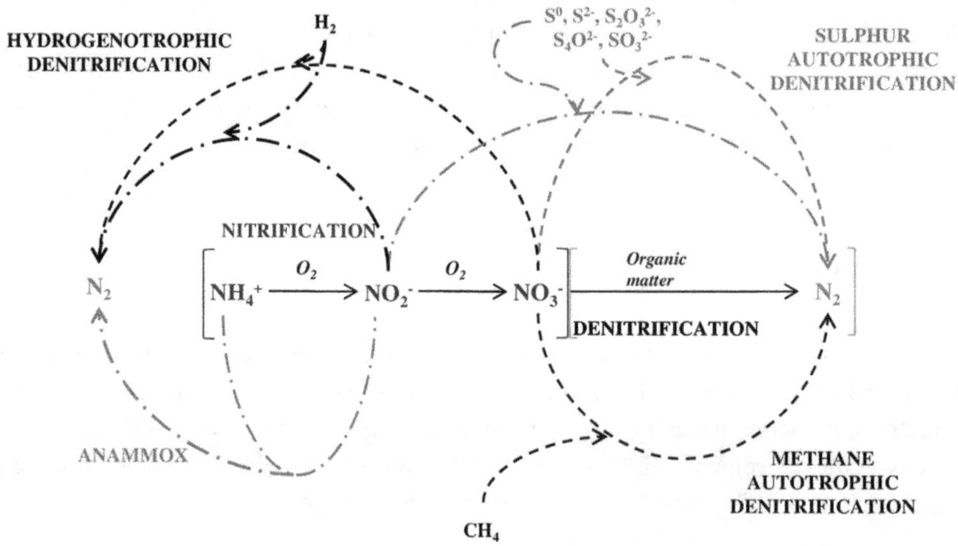

The anammox (Anaerobic Ammonium Oxidation) process is carried out by a group of autotrophic bacteria capable of oxidizing ammonia to nitrogen gas using nitrite as electron acceptor, without providing organic matter and oxygen (Equation 4). However, this chapter was not considered the anammox process, which is addressed in a specific chapter of this book.

$$NH_4^+ + 1.3NO_2^- + 0.066HCO_3^- + 0.13H^+ \rightarrow N_2 + 0.26NO_3^- + 0.066CH_2O_{0.5}N_{0.15} + 2H_2O \quad (4)$$

The goal of this present chapter is to describe and evaluate the use of autotrophic denitrification process mainly with sulfur compounds and hydrogen. These processes were finally to analyze under a technical and economic perspective for the treatment of industrial effluents such as coming from anaerobic digestion. In the next sections, both processes and technologies related to nitrogen removal using autotrophic denitrification are described and their possible applications are evaluated.

## AUTOTROPHIC DENITRIFICATION

The autotrophic denitrification is the biological process for the reduction of $NO_3^-$ and/or $NO_2^-$ to $N_2$, using inorganic compounds (sulfur compounds, $CH_4$ or $H_2$) as electron donors to remove nitrogen (See Equations 1, 2, 3, and Figure 1). When sulfur compounds, methane or hydrogen are used as electron donors to remove nitrogen, all ammonia contained in the anaerobic effluent should be previously oxidized into nitrate or nitrite by means of conventional nitrification or partial nitrification.

The use of inorganic electron donors can reduce operating costs of autotrophic denitrification over the use of organic compounds in heterotrophic denitrification. Taking into account the market prices of methanol (generally used as an external carbon source) and for example, elemental sulfur ($S^0$) and the stoichiometry of both processes, operational costs of autotrophic denitrification is around half of those of heterotrophic denitrification (Fajardo *et al.,* 2012).

The autotrophic denitrification with sulfur compounds (for example Thiosulfate ($S_2O_3^-$), Sulfide ($S^{-2}$) or elemental Sulfur ($S^0$)) has been commonly studied. The elemental sulfur used is of interest for bioremediation of groundwater, because its low-solubility prevents a possible secondary pollution (see Equation 1). When the elemental sulfur is used as electron donor the rate of the process is lower than does that obtained with thiosulfate or sulfite, due to the low solubility of $S^0$ (Sierra-Alvarez *et al.,* 2007).

The principal advantages of this process are the reduction of operational costs (with $S^0$: 0.3 US$/kg $NO_3^-$-$N_{removed}$) and the low production of sludge in comparison with conventional denitrification using organic compounds (with methanol: 1.8 – 3.6 US$/kg $NO_3^-$-$N_{removed}$) (Fajardo *et al.*, 2012; Park & Yoo, 2009).

The main drawback of using sulfur compounds as an electron donor could be related to the inhibitory effects of sulfide on the denitrification process. To avoid this inhibitory effect, the concentration of sulfide should be maintained at low levels. This could be achieved by applying specific sulfide loading rates lower than the specific sulfide removal rate of the biomass (Fajardo *et al.*, 2012) or maintaining an influent S/N ratio lower than the stoichiometric ratio. Another disadvantage of this process could be the secondary pollution, due to the formation of sulfate ($SO_4^{2-}$) at high concentrations, being necessary the control of nitrate concentration in the wastewater (< 80 mg $NO_3^-$-N/L), among others.

The autotrophic denitrification with methane or hydrogen (see Equations 2 and 3) are other inorganic electron donors that can be studied. Research conducted recently, indicated that the use of these compounds avoids the appearance of secondary pollution in comparison with sulfur autotrophic denitrification, allowing that the process can be a sustainable alternative for nitrate removal. Additionally, the costs of the autotrophic denitrification with $H_2$ or $CH_4$ as electron donor may be reduced if these were generated from waste anaerobic digestion process, since both compounds are produced under these conditions (Modin *et al.,* 2007; Rajapakse & Scutt, 1999; Waki *et al.*, 2009).

If an anaerobic treatment is used to remove organic matter from wastewater, the dissolved methane in the effluent is expected to be approximately 12-20 mg $CH_4$/L (equilibrium with biogas 65% $CH_4$). Since it has a strong greenhouse effect, it should be removed to avoid its stripping in the aerobic tank (nitrification). In this way, a three stage system consisting of an anaerobic digester, a denitrifying reactor and a nitrifying reactor could be applied in a predenitrifying configuration. Nitrate generated in the nitrifying reactor would be recycled to the denitrifying unit to carry out the anoxic oxidation of methane (Islas-Lima *et al.,* 2004; Modin *et al.,* 2007). According to the stoichiometry of the reaction (see Equation 2), 1.4 mg $NO_3^-$-N/mg $CH_4$ could be removed in this way (Thalasso *et al.*, 1997). As effluent from anaerobic sewage treatment plants contains 50 mg N/L, 35.7 mg $CH_4$/L are needed to remove all nitrate generated. This implies that a considerable fraction of the $CH_4$ obtained should be consumed during denitrification.

Also the autotrophic denitrification with hydrogen can be an interesting alternative, due to its high selectivity by nitrate and low sludge production, without secondary pollution (Karanasios *et al.,* 2010). Hydrogen gas chemical properties, such as high affinity for oxygen under certain environmental conditions, facilitate its high flammability and explosive potential. Therefore, caution is required when handling this gas. The hydrogen gas has a low water solubility of 1.6 mg/L at 20 °C (Vasiliadou *et al.*, 2006a). To improve the solubility of hydrogen in the wastewater the technologies used for denitrifying are based on hollow fiber membranes (Rezania *et al.*, 2006; Xia *et al.*, 2009; Zhang *et al.*, 2009). The environment factors affecting the autotrophic denitrification with hydrogen and the technologies employed for its application have been few studied.

These autotrophic denitrification processes are described below.

## SULFUR AUTOTROPHIC DENITRIFICATION

The sulfur autotrophic denitrification process is the reduction of $NO_3^-$ and/or $NO_2^-$ to $N_2$, with sulfur compounds ($S_2O_3^{2-}$; $S^{2-}$; S°; $S_4O_6^{2-}$; $SO_4^{2-}$) as electrons donors, $CO_2$ and $HCO_3^-$ as carbon source and by the action of sulfur oxidizing bacteria (SOB) (Fernández *et al.,* 2006; Koenig *et al.,* 2005).

### Factors Controlling

There are some basic operational parameters to consider during the application of autotrophic denitrification for the treatment of wastewaters containing sulfur and nitrogen compounds such as:

1. **pH:** As it is the case with all microorganisms, the autotrophic bacteria have an optimum pH range of operation. In the case of *Thiobacillus denitrificans* it is 7-8 (Claus & Kutzner, 1985; Oh *et al.*, 2000). In this range of pH values the end products of denitrification are $N_2$ and sulfate while at pH values below 7 the denitrification is incomplete and intermediate products such as nitrite and/or

elemental sulfur are detected. At pH values under 6 or over 9 the complete inhibition of denitrification is observed (Oh *et al.*, 2000; Moon *et al.*, 2004).

2. **Temperature:** Sulfur oxidizing bacteria have been found in mesophilic environments (25-35 °C), their optimum value of temperature being around 35 °C. When the temperature is higher than 40 °C (Oh *et al.*, 2000) or lower than 15 °C (Yamamoto-Ikemoto *et al.*, 2000), the autotrophic denitrification rate is negligible.
3. **Alkalinity:** A source of alkalinity as $HCO_3^-$ is necessary to neutralize the protons produced during the biological reaction and to provide a carbon source for bacterial growth. According to equations 5, 6 and 7, alkalinity (as $CaCO_3$) consumption per gram of $NO_3^-$-N removed (alkalinity consumption ratio) is 2.49, 0.93 and 4.57 when $S_2O_3^{2-}$, $H_2S$ and S°, respectively, are used.

$$\begin{aligned} &0.844S_2O_3^{2-} + NO_3^- + 0.347CO_2 + 0.086HCO_3^- + 0.086NH_4^+ + 0.434H_2O \\ &\quad \rightarrow 1.689SO_4^{2-} + 0.5N_2 + 0.086C_5H_7O_2N + 0.697H^+ \end{aligned} \tag{5}$$

$$\begin{aligned} &0.421H_2S + 0.421HS^- + NO_3^- + 0.346CO_2 + 0.086HCO_3^- + 0.086NH_4^+ \\ &\quad \rightarrow 0.842SO_4^{2-} + 0.5N_2 + 0.086C_5H_7O_2N + 0.434H_2O + 0.262H^+ \end{aligned} \tag{6}$$

$$\begin{aligned} &1.1S^0 + NO_3^- + 0.76H_2O + 0.4CO_2 + 0.08NH_4^+ \\ &\quad \rightarrow 1.1SO_4^{2-} + 0.5N_2 + 0.08C_5H_7O_2N + 1.28H^+ \end{aligned} \tag{7}$$

Many research works have been performed using calcite ($CaCO_3$), as alkalinity source for autotrophic denitrification, in packed reactors containing sulfur as electrons donor. In these cases, the alkalinity value ranged between 85–95 mg $CaCO_3$/L at pH values of 6.9-7.5 (Kim & Bae, 2000; Moon *et al.*, 2004).

4. **S/N Ratio:** It has been observed that the S/N ratio of the feeding plays an important role on the autotrophic denitrification. Oh et al. (2000) tested different S/N ratios with thiosulfate as electron donor and nitrate as electron acceptor and they found that at ratios below of 6.51 g S/g N (sulfur limitation) the denitrification was only carried out to nitrite. In the case of S/N ratios higher than the stoichiometric one, the sulfur compound is only oxidized to elemental sulfur (Gadekar *et al.*, 2006). The control of the inlet S/N ratio is required to drive the process to the products of interest.
5. **Sulfur Compounds:** The denitrification rate depends on the oxidation state of the sulfur compound present in the wastewater (Table 1). The oxidation rate of thiosulfate is around 212 mg $NO_3^-$-N/g VSS·d, which is ten times higher than that of 22 mg $NO_3^-$-N/g VSS·d reported elsewhere using elemental sulfur (Beristain *et al.*, 2006).
6. **Oxygen:** Oxygen and nitrate are electron acceptors for the oxidation of sulfur compounds. The oxidation of sulfur compounds in the presence of oxygen is thermodynamically more favored than the oxidation using nitrate. Therefore, its presence should be avoided. Several research works agree that the minimum concentration of dissolved oxygen which does not cause inhibition on autotrophic denitrification is between of 0.1-0.3 mg $O_2$/L. Above these concentrations denitrification is inhibited (Gu *et al.*, 2004; Kimura *et al.*, 2002; Sublette *et al.*, 1997).

*Table 1. Oxidation-reduction reactions with different sulfur compounds*

| Reaction | ΔG°' (kJ·mol⁻¹) | Reference |
|---|---|---|
| $1.25S^{2-} + 2NO_3^- + 2H^+ \rightarrow 1.25SO_4^{2-} + N_2 + H_2O$ | -972.8 | Reyes-Ávila *et al.* (2004) |
| $5S^0 + 6NO_3^- + 8H_2O \rightarrow 5H_2SO_4 + 3N_2 + 6OH^-$ | -1834.0 | Wang *et al.* (2005) |
| $S_2O_3^{2-} + 1.6NO_3^- + 0.2H_2O \rightarrow 2SO_4^{2-} + 0.8N_2 + 0.4H^+$ | -765.7 | Beristain *et al.* (2006) |

7. **Presence of Inhibitory Compounds:** It has been reported the inhibition of the denitrification by nitrate at concentrations of 660 mg $NO_3^-$ -N/L, while nitrite appears to be a strong inhibitor of the denitrification even at low concentrations (36-60 mg $NO_3^-$-N/L) (Oh *et al.*, 2002). The inhibition of denitrification by heavy metals, such as zinc (Zn) and copper (Cu) at concentrations of 0.5 and 1.0 mg/L, respectively, has also been reported (Claus & Kutzner, 1985; Krishnakumar & Manilal, 1999; Moon *et al.*, 2006, 2008; Oh *et al.*, 2000). The organic matter has no inhibitory effect on the process, but it affects the oxidation of sulfur species, decreasing the formation of sulfate (Kim & Son, 2000; Oh *et al.*, 2002). The sulfate is a product of the process and it has been reported to provoke partial inhibition at concentrations of 500 mg $SO_4^{-2}$-S/L and total activity depletion at 6400 mg $SO_4^{-2}$-S/L (Campos *et al.*, 2008; Claus & Kutzner, 1985).

## Applied Technologies

Autotrophic denitrification has been extensively studied in laboratory scale systems. Several kinds of reactors have been used such as reactors packed with sulfur or sulfur/limestone (Figure 2a), biofilm reactors, continuous stirred tank reactors (CSTR) (Figure 2b), UASB, Upflow anoxic filter (UAF) and activated sludge systems. In many cases, high removal efficiencies of both nitrate and sulfur species were obtained (Table 2).

The first reported research with autotrophic denitrification has been performed in a packed bed reactor with S° and S°/limestone known as Sulfur:limestone autotrophic denitrification reactor (SLAD) (Koenig & Liu, 2001; Kuai & Verstraete, 1999; Zhang & Lampe, 1999). In this system both sulfur and limestone are used as a carrier for biomass and limestone also provides alkalinity to maintain a suitable value of pH (Kim *et al.*, 2004; Zhang & Lampe, 1999). Nitrogen loading rates (NLR) up to 2000 mg $NO_3^-$-N/L·d can be treated by this kind of systems (Flere, 1999; Kim *et al.*, 2004; Koenig & Liu 1996).

However, the packed bed reactors can present clogging problems due to the excess of biomass which provokes the gas entrapment, limiting the mass transfer of nitrate from bulk liquid to biofilm (Flere, 1999; Kim *et al.*, 2004). On the other hand, the size of the elemental sulfur particles affects both nitrogen loading rate removed and nitrate removal efficiency. Therefore, it is recommended to use particles sizes between 2 and 5 mm to optimize the performance of the system (Koenig & Liu, 1996, 2001; Moon *et al.*, 2004, 2006, 2008; Sierra-Alvarez *et al.*, 2007).

Most of the studies carried out using thiosulfate were performed in batch reactors to obtain the kinetic parameters of the biomass (Campos *et al.*, 2008; Claus & Kutzner, 1985; Justin & Kelly, 1978; Oh *et*

*Figure 2. Configurations of denitrifying autotrophic reactors: a) packed-bed reactor (S°/limestone)) and b) CSTR*
*(a. Adapted from: Moon et al., 2004)*
*(b. Adapted from: Gadekar et al., 2006)*

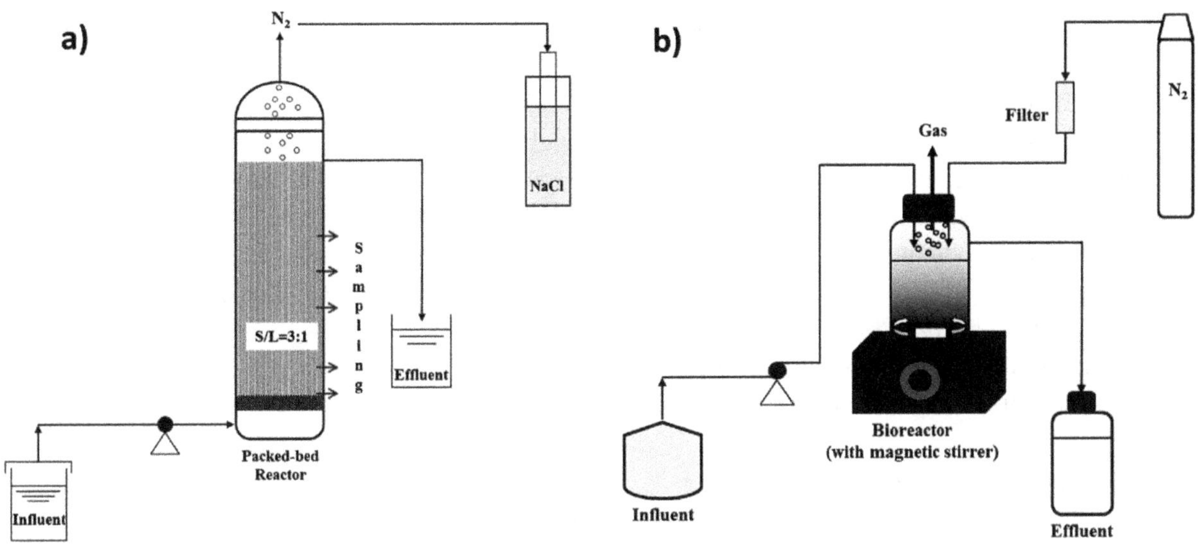


*Table 2. Sulfur autotrophic denitrifying reactors*

| Reactor | Bacterial population | Electron donor | Electron acceptor | Product | Removal efficiency | Reference |
|---|---|---|---|---|---|---|
| CSTR | *Thiomicrospira* sp. | $HS^-$ | $NO_3^-$ | $SO_4^{2-}$, $NO_2^-$ | 100% $HS^-$ | Gadekar *et al.* (2006) |
| CSTR | *T. denitrificans* | $S^{2-}$ | $NO_3^-$ | S° | 100% $S^{2-}$, 95% $NO_3^-$ | Wang *et al.* (2005) |
| Fed-Batch | Enriched sludge | $H_2S$, $S_2O_3^{2-}$, S° | $NO_3^-$ | $SO_4^{2-}$, $N_2$, $NO_2^-$ | 100% N and S | Beristain *et al.* (2006) |
| Packed bed reactor (S°) | Enriched sludge | S° | $NO_3^-$ | $SO_4^{2-}$ | 95% $NO_3^-$ | Soares (2002) |
| Membrane reactor | Enriched sludge | S° | $NO_3^-$ | $SO_4^{2-}$ | 80% $NO_3^-$ | Kimura *et al.* (2002) |
| Fluidized bed reactor (S°) | Enriched sludge | S° | $NO_3^-$ | $SO_4^{2-}$ | 90% $NO_3^-$ | Kim *et al.* (2004) |
| UAF | Enriched sludge | $HS^-$ | $NO_3^-$ | $SO_4^{2-}$ | 100% $HS^-$, $NO_3^-$ | Kleerebezem & Méndez (2002) |
| Packed bed reactor (S°) | *T. denitrificans* | S° | $NO_3^-$ | $SO_4^{2-}$, $H_2S$, $NO_2^-$ | 97% $NO_3^-$ | Gu *et al.* (2004) |
| Packed bed reactor (S°) | *T. denitrificans* | S° | $NO_3^-$ | $SO_4^{2-}$ | 100% $NO_3^-$ | Moon *et al.* (2004) |
| CSTR | Bioaugmented *T. denitrificans* | $HS^-$ | $NO_3^-$ | $SO_4^{2-}$ | 100% $HS^-$ | Manconi *et al.* (2007) |
| CSTR | Bioaugmented *T. denitrificans* | $S_2O_3^{2-}$ | $NO_3^-$ | $SO_4^{2-}$ | 95% $S_2O_3^{2-}$, 100% $NO_3^-$ | Manconi *et al.* (2007) |
| UASB | Methanogenic sludge | $HS^-$ | $NO_2^-$ | $SO_4^{2-}$,S°, $N_2$ | 98% $HS^-$, 82% $NO_2^-$ | Mahmood *et al.* (2008) |
| UAF | -- | $S_2O_3^{2-}$ | $NO_3^-$ | $SO_4^{2-}$,S°, $N_2$ | 90% $NO_3^-$ | Yamamoto-Ikemoto *et al.* (2000) |

*al.*, 2000). There are only a few works where continuous autotrophic denitrifying reactors were operated using this sulfur source. Yamamoto-Ikemoto *et al.* (2000) and Fernández *et al.* (2008) were able to treat up to 6.3 g $NO_3^-$- N/L·d with nitrogen removal efficiencies of 90% using an UASB system, while Manconi *et al.* (2007) treated a NLR of 0.49 g $NO_3^-$- N/L·d using an activated sludge system.

Autotrophic denitrification, using sulfide as sulfur source, was widely studied with different kinds of reactors: activated sludge reactors (Manconi *et al.*, 2006, 2007); UASB (Fernández *et al.*, 2008; Jing *et al.*, 2009; Mahmood *et al.*, 2007, 2008); CSTRs (Gadekar *et al.*, 2006; Reyes-Ávila *et al.*, 2004; Vaiopoulou *et al.*, 2005; Wang *et al.*, 2005); biofilm reactors (Kleerebezem & Méndez, 2002; Ma *et al.*, 2006); and sequential batch reactors (Pérez *et al.*, 2007). In these works the NLR applied ranged between 0.1 and 0.6 g $NO_3^-$- N/L·d while the SLR ranged between 0.3 – 3.25 g $S^{2-}$/L·d. Although high sulfide removal efficiencies can be obtained in these systems, their main disadvantage is the nitrite accumulation when high sulfide concentrations are tested which provokes poor nitrogen removal efficiencies (Mahmood *et al.*, 2007; Manconi *et al.*, 2007).

The anaerobic treatment of industrial effluents containing sulfate (canneries, petrochemical industries, tanneries, etc.) implies generation of sulfide. This compound presents some problems such as:

1. Odor problems and toxicity;
2. Decrease of organic matter removal efficiency and, therefore, less methane generated;
3. Corrosion problems;
4. The need for biogas conditioning and post-treatment of effluents.

A simple method to remove sulfide is autotrophic denitrification and can be applied by using the ANANOX (Anaerobic-Anoxic-Oxic) technology (Garuti *et al.*, 2001).

This technology is based on a two units configuration. The first unit is an anaerobic reactor with three compartments (2 anaerobic + 1 anoxic) containing flocculant sludge (Figure 3). The second unit is a conventional activated sludge system with a settler. In the first unit, anaerobic digestion of organic

*Figure 3. ANANOX process*

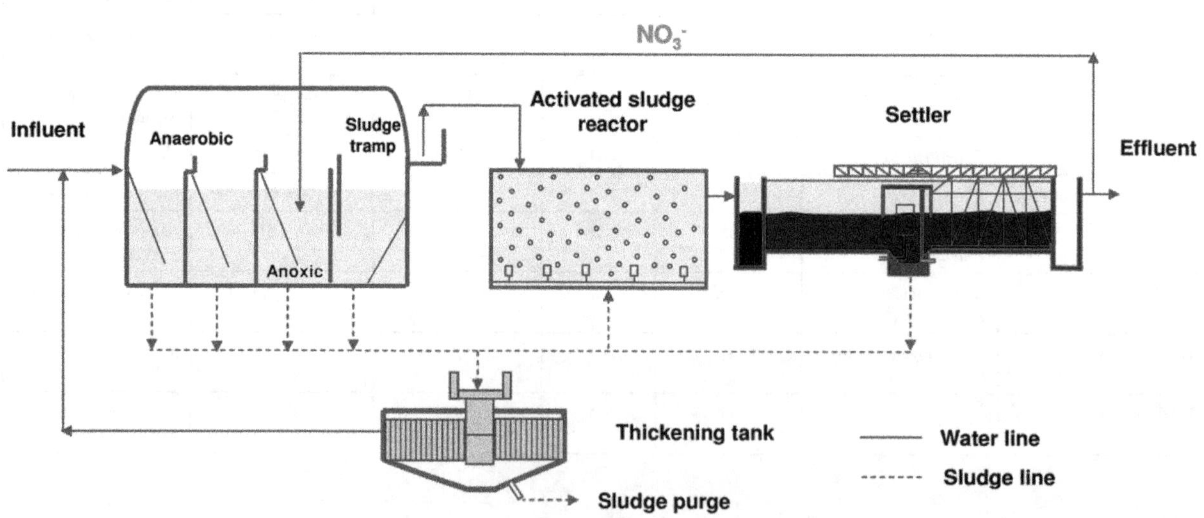

matter and sulfate reduction into sulfide are carried out. During autotrophic denitrification, sulfide is again oxidized into sulfate with the nitrate coming from the effluent recirculated.

Kleerebezem and Méndez (2002) also proposed the use of sulfide to remove nitrate during the post-treatment of fish cannery effluents treated by anaerobic digestion. The proposed configuration presented several advantages compared to the conventional configuration (Figure 4) such as:

1. $H_2S$ removal from biogas;
2. No requirement of recirculation between nitrifying and denitrifying reactors;
3. Autotrophic denitrification produces less sludge and
4. It is no necessary to control the amount of organic matter in the effluent of the anaerobic digester to carry out heterotrophic denitrification.

*Figure 4. a) Conventional treatment of fish cannery effluents and b) Treatment of fish cannery effluents by applying autotrophic denitrification*
*(a. Adapted from: Fajardo et al., 2014)*
*(b. Adapted from: Fajardo et al., 2014)*

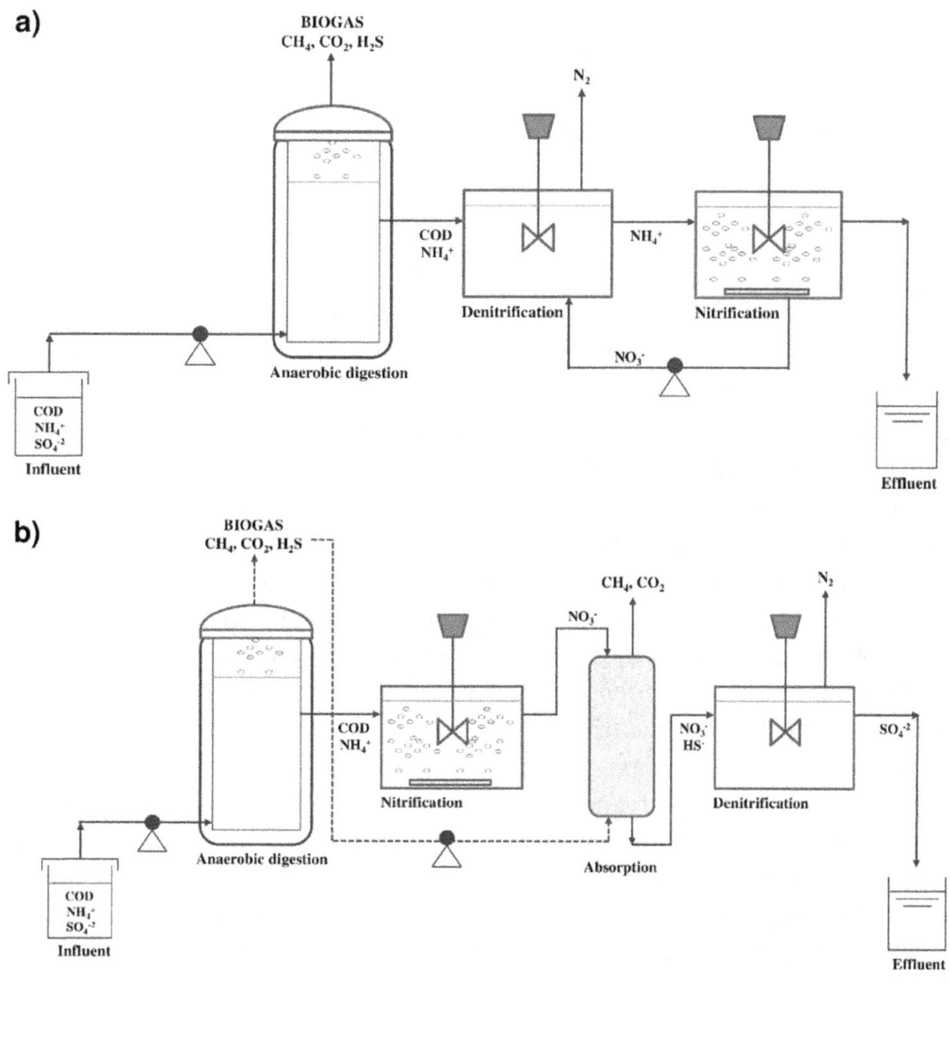

Vaiopoulou *et al.* (2005) applied the autotrophic denitrification process to remove both nitrate and sulfide from effluents of a petrochemical industry. According to the satisfactory obtained results during the operation of the pilot-scale plant, a new stage in the wastewater processing plant was introduced, reducing the $CO_2$ consumption by stripping of $H_2S$ from 2000 $m^3/h$ to 600 $m^3/h$ and obtaining considerable energy and financial savings.

## HYDROGEN AUTOTROPHIC DENITRIFICATION

The hydrogen autotrophic denitrification process or hydrogenotrophic denitrification is the reduction of $NO_3^-$ to $N_2$, where the hydrogen gas is used as an electron donor, while $CO_2$ and $HCO_3^-$ as a carbon source (Epsztein *et al.*, 2016; Karanasios *et al.*, 2010). This process, unlike of the sulfur autotrophic denitrification, has been applied mainly in the groundwater treatment and drinking water, and aquaculture wastewater in minor order. The application of this process at industrial scale for the wastewater treatment has been few studied, due to chemical properties of hydrogen gas (Karanasios *et al.*, 2010; Smith *et al.*, 1994; Vasiliadou *et al.*, 2006a).

### Factors Controlling

There are some parameters to consider during the application of hydrogenotrophic denitrification for the treatment of wastewaters containing nitrogen compounds such as:

1. **Nitrate Concentration:** The effects of nitrate concentration on hydrogenotrophic denitrification process have been little studied, found contradictory opinions among the various researchers. Park *et al.* (2005b) were determined that the rate of nitrate reduction increases by enhancing the initial nitrate concentrations between 20 and 492 mg $NO_3^-$-N/L in the wastewater. These authors also found a progressive nitrite accumulation when initial nitrate concentration is increased. On the other hand, Zhou *et al.* (2007) found that with a nitrate concentration of 10 mg $NO_3^-$-N/L the complete removal of nitrate can be obtained. While at concentrations of 30 mg $NO_3^-$-N/L is observed the accumulation of nitrite and the total inhibition of the hydrogenotrophic denitrification. However, this inhibition has not been clearly related to the presence of nitrite in the system. In that regard, Vasiliadou *et al.* (2006a) found that the hydrogenotrophic denitrification can be inhibited at nitrate concentrations of 40 mg $NO_3^-$-N/L, without observing nitrite accumulation in the system.
2. **Hydrogen Concentration:** Chang *et al.* (1999) reported that the critical limit for the concentration of hydrogen dissolved, so that the process has successful, is of 0.2 mg $H_2$/L. When the hydrogen concentration is less than 0.2 mg/L, the hydrogenotrophic denitrification is incomplete, increasing the nitrite concentrations in the system. This is because the reductases enzymes of nitrite and nitrate are inhibited to a hydrogen concentration of less than 0.2 and 0.1 mg $H_2$/L, respectively. Also, Karanasios *et al.* (2011) reported that can be achieved complete removal of nitrate with hydrogen concentrations between 0.4 and 0.8 mg $H_2$/L.
3. **pH:** The optimum range of pH for performing the hydrogenotrophic denitrification is between 7.6 and 8.6 (Ghafari *et al.*, 2009; Rezania *et al.*, 2005). Zhou *et al.* (2007) decreased this range to 7.5 and 7.6, considering the different cultures of hydrogenotrophic microorganisms and the variability of the operating conditions used in the systems. According to this author, when the pH is higher than 7.6, observed inhibition or accumulation of nitrite. Additionally, Rezania *et al.* (2005) indicated

that an increase of pH above 8.6 can cause the nitrite accumulation and a significant decrease in the rate removal of nitrate. In the hydrogenotrophic denitrification to maintain the optimum pH, the most feasible option is the application of carbon ($CO_2$) (Tang *et al.*, 2011).

4. **Temperature:** Several authors (Komori & Sakakibara, 2008; Rezania *et al.*, 2005; Vasiliadou *et al.*, 2009b; Zhou *et al.*, 2007) indicated that the increase of temperature down to 35 °C can have a positive effect in the process, favoring the increase in the rate of hydrogenotrophic denitrification (> 2 kg N-$NO_3^-$/m³·d). The hydrogenotrophic denitrification can occur at a temperature between 2 and 50 °C, because these microorganisms can survive in extreme environmental conditions (Brady & Weil, 2002). Zhou *et al.* (2007) suggest the operation of this process at temperatures above 25 °C to prevent the nitrite accumulation.

## Applied Technologies

Hydrogenotrophic denitrification process has been used for the treatment of wastewater and drinking water. The most applied technology to carry out this process is the biofilm reactor, this to retain the specific biomass, due to the low yield of the hydrogenotrophic denitrifying bacteria. The reactors most used in the process are: Packed Bed Reactor (PBR), Fluidized Bed Reactor (FBR), Membrane Biofilm Reactor (MBR), Glass Beds Reactor (GBR), and Sequential Batch Reactor (SBR), as it is shown in Table 3.

*Table 3. Hydrogenotrophic denitrification reactors*

| Reactor | Substrate | $NO_3^-$-N $_{In}$ (mg/L) | Rate denitrification (kg N/m³·d) | Removal efficiency $NO_3^-$-N (%) | HRT (h) | Reference |
|---|---|---|---|---|---|---|
| HFMBR | Synthetic | 10-12.5 | 0.23-0.37 | 75-92 | 0.7 | Lee & Rittmann (2000) |
| HFMB | Synthetic | 145 | 0.77 | 99 | 4.1 | Ergas & Reuss (2001) |
| HFMBR | Synthetic | 5-15 | 0.25-0.55 | 39-98 | 0.7 | Lee & Rittmann (2002) |
| -- | Synthetic | 20-40 | 2.70-5.30 (g N/m²·d) | 58-96 | -- | Mansell & Schroeder (2002) |
| HFMBR | Synthetic | 12-72 | 0.02-0.19 | 72-100 | 9.0-12.0 | Mo *et al.* (2005) |
| MBR | Wastewater | 33 | 0.14 | 100 | 3.0 | Rezania *et al.* (2006) |
| MBfR | Synthetic | 5-50 | 0.41 | >90 | 0.6-3.8 | Zhang *et al.* (2009) |
| MBfR | Synthetic | 2.5-20 | -- | -- | 4.7 | Xia *et al.* (2011) |
| HF MBfR | Synthetic | 30-150 | -- | 66 | 6.0 | Shin *et al.* (2005) |
| GBR | Synthetic | 20-150 | -- | -- | -- | Park *et al.* (2005a) |
| Discontinuous | Synthetic | 80 | 0.08 | -- | -- | Vasiliadou *et al.* (2006a) |
| MBfR | Synthetic | 17 | 0.93–1.20 (g/m²·d) | -- | 4.0 | Celmer *et al.* (2008) |
| SBR | Synthetic | 20 | 0.03 – 0.11 | -- | 6.0-15.0 | Ghafari *et al.* (2010) |
| SPEME | Synthetic | 21-95 | 2.16 | 50-60 | 1.0 | Kamori *et al.* (2008) |
| PBR | Synthetic | -- | 2.40 | -- | 2.0 | Lee *et al.* (2010) |
| MBfR | Synthetic | 40-50 | -- | >80 | 1.5–6.7 | Sahu *et al.* (2009) |
| MBfR | Synthetic | 100 | 6.58 (g/m²·d) | 99 | -- | Terada *et al.* (2006) |

(--): without information; SBR: Sequencing Batch Reactor; GBR: Glass Beads Reactor.

In PBRs, for carry out the hydrogenotrophic denitrification, have been used as support to retain the specific biomass both plastic materials (polyurethane, propylene) and inorganic materials (sand, ceramic, glass). Although in this type of reactor the mass internal transfer and external resistances are high, nitrate removal rates can be achieved with values around of 0.2 and 0.3 kg $NO_3^-$-N/m$^3$·d (Park *et al.*, 2005b). These values are higher than those obtained with suspended biomass systems (0.1 kg $NO_3^-$-N/m$^3$·d) (Ghafari *et al.*, 2010), due to the greater amount of retained biomass. The main limitations associated with the use of PBR system are:

1. The occlusions and flow lines that cause a decrease in the efficiency to remove nitrogen in the reactor, and
2. Uncontrolled detachment of biofilm generating secondary contamination in the treated water.

The use of FBR in hydrogenotrophic denitrification could avoid the problems associated with the use of PBR system (occlusions and flow lines). Although high rates of denitrification (0.6 to 2.2 kg $NO_3^-$-N/m$^3$·d) are achieved with FBR, it requires fluidizing the bed which means operate with high speeds of upflow, and with a short hydraulic retention time (HRT) (Komori & Sakakibara, 2008; Kurt *et al.*, 1987). According to Karanasios *et al.* (2010), this causes the loss efficiency of nitrate removal.

In the reactors described above, the hydrogen gas is introduced into the system by bubbling. This leads to a low rate of mass transfer in the reactor, due to the low solubility of hydrogen gas, and possible accumulation in closed headspace, which could create an explosive environment. To avoid these problems some researchers have proposed the use of membranes permeable to gas to improve the supply of hydrogen (McAdam & Judd, 2006; Terada *et al.*, 2006). These membranes are used both to diffuse hydrogen gas in the liquid phase, as give support for the biomass. Therefore, the selection of the membrane is a critical factor for the efficiency of this technology in the removal of nitrate by hydrogenotrophic denitrification. The gas permeable membranes are mainly composed of materials such as: polypropylene, polysulfone, silicone cured with platinum and silicone coated with iron-nickel (Ergas & Reuss, 2001; Shin *et al.*, 2005; Smith *et al.*, 2007). The advantage of membranes system in hydrogenotrophic denitrification is the reduction of the costs of hydrogen supply because it uses practically the 100% of $H_2$ supplied.

Also, the retention time is minimized due to the diffusion countercurrent, allowing high flows of nitrate and hydrogen in the system. The membranes most common used in the process are the hollow fiber. The applied technologies in hydrogenotrophic denitrification with these systems are: Membrane Biofilm Reactor hollow-fiber membrane (HFM), hollow-fiber membrane biofilm (HFMB), hollow-fiber membrane biofilm reactor (HFMBR or HFMBfR) (Ergas & Reuss, 2001; Lee *et al.,* 2000; Rezania *et al.,* 2006; Xia *et al,*. 2011).

These have a high specific surface area for the transfer of the hydrogen and biofilm formation, and can thus achieve high rates of nitrate removal (up to 23.5 kg $NO_3^-$-N/m$^3$·d) (Sahu *et al.,* 2009).

The efficiency of these systems to remove nitrogen can range from 39 to 100% depending on the operational conditions as: initial nitrate concentration, HRT, and others (See Table 3).

However the biofilm formed directly on the membranes is highly exposed to fouling. This fouling is caused by the production of exopolymers from the biofilm and by precipitation of minerals during the denitrification process. This is the main disadvantage in the operation of the membrane system. The progressive fouling membrane can reduce the capability hydrogen transfer. This provokes the limitation on the outermost part of the biofilm, causing the partial loss of the activity of denitrifying biomass and even their detachment (Celmer *et al.*, 2008). The fouling membrane can be mitigated by performing regular cleaning, which increases operating costs of the treatment (Tang *et al.*, 2012).

## METHANE AUTOTROPHIC DENITRIFICATION

The methane autotrophic denitrification or methanotrophic denitrification has been few studied. This process uses methane as a carbon source for denitrification, which is oxidized under aerobic or anaerobic conditions (Modin *et al.*, 2007, 2010). The aerobic methane oxidation (AME) has been the most known process (Rhee & Fuhs, 1978) and it is carried out mainly by bacteria, while that the anaerobic methane oxidation (ANME) process has been few studied (Islas-Lima *et al.*, 2004, Modin *et al.*, 2007). The rate of aerobic methane oxidation with coupling the denitrification can be up to two times higher in order of magnitude than under anaerobic conditions (Noroi & Thamdrup, 2014). Some authors indicate that the process ANME is carried out by a microbial consortium between bacteria and archaea (Ettwing *et al.*, 2010; Noroi & Thamdrup, 2014). While others authors suggested the existence of specific anaerobic bacteria, as *Candidatus Methylomirabilis oxyfera* present in natural environment (sediments, soil) (Ettwing *et al.*, 2010; Serrano-Silva *et al.*, 2014), or archaea, as *Candidatus Methanoperedens nitroreducens* (Cai *et al.*, 2015; Haroon *et al.*, 2013), that have been found recently. These microorganisms have been described as nitrite or nitrate dependent, respectively, with slow growth and low specific biomass activity (0.09-0.4 mol $CH_4$/g cell·d by *M. oxyfera*) (He *et al.*, 2015a), where methane is used as a donor electron and nitrite or nitrate as an acceptor of electrons under anaerobic conditions, process known as Nitrogen-Dependent Anaerobic Methane Oxidation (N-DAMO) (Cai *et al.*, 2015; Haroon *et al.*, 2013; Luesken *et al.*, 2011; Modín *et al.*, 2010; Waki *et al.*, 2009).

Despite this, it is known that in the anaerobic methane oxidation coupled to denitrification (ANME-D), the bacteria and/or archaea can perform first the methane oxidation to carbon dioxide, following of the reduction of nitrite (Equation 8) or nitrate (Equation 3) to $N_2$ (Modin *et al.*, 2007; Serrano-Silva *et al.*, 2014).

$$3CH_4 + 8NO_2^- + 8H^+ \rightarrow 3CO_2 + 4N_2 + 10H_2O(\Delta G: -928 \quad kJ/mol) \tag{8}$$

This process plays an important role in the natural control of the atmospheric emissions of methane from the oceans (2%) (Hu *et al.*, 2009; Luesken *et al.*, 2011). Because of this, the few existing studies of this process are focused on sediments or natural environment (Bhattacharjee *et al.*, 2016; Ettwing *et al.*, 2010).

### Factors Controlling

The factors controlling ANME-D process are poorly known, in contrast to the processes described above. Even so, the factors that may be important taking into account the natural environment in ANME-D microorganisms would be: the temperature, pH, salinity, C/N ratio, nitrite and medium (the presence of trace metals).

1. **Temperature:** The microorganisms responsible for the ANME-D (bacteria or archaea) process have a very slow growth (for example methane oxidizing bacteria doubling time is 14-25 days) (He *et al.*, 2016), due to its low biomass productivity (Bhattacharjee *et al.*, 2016). This affects the total denitrification rate. The temperature is an important factor in the efficiency of the process (He *et al.*, 2015a; Hu *et al.*, 2009; Rajapakse & Scutt, 1999; Serrano-Silva *et al.*, 2014; Shen *et al.*, 2012). Rajapakse and Scutt (1999) indicate that specific growth rate of methanotrophic denitrifying is reduced by 40%, due to the decrease of the temperature from 35 ° to 20 ºC. He *et al.* (2015a) observed that in the biomass enriched by *M. oxyfera* on anaerobic methane oxidation coupled to nitrite reduction, the specific activity at temperatures between 15 and 35 ° C have exponential growth. When the temperature is higher (> 35 ºC), the activity decreases dramatically by 88% (He *et al.*, 2015a). Several authors (Cai *et al.*, 2015; Hu *et al.*, 2009; Islas-Lima *et al.*, 2004) indicate that the optimum temperature range of the process is between 30-35 ºC. While this process could also be at ambient temperature (Shen *et al.*, 2012), reducing the costs of the treatment wastewater.
2. **pH:** Regarding pH, the research works indicated that the process is performed between 7-8 (Cai *et al.*, 2015; Islas-Lima *et al.*, 2004; Serrano-Silva *et al.*, 2014). He *et al.*, (2015a) studied the effect of pH (6 – 9) on N-DAMO specific activity. According to He *et al.* (2015a) this specific activity can be increasing at low pH (< 7.6 optimal value). When the pH of the system is greater than 7.6 the activity begins to decrease, noting a reduction of up to a 50% at pH 9. In ANME-D the controlling of pH is important both in the cultivation or enrichment of the microorganism (He *et al.*, 2016).
3. **Salinity:** The inoculum used to carry out different research works of the ANME-D process, comes mainly from sediments (marine or freshwater), where the oxidation of methane has an important role (Islas-Lima *et al.*, 2004; Raghoebarsing *et al.*, 2006; Waki *et al.*, 2009). This would indicate that some microorganisms (mainly bacteria) depending on their natural condition and origin could tolerate high salt concentrations (Serrano-Silva *et al.*, 2014). Some authors (He *et al.*, 2015a; Serrano-Silva *et al.*, 2014) assessed the effect of the increase in the salt concentration in the activity of the microorganisms using freshwater sediments, where *M. oxyfera* is the dominant organism (60%). He *et al.* (2015a) observed that to increase the salt concentrations from 0 to 15 g NaCl/L in the medium, the specific activity decreases. Since the increased cell osmotic pressure by high salinity would be responsible for the inhibition of microorganisms (He *et al.*, 2015a). Despite this, the microorganisms would be able to adapt at high salt concentrations recovering its activity after 90 days of incubation (He *et al.*, 2015a). This would indicate that the process has an origin in saline environments.
4. **C/N Ratio:** The C/N ratio in the process would have an effect on the selection of microorganisms from the inoculum (Islas-Lima *et al.*, 2004) and in the efficiency of the system applied as treatment. The main disadvantage of the use of methane as a carbon source in autotrophic denitrifying processes is its low solubility, as well as hydrogen in hydrogenotrophic denitrification process (Bhattacharjee *et al.*, 2016; Islas-Lima *et al.*, 2004). Generally, the process is applied to wastewaters with low C/N ratio, as the effluents from the anaerobic digestion (Islas-Lima *et al.*, 2004). According to the stoichiometry of the process, the $CH_4$/N ratio could be 0.6 or 0.4 depending on the acceptor of electrons used (nitrate or nitrite, respectively) (See Equations 3 and 8) (Islas-Lima *et al.*, 2004; Raghoebarsing *et al.*, 2006). With this, some authors have estimated the theoretical denitrification rate between 0.2-2.5 mg N/g VSS·h (Islas-Lima *et al.*, 2004).
5. **Nitrite:** Some authors indicate that the nitrite accumulation in the system would inhibit at microorganisms ANME-D, although of the metabolic mechanism is unknown (Cui *et al.*, 2015). While

others indicate that the presence of nitrite has an effect on the selection of predominant microorganisms (bacteria or archaea) in the ANME-D process (Cai *et al.,* 2015; Haroon *et al.,* 2013; Hu *et al.,* 2011). However, recent research on bioreactors would indicate that there is an interaction between the populations of microorganisms (bacteria-archaea) in systems fed with methane (Haroon *et al.,* 2013). *M. nitroreducens* (ANME archaea) is capable of reducing the nitrate to nitrite, and the latter is reduced to nitrogen gas by *M. oxyfera* (NC10 group Bacteria) both using the methane as a carbon source (Cui *et al.,* 2015; Haroon *et al.,* 2013) (See Figure 5). Cui *et al.* (2015) observed that *M. oxyfera* prefer to utilize nitrite as a substrate in comparison with nitrate. However, high contents of nitrite (> 1.4 mg $NO_2^-$-N/L) could cause the inhibition of *M. oxyfera* (Hu *et al.,* 2011).

6. **Medium: Presence of Trace Metals:** The presence of trace metals as nutrients in the process would have an important effect on the growth of organisms (Bhattacharjee *et al.,* 2016). He *et al.* (2015b) observed that the presence of trace metals as iron (Fe (II): 20 µmol/L) and copper (Cu (II): 10 µmol/L) have a significant effect on promoting the biomass activity and the growth of *Candidatus Methylomirabilis oxyfera.* The presence of copper activates the enzymes responsible for the oxidation of methane, while iron enzymes act on the reduction of nitrite (He *et al.,* 2015b). Copper concentrations over 25 µmol/L can inhibit this process (He *et al.*, 2015b).

## Applied Technologies

The implementation of this process has focused until now on scale laboratory. In wastewater treatment the methane could be considered as a relatively inexpensive electron donor carbon source to denitrification due favorable thermodynamics conditions (Bannert *et al.*, 2012). Experimental conversion rates of methane and nitrite by *M. oxyfera* have a stoichiometry methane to nitrite ratio of $3CH_4$:$8NO_2^-$. On the other hand, ANME-D can be chosen as a competitive process in biological denitrification for wastewa-

*Figure 5. Microbial populations A) Co- culture Archaea-Bacteria and B) predominant bacteria population*

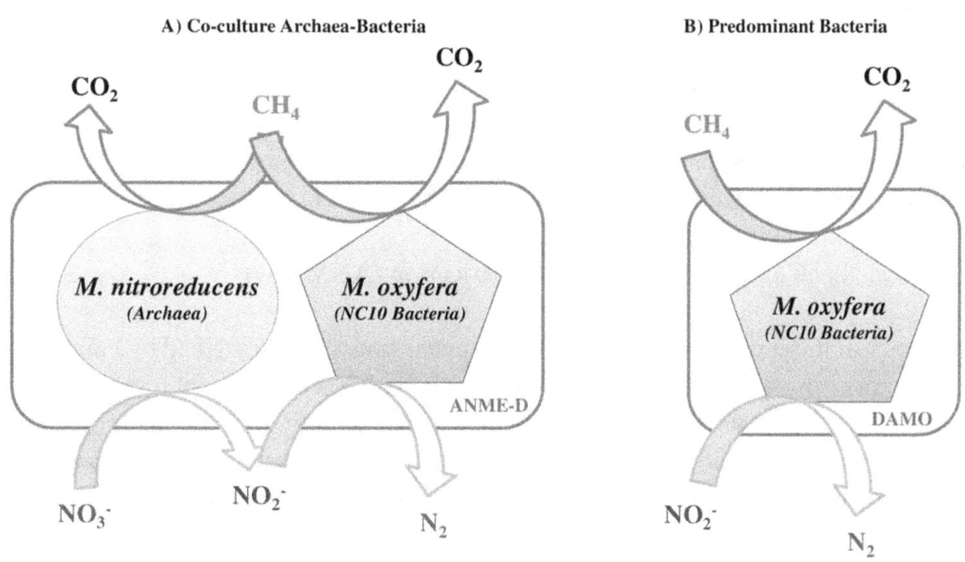

ter treatment for both mainstream and sidestream nitrogen removal (Shen *et al.,* 2012). Regarding the specificity of the methane elimination or nitrite/nitrate elimination of the system, is necessary known if N-DAMO microorganisms survive by methane oxidation or by N-reduction (e.g for *M. oxyfera*, both, $NO_2^-$ and $CH_4$ are essential for their survivor). This is useful to establish the reaction capacities of N-DAMO biotechnologies and to design strategies to reach a high rate of reactions to remove all contaminants presents. This process would be used for the nitrogen removal from drinking water, wastewater (anaerobically treated wastewater, aquaculture, and municipal), groundwater or landfill leachate (Modin *et al.*, 2007; Waki *et al.,* 2009).

In the case of anaerobically treated wastewater that has a high concentration of ammonium, it must combine with partial or conventional nitrification. Recent research works indicate that the ANME-D process can be applied with co-cultures anammox, carrying out simultaneous nitrate, nitrite and ammonium removal (Haroon *et al.,* 2013). The low solubility of methane as well as the slow growth of microorganisms that perform ANME-D can be limiting their application at greater scale, since it requires technologies that will enable to assist in this process (Hu *et al.,* 2009, 2011). It is the case of SBR systems and membrane reactor that are the most studied (See Table 4). Cai *et al.* (2015) indicate that the MBR systems favor the growth of the bacteria-archaea ANME-D consortium in contrast to SBR systems, especially in the enrichment of the biomass (> 8 months onset of apparent activity) (Hu *et al.,* 2009). The above due to hollow fiber membranes are employed to improve the diffusion of methane and surface for the development of microorganisms. The operation of these systems require higher HRT (> 25 d) (Cai *et al.,* 2015). On the other hand, the higher rate of nitrogen removal is achieved in MBR system (0.68 kg $N/m^3 \cdot d$) used as inoculum co-culture of anammox and ANME-D (bacteria-archaea) (Cai *et al.*, 2015). More studies are required that permit to know the metabolisms of the process, as well as the factors affecting for their application to industrial level for different wastewaters.

## COMPARISON OF NITRATE REMOVAL PROCESS

The implementation of different advanced processes for biological nitrogen removal from industrial effluents with low COD/N (< 5), previously oxidizing ammonia via nitrification, were studied from an economic perspective. The processes evaluated are:

1. Heterotrophic denitrification with methanol (D+MeOH) as organic carbon source;
2. Sulfur autotrophic denitrification using elemental sulfur ($S^0D$) as electron donor;
3. Hydrogenotrophic denitrification ($H_2D$); and
4. Anaerobic methane oxidation coupling denitrification (ANME-D).

Operating costs of these processes were estimated according to the described by Fajardo *et al.* (2013), having as key factors: consumption energy due to aeration required to carry out previous the nitrification process, the costs associated with the electron donor source, and the disposal of sludge produced. This estimation was based on the removal of a 1 kg of nitrogen in wastewater. To estimate the costs associated with aeration, was used as average the consumption of power 1 kW h/kg $O_2$ transferred in the nitrification process, and an electricity price 0.160 US$/kW h. For the processes of heterotrophic denitrification with methanol and sulfur autotrophic denitrification, was estimated the amount required for each electrons donor external means stoichiometric reactions with nitrite. Stoichiometric reactions was obtained

*Table 4. Anaerobic methane oxidation: denitrification reactors*

| Reactor | Substrate | Inoculum | $NO_3^-$-$N_{in}$ (mg/L) | Rate denitrification (kg N/m³·d) | Removal efficiency $NO_3^-$-N (%) | Temperature (°C) | pH | Reference |
|---|---|---|---|---|---|---|---|---|
| Batch Fixed bed (sand bed) with support biomass | Synthetic | *Methanotrophic* mixed culture | 30 | 0.260 | 93(1) | 14.5-24.0 | -- | Rajapakse & Scutt (1999) |
| Batch assays | Synthetic | Mixed culture | 35(2) | 0.23 to 4.9x10⁻³ (g $NO_3^-$-N/g VSS·d)(3) | >95 | 30 | -- | Islas-Lima *et al.* (2004) |
| Gravel-Packed columns | Synthetic | Anaerobic culture | 30-35 | 0.02 0.03(4) | 90 | -- | -- | Sollo *et al.* (1976) |
| Sequencing batch reactor (SBR) | Synthetic | Anaerobic freshwater sediment (enrichment culture Bacterial and Archaea) | -- | 0.06 (g $NO_3^-$-N/g VSS·d) | -- | 25 | -- | Raghoebarsing *et al.* (2006) |
| SBR | Synthetic | Mixed culture (freshwater sediment, digester sludge, activated sludge) | -- | 0.07 (mmol $NO_3^-$-N /L·d) | -- | 22 | -- | Hu *et al.* (2009) |
| SBR | Synthetic | Mixed culture (freshwater sediment, digester sludge, activated sludge) | -- | 2.0 (mmol $NO_3^-$-N /L·d) | -- | 35 | -- | Hu *et al.* (2009) |
| MBfR | Synthetic | Co-culture DAMO-anammox | 600-1000 | 0.68(5)(6) | >95 | 22 | 7-8 | Cai *et al.* (2015) |
| SBR | Synthetic | n-damo culture enriched from paddy soil | 2 mmol/L | 0.69–6.42 (μmol/L·h) | | 15-45 | 6-9 | He *et al.* (2015a) |
| I-SBR(7) | Synthetic | River sediment enrichment | 7(8) | 2.88 (mmol $NO_2^-$-N /L·d) | | 35 | 7.5 | Bhattacharjee *et al.* (2016) |

(1): 0.6 m/h water filtration rates; (2): methane partial pressures ranging from 1.8 to 35.7 kPa; (3): methane partial pressure > 8.8 kPa; (4): value obtained in an activated carbon fluidized bed reactor; (5): nitrate removal rate using DAMO process by archaea; (6): HRT: 3 to 1.5 d; (7): Intermittently-fed SBR, with cycle 6 day to HRT of 29 days; (8): $NO_2^-$-N/$NO_3^-$-N ratio influent: 2.28.

values of: 1.4 kg $Methanol_{consumed}$/kg $NO_2^-$-N and 1.5 kg $S^0_{consumed}$/kg $NO_2^-$-N. The costs associated with the amount substrate were determined using as prices reference 0.73 US$/kg Methanol and 0.11 US$/kg $S^0$ (Fajardo *et al.,* 2013). The amount of sludge produced in each process was estimated from the kinetic parameters (specific productivity of microorganisms) which are shown in Table 5. Costs related to the disposal of sludge were estimated assuming a VSS/TSS ratio in the sludge of 0.8 g/g (Ekama *et al.,* 2004) and a cost per available of 272 US$/Ton TSS.

The results of efficiency to nitrogen removal and operational costs obtained for each biological process are shown in Table 6.

*Table 5. Specific productivity of microorganisms associated to nitrogen removal*

| Microorganisms | Specific Productivity | Reference |
|---|---|---|
| Ammonia oxidizing bacteria | 0.12 g VSS/g $NH_4^+$ | Wiesmann (1994) |
| Heterotrophic denitrifying bacteria (using methanol) | 0.60 g VSS/g $NO_3^-$ | Foglar & Briski (2003) |
| Sulfur oxidizing bacteria (using S°) | 0.30 g VSS/g $NO_2^-$ | Fajardo *et al.* (2013) |
| Methanotrophic denitrifying bacteria | 0.01-0.13 g VSS/g $NO_2^-$ | Kampman *et al.* (2012) |
| Hydrogenotrophic denitrifying bacteria | 0.22 g VSS/g $NO_3^-$ | Karanasios *et al.*(2010) |

*Table 6. Operational costs of each process (reference: 1 kg Nitrogen$_{removed}$)*

| Process | Efficiency Removal Nitrogen (%) | Operational Costs (US$) | | | |
|---|---|---|---|---|---|
| | | Aeration | Substrate | Sludge | Total Costs |
| Nitri-D+MeOH | 20-45(1) | 0.540 | 1.010 | 0.163 | 1.713 |
| Nitri-S°D | 13-33(1) | 0.540 | 0.163 | 0.150 | 0.853 |
| Nitri-ANME-D | 70-80(2) | 0.540 | -- | 0.003-0.044 | 0.543-0.584 |
| Nitri-$H_2$D | 39-100(3) | 0.540 | -- | 0.075 | 0.615 |

*(1) Fajardo et al., 2013; (2) Kampman et al., 2012; (3) Lee & Rittmann, 2002*

According to Table 5, with the combination of the processes of nitrification-sulfur autotrophic denitrification with S° a reduction in the operating costs of about 50% is observed compared with the previous process, although the nitrogen removal efficiency is low (<33%). On the other hand, it is necessary in the process to control the formartion of sulfate to avoid secondary pollution. However, the combinations of nitrification-autotrophic denitrification using methane or hydrogen (from anaerobic digestion) as electron donor sources, allow reducing the costs related to the substrate and generating removal efficiencies above 70-80%.

## CONCLUSION

Autotrophic denitrification with sulfur is the process most know and applied for the treatment of wastewater with low C/N ratio. The production of sulfates in the process is the key factor to avoid a possible secondary pollution. The use of $H_2$ or $CH_4$ as electron donors in the autotrophic denitrification would be more sustainable alternatives nowadays to remove the nitrogen in wastewater with low C/N ratio. The application to scale industry depends on knowing what factors are those who control these processes. This is in relation to favor the growth of specific microorganisms (slowly growth) and to improve the transfer and diffusion of the $H_2$ or $CH_4$, this due to its low solubility. This trend shows that the membrane reactors would be the most suitable alternative, at leads in laboratory research.

Alternatives efficient and cost-effectives to nitrogen removal coming from effluent anaerobic digestion are the combination of nitrification with ANME-D following of $H_2$D.

The high operational cost is related to the consumption of electricity due to aeration during oxidation of ammonium to nitrite or nitrate (nitrification) as a requirement for the autotrophic denitrification in effluent anaerobic.

## ACKNOWLEDGMENT

The authors thank to: FONDECYT 3140276 of the Chilean Government; PAI-MEC-CONICYT (Folio n°80140011); Fondo de Desarrollo Disciplinario-UPLA (18105); PROMEP (UAM-EXB-128) of Secretaria de Educación Pública of México; GRAIL 613667 (FP7-KBBE-2013), Galician Competitive Research Group (GRC2013-032), CRETUS Strategic Partnership (AGRUP2015/02), and Spanish Red_Novedar: CTQ2014-51693-REDC.

This chapter is especially dedicated to Gonzalo Ruiz-Filippi (1972-2016), for his contribution to the research and development of Science in Chile, both your *alma mater* the Pontificia Universidad Católica de Valparaíso (Chile) and the University of Santiago de Compostela (Spain). Thank you for everything!

## REFERENCES

Ahn, Y. H. (2006). Sustainable nitrogen elimination biotechnologies: A review. *Process Biochemistry*, *41*(8), 1709–1721. doi:10.1016/j.procbio.2006.03.033

Bannert, A., Bogen, C., Esperschütz, J., Koubová, A., Buegger, F., Fischer, D., & Schloter, M. et al. (2012). Anaerobic oxidation of methane in grassland soils used for cattle husbandry. *Biogeosciences*, *9*(10), 3891–3899. doi:10.5194/bg-9-3891-2012

Beristain, C. R., Sierra-Alvarez, R., Rowlette, P., Razo-Flores, E., Gómez, J., & Field, J. (2006). Sulfide Oxidation Under Chemolitoautotrophic Denitrifying Conditions. *Biotechnology and Bioengineering*, *95*(6), 1148–1156. doi:10.1002/bit.21084 PMID:16807929

Bhattacharjee, A. S., Motlagh, A. M., Jetten, M. S. M., & Goel, R. (2016). Methane dependent denitrification- from ecosystem to laboratory-scale enrichment for engineering applications. *Water Research*, *99*, 244–252. doi:10.1016/j.watres.2016.04.070 PMID:27176548

Brady, N. C., & Weil, R. R. (2002). *The Nature and Properties of Soils* (13th ed.). Prentice Hall.

Cai, C., Hu, S., Guo, J., Shi, Y., Xie, G.-J., & Yuan, Z. (2015). Nitrate reduction by denitrifying anaerobic methane oxidizing microorganisms can reach a practically useful rate. *Water Research*, *87*, 211–217. doi:10.1016/j.watres.2015.09.026 PMID:26414889

Campos, J. L., Carvalho, S., Portela, R., Mosquera-Corral, A., & Méndez, R. (2008). Kinetics of denitrification using sulphur compounds: Effects of S/N ratio, endogenous and exogenous compounds. *Bioresource Technology*, *99*(5), 1293–1299. doi:10.1016/j.biortech.2007.02.007 PMID:17399983

Campos, J. L., Vázquez-Padín, J. R., Figueroa, M., Fajardo, C., Mosquera-Corral, A., & Méndez, R. (2010). Novel biological nitrogen-removal processes: applications and perspectives. In K.W. Canton (Ed.), Fluid Waste Disposal (pp. 153-181). Academic Press.

Celmer, D., Oleszkiewicz, J. A., & Cicek, N. (2008). Impact of shear force on the biofilm structure and performance of a membrane biofilm reactor for tertiary hydrogen-driven denitrification of municipal wastewater. *Water Research*, *42*(12), 3057–3065. doi:10.1016/j.watres.2008.02.031 PMID:18396310

Chang, C. C., Tseng, S. K., & Huang, H. K. (1999). Hydrogenotrophic denitrification with immobilized *Alcalogenes eutrophus* for drinking water treatment. *Bioresource Technology*, *69*(1), 53–58. doi:10.1016/S0960-8524(98)00168-0

Claus, G., & Kutzner, J. (1985). Physiology and kinetics of autotrophic denitrification by Thiobacillus denitrificans. *Applied Microbiology and Biotechnology*, *22*, 283–288.

Cui, M., Ma, A., Qi, H., Zhuang, X., & Zhuang, G. (2015). Anaerobic oxidation of methane: An «active» microbial process. *Microbiology Open*, *4*(1), 1–11. doi:10.1002/mbo3.232 PMID:25530008

Ekama, G. A., & Wentzel, M. C. (2004). Modelling inorganic material in activated sludge systems. *Water S.A.*, *30*(2), 153–174. doi:10.4314/wsa.v30i2.5060

Epsztein, R., Beliavski, M., Tarre, S., & Green, M. (2016). High-rate hydrogenotrophic denitrification in a pressurized reactor. *Chemical Engineering Journal*, *286*, 578–584. doi:10.1016/j.cej.2015.11.004

Ergas, S. J., & Reuss, A. F. (2001). Hydrogenotrophic denitrification of drinking water using a hollow fibre membrane bioreactor. *Journal of Water Supply: Research & Technology - Aqua*, *50*, 161–171.

Ettwig, K. F., Butler, M. K., Le Paslier, D., Pelletier, E., Mangenot, S., Kuypers, M. M. M., & Strous, M. et al. (2010). Nitrite-driven anaerobic methane oxidation by oxygenic bacteria. *Nature*, *464*(7288), 543–548. doi:10.1038/nature08883 PMID:20336137

Fajardo, C., Mora, M., Fernández, I., Mosquera-Corral, A., Campos, J. L., & Méndez, R. (2014). Cross effect of temperature, pH and free ammonia on autotrophic denitrification process with sulphide as electron donor. *Chemosphere*, *97*, 10–15. doi:10.1016/j.chemosphere.2013.10.028 PMID:24216266

Fajardo, C., Mosquera-Corral, A., Campos, J. L., & Méndez, R. (2012). Autotrophic denitrification with sulphide in a sequencing batch reactor. *Journal of Environmental Management*, *113*, 552–556. doi:10.1016/j.jenvman.2012.03.018 PMID:22704254

Fajardo, C., Mosquera-Corral, A., Campos, J. L., & Méndez, R. (2013). Post-treatment of fish canning effluents by sequential nitrification and autotrophic denitrification processes. *Process Biochemistry*, *48*(9), 1368–1374. doi:10.1016/j.procbio.2013.06.005

Fernández, N., Gómez, R., Amils, R., Sierra-Alvarez, R., Field, J. A., & Sanz, J. L. (2006). Microbial and structural aspects of granular sludge from autotrophic denitrifying reactors. *Water Science and Technology*, *54*(2), 11–17. doi:10.2166/wst.2006.480 PMID:16939078

Fernández, N., Sierra-Alvarez, R., Field, J. A., Amils, R., & Sanz, J. L. (2008). Microbial community dynamics in a chemolithotrophic denitrification reactor inoculated with methanogenic granular sludge. *Chemosphere*, *70*(3), 462–474. doi:10.1016/j.chemosphere.2007.06.062 PMID:17689587

Flere, J. M., & Zhang, T. C. (1999). Nitrate removal with sulphur-limestone autotrophic denitrification process. *Journal of Environmental Engineering*, *125*(8), 721–729. doi:10.1061/(ASCE)0733-9372(1999)125:8(721)

Foglar, L., & Briški, F. (2003). Wastewater denitrification process-the influence of methanol and kinetic analysis. *Process Biochemistry*, *39*(1), 95–103. doi:10.1016/S0032-9592(02)00318-7

Gadekar, S., Nemati, M., & Hill, G. A. (2006). Batch and continuous bio-oxidation of sulphide by *Thiomicrospira* sp. CVO: Reaction kinetics and stoichiometry. *Water Research*, *40*(12), 2436–2446. doi:10.1016/j.watres.2006.04.007 PMID:16730776

Garuti, G., Giordano, A., & Pirozzoli, F. (2001). Full-scale ANANOX system performance. *Water S.A.*, *27*, 189–197.

Ghafari, S., Hasan, M., & Aroua, M. K. (2008). Bio-electrochemical removal of nitrate from water and wastewater-A review. *Bioresource Technology*, *99*(10), 3965–3974. doi:10.1016/j.biortech.2007.05.026 PMID:17600700

Ghafari, S., Hasan, M., & Aroua, M. K. (2009). Effect of carbon dioxide and bicarbonate as inorganic carbon sources on growth and adaptation of hydrogenotrophic denitrifying bacteria. *Journal of Hazardous Materials*, *162*(2-3), 1507–1513. doi:10.1016/j.jhazmat.2008.06.039 PMID:18639979

Ghafari, S., Hasan, M., & Aroua, M. K. (2010). A kinetic study of autohydrogenotrophic denitrification at the optimum pH and sodium bicarbonate dose. *Bioresource Technology*, *101*(7), 2236–2242. doi:10.1016/j.biortech.2009.11.068 PMID:20015639

Gu, J. D., Qiu, W., Koenig, A., & Fan, Y. (2004). Removal of high $NO_3^-$ concentration in saline water through autotrophic denitrification by the bacterium *Thiobacillus denitrificans* strain MP. *Water Science and Technology*, *49*(5-6), 105–112. PMID:15137413

Haroon, M., Hu, S., Shi, Y., Imelfort, M., Keller, J., Hugenholtz, P., & Tyson, G. W. et al. (2013). Anaerobic oxidation of methane coupled to nitrate reduction in a novel archaeal lineage. *Nature*, *500*(7464), 567–570. doi:10.1038/nature12375 PMID:23892779

He, Z., Geng, S., Pan, Y., Cai, C., Wang, J., Wang, L., & Hu, B. et al. (2015b). Improvement of the trace metal composition of medium for nitrite-dependent anaerobic methane oxidation bacteria: Iron (II) and copper (II) make a difference. *Water Research*, *85*, 235–243. doi:10.1016/j.watres.2015.08.040 PMID:26340061

He, Z., Geng, S., Shen, L., Lou, L., Zheng, P., Xu, X., & Hu, B. (2015a). The short- and long-term effects of environmental conditions on anaerobic methane oxidation coupled to nitrite reduction. *Water Research*, *68*, 554–562. doi:10.1016/j.watres.2014.09.055 PMID:25462761

He, Z., Geng, S., Wang, L., Cai, C., Wang, J., Liu, S., & Hu, B. et al. (2016). Improvement of mineral nutrient concentrations and pH control for the nitrite-dependent anaerobic methane oxidation process. *Separation and Purification Technology*, *162*, 148–153. doi:10.1016/j.seppur.2016.02.016

Hu, S., Zeng, R. J., Burow, L. C., Lant, P., Keller, J., & Yuan, Z. (2009). Enrichment of denitrifying anaerobic methane oxidizing microorganisms. *Environmental Microbiology Reports*, *1*(5), 377–384. doi:10.1111/j.1758-2229.2009.00083.x PMID:23765890

Hu, S., Zeng, R. J., Keller, J., Lant, P. A., & Yuan, Z. (2011). Effect of nitrate and nitrite on the selection of microorganisms in the denitrifying anaerobic methane oxidation process. *Environmental Microbiology Reports*, *3*(3), 315–319. doi:10.1111/j.1758-2229.2010.00227.x PMID:23761277

Islas-Lima, S., Thalasso, F., & Gómez-Hernández, J. (2004). Evidence of anoxic methane oxidation coupled to denitrification. *Water Research*, *38*(1), 13–16. doi:10.1016/j.watres.2003.08.024 PMID:14630098

Jing, C., Ping, Z., & Mahmood, Q. (2009). Simultaneous sulfide and nitrate removal in anaerobic reactor under shock loading. *Bioresource Technology*, *100*(12), 3010–3014. doi:10.1016/j.biortech.2008.12.041 PMID:19231174

Justin, P., & Kelly, D. P. (1978). Growth kinetics of *Thiobacillus denitrificans* in anaerobic and aerobic chemostat culture. *Journal of General Microbiology*, *107*(1), 123–130. doi:10.1099/00221287-107-1-123 PMID:24083

Kampman, C., Hendrickx, T. L. G., Luesken, F. A., van Alen, T. A., Op den Camp, H. J. M., Jetten, M. S. M., & Temmink, H. et al. (2012). Enrichment of denitrifying methanotrophic bacteria for application after direct low-temperature anaerobic sewage treatment. *Journal of Hazardous Materials*, *227-228*, 164–171. doi:10.1016/j.jhazmat.2012.05.032 PMID:22657102

Karanasios, K. A., Michailidis, M. K., Vasiliadou, I. A., Pavlou, S., & Vayenas, D. V. (2011). Potable water hydrogenotrophic denitrification in packed-bed bioreactors coupled with a solar-electrolysis hydrogen production system. *Desalination and Water Treatment*, *33*(1-3), 86–96. doi:10.5004/dwt.2011.2614

Karanasios, K. A., Vasiliadou, I. A., Pavlou, S., & Vayenas, D. V. (2010). Hydrogenotrophic denitrification of potable water: A review. *Journal of Hazardous Materials*, *180*(1-3), 20–37. doi:10.1016/j.jhazmat.2010.04.090 PMID:20471745

Kim, E.-W., & Bae, J.-H. (2000). Alkalinity requirements and the possibility of simultaneous heterotrophic denitrification during sulphur-utilizing autotrophic denitrification. *Water Science and Technology*, *42*(3-4), 233–238.

Kim, H.-R., Lee, I.-S. & Bae, J.-H. (2004). Performance of a sulphur-utilizing fluidized bed reactor for post-denitrification. *Process Biotechnology*, *39*, 1591-1597.

Kim, I. S., & Son, J.-H. (2000). Impact of COD/N/S ratio on denitrification by the mixed culture of sulphate reducing bacteria and sulphur denitrifying bacteria. *Water Science and Technology*, *42*(3-4), 69–76.

Kimura, K., Nakamura, M., & Watanabe, Y. (2002). Nitrate removal by a combination of elemental sulphur-based denitrification and membrane filtration. *Water Research*, *36*(7), 1758–1766. doi:10.1016/S0043-1354(01)00376-1 PMID:12044075

Kleerebezem, R., & Méndez, R. (2002). Autotrophic denitrification for combined hydrogen sulphide removal from and post-denitrification. *Water Science and Technology*, *45*(10), 349–356. PMID:12188569

Koenig, A., & Liu, H. (1996). Autotrophic denitrification of landfill leachate using elemental sulphur. *Water Science and Technology*, *34*(5-6), 469–476. doi:10.1016/0273-1223(96)00680-4

Koenig, A., & Liu, L. H. (2001). Kinetic model of autotrophic denitrification in sulphur packed-bed reactors. *Water Research*, *35*(8), 1969–1978. doi:10.1016/S0043-1354(00)00483-8 PMID:11337843

Koenig, A., Tong, Z., Liu, L., & Fang, H. P. (2005). Microbial community and biochemistry process in autosulfurotrophic denitrifying biofilm. *Chemosphere*, *58*(8), 1041–1047. doi:10.1016/j.chemosphere.2004.09.040 PMID:15664612

Komori, M., & Sakakibara, Y. (2008). High-rate hydrogenotrophic denitrification in a fluidized-bed biofilm reactor using solid-polymer-electrolyte membrane electrode (SPEME). *Water Science and Technology*, *58*(7), 1441–1446. doi:10.2166/wst.2008.725 PMID:18957757

Krishnakumar, B., & Manilal, V. B. (1999). Bacterial oxidation of sulphide under denitrifying condition. *Biotechnology Letters*, *21*(5), 437–440. doi:10.1023/A:1005584920800

Kuai, L., & Verstraete, W. (1999). Autotrophic denitrification with elemental sulphur in small-scale wastewater treatment facilities. *Environmental Technology*, *20*(2), 201–209. doi:10.1080/09593332008616809

Kurt, M., Dunn, J., & Bourne, J. R. (1987). Biological denitrification of drinking water using autotrophic organisms with $H_2$ in a fluidized-bed biofilm reactor. *Biotechnology and Bioengineering*, *29*(4), 493–501. doi:10.1002/bit.260290414 PMID:18576478

Lee, J., Lee, K., Park, K., & Maeng, S. (2010). Hydrogenotrophic denitrification in a packed bed reactor: Effects of hydrogen-to-water flow rate ratio. *Bioresource Technology*, *101*(11), 3940–3947. doi:10.1016/j.biortech.2010.01.022 PMID:20144861

Lee, K. C., & Rittmann, B. E. (2000). A novel hollow-fibre membrane biofilm reactor for autohydrogenotrophic denitrification of drinking water. *Water Science and Technology*, *41*, 219–226.

Lee, K. C., & Rittmann, B. E. (2002). Applying a novel autohydrogenotrophic hollow-fiber membrane biofilm reactor for denitrification of drinking water. *Water Research*, *36*(8), 2040–2052. doi:10.1016/S0043-1354(01)00425-0 PMID:12092579

Luesken, F. A., van Alen, T. A., van der Biezen, E., Frijters, C., Toonen, G., Kampman, C., & Jetten, M. S. M. et al. (2011). Diversity and enrichment of nitrite-dependent anaerobic methane oxidizing bacteria from wastewater sludge. *Environmental Biotechnology*, *92*(4), 845–854. doi:10.1007/s00253-011-3361-9 PMID:21667086

Ma, Y.-L., Yang, B.-L., & Zhao, J.-L. (2006). Removal of $H_2S$ by *Thiobacillus denitrificans* immobilized on different matrices. *Bioresource Technology*, *97*(16), 2041–2046. doi:10.1016/j.biortech.2005.09.023 PMID:16289624

Mahmood, Q., Zheng, P., Cai, J., Wu, D., Hu, B., & Li, J. (2007). Anoxic sulfide biooxidation using nitrite as electron acceptor. *Journal of Hazardous Materials*, *147*(1-2), 249–256. doi:10.1016/j.jhazmat.2007.01.002 PMID:17261349

Mahmood, Q., Zheng, P., Hayat, Y., Islam, E., Wu, D., & Ren-cun, J. (2008). Effect of pH on anoxic sulfide oxidizing reactor performance. *Bioresource Technology*, *99*(8), 3291–3296. doi:10.1016/j.biortech.2007.07.006 PMID:17702569

Manconi, I., Carucci, A., & Lens, P. (2007). Combined removal of sulfur compounds and nitrate by autotrophic denitrification in bioaugmented activated sludge system. *Biotechnology and Bioengineering*, *98*(3), 551–560. doi:10.1002/bit.21383 PMID:17724757

Manconi, I., Carucci, A., Lens, P., & Rossetti, S. (2006). Simultaneous biological removal of sulphide and nitrate by autotrophic denitrification in an activated sludge system. *Water Science and Technology, 53*(12), 91–99. doi:10.2166/wst.2006.410 PMID:16889245

Mansell, B. O., & Schroeder, E. D. (2002). Hydrogenotrophic denitrification in a microporous membrane bioreactor. *Water Research, 36*(19), 4683–4690. doi:10.1016/S0043-1354(02)00197-5 PMID:12448509

McAdam, E. J., & Judd, S. J. (2006). A review of membrane bioreactor potential for nitrate removal from drinking water. *Destilation, 196*, 135–148.

Mo, H., Oleszkiewicz, J.A., Cicek, N., Rezania, B. (2005). Incorporating membrane gas diffusion into a membrane bioreactor for hydrogenotrophic denitrification of groundwater. *Water Science and Technology, 51*, 357-364.

Modin, O., Fukushi, K., Nakajima, F., & Yamamoto, K. (2010). Nitrate removal and biofilm characteristics in methanotrophic membrane biofilm reactors with various gas supply regimes. *Water Research, 44*(1), 85–96. doi:10.1016/j.watres.2009.09.009 PMID:19781736

Modin, O., Fukushi, K., & Yamamoto, K. (2007). Denitrification with methane as external carbon source. *Water Research, 41*(12), 2726–2738. doi:10.1016/j.watres.2007.02.053 PMID:17433401

Moon, H. S., Ahn, K.-H., Lee, S., Nam, K., & Kim, J. Y. (2004). Use of autotrophic sulphur-oxidizers to remove nitrate from bank filtrate in a permeable reactive barrier system. *Environmental Pollution, 129*(3), 499–507. doi:10.1016/j.envpol.2003.11.004 PMID:15016470

Moon, H. S., Chang, S. W., Nam, K., Choe, J., & Kim, J. Y. (2006). Effect of reactive media composition and co-contaminants on sulfur-based autotrophic denitrification. *Environmental Pollution, 144*(3), 802–807. doi:10.1016/j.envpol.2006.02.020 PMID:16632130

Moon, H. S., Shin, D. Y., Nam, K., & Kim, J. Y. (2008). A long-term performance test on an autotrophic denitrification column for application as a permeable reactive barrier. *Chemosfere, 73*(5), 723–728. doi:10.1016/j.chemosphere.2008.06.065 PMID:18722639

Noroi, K. (2014). Nitrate-dependent anaerobic methane oxidation in a freshwater sediment. *Geochimica et Cosmochimica Acta, 132*, 141–150. doi:10.1016/j.gca.2014.01.032

Oh, S. E., Bum, M. S., Yoo, Y. B., Zubair, A., & Kim, I. S. (2002). Nitrate removal by simultaneous sulphur utilizing autotrophic and heterotrophic denitrification under different organic and alkalinity condition: Batch experiments. *Water Science and Technology, 47*(1), 237–244. PMID:12578200

Oh, S.-E., Kim, K.-S., Choi, H.-C., Cho, J., & Kim, I. S. (2000). Kinetics and physiological characteristics of autotrophic denitrifying sulphur bacteria. *Water Research & Technology, 42*(3-4), 959–968.

Park, H. I., Choi, Y. J., & Pak, D. (2005a). Autohydrogenotrophic denitrifying microbial community in a glass beads biofilm reactor. *Biotechnology Letters, 27*(13), 949–953. doi:10.1007/s10529-005-7654-x PMID:16091891

Park, H. I., Kim, D. K., Choi, Y., & Pak, D. (2005b). Nitrate reduction using an electrode as direct electron donor in a biofilm-electrode reactor. *Process Biochemistry, 40*(10), 3383–3388. doi:10.1016/j.procbio.2005.03.017

Park, J. Y., & Yoo, Y. J. (2009). Biological nitrate removal in industrial wastewater treatment: Which electron donor we can choose. *Applied Microbiology and Biotechnology*, *82*(3), 415–429. doi:10.1007/s00253-008-1799-1 PMID:19148639

Pérez, R., Galí, A., Dosta, J., & Mata-Álvarez, J. (2007). Biological nitrogen removal (BNR) using sulfides for autotrophic denitrification in a sequencing batch reactor (SBR) to treat reject water. *Industrial & Engineering Chemistry Research*, *46*(21), 6646–6649. doi:10.1021/ie061496h

Raghoebarsing, A. A., Pol, A., van de Pas-Schoonen, K. T., Smolders, A. J. P., Ettwig, K. F., Rijpstra, I. C., & Strous, M. et al. (2006). A microbial consortium couples anaerobic methane oxidation to denitrification. *Nature*, *440*(7086), 918–921. doi:10.1038/nature04617 PMID:16612380

Rajapakse, J. P., & Scutt, J. E. (1999). Denitrification with natural gas and various new growth media. *Water Research*, *33*(18), 3723–3734. doi:10.1016/S0043-1354(99)00088-3

Reyes-Ávila, J., Flores-Razo, E., & Gómez, J. (2004). Simultaneous biological removal of nitrogen, carbon and sulfur by denitrification. *Water Research*, *38*(14-15), 3313–3321. doi:10.1016/j.watres.2004.04.035 PMID:15276748

Rezania, B., Cicek, N., & Oleszkiewicz, J. A. (2005). Kinetics of hydrogen-dependent denitrification under varying pH and temperature conditions. *Biotechnology and Bioengineering*, *92*(7), 900–906. doi:10.1002/bit.20664 PMID:16116656

Rezania, B., Oleszkiewicz, J. A., & Cicek, N. (2006). Hydrogen-driven denitrification of wastewater in an anaerobic submerged membrane bioreactor: Potential for water reuse. *Water Science and Technology*, *54*(11), 207–214. doi:10.2166/wst.2006.844 PMID:17302322

Rhee, G. Y., & Fuhs, G. W. (1978). Wastewater denitrification with one carbon compounds as energy source. *Journal - Water Pollution Control Federation*, *50*(9), 2111–2119.

Sahu, A. K., Conneely, T., Nusslein, K., & Ergas, S. J. (2009). Hydrogenotrophic denitrification and perchlorate reduction in ion exchange brines using membrane biofilm reactors. *Biotechnology and Bioengineering*, *104*(3), 483–491. doi:10.1002/bit.22414 PMID:19544384

Serrano-Silva, N., Sarria-Guzmán, Y., Dendooven, L., & Luna-Guido, M. (2014). Methanogenesis and Methanotrophy in soil: A Review. *Pedosphere*, *24*(3), 291–307. doi:10.1016/S1002-0160(14)60016-3

Shen, L.-D., He, Z.-F., Zhu, Q., Chen, D.-Q., Lou, L.-P., Xu, X.-Y., Zheng, P & Hu B.L. (2012) Microbiology, ecology, and application of the nitrite-dependent anaerobic methane oxidation process. *Frontiers in Microbiology*, *3*, 269-273.

Shin, J. H., Sang, B. I., Chung, Y. C., & Choung, Y. K. (2005). The removal of nitrogen using an autotrophic hybrid hollow-fiber membrane biofilm reactor. *Desalination*, *183*(1-3), 447–454. doi:10.1016/j.desal.2005.03.045

Sierra-Álvarez, R., Beristain-Cardoso, R., Salazar, M., Gómez, J., Razo-Flores, E., & Field, J. A. (2007). Chemolithotrophic denitrification with elemental sulphur for groundwater treatment. *Water Research*, *41*(6), 1253–1262. doi:10.1016/j.watres.2006.12.039 PMID:17296214

Smith, D. P., Rector, T., Reid-Black, K., Hummerick, M., Strayer, R., Birmele, M., & Garland, J. L. et al. (2007). Redox control bioreactor: A unique biological water processor. *Biotechnology and Bioengineering, 99*(4), 830–845. doi:10.1002/bit.21617 PMID:17705250

Smith, R. L., Ceazan, M. L., & Brooks, M. H. (1994). Autotrophic, hydrogen-oxidizing, denitrifying bacteria in groundwater, potential agents for bioremediation of nitrate contamination. *Applied and Environmental Microbiology, 60*(6), 1949–1955. PMID:16349284

Soares, M. I. M. (2002). Denitrification of groundwater with elemental sulphur. *Water Research, 36*(5), 1392–1395. doi:10.1016/S0043-1354(01)00326-8 PMID:11905445

Sollo, F. W., Mueller, H. F., & Larson, T. E. (1976). Denitrification of wastewater effluents with methane. *Journal - Water Pollution Control Federation, 48*(7), 1840–1842.

Sublette, K. L., & Sylvester, N. D. (1997). Oxidation of hydrogen sulphide by mixed cultures of *Thiobacillus denitrificans* and heterotrophs. *Biotechnology and Bioengineering, 29*(6), 759–761. doi:10.1002/bit.260290614 PMID:18576512

Tang, Y., Zhou, C., Van Ginkel, S. W., Ontiveros-Valencia, A., Shin, J., & Rittmann, B. E. (2012). Hydrogen permeability of the hollow fibers used in $H_2$-based membrane biofilm reactors. *Journal of Membrane Science, 407–408*, 176–183. doi:10.1016/j.memsci.2012.03.040

Tang, Y., Zhou, C., Ziv-El, M., & Rittmann, B. E. (2011). A pH-control model for heterotrophic and hydrogen-based autotrophic denitrification. *Water Research, 45*(1), 232–240. doi:10.1016/j.watres.2010.07.049 PMID:20705316

Terada, A., Kaku, S., Matsumoto, S., & Tsuneda, S. (2006). Rapid autohydrogenotrophic denitrification by a membrane biofilm reactor equipped with a fibrous support around a gas-permeable membrane. *Biochemical Engineering Journal, 31*(1), 84–91. doi:10.1016/j.bej.2006.06.004

Thalasso, F., Vallecillo, A., García-Encina, P., & Fdz-Polanco, F. (1997). The use of methane as a sole carbon source for wastewater denitrification. *Water Research, 31*(1), 55–60. doi:10.1016/S0043-1354(96)00228-X

Vaiopoulou, E., Melidis, P., & Aivasidis, A. (2005). Sulfide removal in wastewater from petrochemical industries by autotrophic denitrification. *Water Research, 39*(17), 4101–4109. doi:10.1016/j.watres.2005.07.022 PMID:16139866

Vasiliadou, I. A., Pavlou, S., & Vayenas, D. V. (2006a). A kinetic study of hydrogenotrophic denitrification. *Process Biochemistry, 41*(6), 1401–1408. doi:10.1016/j.procbio.2006.02.002

Vasiliadou, I. A., Pavlou, S., & Vayenas, D. V. (2009a). Dynamics of a chemostat with three competitive hydrogen oxidizing denitrifying microbial populations and their efficiency for denitrification. *Ecological Modelling, 220*(8), 1169–1180. doi:10.1016/j.ecolmodel.2009.02.009

Vasiliadou, I. A., Pavlou, S., & Vayenas, D. V. (2009b). Experimental and modelling study of drinking water hydrogenotrophic denitrification in packed-bed reactors. *Journal of Hazardous Materials, 165*(1-3), 812–824. doi:10.1016/j.jhazmat.2008.10.067 PMID:19054612

Vasiliadou, I. A., Siozios, S., Papadas, I. T., Bourtzis, K., Pavlou, S., & Vayenas, D. V. (2006b). Kinetics of pure cultures of hydrogen-oxidizing denitrifying bacteria and modeling of the interactions among them in mixed cultures. *Biotechnology and Bioengineering*, *95*(3), 513–525. doi:10.1002/bit.21031 PMID:16758460

Waki, M., Yasuda, T., Yokoyama, H., Hanajima, D., Ogino, A., Suzuki, K., & Tanaka, Y. et al. (2009). Nitrogen removal by co-occurring methane oxidation, denitrification, aerobic ammonium oxidation, and anammox. *Applied Microbiology and Biotechnology*, *84*(5), 977–985. doi:10.1007/s00253-009-2112-7 PMID:19603162

Wang, A., Du, D., Ren, N., & van Groenestijn, J. W. (2005). An innovative process of simultaneous desulfurization and denitrication by *Thiobacillus denitrificans*. *Journal of Environmental Science and Health*, *40*(10), 1939–1949. doi:10.1080/10934520500184590 PMID:16194914

Wiesmann, U. (1994). Biological nitrogen removal from wastewater. In A. Fletcher (Ed.), *Advances in biochemical engineering biotechnology* (Vol. 51, pp. 113–154). Berlin: Spinger Verlag.

Xia, S., Zhang, Y. H., & Zhong, F. H. (2009). A continuous stirred hydrogen-based polyvinyl chloride membrane biofilm reactor for the treatment of nitrate contaminated drinking water. *Bioresource Technology*, *100*(24), 6223–6228. doi:10.1016/j.biortech.2009.07.002 PMID:19656675

Xia, S., Zhang, Z., Zhong, F., & Zhang, J. (2011). High efficiency removal of 2-chlorophenol from drinking water by a hydrogen-based polyvinyl chloride membrane biofilm reactor. *Journal of Hazardous Materials*, *186*(2-3), 1367–1373. doi:10.1016/j.jhazmat.2010.12.023 PMID:21215515

Yamamoto-Ikemoto, R., Komori, T., Nomura, M., Ide, Y., & Matsukami, T. (2000). Nitrogen removal from hydroponic culture wastewater by autotrophic denitrification using thiosulfate. *Water Science and Technology*, *42*(3-4), 369–376.

Zhang, T. C., & Lampe, D. G. (1999). Sulfur: limestone autotrophic denitrification processes for treatment of nitrate-contaminated water: batch experiments. *Water Research*, *33*(3), 599–608. doi:10.1016/S0043-1354(98)00281-4

Zhang, Y., Zhong, F., Xia, S., Wang, X., & Li, J. (2009). Autohydrogenotrophic denitrification of drinking water using a polyvinyl chloride hollow fiber membrane biofilm reactor. *Journal of Hazardous Materials*, *170*(1), 203–209. doi:10.1016/j.jhazmat.2009.04.114 PMID:19473764

Zhou, M., Fu, W., Gu, H., & Lei, L. (2007). Nitrate removal from groundwater by a novel three-dimensional electrode biofilm reactor. *Electrochimica Acta*, *52*(19), 6052–6059. doi:10.1016/j.electacta.2007.03.064

# Chapter 7
# Modelling and Control of Nitrogen and Phosphorus Removing Systems

**Borja Valverde-Pérez**
*Technical University of Denmark, Denmark*

**Anna Katrine Vangsgaard**
*Krüger Veolia, Denmark*

**Xavier Flores-Alsina**
*Technical University of Denmark, Denmark*

**Miguel Mauricio-Iglesias**
*Universidade de Santiago de Compostela, Spain*

## ABSTRACT

*A wide variety of technologies addressing nutrient removal in wastewater treatment have been developed in the latest years. In order to be able to design, manage, operate, optimise and benchmark these novel technologies with potentially existing standard technologies, detailed modelling of nutrient removal becomes essential. Nutrient removal is prominently considered, with different degrees of complexity, in the commonly used Activated Sludge Models (ASM). However, the description of nutrient-related compounds in these standard models is not sufficiently detailed for novel and not so novel technologies. Furthermore, the environmental evaluation of a process, which includes an estimation of the greenhouse gas release, would require a more complex process to represent nitrogen removal. This chapter reviews the current state of nutrient removal modelling, with special focus on the needs created by emerging technologies in nitrogen and phosphorus removal, paying special attention to the objectives that must be evaluated for each technology.*

## INTRODUCTION

Mathematical models can serve as essential tools to help understand governing mechanisms in wastewater treatment processes and therefore modelling is central in current procedures of process design and optimisation. In particular, dynamic models, i.e. where variables change with time, are especially relevant in municipal wastewater treatment as input variations caused by changes in influent flowrate and influent concentration are practically unavoidable. Mathematical models can also be used to evaluate alternative options of plant or reactor layouts, to test operational strategies for a smaller cost than experimental test-

DOI: 10.4018/978-1-5225-1037-6.ch007

ing, or to narrow down the experimental range to be explored. Indeed, building up a sound model for a real wastewater treatment plant is a cost-intensive task as it requires staff with expertise from different fields such as reactor engineering, chemical and biochemical reaction kinetics, transport phenomena and control engineering as well as detailed knowledge about the actual plant. However, the costs of model development can be quickly compensated by an increased efficiency leading to reduced operating costs or minimized construction costs due to models allowing smaller safety margins in the design phase.

In order to facilitate and enhance the model construction and use in biological wastewater treatment, the IAWPRC (now IWA) formed a task group dedicated to provide a modelling platform in 1983. The result was a model for a single-sludge treatment featuring COD oxidation and nitrogen removal by nitrification and denitrification (Grady et al.,1986). The model was evaluated and revised later by Henze et al. (1987) resulting in the first version of the widely used activated sludge model no. 1 (ASM1). The model is based on a mechanistic interpretation of the behaviour of the microbial groups catalysing the process reactions considered in the system, as well as detailed characterization of influent wastewater composition with respect to different fractions of COD and nitrogen. Enhanced biological phosphorus removal (EBPR) was later added to a revised version of the ASM1, giving place to the ASM2d (Henze et al., 1999). A new version of the ASM1 including a better description of COD storage by heterotrophs gave place to the ASM3, although the description of removal of nitrogen did not change in this model (Gujer et al., 1999). Since then, numerous activated sludge models have been developed and evaluated, addressing different levels of detail, such as the two step nitrification and denitrification or models for phosphorus accumulating organisms (PAOs). Furthermore, a number of processes have been developed in the last two decades featuring novel processes for nutrient removal such as anaerobic ammonium oxidation (anammox) for which, new models have been quickly proposed. The models for nitrogen and phosphorus removal are respectively presented and reviewed in sections 2 and 3. The importance of pH modelling and its challenges is summarised in section 4. Finally, in section 5, the generic challenges of control in wastewater treatment plants (WWTP) are presented together and the particular control structures applied in nutrient removal.

## MODELS FOR NITROGEN REMOVAL

Nitrogen removal processes have been largely dominated by nitrification-denitrification processes which are properly described with the ASM models. In the last twenty years, the development of new autotrophic nitrogen removal processes led to the parallel creation of new models thereof. Additionally, increasing environmental concerns have increased the attention to models that include the production of $N_2O$ and other intermediates of nitrification and denitrification processes in order to mitigate its release.

### Nitrification

The ASM1 includes one-step nitrification (ammonium is converted to nitrate) and one-step denitrification (nitrate is used as electron acceptor, and converted to nitrogen gas) processes.

One-step nitrification: $NH_4^+ + 2\ O_2 \rightarrow NO_3^- + H_2O + 2\ H^+$

Despite some obvious simplifications and shortcomings, the ASM1 is still one of the most frequently used activated sludge models, and is therefore also considered as the reference model here. The reaction rate proposed to model one-step nitrification depends on the substrates and also takes into account the consumption of alkalinity (Equation 1).

$$\rho_{Nit} = \mu_{Aut} \frac{S_O}{K_{Aut,O} + S_O} \cdot \frac{S_{Alk}}{K_{Aut,\,Alk} + S_{Alk}} \cdot \frac{S_{NH}}{K_{Aut,\,NH} + S_{NH}} \cdot X_{Aut} \tag{1}$$

where $\rho_{Nit}$ is the growth rate of the autotrophic nitrifying organisms (e.g. in g biomass $d^{-1}$ $m^{-3}$), $\mu_{Aut}$ is their maximum growth rate, $K_{Aut,\,i}$ stands for the affinity constants with respect to substrate *i*, $X_{Aut}$ is the concentration of autotrophic nitrifying organisms and subindexes *O*, *Alk* and *NH* stand for dissolved oxygen, alkalinity and ammonia respectively.

The Activated Sludge Model no. 3 (ASM3) (Henze et al., 2000) introduced a number of significant changes in the model structure that corrected for ASM1 deficiencies. The death-regeneration principle was replaced by the endogenous respiration concept, and – in line with reported experimental observations (Nowak et al., 1995; Siegrist et al., 1999) – biomass decay in ASM3 is electron acceptor depending, which should lead to more realistic simulation results for plants that have a significant anaerobic volume. However, the ASM3 also introduced a significant increase of the number of model parameters, leading to a more complex calibration procedure (Guisasola et al., 2007a), which might partly explain why this model never became as popular as the ASM1.

Along the years, it has been quite common to extend the ASM1 or any other activated sludge model with two-step nitrification, for example to simulate plants with significant nitrite build-up. A detailed review of modelling concepts to describe nitrite as an intermediate in activated sludge systems and other nitrification/denitrification process intermediates can be found in Sin et al. (2008).

Two-step nitrification:

$NH_4^+ + 1.5\ O_2 \rightarrow NO_2^- + H_2O + 2\ H^+$

$NO_2^- + 0.5\ O_2 \rightarrow NO_3^-$

In two-step nitrification models, ammonium is first converted to nitrite by ammonium oxidizing bacteria (AOB) using oxygen as electron acceptor, whereas nitrite is oxidized further to nitrate by nitrite oxidizing bacteria (NOB) also utilizing oxygen as electron acceptor (see Figure 1). Several sets of reaction rates are proposed to model two-step nitrification with no universal consensus to date. The possible reaction rates may differ on which true substrate is used for the rate equation, the inclusion of alkalinity limitations or the extent of inhibition. The following equations (Equation 2 and 3) cover the main points of agreement:

$$\rho_{AOB} = \mu_{AOB} \frac{S_O}{K_{AOB,O} + S_O} \cdot \frac{K_{I,AOB,\,HNO2}}{K_{I,AOB,\,HNO2} + S_{HNO2}} \cdot \frac{S_{NH}}{K_{AOB,\,NH} + S_{NH}} \cdot X_{AOB} \tag{2}$$

*Figure 1. The inorganic nitrogen cycle indicating biological reactions converting nitrogen. The number in between the parentheses behind the compounds indicate the oxidation state of the nitrogen atom.*

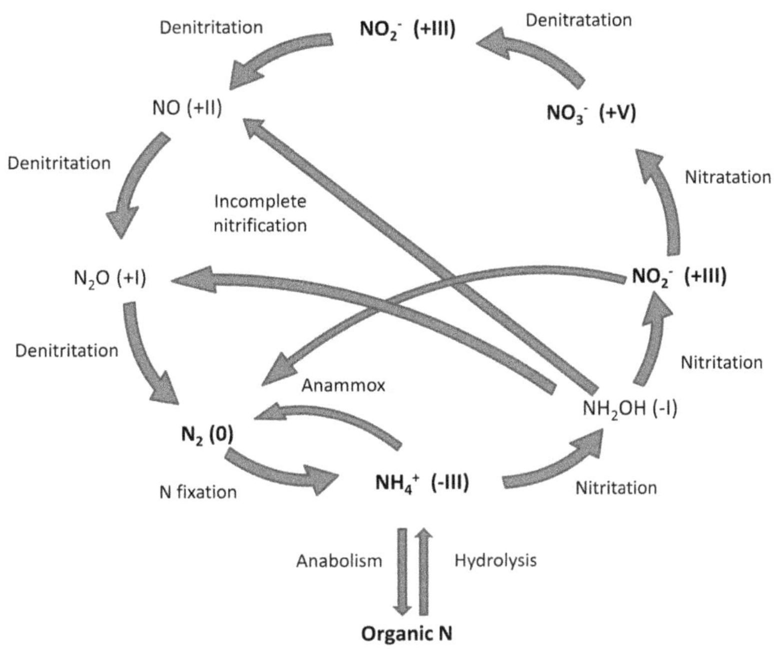


$$\rho_{NOB} = \mu_{NOB} \frac{S_O}{K_{NOB,O} + S_O} \cdot \frac{S_{HNO2}}{K_{NOB,HNO2} + S_{HNO2}} \cdot \frac{K_{I,NOB,HNO2}}{K_{I,NOB,HNO2} + S_{HNO2}} \cdot X_{NOB} \quad (3)$$

where $\rho$ is the growth rate of biomass (e.g. in g biomass $d^{-1}$ $m^{-3}$), $\mu$ is the maximum growth rate, $K$ stands for the affinity constants with respect to substrate, $K_I$ is the inhibition constant, $X$ is the concentration of biomass and subindexes *AOB, NOB, O*, $HNO_2$ and *NH* stand for ammonium oxidizing bacteria, nitrite oxidizing bacteria, dissolved oxygen, nitrous acid and ammonia respectively.

In literature, a number of model structures are reported together with different sets of parameter values to describe two-step nitrification models (e.g. substrate inhibition, product inhibition, etc.) as discussed by Sin et al. (2008).

## Denitrification

Similar to the nitrification process, denitrification is usually modelled as a one-step process that represents a major simplification of the four biochemical steps necessary to reduce nitrate to nitrogen gas (Figure 1). A unifying model structure to describe the denitrification process has not been reached yet. Denitrification is known to be carried out by a number of denitrifying heterotrophic microorganisms which can perform some or all of the biochemical steps in denitrification. Hence, some microorganisms can catalyse all the steps leading to complete denitrification or different steps may be carried out by different microorganisms leading to partial denitrification (Knowles, 1982; Almeida et al., 1995). Whether partial denitrification is observed or not will depend on operational conditions and sludge history. These factors determine which denitrifying microorganisms will be present, and thereby the required model structure/parameter values. Likewise, the influence of operational conditions explains partly the differing observations made about denitrification, and as a consequence the range of model structure proposals available in the literature (von Schulthess & Gujer, 1996; Sin et al., 2008). Some examples of ASM extensions using several denitrification steps (2, 3 or more) include von Schulthess et al. (2006), Hiatt and Grady (2008), Iacopozzi et al. (2008) and Sin et al. (2008).

Extended models in the last years focus particularly on $N_2O$, due to its immense global warming potential. In addition to the partial denitrification, $N_2O$ can also be formed and released during incomplete nitrification or via autotrophic denitrification, whereby AOB use $NO_2^-$ as terminal electron acceptor (Kampschreur et al., 2009; Figure 1). In general, modelling of greenhouse gas emissions is an area of intense research, since it is important for assessing the overall carbon footprint of a technology or operational strategy, whereas focus previously has been mainly on operational costs and removal efficiency. Unfortunately, a unified model that describes all mechanisms independently is still a topic of discussion (Ni & Yuan, 2015).

## Anaerobic Ammonium Oxidation (Anammox)

In the anammox process, ammonium is oxidized by using nitrite as electron acceptor, to form nitrogen gas (Figure 1) and nitrate as by-product. This process is performed by anaerobic ammonium oxidizing bacteria (AnAOB). A simplified version of the process stoichiometry is:

$NH_4^+ + NO_2^- \rightarrow N_2 + 2\,H_2O$

with the following growth rate (Equation 4):

$$\rho_{AnAOB} = \mu_{AnAOB} \frac{S_{NH}}{K_{AnAOB,NH} + S_{NH}} \cdot \frac{S_{HNO2}}{K_{AnAOB,HNO2} + S_{HNO2}} \cdot \frac{K_{I,O}}{K_{I,O} + S_O} \cdot X_{AnAOB} \qquad (4)$$

where $\rho$ is the growth rate of biomass (g biomass $d^{-1}$ $m^{-3}$),, $\mu$ is the maximum growth rate, $K$ stands for the affinity constants with respect to substrate, $K_I$ is the inhibition constant, $X$ is the concentration of biomass and subindexes *AnAOB O*, $HNO_2$ and *NH* stand for anaerobic ammonium oxidizing bacteria, dissolved oxygen, nitrous acid and ammonia respectively.

In practice, the stoichiometry of ammonium to nitrite is 1 to 1.14. Most of the nitrogen is converted to $N_2$, but about 6-7% of the converted N can be found as nitrate. The remaining nitrogen is incorporated in new biomass that is produced during growth.

AnAOB are extremely slow growing with a doubling time of approximately 11 days (Strous et al., 1998). They are very sensitive toward certain compounds and are inhibited by oxygen and nitrite (Strous et al., 1999). Since the process is performed by an obligate anoxic microorganism, oxygen has an inhibiting effect on AnAOB already at a concentration of 0.2 mg $O_2$ $L^{-1}$ (Jung et al., 2007). However, it has been found that AnAOB can recover their activity after exposure to low $O_2$ concentrations (Strous et al., 1997; Egli et al., 2001), thus the inhibition is probably reversible. Different model structure and kinetic parameter values have been presented (Strous et al., 1998; Strous et al., 1999; Koch et al., 2000; Hao et al., 2002a,b; Dapena-Mora et al., 2004). The multiplicity of models is partly caused by the diversity of the anammox-performing microorganisms, partly due to the reactor/experimentation layout of the anammox application.

Models including the anammox process depend on the process layout. It can be put in practice either in a two-stage configuration (nitrification and anammox) or as a simultaneous process in a single stage, where nitrification and anammox are occurring in the same process step. The latter case requires a space- or time-distributed model (or both) so that different environmental conditions are possible in a single reactor.

## Two-Stage Complete Autotrophic Nitrogen Removal Modelling

As a first approach, many studies have modelled two-step nitrification as an extension of the ASM1 or ASM3, and then, investigated how to stimulate and achieve partial nitrification or nitritation, i.e. promote AOB growth and while suppressing NOB growth (see Nitrification chapter). Picioreanu et al. (1997) investigated the effect of oxygen concentration in a nitrifying biofilm reactor, Hellinga et al. (1999) modelled a SHARON (single reactor system for high activity ammonium removal over nitrite) reactor for process design calculations at a full-scale plant, Wyffels et al. (2004) and Pambrun et al. (2006) modelled nitrification in a membrane bioreactor (MBR) and in an SBR (sequenced batch reactor), respectively, both with the objective of obtaining optimal operational conditions for partial nitrification, while Bernet et al. (2005) constructed a steady-state biofilm model with homogenous AOB and NOB distribution in order to design a control system ensuring nitrite accumulation.

## Single-Stage Complete Autotrophic Nitrogen Removal Modelling

The study of Koch et al. (2000) was amongst the first to mathematically describe a single-stage complete autotrophic nitrogen removal (CANR) configuration by modelling a biofilm based RBC (rotating biological contactor) removing nitrogen without organic carbon supply. Under the operational conditions investigated it was found that diffusion of nitrite to the anoxic AnAOB containing parts of the biofilm was the limiting process for nitrogen removal in the system. Later, the effect of temperature, influent ammonium concentration, and influent flow rate (Hao et al., 2002b), plus bulk oxygen concentration and biofilm thickness (Hao et al., 2002a) were investigated by modelling of the CANON (complete autotrophic nitrogen removal over nitrite) process as a 1-D flat-sheet biofilm. It should be noted that both the model of Koch et al. (2000) and the model of Hao et al. (2002a and 2002b) did not consider the presence of heterotrophic denitrifiers and did not explicitly include external mass transfer resistance. However, the study of Hao et al. (2004) investigated the impact of the presence of heterotrophic denitrifying microorganisms in the biofilm in cases where biodegradable organic carbon was present in the influent. This was extended by the study of Lackner et al. (2008) which looked into the effect of denitrification and COD in co-diffusion and counter-diffusion biofilm systems. The counter-diffusion system was a membrane aerated biofilm reactor (MABR) in which the oxygen was supplied through a membrane acting as biofilm substratum (Terada et al., 2007). The study found that the surface loading ratio of ammonium to oxygen was determining the performance of the N removal. The latest development within modelling of single-stage CANR is the modelling of granular sludge reactors (Volcke et al., 2010; Vangsgaard et al., 2012).

## MODELS FOR PHOSPHORUS REMOVAL

Biological phosphorus removal has been modeled using two different model families: Activated Sludge Models (Henze et al., 2000) and metabolic models (e.g. Oehmen et al., 2010a). The ASM families provide a set of default values which are case specific and must be re-estimated for each enhanced biological phosphorus removal (EBPR). On the contrary, yield coefficients are estimated in metabolic models based on the theoretical balances over substrate, energy and reducing power, thereby reducing the calibration efforts (Oehmen et al., 2007). In this chapter we present the most representative models of each family, introducing only those biochemical processes related to phosphorus accumulating organisms (PAO) and glycogen accumulating organisms (GAO).

### Activated Sludge Modelling Framework for Biological Phosphorus Removal

The most common description for modelling biological phosphorus removal was first reported by Gujer et al. (1995), where ASM1 was extended to ASM2 to include the activity of PAO. This model was afterwards extended with PAO denitrifying capacity in the ASM2d (Henze et al., 1999). Default stoichiometric and kinetic parameters were proposed and since then, many investigations have been conducted to find optimal parameter values in different scenarios. ASM2d was further expanded by considering electron dependency in decay rates (Gernaey and Jorgensen, 2004), more in accordance with the principles stated in ASM3 (Henze et al., 2000). Additionally, a model extension for biological phosphorus removal, similar to the ASM2d, was developed for the ASM3 by Rieger et al. (2001). In this section, we will focus on the description of the ASM2d, which is the most widely used model for biological phosphorus removal (Hauduc et al., 2009), and its comparison with other models and extensions available in the ASM framework.

Under anaerobic conditions PAO may release phosphate, ($S_{PO4}$), by hydrolysing the internal stored polyphosphate ($X_{PP}$). This process releases energy that PAO use to take up volatile fatty acids, which are considered to be acetate ($S_A$) in the ASM2d framework, and store them as polyhydroxialkanoates ($X_{PHA}$). As PHA storage may also occur under aerobic and anaerobic conditions there is no inhibition term included in the process rate equation (Equation 5).

$$\rho_{PHA} = q_{PHA} \cdot \frac{S_A}{K_A + S_A} \cdot \frac{S_{Alk}}{K_{Alk} + S_{Alk}} \cdot \frac{X_{PP} / X_{PAO}}{K_{PP} + X_{PP} / X_{PAO}} \cdot X_{PAO} \tag{5}$$

where $\rho$ is the storage rate of PHA, $q_{PHA}$ is the maximum PHA storage rate and $K$ stands for the affinity constants with respect to substrate. Under aerobic conditions PAO store back the soluble phosphorus ($S_{PO}$) as polyphosphate ($X_{PP}$). The energy required for this process is provided via hydrolysis of the anaerobically stored PHA. PAO have a maximum storage capacity beyond which they are not able to store more phosphorus. This process is modelled by Equation 6.

$$\rho_{PP} = q_{PP} \cdot \frac{S_O}{K_O + S_O} \cdot \frac{S_{PO}}{K_{PO} + S_{PO}} \cdot \frac{S_{Alk}}{K_{Alk} + S_{Alk}} \cdot \frac{X_{PHA} / X_{PAO}}{K_{PHA} + X_{PHA} / X_{PAO}} \cdot \frac{K_{MAX} - X_{PP} / X_{PAO}}{K_{iPP} + K_{MAX} - X_{PHA} / X_{PAO}} \cdot X_{PAO} \tag{6}$$

Phosphate storage can also take place under anoxic conditions. The process rate equation (Equation 7) is similar, additionally including an oxygen inhibition term (with inhibition constant $K_{IO}$) and an anoxic correction factor ($\eta_{NO}$). The correction factor aims to account for the fact that not all PAO are able to denitrify (Mielczarek et al., 2013).

$$\rho_{PP,Anox} = \rho_{PP} \cdot \eta_{NO} \cdot \frac{S_{NO}}{K_{NO} + S_{NO}} \cdot \frac{K_{IO}}{S_O} \tag{7}$$

PAO can grow under aerobic conditions at the expense of the internally stored PHAs and external soluble nutrients (i.e. $S_{NH}$ and $S_{PO}$). It is known that PAO can grow on soluble organic substrates. However, it is considered that soluble organics are normally taken up in the anaerobic phase of the EBPR systems and consequently this process is not considered in the model. Similarly to the PHA storage process, PAO clades can also grow under anoxic conditions (Equation 8 and 9). This process is modelled using the same correction factor as in Equation 7.

$$\rho_{PAO} = \mu_{PAO} \cdot \frac{S_O}{K_O + S_O} \cdot \frac{S_{NH}}{K_{NH} + S_{NH}} \cdot \frac{S_{PO}}{K_{PO} + S_{PO}} \cdot \frac{S_{Alk}}{K_{Alk} + S_{Alk}} \cdot \frac{X_{PHA} / X_{PAO}}{K_{PHA} + X_{PHA} / X_{PAO}} \cdot X_{PAO} \tag{8}$$

$$\rho_{PAO,Anox} = \rho_{PAO} \cdot \eta_{NO} \cdot \frac{S_{NO}}{K_{NO} + S_{NO}} \cdot \frac{K_{IO}}{S_O} \tag{9}$$

Finally, the ASM2d also includes the lysis of PAO and their stored products.

Despite its diffusion, ASM2d presents a number of shortcomings when applied to given processes that have fostered the development of extensions or different modelling formulations. For instance, ASM2d neglects the use of glycogen as a reducing power source by PAO under anaerobic conditions (Hauduc et al., 2013). If glycogen is not considered, the model overestimates substrate storage, as volatile fatty acids (VFA) assimilation may cease once the internal glycogen storage is depleted (Mino et al., 1987). Some models have included glycogen as a state variable: under anaerobic conditions glycogen is oxidised to carbon dioxide and acts as electron donor whereas under aerobic conditions glycogen is replenished via PHA (Manga et al., 2001; Serralta et al., 2004a). Reducing power can be also produced via tricarboxylic acid (TCA) cycle, without using the glycogen storage as modelled by Comeau et al., (1986). Other authors have tried to correct the fact that all decay processes rates are the same regardless the redox environment in ASM2d. Flores-Alsina et al. (2012) developed a model extension whereby the decay rate is dependent on the electron acceptor. The model predictions were significantly different to those from original ASM2d, especially when a reactive settler model was considered.

Under anaerobic conditions PAO may compete for organic substrates with GAO, which do not contribute to phosphorus removal. As a consequence, EBPR performance may be deteriorated and models that do not account with GAO will fail predicting this behaviour (Hauduc et al., 2013). The ASM2d has been extended to include GAO and their related biochemical processes by Manga et al. (2001), Yagci et al.(2004) and Schuler (2005).

The implementation of biological phosphorus and nitrogen removal in a real WWTP presents complex interactions that must be taken into account in the model description to avoid a potential process failure.

The presence of nitrate or nitrite in the anaerobic zone is one of the most reported causes of phosphorus removal problems in WWTP. On the one hand, the competition between heterotrophic organisms and PAO for the carbon source under anoxic conditions may lead to the outcompetition of the second group, thereby stopping the biological phosphorus removal (Guerrero et al., 2011a). However, ASM2d does not consider nitrite as a state variable despite the fact that nitrite is a key intermediate for describing the anoxic COD consumption accurately (section 2). Therefore, several models have been reported to describe the effects of nitrite on the EBPR process in both real WWTPs and lab-scale studies (Wett & Rauch, 2003; Sin & Vanrolleghem, 2006). The PAO are modelled with different denitrifying capabilities depending on whether nitrite or nitrate is denitrified (Meinhold et al., 1999a; Saito et al., 2004; Guerrero et al., 2011b). On the other hand, nitrite has been reported as an inhibitor of phosphorus uptake under aerobic and anoxic conditions (Meinhold et al., 1999b; Saito et al., 2004). Sin et al. (2007b) included a reversible inhibition term for phosphorus uptake under aerobic conditions in the presence of nitrite.

## Metabolic Models for Biological Phosphorus Removal

Metabolic models are derived theoretically through substrate, energy and electron balances (Oehmen et al., 2007). Smolders et al. (1994a) developed the first metabolic model to describe the anaerobic metabolism of acetate by PAO. This model was further developed to include the aerobic (Smolders et al., 1994b) and anoxic (Kuba et al., 1996) metabolic processes. Brdjanovic et al. (2000) and Meijer (2004) proposed the so-called TUDP (Technical University of Delft Phosphorous) model by combining the metabolic model for denitrifying and non-denitrifying PAO with the ASM1 model. Contrary to ASM2/ASM2d, all the organic components of the metabolism of PAO are also considered in the TUDP model (e.g. glycogen). This model has been fully validated in pilot plants (Brdjanovic et al., 2000).

More recent studies have included competition between PAO and GAO for acetate (Filipe et al., 2001) and propionate (Oehmen et al., 2005 and 2006). In a follow-up, Oehmen et al. (2010a; 2010b) included new metabolic aspects to the description of PAO-GAO competition that potentially improve the capacity to predict a wider range of biological phosphorus removal conditions. These novel models have been validated with full-scale data (e.g. Lanham et al., 2014). Models with metabolic description are also SUITABLE to model the impact of environmental variables such as pH, temperature or oxygen levels on GAO-PAO competition. Thus, Lopez-Vazquez et al. (2009) presented a model including the effect of temperature, pH and carbon source, whilst the effect of oxygen was included by Carvalheira et al. (2014).

## Models for Chemical Phosphorus Removal

Phosphorus precipitation, for instance by addition of ferric chloride, is also highly important to model phosphorus removal in a WWTP. Although chemical phosphorus removal was already described in the ASM2, only ferric-phosphate precipitation was taken into account (Henze et al., 1999). Maurer et al. (1999), Barat et al. (2011) and Lizarralde et al. (2015)proposed several models to include the precipitation of different compounds, such as struvite ($MgNH_4PO_4$) hydroxydicalcium phosphate ($Ca_2HPO_4(OH)_2$, hydroxyapatite ($Ca_5(PO_4)_3OH$) or calcium phosphate ($Ca_3(PO_4)_2$). The challenge on describing accurately phosphorus precipitation is the large number of species that can play a role as common ions or have an impact on pH, and thereby on compound speciation (Hauduc et al., 2015). To address this issue, generalized physicochemical model for the whole WWTP, and including phosphorus precipitation processes, have been proposed by Jones et al. (2007) and Batstone et al. (2012). This framework has been expanded recently by Solon et al. (2015), Flores-Alsina et al. (2015), Kazadi-Mbamba et al. (2015a and 2015b) where an aqueous-phase plus general precipitation modules have been proposed and were experimentally validated.

## Future Trends on Biological Phosphorus Removal

Conventionally, the arrangement of biological and physical unit processes in wastewater treatment systems is used to remove organic carbon, nitrogen and phosphorus. Current research promotes a paradigm shift, whereby wastewater is considered as a source of energy, nutrients and water and not only as a source of pollution. Particular scientific interest focuses on the development of cost-effective ways to recover nitrogen and phosphorus from wastewater as slow-leaching forms for further application as fertilizer (Carballa et al., 2009; Yuan et al., 2012). Phosphorus recovery as a salt, like struvite or calcium phosphate, has been widely study (de Bashan and Bashan, 2004). However, conventional models need to be updated to properly predict the physicochemical reactions involved in the precipitation processes (section 3.3 and 4).

EBPR systems concentrate low strength streams to facilitate the precipitation process (Yuan et al., 2012) and therefore, the use of models for design and optimisation of EBPR plays a crucial role in resource recovery. An obstacle in model-based design and optimisation is their relatively narrow range of validity, caused by the hypotheses needed for model development. For instance, Valverde-Pérez et al. (2015) carried out a model-based design of enhanced biological phosphorus removal and recovery (EBP2R) but the experimental validation proved that filamentous bacteria compromised the process performance. Actually, filamentous bacteria are frequently found in EBPR systems, compromising the effluent quality and deteriorating sludge settleability (Mielczarek et al., 2012; Wágner et al., 2015) but are not included in the models available in literature.

A different means to recover phosphorus is precipitation after anaerobic digestion (Yuan et al., 2012). However, the anaerobic digestion model 1 (ADM1, Batstone et al., 2001) is not adequate to describe plant wide phosphorus transformations. In effect, the physicochemical formulation in the model does not cover complex effects such as ion activity and ion pairing, which strongly influence phosphate availability, and hence its partitioning in precipitates (Batstone et al., 2012, Flores-Alsina et al., 2016). As a consequence, plant-wide phosphorus description requires continuous ionic strength tracking, extensive consideration of non-ideality (including ion activities instead of molar concentrations) and finally complexation and precipitation processes (Solon et al., 2011, Flores-Alsina et al., 2015, Kazadi-Mbamba et al., 2015b). Plant wide application can be found in Lizarralde et al. (2015).

# PH MODELLING IN NUTRIENT REMOVAL SYSTEMS

Chemical and biological processes are directly influenced by the pH through substrate and inhibitor speciation. The structure and stability of bacterial cells are also impacted, which in turn impacts the activity level and process rates (Van Hulle et al., 2007). In ASM1 and ASM2d, it is assumed that pH is constant and near neutrality. The inclusion of alkalinity, as one of the state variables in the models, allows one to detect possible pH problems. For some reactions, specific functions can be added to the model to describe direct inhibitory pH effects on biological activity, as illustrated by Hellinga et al. (1999) for nitrification. A sounder and more accurate way of describing the effect of pH is to include the speciation of the modelled compounds and differentiating the metabolic role of the ionised and unionised species. As an illustration, Anthonisen et al. (1976) were among the first to propose $NH_3$ as substrate for AOB rather than $NH_4^+$ or total ammonium nitrogen (TAN=$NH_4^+ + NH_3$). The identification of the true substrate and inhibitory form has been, and still remains, a point of discussion (Sin et al., 2008) which influences how affinity constants and reaction rates are reported in nitrogen conversion processes.

Regarding phosphorus, high pH may induce chemical phosphorus precipitation, thereby reducing the phosphorus availability for PAO. For this reason, it is recommended to include a pH model, together with a precipitation model, to predict phosphorus concentrations at pH above 7.5 (Hauduc et al., 2013). Similarly to biological nitrogen removal processes, pH may affect the biological activity by directly inhibiting microbial growth or by affecting the speciation of some compounds. As an example, free nitrous acid has been demonstrated to be the real inhibitor of phosphorus uptake by PAO (Pijuan et al., 2010). High pH increases the energy demand for VFA uptake by PAO, which is compensated by increasing the phosphorus release per mole of VFA (Oehmen et al., 2007). To overcome this situation, the phosphorus release yield is modelled as function of the pH (Smolders et al., 1994a; Filipe et al., 2001). Given the importance of pH, Serralta et al. (2004a) developed a pH model extension for the ASM2d.

Modelling the flow rates and composition of gas stripping is also greatly impacted by pH. In particular, $CO_2$ stripping has a direct impact on the alkalinity inventory and as a consequence, impacts the pH of the medium. At the same time, as only neutral molecules can be stripped, ammonia stripping is negligible at pH values well below ammonium pKa (9.25). The ASM framework has been extended with expressions capturing the effect of stripping as illustrated by Sin and Vanrolleghem (2007).

Traditionally, pH has been modelled by solving a charge balance that was closed by estimating the proton concentration in the water (Hellinga et al., 1999). This approach was extended in a dynamic simulation by Vangsgaard et al. (2013) by formulating a set of equations describing the mass balances and equilibrium equations of weak acids and bases. These set of nonlinear algebraic equations was closed with the charge balance that allowed one to find the proton concentration in the medium via multi-dimensional Newton-Raphson method. During the last couple of years special attention is placed on improving the description of the acid-base system by including the effect of ionic strength and developing of improved numerics (Solon et al., 2015; Flores-Alsina et al., 2015).

## CONTROL DESIGN IN WASTEWATER TREATMENT

The automatic control of wastewater treatment processes is characterised by the need to keep discharge limits while receiving continuous load variations from the influent (in flow rate, composition, temperature, etc.). Therefore, most controllers implemented in WWTP aim at regulating the system, i.e. to keep the process close to a desired state by rejecting the disturbances introduced by the influent variations while keeping the effluent requirements. More advanced controllers that aim at supervising or optimising the process, i.e. to automatically drive the process towards an optimal state, are rarely seen in WWTP with a few relevant exceptions that are mentioned below. Advanced control schemes such as model predictive control (MPC) are usually hindered by the low accuracy of models describing the microbial metabolism, highly nonlinear behaviour, interactive dynamics, model uncertainty, and by the complexity of such controllers (Olsson, 2011). In this respect, the simplicity of a controller is an important characteristic in a bioreactor since it is likely that frequent maintenance will be needed as a result of variations in the feedstock, seasonal conditions and even because of microbial evolution. Hence, a trade-off must be achieved between efficient control and monitoring tools on the one hand, and simplicity on the other hand in order to ensure the success of the control scheme.

## Control of Ammonium in Nitrification-Denitrification Processes

The use of feedback control to regulate the level of ammonium in the effluent is arguably the most spread control structure for nutrient removal in WWTPs. In general, this control structure consists of manipulating the aeration flow rate in the aerated tanks. The simplest feedback controller tries to keep a constant dissolved oxygen (DO) level. The DO setpoint is chosen as a trade-off: it must be high enough to ensure the nitrification of all the ammonium in the influent while a too high level would lead to wasting energy caused by aeration excess and can even hamper denitrification.

The most obvious upgrade to the ammonium control loop is to set up a cascade-control whereby the concentration of ammonium is used to fix the setpoint of DO (Figure 2). The advantage of this cascaded control is that it adjusts the DO setpoint to the changes in ammonium load, thereby providing savings in aeration energy. On the other hand, such a controller requires ammonium sensors which require higher maintenance than DO or pH sensors, while having a comparatively worse performance in terms of drift. In the cascaded controller configuration, the ammonium feedback loop (generically called the master loop) is tuned more slowly than the DO feedback loop (the slave loop). Ammonium dynamics are generally caused by load variations; therefore, tuning the controllers with a 1 day response rate is generally enough.

The cascade ammonium loop can be considered a supervision/optimising loop given that, besides the regulation of the process, aims at minimizing the energy consumption. In a full-scale study, Åmand et al. (2013) showed that energy savings of up to 7% to 19% were possible. However, the authors pointed out that the proper implementation of the cascaded controller required considerable time (tuning, anti-windup management) and that ammonium sensors required frequent maintenance.

*Figure 2. Control structures for ammonium removal. The regulatory loop is highlighted within the blue dash-dotted line. The supervisory loop, which acts as the master loop in a cascade configuration, is limited by the red dash-dotted line*

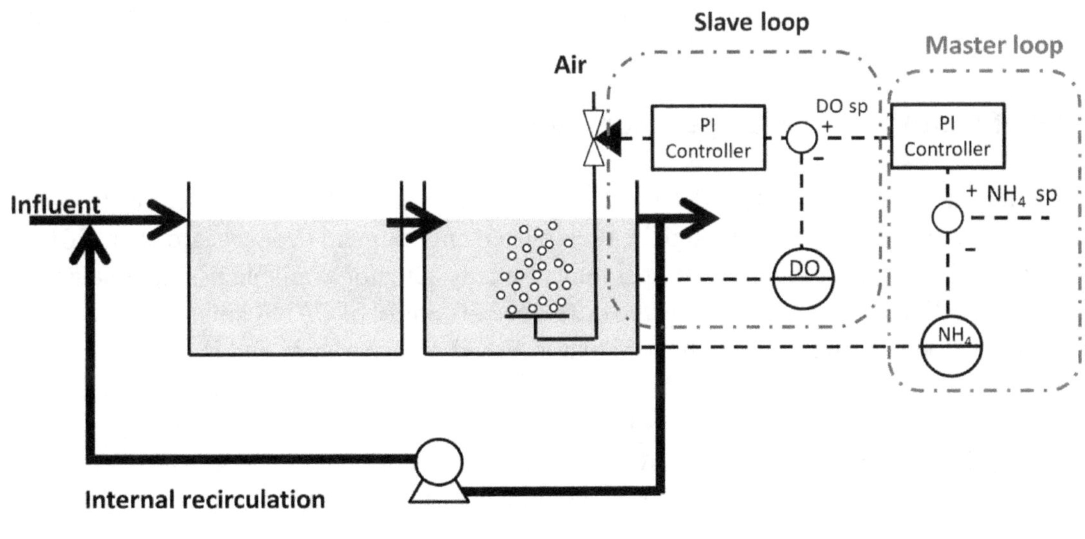

## Control of Nitrate in Pre-Denitrifying Processes

Nitrate removal in pre-denitrification tanks takes place as heterotrophic bacteria oxidize COD by using nitrate as an electron acceptor in an anoxic environment. Therefore, the two main actions that are used to control nitrate removal are, precisely, to increase the amount of nitrate in the tank and/or to increase the available COD. In the first case, the amount of nitrate in the anoxic tanks can be regulated by manipulating the internal recirculation flowrate so that as much COD as possible is used in nitrate reduction. This strategy relies in ensuring that the nitrification process is working properly. In such a case, it is economically attractive as only pumping energy costs are increased. However, in plants with an unfavourable COD to N ratio, addition of an external carbon source in the anoxic part can be the most effective solution. The drawbacks of this strategy are that carbon addition may represent a significant increase of operating costs and, excess of carbon dosing may deteriorate the effluent quality.

The two mentioned control strategies can be implemented in a rather similar way (Yong et al., 2006). The controlled variable is frequently the nitrate concentration at the end of the anoxic area (with a set-point of 1-3 mgN/L). The use of the recirculation flowrate as a manipulated variable imposes tuning the controller rather as changing the nitrate concentration by recirculation introduces a considerable delay in the loop. However, the typical influent dynamics do not feature fast changes of the COD to N ratio and this is rarely a problem. Actually, selecting which control strategy best fits a given WWTP is mainly done on the basis of the COD to N ratio: high COD to N ratios naturally preferring the use of internal recirculation as a manipulated variable to maximize the use of COD and low COD to N ratios using the addition of external carbon to supplement the lack of COD for denitrification. For cases where the COD to N ratio can be expected to vary notably along the year, Yong et al. (2006) proposed several integrated control structure that combine both characteristics. The common idea to these integrated controllers is to use two distinct controlled variables: the nitrate concentration at the outlet of the anoxic part and at the outlet of the aerobic part. Thus, it is possible to set up two independent controllers that manipulate the recirculation flowrate and the addition of external carbon. In their study, they benchmarked several integrated control structures and the proposed the following pairing: nitrate in the anoxic part controlled by the addition of external carbon and nitrate in the aerobic part controlled by the recirculation flowrate.

Supervisory control structures have also been proposed for nitrate removal, including model predictive control (Liu & Yoo, 2011) or, together with ammonium control, model predictive control with feedforward action (Stare et al., 2007; Santín et al., 2015).

## Control of Autotrophic Nitrogen Removal Processes

The operation of a single-stage CANR system is characterized by the fact that competing microbial groups coexist in the same vessel although in a segregated environment (Vangsgaard et al., 2014). Apart from the regulation of pH, mixing and temperature, the only actuator available in single-stage CANR is generally the aeration flowrate. Therefore, the difference in control of CANR single stage systems relies on how to relate the measurements with the actuator. Based on experience and insight on the microbial metabolism, Vlaeminck et al. (2012) proposed a full strategy relating the process objectives (e.g. maximize nitrogen removal, minimize risk of $N_2O$ and NO production…) with the operating conditions. In the same way, Mutlu et al. (2013) proposed an operation protocol based on a decision tree, to facilitate start-up and operation of single stage CANR. This simple protocol, which is based on calculating ratios

*Figure 3. Control structures for nitrate removal for two possible manipulated variables (MV)*

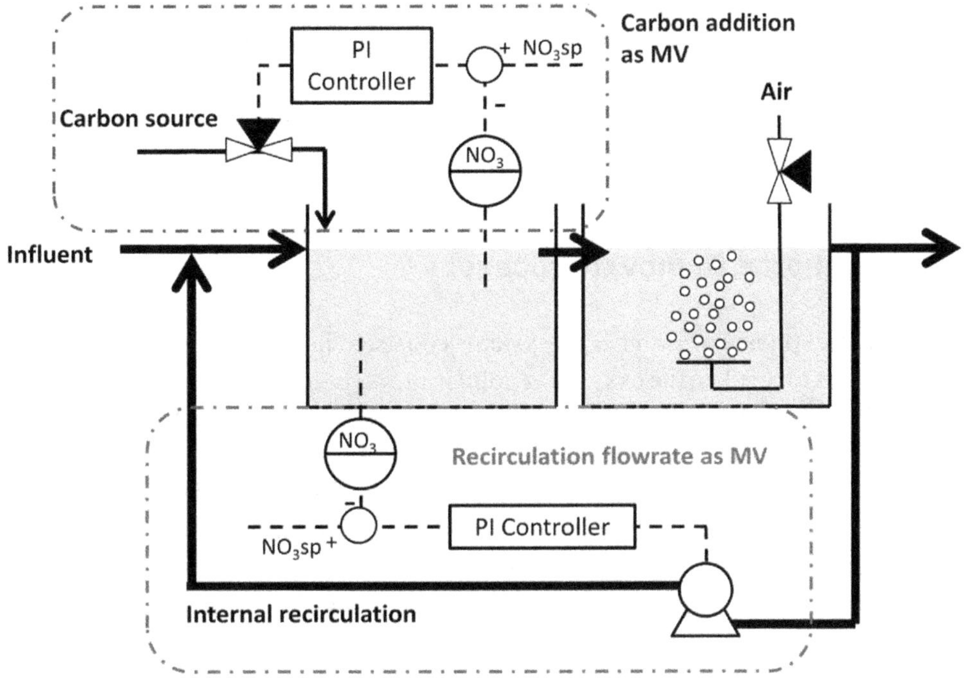


of the nitrogen species, was translated into a fuzzy automatic controller and diagnosis system by Boiocchi et al. (2015) achieving in an *in-silico* study high and stable nitrogen removal efficiency.

As the aeration flowrate is the most common manipulated variable in CANR, dissolved oxygen (DO) can be seen as the natural controlled variable. Vazquez-Padin et al. (2014) claim in a granted patent that it is possible to control single-stage CANR with a constant DO setpoint higher than 0.8 mg $L^{-1}$. The rationale behind this controller is that once that the granules are developed and ammonium oxidizing bacteria (AOB) are established in the granule crust, AOB will deplete enough oxygen to ensure that an anoxic environment is still present in the granule core. To go beyond the process regulation and in order to maximize the nitrogen removal efficiency, Trollberg et al. (2014) proposed instead the use of extremum seeking control systems to maximize the nitrogen removal efficiency. Extremum seeking control is a model-free adaptive controller that tries to improve the operation of the process automatically by recording the changes in a defined objective function (in this case nitrogen removal efficiency) during operation. In the search of a control structure to maximize the nitrogen removal efficiency, Vangsgaard et al. (2014) tested by simulation three control structures, all of them using the aeration flowrate as a manipulated variable: 1) using a feedforward loop that measures the ammonium load in the influent and tries to keep a constant nitrogen/oxygen loading ratio; 2) measuring the nitrogen species in the reactor, calculating the nitrogen removal efficiency and using a feedback loop to maximize the efficiency and; 3) combining the two previous control strategies in order to provide fast feedforward response and model mismatch correction by feedback. This last control structure was implemented in a 4L reactor by the same authors (Mauricio-Iglesias et al., 2015) who found it able to achieve nitrogen removal efficiencies close to 90% despite disturbances in ammonium loading ratio.

CANR control in two-stage configuration has been thoroughly tackled by Volcke et al. (2006 and 2007) who showed that aeration could be used to regulate the nitrite/ammonium ratio in the first reactor. In these contributions, the DO setpoint is set by a cascade control structure where the master loop tries to adjust the nitrite/ammonium ratio to the stoichiometric theoretical optimum (1.3). Thus, the influent to the second reactor (anammox) can achieve a maximum conversion to nitrogen. Such a design was confirmed by Valverde-Perez et al. (2016a) who produced an optimal control for the two-stage CANR based on the measurement of the nitrogen species and the use of two cascaded controllers.

## Control of Phosphorus Removal Processes

First attempts to control phosphorus in EBPR systems consisted in control loops based on proportional-integral (PI) controllers to regulate the oxygen set point in the aerobic reactors to achieve good phosphorus removal (see e.g. Gernaey & Jørgersen, 2004). However, the proposed control loop leads to poor nutrient removal due to interactions with ammonia effluent control (Machado et al., 2009). Consequently, many studies focused on the control of nitrogen discharges, while acceptable phosphorus removal was achieved by optimizing the set points of the controllers regulating the nitrification-denitrification process (Guerrero et al., 2011c; Guerrero et al., 2014). Alternatively, some studies proposed to control the phosphorus release in the anaerobic reactors by manipulating the recirculation from the bottom of the clarifier and thereby manipulating the nitrate recycling to the anaerobic reactors (Machado et al., 2009; Machado et al., 2015). The nitrate concentration in the anaerobic reactors drives the competition for VFA between PAO and denitrifying bacteria, which affects the phosphorus release (Ostace et al., 2013).

Chemical dosing is a common practice in WWTP to control nutrient removal (Olsson 2012). In the case of phosphorus removal, metal salts (e.g. $FeClSO_4$) can be dosed to precipitate the phosphate that cannot be removed biologically (Devisscher et al., 2002). Other option is the use of an external carbon source (e.g. acetate) in the anaerobic reactors to ensure the PHA storage by PAO, thereby allowing their growth and polyphosphate storage at the aerobic stage (Ostace et al., 2013). This strategy has been successfully applied to pilot scale systems using glycerol as external carbon source (Guerrero et al., 2015). However, the addition of chemicals adds an environmental cost that can compromise the sustainability of the EBPR process (Corominas et al., 2013). Yuan et al. (2008) proposed to optimize the microbial communities on the reactor to reduce the chemical demand, strategy that has been successfully applied by Valverde-Pérez et al. (2016b) to design control structures for phosphorus recovery and removal.

Other option to ensure that PAO take up and store enough organics is to monitor the phosphorus release. Usually, for sequenced batch reactors, this task can be performed by monitoring conductivity or pH (Maurer & Gujer, 1995). Shifts on pH trends along a batch can indicate the end of denitrification, the beginning of phosphorus release, carbon dioxide stripping or even out competition of PAO by GAO (Serralta et al., 2004b; Spagni et al., 2001). Conductivity in the anaerobic phase increases with phosphorus release until a plateau is reached, indicating the depletion of COD or polyphosphate storage. During the aerobic phase conductivity decreases until all phosphate is removed. By monitoring these parameters the length of each phase of the SBR can be optimized achieving fast removal rates and ensuring good effluent quality (Aguado et al., 2006; Guisasola et al., 2007b). A simpler approach was implemented by Ingildsen et al. (2006), whereby the length of the anaerobic period was extended to enhance PHA accumulation whenever ammonia was removed faster than phosphate in the aerobic period.

## Fuzzy Control and Diagnosis of Nutrient Removal Processes

Fuzzy decision methods have been used for diagnosis and control as a means to reconcile the use of expert knowledge and quantitative models (Honda and Kobayashi 2000). In fuzzy control, control laws are expressed in linguistic terms which can be converted into mathematical expression. Likewise, fuzzy decision systems can provide a diagnosis of the process in common linguistic terms from the available measurements (Puñal et al., 2001). Hence, fuzzy controller and fuzzy diagnosis modules are intrinsically easy to understand and to adapt in function of control performance requirements. Boiocchi et al. (2015) proposed a setup composed of fuzzy controller and diagnosis for a CANR process. They converted the diagnosis tree presented by Mutlu et al. (2013) into a fuzzy logic diagnosis module. In this setup (Figure 4) the fuzzy logic diagnosis system converts the measurements into a diagnosis with four linguistic variables: "*Performance*", "*Nitritration*", "*Nitratation*" and "*Anaerobic ammonium oxidizing bacteria activity*". These four results of diagnosis can be read by the process operators or fed to an automatic fuzzy logic controller that acts by manipulating the aeration flowrate. Although it is possible to design a mathematically equivalent setup with one single module, the two-module architecture provides a better insight of the diagnosis/control and makes the tuning, upgrade or retrofitting much easier.

## Nutrient Removal Control with Soft Sensors

Control in nutrient removal, in particular ammonium control by aeration is well known and widely applied in large WWTPs. In contrast, ammonium control is not necessarily economically attractive in small WWTPs as energy and cost saving do not compensate the costs of installation and maintenance of ammonium measuring sensors. Soft sensors have been tested as an alternative to implement ammonium control using more widespread and robust sensors such as DO and pH meters. A soft sensor is a described as an input-output model that relates easily measurable variables and signals with primary variables which are hard or costly to measure (Haimi et al., 2013). Soft sensors were actually applied and implemented in a number of cases. Bongards (2001), in one of the first cases of real application of soft sensors, used the influent flow-rate and conductivity, and DO in the aerated nitrification tank as input

*Figure 4. Structure of fuzzy logic diagnosis (FLD) providing four diagnosis items to the fuzzy logic controller (FLC) that subsequently manipulates the flowrate in a complete autotrophic nitrogen removal reactor (Adapted from Boiocchi et al., 2015)*

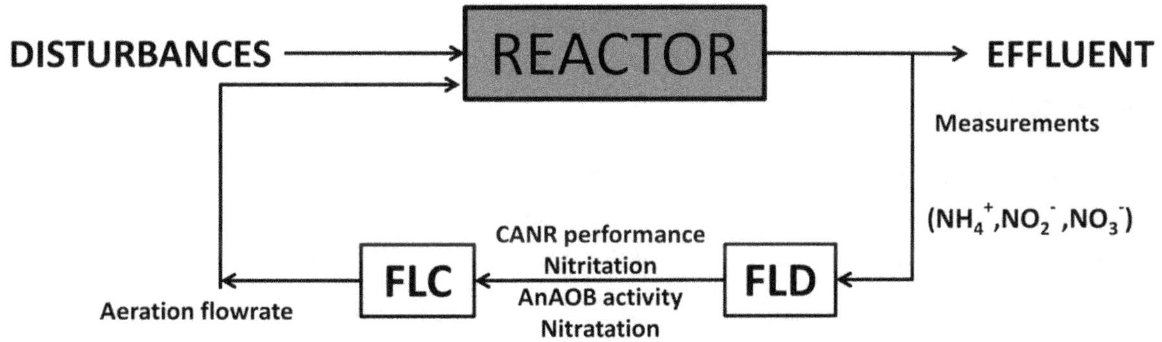

variables in order to estimate the ammonium and nitrate concentrations. A fuzzy controller was set to regulate ammonium and nitrate concentration by adjusting the aeration flowrate. Recently, Thürlimann et al. (2015) used only two pH sensors (placed at the entrance and the end of the aerated tanks) to estimate the ammonium load to a small municipal WWTP in Switzerland. The goal was, distinguishing between high and low ammonium loads, to assign an oxygen setpoint to each of the two states. Implementing a controller for aeration, they were able to achieve savings of over 10% in energy duty for aeration.

## CONCLUSION

As nutrient removal processes become increasingly complex, mathematical models appear as an essential tool to assist in WWTP design and to provide valuable process understanding. Starting from the very popular ASM family models, there have been contributions in the latest years aiming at include novel processes and/or answer to new needs and concerns such as the release of greenhouse gases or the recovery of phosphorus. For instance, the precipitation of phosphorus salts requires detailed physico-chemical models describing ion interactions and precipitation and efficient numerical methods to solve the resulting models.

Mathematical models of nutrient removal systems are also applied in control design as they provide information about process dynamics and potential loop interactions. Simulation platforms count with generally accepted nutrient removal models and can be used to test and demonstrate the design of a control system *in silico*. For complex technologies where the models are i) not accurate enough for dynamic simulation or ii) too complex for an efficient solving procedure, data-driven solutions such as fuzzy control or the use of soft sensors have gained acceptance.

## ACKNOWLEDGMENT

Dr. Mauricio-Iglesias gratefully acknowledges the financial support provided by the People Program (Marie Curie Actions) of the European Union's Seventh Framework Programme FP7/2007-2013 under REA agreement 627475 (GREENCOST). Dr. Mauricio-Iglesias belongs to the Galician Competitive Research Group GRC 2013-032, programme co-funded by FEDER. Dr Flores-Alsina gratefully acknowledges the financial support of the collaborative international consortium WATERJPI2015 WATINTECH of the Water Challenges for a Changing World Joint Programming Initiative (Water JPI) 2015 call. Dr. B. Valverde-Pérez thanks the Integrated Water Technology (InWaTech) project between the Technical University of Denmark and the Korean Advanced Institute of Science and Technology for the financial support.

## REFERENCES

Aguado, D., Montoya, T., Ferrer, J., & Seco, A. (2006). Relating ions concentration variations to conductivity variations in a sequencing batch reactor operated for enhanced biological phosphorus removal. *Environmental Modelling & Software*, *21*(6), 845–851. doi:10.1016/j.envsoft.2005.03.004

Almeida, J. S., Reis, M. A. M., & Carrondo, M. J. T. (1995). Competition between nitrate and nitrite reduction in denitrification by Pseudomonas fluorescens. *Biotechnology and Bioengineering*, *46*, 474–484. PMID:18623340

Åmand, L., Laurell, C., Stark-Fujii, K., Thunberg, A., & Carlsson, B. (2014). Lessons learnt from evaluating full-scale ammonium. *Water Science and Technology*, *69*(7), 1573. doi:10.2166/wst.2014.061 PMID:24718353

Anthonisen, A. C., Loehr, R. C., Prakasam, T. B. S., & Srinath, E. G. (1976). Inhibition of nitrification by ammonium and nitrous-acid. *Journal - Water Pollution Control Federation*, *48*, 835–852. PMID:948105

Barat, R., Montoya, T., Seco, A., & Ferrer, J. (2011). Modelling biological and chemically induced precipitation of calcium phosphate in enhanced biological phosphorus removal systems. *Water Research*, *45*(12), 3744–3752. doi:10.1016/j.watres.2011.04.028 PMID:21565382

Batstone, D. J., Amerlinck, Y., Ekama, G., Goel, R., Grau, P., Johnson, B., & Volcke, E. et al. (2012). Towards a generalised physicochemical framework. *Water Science and Technology*, *66*(6), 1147–1161. doi:10.2166/wst.2012.300 PMID:22828290

Batstone, D. J., Keller, J., Angelidaki, I., Kalyuzhnyi, S. V., Pavlostathis, S. G., Rozzi, A., & Vavilin, V. A. et al. (2001). The IWA Anaerobic Digestion Model No 1 (ADM1). *Water Science and Technology*, *45*(10), 65–73. PMID:12188579

Bernet, N., Sanchez, O., Cesbron, D., Steyer, J. P., & Delgenes, J. P. (2005). Modeling and control of nitrite accumulation in a nitrifying biofilm reactor. *Biochemical Engineering Journal*, *24*(2), 173–183. doi:10.1016/j.bej.2005.02.002

Boiocchi, R., Mauricio-Iglesias, M., Vangsgaard, A. K., Gernaey, K. V., & Sin, G. (2015). Aeration control by monitoring the microbiological activity using fuzzy logic diagnosis and control. Application to a complete autotrophic nitrogen removal reactor. *Journal of Process Control*, *30*, 22–33. doi:10.1016/j.jprocont.2014.10.011

Bongards, M. (2001). Improving the efficiency of a wastewater treatment plant by fuzzy control and neural networks. *Water Science and Technology*, *43*(11), 189–196. PMID:11443962

Brdjanovic, D., van Loosdrecht, M. C. M., Versteeg, P., Hooijmans, C. M., Alaerts, G. J., & Heijnen, J. J. (2000). Modelling COD, N and P removal in a full-scale WWTP Haarlem Waarderpolder. *Water Research*, *34*(3), 846–858. doi:10.1016/S0043-1354(99)00219-5

Carballa, M., Moerman, W., De Windt, W., Grootaerd, H., & Vestraete, W. (2008). Strategies to optimize phosphate removal from industrial anaerobic effluents by magnesium ammonium phosphate (MAP) production. *Journal of Chemical Technology and Biotechnology (Oxford, Oxfordshire)*, *84*(1), 63–68. doi:10.1002/jctb.2006

Carvalheira, M., Oehmen, A., Carvalho, G., Eusébio, M., & Reis, M. A. M. (2014). The impact of aeration on the competition between polyphosphate accumulating organisms and glycogen accumulating organisms. *Water Research*, *66*, 296–307. doi:10.1016/j.watres.2014.08.033 PMID:25222333

Comeau, Y., Hall, K. J., Hancock, R. E. W., & Oldham, W. K. (1986). Biochemical model for enhanced biological phosphorus removal. *Water Research*, *20*(12), 1511–1521. doi:10.1016/0043-1354(86)90115-6

Corominas, L., Larsen, H. F., Flores-Alsina, X., & Vanrolleghem, P. A. (2013). Including life cycle assessment for decision-making in controlling wastewater nutrient removal systems. *Journal of Environmental Management*, *128*, 759–767. doi:10.1016/j.jenvman.2013.06.002 PMID:23856224

Dapena-Mora, A., van Hulle, S. W. H., Campos, J. L., Mendez, R., Vanrolleghem, P. A., & Jetten, M. (2004). Enrichment of Anammox biomass from municipal activated sludge: Experimental and modelling results. *Journal of Chemical Technology and Biotechnology (Oxford, Oxfordshire)*, *79*(12), 1421–1428. doi:10.1002/jctb.1148

de-Bhasan, L. E., & Bashan, Y. (2004). Recent advances in removing phosphorus from wastewater and its future use as fertilizer (1997-2003). *Water Research*, *8*(19), 4222–4246. doi:10.1016/j.watres.2004.07.014 PMID:15491670

Devisscher, M., Bogaert, H., Bixio, D., Van de Velde, J., & Thoeye, C. (2002). Feasibility of automatic chemicals dosage control – a full-scale evaluation. *Water Science and Technology*, *45*(4-5), 445–452. PMID:11936665

Egli, K., Fanger, U., Alvarez, P. J. J., Siegrist, H., van der Meer, J. R., & Zehnder, A. J. B. (2001). Enrichment and characterization of an anammox bacterium from a rotating biological contactor treating ammonium-reach leachate. *Archives of Microbiology*, *175*(3), 198–207. doi:10.1007/s002030100255 PMID:11357512

Filipe, C. D. M., Daigger, G. T., & Grady, C. P. L. Jr. (2001). Stoichiometry and kinetics of acetate uptake under anaerobic conditions by an enriched culture of phosphorus-accumulating organisms at different pHs. *Biotechnology and Bioengineering*, *76*(1), 32–43. doi:10.1002/bit.1023 PMID:11400104

Flores-Alsina, X., Gernaey, K. V., & Jeppsson, U. (2012). Benchmarking biological nutrient removal in wastewater treatment plants: Influence of mathematical model assumptions. *Water Science and Technology*, *65*(8), 1496–1505. doi:10.2166/wst.2012.039 PMID:22466599

Flores-Alsina, X., Kazadi-Mbama, C., Solon, K., Vrecko, D., Tait, S., Batstone, D., & Gernaey, K. V. et al. (2015). A plant wide aqueous phase chemistry module describing pH variations and ion speciation/pairing in wastewater treatment models. *Water Research*, *85*, 255–265. doi:10.1016/j.watres.2015.07.014 PMID:26342179

Flores-Alsina, X., Solon, K., Kazadi-Mbamba, C., Tait, S., Jeppsson, U., Gernaey, K. V., & Batstone, D. J. (2016). Modelling phosphorus (P), sulfur (S) and iron (Fe) interactions for dynamic simulations of anaerobic digestion processes. *Water Research*, *95*, 370–382. doi:10.1016/j.watres.2016.03.012 PMID:27107338

Gernaey, K. V., & Jørgensen, S. B. (2004). Benchmarking combined biological phosphorus and nitrogen removal wastewater treatment processes. *Control Engineering Practice*, *12*(3), 357–373. doi:10.1016/S0967-0661(03)00080-7

Grady, C. P. L., Gujer, W., Henze, M., Marais, G. V. R., & Matsuo, T. (1986). A model for single-sludge wastewater treatment systems. *Water Science and Technology*, *16*(6), 47–61.

Guerrero, J., Guisasola, A., & Baeza, J. A. (2011b). The nature of the carbon source rules the competition between PAO and denitrifiers in systems for simultaneous biological nitrogen and phosphorus removal. *Water Research*, *45*(16), 4793–4802. doi:10.1016/j.watres.2011.06.019 PMID:21774957

Guerrero, J., Guisasola, A., & Baeza, J. A. (2014). A novel control strategy for efficient biological phosphorus removal with carbon-limited wastewaters. *Water Science and Technology*, *70*(4), 691–697. doi:10.2166/wst.2014.280 PMID:25116500

Guerrero, J., Guisasola, A., & Baeza, J. A. (2015). Controlled crude glycerol dosage to prevent EBPR failures in C/N/P removal WWTPs. *Chemical Engineering Journal*, *271*, 114–127. doi:10.1016/j.cej.2015.02.062

Guerrero, J., Guisasola, A., Comas, J., Rodríguez-Roda, I., & Baeza, J. A. (2011a). Multi-criteria selection of optimum WWTP control setpoints base on microbiology-related failures, effluent quality and operating costs. *Chemical Engineering Journal*, *188*, 23–29. doi:10.1016/j.cej.2012.01.115

Guerrero, J., Guisasola, A., Vilanova, R., & Baeza, J. A. (2011c). Improving the performance of a WWTP control system by model-based setpoint optimization. *Environmental Modelling & Software*, *26*(4), 492–497. doi:10.1016/j.envsoft.2010.10.012

Guisasola, A., Petzet, S., Baeza, J. A., Carrera, J., & Lafuente, J. (2007). Inorganic carbon limitations on nitrification: Experimental assessment and modeling. *Water Research*, *41*(2), 277–286. doi:10.1016/j.watres.2006.10.030 PMID:17169397

Guisasola, A., Vargas, M., Marcelino, M., Lafuente, J., Casas, C., & Baeza, J. A. (2007b). On-line monitoring of the enhanced biological phosphorus removal process using respirometry and titrimetry. *Biochemical Engineering Journal*, *35*(3), 371–379. doi:10.1016/j.bej.2007.02.001

Gujer, W., Henze, M., Mino, T., Matsuo, T., Wentzel, M. C., & Marais, G. R. (1995). The Activated Sludge Model No. 2: Biological phosphorus removal. *Water Science and Technology*, *31*(2), 1–11. doi:10.1016/0273-1223(95)00175-M

Gujer, W., Henze, M., Mino, T., & van Loosdrecht, M. C. M. (1999). Activated sludge model No. 3. *Water Science and Technology*, *39*(1), 183–193. doi:10.1016/S0273-1223(98)00785-9

Haimi, H., Mulas, M., Corona, F., & Vahala, E. (2013). Data-derived soft-sensors for biological wastewater treatment plants: An overview. *Environmental Modelling & Software*, *47*, 88–107. doi:10.1016/j.envsoft.2013.05.009

Hao, X. D., Heijnen, J. J., & van Loosdrecht, M. C. M. (2002a). Sensitivity analysis of a biofilm model describing a one-stage completely autotrophic nitrogen removal (CANON) process. *Biotechnology and Bioengineering*, *77*(3), 266–277. doi:10.1002/bit.10105 PMID:11753935

Hao, X.D., Heijnen, J.J., & van Loosdrecht, M.C.M. (2002b) Model-based evaluation of temperature and inflow variations on a partial nitrification-ANAMMOX biofilm process. *Water Research, 36*, 4839-4849.

Hao, X. D., & van Loosdrecht, M. C. M. (2004). Model-based evaluation of COD influence on a partial nitrification-Anammox biofilm (CANON) process. *Water Science and Technology*, *49*(11-12), 83–90. PMID:15303727

Hauduc, H., Guillot, S., Rieger, L., Ohtsuki, T., Shaw, A., Takács, I., & Winkler, S. (2009). Activated sludge modelling practice: An international survey. *Water Science and Technology*, *60*(8), 1943–1951. doi:10.2166/wst.2009.223 PMID:19844041

Hauduc, H., Rieger, L., Oehmen, A., van Loosdrecht, M. C. M., Comeau, Y., Héduit, A., & Guillot, S. et al. (2013). Critical review of activated sludge modeling: State of process knowledge, modeling concepts and limitations. *Biotechnology and Bioengineering*, *1*(1), 24–46. doi:10.1002/bit.24624 PMID:22886494

Hauduc, H., Takács, I., Smith, S., Szabo, A., Murthy, S., Daigger, G. T., & Spérandio, M. (2015). A dynamic physicochemical model for chemical phosphorus removal. *Water Research*, *73*, 157–170. doi:10.1016/j.watres.2014.12.053 PMID:25655322

Hellinga, C., van Loosdrecht, M. C. M., & Heijnen, J. J. (1999). Model based design of a novel process for nitrogen removal from concentrated flows. *Mathematical and Computer Modelling*, *5*(4), 351–371.

Henze, M., Grady, C. P. L., Jr., Gujer, W., Marais, G. R., & Matsuo, T. (1987). *Activated Sludge Model No. 1*. IAWQ Scientific and Technical Report No. 1.

Henze, M., Gujer, W., Mino, T., Matsuo, T., & van Loosdrecht, M. C. M. (2000). *Activated sludge models ASM1, ASM2, ASM2d and ASM3*. IWA Publishing.

Henze, M., Gujer, W., Mino, T., Matsuo, T., Wentzel, M. C., Marais, G. R., & van Loosdrecht, M. C. M. (1999). Activated sludge model No. 2D, ASM2D. *Water Science and Technology*, *39*(1), 165–182. doi:10.1016/S0273-1223(98)00829-4

Hiatt, W. C., & Grady, C. P. L. Jr. (2008). An updated process model for carbon oxidation, nitrification, and denitrification. *Water Environment Research*, *80*(11), 2145–2156. doi:10.2175/106143008X304776 PMID:19024730

Honda, H., & Kobayashi, T. (2000). Fuzzy control of bioprocess. *Journal of Bioscience and Bioengineering*, *89*(5), 401–408. doi:10.1016/S1389-1723(00)89087-8 PMID:16232768

Iacopozzi, I., Innocenti, V., Marili-Libelli, S., & Giusti, E. (2007). A modified Activated Sludge Model No. 3 (ASM3) with two-step nitrification-denitrification. *Environmental Modelling & Software*, *22*(6), 847–861. doi:10.1016/j.envsoft.2006.05.009

Ingildsen, P., Rose, C., Gernaey, K. V., Nielsen, M. K., Guidal, T., & Jacobsen, B. N. (2006). Modelling and control strategy testing of biological and chemical phosphorus removal at Avedøre WWTP. *Water Science and Technology*, *53*(4-5), 105–113.

Jones, R. M., Dold, P. L., Takács, I., Chapman, K., Wett, B., Murphy, S., & O'Shaughnessy, M. (2007). Simulation for operation and control of reject water treatment processes. In *Proc. WEFTEC2007*.

Jung, J. Y., Kang, S. H., Chung, Y. C., & Ahn, D. H. (2007). Factors affecting the activity of anammox bacteria during start up in the continuous culture reactor. *Water Science and Technology*, *55*(1-2), 459–468. doi:10.2166/wst.2007.023 PMID:17305171

Kampschreur, M. J., Temmink, H., Kleerebezem, R., Jetten, M. S. M., & van Loosdrecht, M. C. M. (2009). Nitrous oxide emission during wastewater treatment. *Water Research*, *43*(17), 4093–4103. doi:10.1016/j.watres.2009.03.001 PMID:19666183

Kazadi Mbamaba, C., Flores-Alsina, X., Batstone, D., & Tait, S. (2015a). A generalised chemical precipitation modelling approach in wastewater treatment applied to calcite. *Water Research*, *68*, 342–353. doi:10.1016/j.watres.2014.10.011 PMID:25462741

Kazadi Mbamaba, C., Flores-Alsina, X., Batstone, D., & Tait, S. (2015b). A systematic study of multiple minerals precipitation modelling in wastewater treatment. *Water Research*, *85*, 359–370. doi:10.1016/j.watres.2015.08.041 PMID:26360229

Knowles, R. (1982). Denitrification. *Microbiological Reviews*, *46*, 43–70. PMID:7045624

Koch, G., Egli, K., van der Meer, J.R., & Siegrist, H. (2000). Mathematical modeling of autotrophic denitrification in a nitrifying biofilm of a rotating biological contactor. *Water Science & Technology*, *41*, 191-198.

Kuba, T., Murnleitner, E., van Loosdrecht, M. C. M., & Heijen, J. J. (1996). A metabolic model for biological phosphorus removal by denitrifying organisms. *Biotechnology and Bioengineering*, *52*(6), 685–695. doi:10.1002/(SICI)1097-0290(19961220)52:6<685::AID-BIT6>3.3.CO;2-M PMID:18629947

Lackner, S., Terada, A., & Smets, B. F. (2008). Heterotrophic activity compromises autotrophic nitrogen removal in membrane-aerated biofilms: Results of a modeling study. *Water Research*, *42*(4-5), 1102–1112. doi:10.1016/j.watres.2007.08.025 PMID:17915280

Lanham, A. B., Oehmen, A., Saunders, A. M., Carvalho, G., Nielsen, P. H., & Reis, M. A. M. (2014). Metabolic modelling of full-scale enhanced biological phosphorus removal sludge. *Water Research*, *66*, 283–295. doi:10.1016/j.watres.2014.08.036 PMID:25222332

Liu, H., & Yoo, C. (2011). Performance assessment of cascade controllers for nitrate control in a wastewater treatment process. *Korean Journal of Chemical Engineering*, *28*(3), 667–673. doi:10.1007/s11814-010-0442-x

Lizarralde, I., Fernandez-Arevalo, T., Brouckaert, C., Vanrolleghem, P. A., Ikumi, D., Ekama, D., & Grau, P. et al. (2015). A new general methodology for incorporating physico-chemical transformations into multiphase wastewater treatment process models. *Water Research*, *74*, 239–256. doi:10.1016/j.watres.2015.01.031 PMID:25746499

Lopez-Vazquez, C. M., Oehmen, A., Hooijmans, C. M., Brdjanovic, D., Gijzen, H. J., Yuan, Z., & van Loosdrecht, M. C. M. (2009). Modelling the PAO-GAO competition: Effects of carbon source, pH and temperature. *Water Research*, *43*(2), 450–462. doi:10.1016/j.watres.2008.10.032 PMID:19022471

Machado, V. C., Gabriel, D., Lafuente, J., & Baeza, J. A. (2009). Cost and effluent quality controllers design based on the relative gain array for a nutrient removal WWTP. *Water Research*, *43*(20), 5129–5141. doi:10.1016/j.watres.2009.08.011 PMID:19815250

Machado, V. C., Lafuente, J., & Baeza, J. A. (2015). Model-based control structure design of a full-scale WWTP under the retrofitting process. *Water Science and Technology*, *71*(11), 1661–1671. doi:10.2166/wst.2015.140 PMID:26038931

Manga, J., Ferrer, J., Garcia-Usach, F., & Seco, A. (2001). A modification to the Activated Sludge Model No. 2 based on the competition between phosphorus-accumulating organisms and glycogen-accumulating organisms. *Water Science and Technology*, *43*(11), 161–171. PMID:11443958

Maurer, M., Abramovich, D., Siegrist, H., & Gujer, W. (1999). Kinetics of biologically induced phosphorus precipitation in wastewater treatment. *Water Research*, *33*(2), 484–493. doi:10.1016/S0043-1354(98)00221-8

Maurer, M., & Gujer, W. (1995). Monitoring of microbial phosphorus release in batch experiments using electric conductivity. *Water Research*, *29*(11), 2613–2617. doi:10.1016/0043-1354(95)00146-C

Mauricio-Iglesias, M., Vangsgaard, A. K., Gernaey, K. V., Smets, B. F., & Sin, G. (2015). A novel control strategy for single-stage autotrophic nitrogen removal in SBR. *Chemical Engineering Journal*, *260*, 64–73. doi:10.1016/j.cej.2014.08.069

Meijer, S. C. F. (2004). *Theoretical and practical aspects of modeling activated sludge processes*. (PhD Thesis). Delft University of Technology, The Netherlands.

Meinhold, J., Arnold, E., & Isaacs, S. (1999b). Effect of nitrite on anoxic phosphate uptake in biological phosphorus removal activated sludge. *Water Research*, *33*(8), 1871–1883. doi:10.1016/S0043-1354(98)00411-4

Meinhold, J., Filipe, C. D. M., Digger, G. T., & Isaacs, S. (1999a). Characterization of the denitrifying fraction of phosphate accumulating organisms in biological phosphate removal. *Water Science and Technology*, *39*(1), 31–42. doi:10.1016/S0273-1223(98)00773-2

Mielczarek, A. T., Kragelund, C., Eriksen, P. S., & Nielsen, P. H. (2012). Population dynamics of filamentous bacteria in Danish wastewater treatment plants with nutrient removal. *Water Research*, *46*(12), 3781–3795. doi:10.1016/j.watres.2012.04.009 PMID:22608099

Mielczarek, A. T., Nguyen, H. T. T., Nielsen, J. L., & Nielsen, P. H. (2013). Population dynamics of bacteria involved in enhanced biological phosphorus removal in Danish wastewater treatment plants. *Water Research*, *47*(4), 1529–1544. doi:10.1016/j.watres.2012.12.003 PMID:23317522

Mino, T., Arun, V., Tsuzuki, Y., & Matsuo, T. (1987). Effect of phosphorus accumulation on acetate metabolism in the biological phosphorus removal process. In R. Ramadori (Ed.), *Biological phosphate removal from wastewaters* (pp. 27–38). Oxford, UK: Pergamon Press. doi:10.1016/B978-0-08-035592-4.50009-2

Mutlu, A. G., Vangsgaard, A. K., Sin, G., & Smets, B. F. (2013). An operation protocol for facilitating start-up of single-stage autotrophic nitrogen removing reactors based on process stoichiometry. *Water Science and Technology*, *68*(3), 514–521. doi:10.2166/wst.2013.157 PMID:23925177

Ni, B. J., & Yuan, Z. (2015). Recent Advances in Mathematical Modeling of Nitrous Oxides Emissions from Wastewater Treatment Processes. *Water Research*, *87*, 336–346. doi:10.1016/j.watres.2015.09.049 PMID:26451976

Nowak, O., Svardal, K., & Schweighofer, P. (1995). The dynamic behaviour of nitrifying activated sludge systems influenced by inhibiting wastewater compounds. *Water Science and Technology*, *31*(2), 115–124. doi:10.1016/0273-1223(95)00185-P

Oehmen, A., Carvalho, G., Lopez-Vazquez, C. M., van Loosdrecht, M. C. M., & Reis, M. A. M. (2010b). Incorporating microbial ecology into the metabolic modelling of polyphosphate accumulating organisms and glycogen accumulating organisms. *Water Research*, *44*(17), 4992–5004. doi:10.1016/j.watres.2010.06.071 PMID:20650504

Oehmen, A., Lemos, P. C., Carvalho, G., Yuan, Z., Keller, J., Blackhall, L. L., & Reis, M. A. M. (2007). Advances in enhanced biological phosphorus removal: From micro to macro scale. *Water Research*, *41*(11), 2271–2300. doi:10.1016/j.watres.2007.02.030 PMID:17434562

Oehmen, A., Lopez-Vazquez, C. M., Carvalho, G., Reis, M. A. M., & van Loosdrecht, M. C. M. (2010a). Modelling the population dynamics and metabolic diversity of organisms relevant in anaerobic/anoxic/aerobic enhanced biological phosphorus removal processes. *Water Research*, *44*(15), 4473–4486. doi:10.1016/j.watres.2010.06.017 PMID:20580055

Oehmen, A., Saunders, A. M., Vives, M. T., Yuan, Z., & Keller, J. (2006). Competition between polyphosphate and glycogen accumulating organisms in enhance biological phosphorus removal systems with acetate and propionate as carbon sources. *Journal of Biotechnology*, *123*(1), 22–32. doi:10.1016/j.jbiotec.2005.10.009 PMID:16293332

Oehmen, A., Yuan, Z. G., Blackall, L. L., & Keller, J. (2005). Comparison of acetate and propionate uptake by polyphosphate accumulating organisms and glycogen accumulating organisms. *Biotechnology and Bioengineering*, *91*(2), 162–168. doi:10.1002/bit.20500 PMID:15892052

Olsson, G. (2011). Automation development in water and wastewater systems. *Environmental Engineering Research*, *16*(4), 197–200.

Olsson, G. (2012). ICA and me – A subjective review. *Water Research*, *46*(6), 1585–1624. doi:10.1016/j.watres.2011.12.054 PMID:22284982

Ostace, G. S., Baeza, J. A., Guerrero, J., Guisasola, A., Cristea, V. M., Agachi, P. S., & Lafuente, J. (2013). Development and economic assessment of different WWTP control strategies for optimal simultaneous removal of carbon, nitrogen and phosphorus. *Computers & Chemical Engineering*, *53*, 164–177. doi:10.1016/j.compchemeng.2013.03.007

Pambrun, V., Paul, E., & Sperandio, M. (2006). Modeling the partial nitrification in sequencing batch reactor for biomass adapted to high ammonium concentrations. *Biotechnology and Bioengineering*, *95*(1), 120–131. doi:10.1002/bit.21008 PMID:16715534

Picioreau, C., van Loosdrecht, M. C. M., & Heijnen, J. J. (1997). Modelling the effect of oxygen concentration on nitrite accumulation in a biofilm airlift suspension reactor. *Water Science and Technology*, *36*(1), 147–156. doi:10.1016/S0273-1223(97)00347-8

Pijuan, M., Ye, L., & Yuan, Z. (2010). Free nitrous acid inhibition on the aerobic metabo-lism of polyphosphate accumulating organisms. *Water Research*, *44*(20), 6063–6072. doi:10.1016/j.watres.2010.07.075 PMID:20709350

Puñal, A., Rodríguez, J., Franco, A., Carrasco, E. F., Roca, E., & Lema, J. M. (2001). Advanced monitoring and control of anaerobic wastewater treatment plants: Diagnosis and supervision by a fuzzy-based expert system. *Water Science and Technology*, *43*(7), 191–198. PMID:11385847

Rieger, L., Koch, G., Kühni, M., Gujer, W., & Siegrist, H. (2001). The EAWAG bio-P module for activated sludge model No. 3. *Water Research*, *35*(16), 3887–3903. doi:10.1016/S0043-1354(01)00110-5 PMID:12230171

Saito, T., Brdjanovic, D., & van Loosdrecht, M. C. M. (2004). Effect of nitrite on phosphate uptake by phosphate accumulating organisms. *Water Research*, *38*(17), 3760–3768. doi:10.1016/j.watres.2004.05.023 PMID:15350428

Santin, I., Pedret, C., & Vilanova, R. (2015). Applying variable dissolved oxygen set point in a two level hierarchical control structure to a wastewater treatment process. *Journal of Process Control*, *28*, 40–55. doi:10.1016/j.jprocont.2015.02.005

Schuler, A. J. (2005). Diversity matters: Dynamic simulation of distributed bacterial states in suspended growth biological wastewater treatment systems. *Biotechnology and Bioengineering*, *91*(1), 62–74. doi:10.1002/bit.20484 PMID:15880520

Serralta, J., Borrás, L., Blanco, C., Barat, R., & Seco, A. (2004b). Monitoring pH and electric conductivity in an EBPR sequencing batch reactor. *Water Science and Technology*, *50*(10), 145–152. PMID:15656307

Serralta, J., Ferrer, J., Borrás, L., & Seco, A. (2004a). An extension of ASM2d including pH calculation. *Water Research*, *38*(19), 4029–4038. doi:10.1016/j.watres.2004.07.009 PMID:15491650

Siegrist, H., Brunner, I., Koch, G., Linh, C. P., & Van Chieu, L. (1999). Reduction of biomass decay rate under anoxic and anaerobic conditions. *Water Science and Technology*, *39*(1), 129–137. doi:10.1016/S0273-1223(98)00782-3

Sin, G., Kaelin, D., Kampschreur, M. J., Takács, I., Wett, B., Gernaey, K. V., & van Loosdrecht, M. C. M. et al. (2008). Modelling nitrite in wastewater treatment systems: A discussion of different modelling concepts. *Water Science and Technology*, *58*(6), 1155–1171. doi:10.2166/wst.2008.485 PMID:18845852

Sin, G., Niville, K., Bachis, G., Jiang, T., Nopens, I., van Hulle, S., & Vanrolleghem, P. A. (2007b). Nitrite effect on the phosphorus uptake activity of phosphate accumulating organisms (PAOs) in pilot-scale SBR and MBR reactors. *Water S.A.*, *34*(2), 249–260.

Sin, G., & Vanrolleghem, P. A. (2006). Evolution of an ASM2d-like model structure due to operational changes of an SBR process. *Water Science and Technology*, *53*(12), 237–245. doi:10.2166/wst.2006.426 PMID:16889260

Sin, G., & Vanrolleghem, P. A. (2007a). Extensions to modeling aerobic carbon degradation using combined respirometric-titrimetric measurements in view of activated sludge model calibration. *Water Research*, *41*(15), 3345–3358. doi:10.1016/j.watres.2007.03.029 PMID:17498768

Smolders, G. J. F., van der Meij, J., van Loosdrecht, M. C. M., & Heijnen, J. J. (1994a). Model of the anaerobic metabolism of the biological phosphorus removal process: Stoichiometry and pH influence. *Biotechnology and Bioengineering*, *43*(6), 461–470. doi:10.1002/bit.260430605 PMID:18615742

Smolders, G. J. F., van der Meij, J., van Loosdrecht, M. C. M., & Heijnen, J. J. (1994b). Stoichiometric model of the aerobic metabolism of the biological phosphorus removal proces. *Biotechnology and Bioengineering*, *44*(7), 837–848. doi:10.1002/bit.260440709 PMID:18618851

Solon, K., Flores-Alsina, X., Kazadi Mbamaba, C., Gernaey, K. V., Tait, S., Batstone, D., & Jeppsson, U. et al. (2015). Effects of ion strength and ion pairing on plant wide modelling on anaerobic digester. *Water Research*, *70*, 235–245. doi:10.1016/j.watres.2014.11.035 PMID:25540837

Spagni, A., Buday, J., Ratini, P., & Bortone, G. (2001). Experimental considerations on monitoring ORP, pH, conductivity and dissolved oxygen in nitrogen and phosphorus biological removal processes. *Water Science and Technology*, *43*(11), 197–204. PMID:11443963

Stare, A., Vrecko, D., Hvala, N., & Strmcnik, S. (2007). Comparison of control strategies for nitrogen removal in an activated sludge process in terms of operating costs: A simulation study. *Water Research*, *41*(9), 2004–2014. doi:10.1016/j.watres.2007.01.029 PMID:17346768

Strous, M., Heijnen, J. J., Kuenen, J. G., & Jetten, M. S. M. (1998). The sequencing batch reactor as a powerful tool for the study of slowly growing anaerobic ammonium-oxidizing microorganisms. *Applied Microbiology and Biotechnology*, *50*(5), 589–596. doi:10.1007/s002530051340

Strous, M., Kuenen, J. G., & Jetten, M. S. M. (1999). Key physiology of anaerobic ammonium oxidation. *Applied and Environmental Microbiology*, *65*, 3248–3250. PMID:10388731

Strous, M., van Gerven, E., Kuenen, J. G., & Jetten, M. S. M. (1997). Effects of aerobic and micro-aerobic conditions on anaerobic ammonium-oxidizing (Anammox) sludge. *Applied and Environmental Microbiology*, *63*(3), 2446–2448. PMID:16535633

Terada, A., Lackner, S., Tsuneda, S., & Smets, B. F. (2007). Redox-stratification controlled biofilm (ReSCoBi) for completely autotrophic nitrogen removal: The effect of co- versus counter-diffusion on reactor performance. *Biotechnology and Bioengineering*, *97*(1), 40–51. doi:10.1002/bit.21213 PMID:17013935

Thürlimann, C. M., Dürrenmatt, D. J., & Villez, K. (2015). Evaluation of Qualitative Trend Analysis as a Tool for Automation. *Computer Aided Chemical Engineering*, *37*, 2531–2536. doi:10.1016/B978-0-444-63576-1.50116-3

Trollberg, O., Carlsson, B., & Jacobsen, E. W. (2014). Extremum seeking control of the CANON process - Existence of multiple stationary solutions. *Journal of Process Control*, *24*(2), 348–356. doi:10.1016/j.jprocont.2013.11.007

Valverde-Pérez, B., Fuentes-Martínez, J. M., Flores-Alsina, X., Gernaey, K. V., Huusom, J. K., & Plósz, B. (in press). Gy., (2016b). Control structure design for resource recovery using the enhanced biological phosphorus removal and recovery (EBP2R) activated sludge process combined with green microalgae cultivation. *Chemical Engineering Journal*. doi:10.1016/j.cej.2016.03.021

Valverde-Pérez, B., Mauricio-Iglesias, M., & Sin, G. (2016a). Systematic design of an optimal control system for the SHARON-Anammox process. *Journal of Process Control*, *39*, 1–10. doi:10.1016/j.jprocont.2015.12.009

Valverde-Pérez, B., Ramin, E., Smets, B. F., & Plósz, B. (2015). EBP2R – An innovative enhanced biological nutrient recovery activated sludge system to produce growth medium for green microalgae cultivation. *Water Research*, *68*, 821–830. doi:10.1016/j.watres.2014.09.027 PMID:25480432

Van Hulle, S. W. H., Volcke, E. I. P., Lopez Teruel, J., Donckels, B., van Loosdrecht, M. C. M., & Vanrolleghem, P. A. (2007). Influence of temperature and pH on the kinetics of the SHARON nitritation process. *Journal of Chemical Technology and Biotechnology (Oxford, Oxfordshire)*, *82*(5), 471–480. doi:10.1002/jctb.1692

Vangsgaard, A. K., Mauricio-Iglesias, M., Gernaey, K. V., & Sin, G. (2014). Development of novel control strategies for single-stage autotrophic nitrogen removal: A process oriented approach. *Computers & Chemical Engineering*, *66*, 71–81. doi:10.1016/j.compchemeng.2014.01.017

Vangsgaard, A. K., Mauricio-Iglesias, M., Gernaey, K. V., Smets, B. F., & Sin, G. (2012). Sensitivity analysis of autotrophic N removal by a granule based bioreactor: Influence of mass transfer versus microbial kinetics. *Bioresource Technology*, *123*, 230–241. doi:10.1016/j.biortech.2012.07.087 PMID:22940324

Vangsgaard, A. K., Mauricio-Iglesias, M., Valverde-Pérez, B., Gernaey, K. V., & Sin, G. (2013). pH variation and influence in an autotrophic nitrogen removing biofilm system using an efficient numerical solution strategy. *Water Science and Technology*, *67*(11), 2608–2615. doi:10.2166/wst.2013.174 PMID:23752396

Vazquez-Padin, J. R., Fernandez, R., Campos, J. L., Mosquera, A., & Mendez, R. (2014). *Method for starting up and controlling a biological process for ammonium removal by the action of autotrophic bacteria in wastewater*. European Patent, EP2740713B1

Vlaeminck, S. E., De Clippeleir, H., & Verstraete, W. (2012). Microbial resource management of one-stage partial nitritation/anammox. *Microbial Biotechnology*, *5*(3), 433–448. doi:10.1111/j.1751-7915.2012.00341.x PMID:22452819

Volcke, E. I. P., Picioreanu, C., De Baets, B., & van Loosdrecht, M. C. M. (2010). Effect of granule size on autotrophic nitrogen removal in a granular sludge reactor. *Environmental Technology*, *31*(11), 1271–1280. doi:10.1080/09593331003702746 PMID:21046957

Volcke, E. I. P., van Loosdrecht, M. C. M., & Vanrolleghem, P. A. (2006). Controlling the nitrite: Ammonium ratio in a SHARON reactor in view of its coupling with an Anammox process. *Water Science and Technology*, *53*(4-5), 45–54. doi:10.2166/wst.2006.109 PMID:16722054

Volcke, E. I. P., van Loosdrecht, M. C. M., & Vanrolleghem, P. A. (2007). Interaction between control and design of a SHARON reactor: Economic considerations in a plant-wide (BSM2) context. *Water Science and Technology*, *56*(9), 117–125. doi:10.2166/wst.2007.607 PMID:18025739

von Schulthess, R., & Gujer, W. (1996). Release of nitrous oxide ($N_2O$) from denitrifying activated sludge: Verification and application of a mathematical model. *Water Research*, *30*(3), 521–530. doi:10.1016/0043-1354(95)00204-9

Wágner, D. S., Ramin, E., Szabo, P., Dechesne, A., & Plósz, B. (2015). *Microthrix parvicella* abundance associates with activated sludge settling velocity and rheology - Quantifying and modelling filamentous bulking. *Water Research*, *78*, 121–132. doi:10.1016/j.watres.2015.04.003 PMID:25935367

Wett, B., & Rauch, W. (2003). The role of inorganic carbon limitation in biological nitrogen removal of extremely ammonia concentrated wastewater. *Water Research*, *37*(5), 1100–1110. doi:10.1016/S0043-1354(02)00440-2 PMID:12553985

Wyffels, S., van Hulle, S. W. H., Boeckx, P., Volcke, E. I. P., van Cleemput, O., Vanrolleghem, P. A., & Verstraete, W. (2004). Modeling and simulation of oxygen-limited nitritation in a membrane-assisted bioreactor (MBR). *Biotechnology and Bioengineering*, *86*(5), 531–542. doi:10.1002/bit.20008 PMID:15129436

Yagci, N., Insel, G., Artan, N., & Orhon, D. (2004). Modelling and calibration of phosphate and glycogen accumulating organisms competition for acetate uptake in a sequencing batch reactor. *Water Science and Technology*, *50*(6), 241–250. PMID:15537013

Yong, M., Yongzhen, P., & Jeppsson, U. (2006). Dynamic evaluation of integrated control strategies for enhanced. *Control Engineering Practice*, *14*(11), 1269–1278. doi:10.1016/j.conengprac.2005.06.018

Yuan, Z., Oehmen, A., Peng, Y., Ma, Y., & Keller, J. (2008). Sludge population optimization in biological nutrient removal wastewater treatment systems through on-line process control: A review. *Reviews in Environmental Science and Biotechnology*, *7*(3), 243–254. doi:10.1007/s11157-008-9134-y

Yuan, Z., Pratt, S., & Batstone, D. (2012). Phosphorus recovery from wastewater through microbial processes. *Current Opinion in Biotechnology*, *23*(6), 878–883. doi:10.1016/j.copbio.2012.08.001 PMID:22922003

## KEY TERMS AND DEFINITIONS

**Activated Sludge Model:** It refers to a class of models aimed at describing the main biological and chemical reactions that take place in activated sludge systems. The most popular are abbreviated as ASM1, ASM2d and ASM3 and have been defined in successive IWA Technical Reports.

**Biological Model:** It is a mathematical model that describes the biological reactions in a process, including substrate uptake, product formation, biomass growth and decay, inhibition.

**Cascade Control:** It is the simplest form of supervisory control, by which the manipulated variable of the supervisory controller is the set point of a regulatory controller. It can additionally be used to improve the regulation of a nonlinear controller.

**Fuzzy Control:** It is a control system based on fuzzy logic. The manipulated variable takes a continuous value (from 0 to 1), after applying a set of qualitative and quantitative rules in response to the measurement of inputs.

**Physicochemical Model:** This term is normally used in wastewater treatment to describe a mathematical model formulated to describe the state of the chemical species in a solution, including the ion strength and pairing, precipitation equilibrium, acid-base equilibrium and vapour-liquid equilibrium.

**Process Rate:** In a biological model, the process rate describes the rate of advancement of a reaction. Process rates are multiplied by the stoichiometric coefficients of all the species taking place in the reaction, including the biomass, in order to determine the reaction rate of each of the species.

**Regulatory Control:** It is the control system or part of the control system with the objective of tracking the set point and rejecting disturbances in order to keep constant the controlled variables.

**Supervisory Control:** It is the control system or part of the control system with the objective of driving the process to the point of desired operation. This point is normally determined by economic and/or regulatory reasons.

# Chapter 8
# Constructed Wetlands for Industrial Wastewater Treatment and Removal of Nutrients

**David de la Varga**
*Sedaqua, Spain*

**Dion van Oirschot**
*Rietland bvba, Belgium*

**Manuel Soto**
*University of A Coruña, Spain*

**Rene Kilian**
*Kilian Water, Denmark*

**Carlos Alberto Arias**
*Aarhus University, Denmark*

**Ana Pascual**
*AIMEN, Spain*

**Juan A. Álvarez**
*AIMEN, Spain*

## ABSTRACT

*Constructed Wetlands (CWs) are low-cost and sustainable systems for wastewater treatment. Traditionally they have been used for urban and domestic wastewater treatment, but in the last two decades, the applications for industrial wastewater treatment increased due to the evolution of the technology and the extended research on the field. Nowadays, CWs have been applied to the treatment of different kind of wastewaters as such as refinery and petrochemical industry effluents, food industry effluents including abattoir, dairy, meat, fruit and vegetables processing industries, distillery and winery effluents, pulp and paper, textile, tannery, aquaculture, steel and mixed industrial effluents. In this chapter, the authors present the main types of CWs, explain how they work and the expected performances, and describe the principal applications of CWs for industrial wastewater treatment with particular attention to suspended solids, organic matter and nutrient removal. A review of these applications as well as some case studies will be discussed.*

DOI: 10.4018/978-1-5225-1037-6.ch008

## INTRODUCTION

In this chapter, the authors refer to the use of constructed wetlands (CWs) for industrial wastewater treatment and their efficiency for nutrient removal. CWs are engineered systems that have been designed and constructed to utilize the natural processes involving wetland vegetation, soils, and their associated microbial assemblages to achieve wastewater treatment (Vymazal, 2014). "Modern treatment wetlands are man-made systems that have been designed to emphasize specific characteristics of wetland ecosystems for improved treatment capacity" (Kadlec & Wallace, 2009). "Besides treatment wetlands", constructed and engineered wetlands can cover a broad range of objectives such as improving biodiversity and environmental conditions related to, wildlife use, irrigation of agriculture lands, improving river water quality, or riverine restoration. Some misleading names have been given to the technology including green filters, biofilters and even sand filters or artificial wetlands. As CWs have evolved with time and applications, other terms like engineered wetlands have appeared that might include the use of devices that upgrade the performance using energy input.

CWs are low-cost and ecofriendly technologies, that take advantage of natural processes to remove pollutants from the water, generally avoiding the use of chemical products and the input of high amounts of external energy. On the other hand, CWs may require a large surface, which is its major drawback. As a result; they are included in the group of extensive technologies for wastewater treatment.

The first research into CWs for wastewater treatment took place in Germany, in the 1950s (e.g., Seidel, 1961), with special focus on phenols removal. From the beginning, the first applications of CWs dealt with urban wastewater, but in the last two or three decades, they have been applied for industrial and agricultural wastewater, as well as stormwater runoff and the treatment of landfill leachates (Vymazal 2011a).

## CONSTRUCTED WETLANDS: HOW DO THEY WORK?

Pollutant removal in natural wetlands takes place due to the combination of physical, chemical and microbial processes. The processes involved in pollutant removal are sedimentation, sorption, precipitation, evapotranspiration, volatilization, photodegradation, diffusion, plant uptake, and microbial degradation (for instance, nitrification, denitrification, sulphate reduction, carbon metabolization, etc.) among others.

### Types of Constructed Wetlands

There are several types of CWs, depending on the hydrology, the type of macrophytic growth and the direction of the flow inside the wetland. As a result, if the hydrology is considered, the principal types of CWs are surface flow (or free water systems) or subsurface flow systems (Figure 1). According to the macrophytic growth, there are emergent, submerged, free-floating and floating leaves. Finally, the direction of flow inside the CW can be vertical, horizontal or mixed flow.

The most widespread CWs are the surface flow systems (FWS), the horizontal subsurface flow systems (HSSF) and the vertical subsurface flow systems (VF). For improving the performance and the removal of pollutants and nutrients, a combination of these systems can be used, known as hybrid systems. The hybrid systems can combine several features in only one or in several sequential steps.

*Figure 1. Schematic representation of types of CWs, depending on the flow regime*

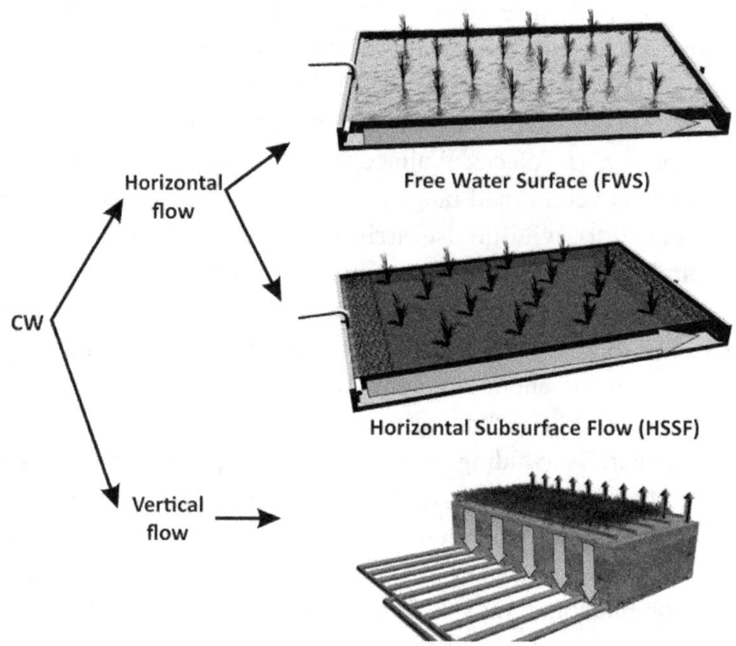


## Free Water Surface Systems (FWS)

The FWS system CWs are shallow beds (often from 20 to 40 cm), planted with aquatic vegetation. The pollutants are removed as wastewater goes exposed to the atmosphere through the planted beds, due to several processes including sedimentation, oxidation, reduction, precipitation, sorption and biological degradation.

This type of FWS is often used for tertiary treatment, to polish effluents from a previous physical-chemical or biological wastewater treatment. The FWS systems have several characteristics:

- Simple to operate and low-cost systems.
- Low removal rates.
- Demand large surface requirements (up to 20 $m^2$/PE[1]).
- Risk of odors forming (if they receive untreated sewage directly) and freezing in winter.
- Can host vectors.

## Horizontal Subsurface Flow Systems (HSSF)

HSSF systems are cells filled with media (from 30 to 60 cm deep) in which aquatic vegetation is planted (Figure 1). This kind of wetland is saturated, and the water column is not exposed to the atmosphere, usually remaining about 5 to 10 cm under the surface of the bed and therefore avoiding fouling odors and the proliferation of vectors.

Water is distributed in the inlet zone by perforated pipes or Thompson channels. The inlet of the wetland is filled with coarse gravel (30-100 mm size), to improve influent distribution and delay the possible clogging in the inlet zones. The bed media is generally gravel (5-20 mm size), where the biofilm develops and at the same time, the media serves as a support for the vegetation. At the outlet zone and the bottom of the bed, a collection pipe is placed to evacuate the treated waters. The pipe is also embedded with coarse gravel to facilitate the effluent collection. After the bed, there must be an outlet sump where a pipe is connected to the collection pipe and can regulate the water level in the wetland by a swivel joint or a flexible pipe.

In a HSSF wastewater goes through the wetland and the media in a horizontal path, coming into contact with aerobic, anoxic and anaerobic zones. The oxygen from the air is transported through the stems and leafs of the vegetation to the rhizosphere, where an aerobic-anoxic microcosm is present (Brix & Schierup, 1990). Normally, the amount of oxygen in HSSF CW treating typical wastewaters is not enough to degrade all the organic matter via aerobic processes much less to allow high dissolved oxygen concentration into the wetland. Therefore the processes responsible for the degradation are mainly anoxic-anaerobic. Some characteristics of this system are the following:

- High resistance to freezing conditions.
- Less surface area needed than FWS (5 $m^2$/PE).
- Higher removal rates than FWS, although this is dependent on operational conditions.
- Risk of clogging problems in the inlet zone.

## Vertical Flow Systems (VF)

The subsurface VF systems are cells filled with coarse sand or fine gravel, usually from 60 to 100 cm deep, and planted with aquatic vegetation (Figure 1). Most VF systems are unsaturated and fed sequentially in short pulses and then a drainage period takes place. The wastewater is loaded homogeneously onto the surface of the wetland, trickles vertically through the filter media and is collected at the bottom by drainage pipes. Additionally, passive aeration pipes from to the drainage pipes to the atmosphere to improve the oxygen transfer to the bed. Due to aeration pipes, pulse loadings and drained periods, the oxygen transfer can reach much higher values, if compared to HSSF or FWS systems. VF CWs have aerobic conditions and therefore higher organic removal rates as well as the capacity to nitrify.

The wastewater distribution system of VF CWs consists of a manifold of pipes installed on the surface of the bed. The media filling the bed can be either by one layer or even several sand layers. At the top and the bottom of the bed, a layer of about 20 cm of coarse gravel is placed to facilitate the distribution and evacuation of the waters.

Some other characteristics of VF CWs are the following:

- Less surface area needed (2-3 $m^2$/PE).
- No clogging problems (if well designed and operated).
- Demands a good influent distribution system.

A modification of VF is commonly called "the French system" or French reed beds (FRB). It is basically a 2-step CW but with some differences, mainly the stratification of the substrate layers and the distribution system. It can treat wastewater directly, without primary treatment. In the FRB the 1-step

VF demands parallel beds that are loaded sequentially and used as primary treatment and the bed retains suspended solids as well as other particulate pollutants. As times goes by and after loading, a layer of stabilized sludge accumulates on top of the wetland. After 8 to 10 years, the sludge layer is removed and can be further composted or applied directly to land as fertilizer. After the first step and according to the effluent needs the vertical bed can be followed by a new set of vertical flow beds or a horizontal flow bed.

The recirculation of treated effluents back to the pumping well or to the sedimentation tank improves the overall system performance and enhances the removal of nitrogen. The recirculation of treated water dilutes and makes the water loaded onto the filter more homogeneous. Additionally, the recirculation of treated water helps maintain the presence of water during vacation periods ensuring the survival of the plants and bacteria/biofilm. The optimal recirculation rate is circa 100%, i.e. the recirculation of half of the effluent back to the pumping well (Arias *et al*, 2011).

### Tidal Flow Constructed Wetlands

Other type of CWs developed during last decade are the tidal-flow constructed wetlands (TFCWs), which have enhanced the organic matter and ammonia removal by overcoming the lack of oxygen in conventional CWs (Sun *et al*., 2005; Wu *et al*., 2011). However, the total nitrogen removal efficiency in TFCWs is not ideal because of the high oxygen content that limits the denitrification processes (Ju *et al*., 2014; Cui *et al*., 2012).

The "tidal flow" principle includes four operational procedures (fill, contact, drain and rest) that constitute the main difference with conventional CWs (Sun *et al*., 2006).

### Engineered Wetlands

As reviewed by Wu *et al*. (2014), different operation strategies and innovative designs can be used in order to intensify the performance of CW systems. These strategies include recirculation, aeration, tidal operation, flow direction reciprocation, earthworm integration, short-term fluctuations in the water table, step-feeding, circular-flow corridor wetlands, towery hybrid CWs, baffled subsurface CWs for the intensifications of the performance.

Aerated wetlands are subsurface flow wetlands, both horizontal and vertical flow, but with artificial aeration implemented in the bed where the air is pumped into the wetland by means of compressors. Forced bed aeration (FBA®) wetlands were patented in USA by Scott Wallace and are gaining ground and are being used more due to their versatility and better capacity to treat industrial waters. Aerated wetlands testing different regimes of aeration "on" and "off", nitrification or denitrification processes can be enhanced.

## Pretreatments for CWs

The most frequently used pretreatments for CWs treating domestic wastewater and, frequently used for treating industrial wastewater, are septic tanks (ST) and Inhoff tanks (IT). Often, these primary treatment technologies are not efficient enough for removing suspended solids, and can promote clogging in CWs, especially when high organic loads are applied. The use of anaerobic digesters and CWs is gaining importance in latest years for treating different industrial wastewaters. Both systems are low-cost, robust and effective in wastewater treatment. According to Álvarez *et al*., (2008) an average and 95 percentile

TSS concentrations of anaerobic treated wastewater were below 60 and 100 mg/L, respectively, for all configurations reported suggesting that anaerobic treatments might be an option when using CWs.

## Removal of Pollutants in CWs

Solids are removed in CWs mainly by physical processes such as sedimentation and filtration. The removal of solids takes place in the inlet zone of HSSF (1/3 or 1/4 of the total length) and in the surface of VF CWs. These processes are favored by the low velocity of the water and the sieving in the interstitial spaces of the granular media inside the CWs.

Organic removal processes are complex and vary, with the interaction of several physical, chemical and biotic mechanisms involved. Particulate organic matter is retained like solids, as described before, and hydrolyzed in simple substrates that can be assimilated by heterotrophic, anaerobic and facultative bacteria. Aerobic degradation of organic matter takes place close to the water surface and the rhizosphere, while facultative and aerobic degradation occurs in the bottom of HSSF and FWS systems. The rest of the bed is dominated by anaerobic organisms and as such anaerobic processes are responsible for further transformations.

Pathogens are removed by complex processes, including filtration, sorption, died off due to environmental conditions and predation.

Under aerobic conditions, heavy metals (HM) retention and accumulation in wetland substrate is mainly due to the formation of metal hydroxides (i.e. Fe and Mn hydroxides; Singer & Stumm, 1970). Under anoxic or anaerobic conditions, precipitation of metal sulphides is the main process contributing to the removal of HMs in CWs. In anaerobic conditions, sulphate reduction leads to the formation of hydrogen sulphide and most HMs react with sulphide to form highly insoluble precipitates (Stumm & Morgan 1981).

## Removal of Nutrients in Constructed Wetlands

The most important mechanism to remove nitrogen from wastewater in CWs is nitrification-denitrification process. The process begins with nitrogen from organic compounds like proteins, transformed both in aerobic and anoxic conditions into ammonia, (ammonification). Nitrification consists on the oxidation of ammonia into nitrite and nitrate by autotrophic bacteria. This process occurs in aerobic conditions. Denitrification is the conversion of nitrate to dinitrogen gas ($N_2$) by heterotrophic bacteria. This process needs organic matter and anoxic conditions.

Recently, the presence of anaerobic ammonium oxidation (anammox) processes has been described in CWs as another path to remove nitrogen (Saeed and Sun, 2012b). Anammox is a reaction that oxidizes ammonium to dinitrogen gas using nitrite as the electron acceptor under anoxic conditions. When partial nitritation ($NH_4^+$ to $NO_2^-$) and anammox is combined, only about one half of ammonium needs to be oxidized to nitrite first, then anammox bacteria use the remaining ammonium as electron donors for autotrophic denitrification of the nitrite produced (van der Star *et al.*, 2007; Kartal *et al.*, 2010). Anammox process consumes 60% less oxygen than nitrification-denitrification cycle and no organic carbon source is required (He *et al.*, 2012).

Other processes involved in nitrogen removal are plant uptake, volatilization, dissimilatory nitrate reduction, and biomass assimilation. The latest process takes place through incorporation of $NH_4^+$ in the heterotrophic biomass to fulfil nutrient requirements. In stationary systems like CWs, biomass accumulation could lead to clogging problems, so biomass assimilation is not expected to be an important process in CWs for removal of nitrogen.

The presence of phosphorus in wastewater is mainly in the forms organic phosphorus, orthophosphate, and polyphosphate. Soluble reactive phosphorus is taken by plants and transformed to biomass, in the form of vegetal tissues and/or organisms. Sorption by sediments, sand or gravel media is the most important phosphorus removal process in CWs. Precipitates and co-precipitates of Al, Fe and Ca occur under certain circumstances mainly depended on pH, but can be redissolved under altered conditions (Kadlec & Wallace, 2009).

## Nitrogen Removal in CWs

If organic matter concentration (carbon source) is sufficient, denitrification might be an important process to remove nitrogen in FWS systems. Unvegetated open water does not promote denitrification, resulting in rate constants about one third of those for vegetated systems (Arheimer & Wittgren, 1994). Smith *et al.* (2000) have shown nitrate removal proportional to the number of shoots in a *Schoenoplectus* spp. planted wetland. These considerations lead to the conclusion that fully vegetated marshes with either emergent or submergent communities are the preferred option for denitrification.

Since HSSF CWs are saturated reactors they have limited nitrification capacity, on the contrary the conditions are best for denitrification processes and therefore total nitrogen removal is effective only if a previous nitrifying system is implemented. However, low loaded HSSF systems can remove influent total nitrogen because near all the nitrified ammonia is effectively removed (nitrification-denitrification), or by the existence of other removal mechanisms.

In VF CWs, ammonia oxidation to nitrate is highly dependent on the organic and nitrogen surface loads and on the VF operation regime, mainly the pulse frequency and volume as well as the duration of drained periods. If well designed and operated, VF systems can reach more than 90% of nitrification, both for urban and industrial wastewater. With aerobic conditions, total nitrogen removal in VF systems is poor or null, due to very limited denitrification rates. Partial recirculation of VF effluent to previous septic tanks or anaerobic reactors can enhance total nitrogen removal (Brix & Arias, 2005; Torrijos *et al.*, 2015).

In TFCWs ammonium cations ($NH_4^+$) are first adsorbed on matrix, pores, and surfaces when wetland cells are flooded. Second, as wetland cells drain, matrix pores are immediately filled with air and the absorbed ammonia is nitrified by bacteria. Finally, in the next flood cycle, nitrate ($NO_3^-$) and nitrite ($NO_2^-$) desorb into bulk water, where they are denitrified into atmospheric nitrogen (Austin, 2006; Chang *et al.*, 2014). Therefore, the flooded time of TFCWs plays a key role in forming effective anoxic conditions that are favorable for reducing oxidized nitrogen and to improve total nitrogen removal (Li *et al*, 2015b).

## The Role of Plants in Constructed Wetlands

Nutrient transformation and sequestration in low-loaded systems, organic matter production and plant uptake of nutrients, as well as root-zone oxygen and organic carbon release systems have been identified as key factors (Brix, 1997; Tanner, 2001; Vymazal, 2011b). According to Tanner (2001), wetland plants promote the enhancement of nutrient removal, mainly by favoring transformations to gaseous forms and sequestration in accumulating organic matter. Recent studies reported that nitrogen removal is usually better for planted systems (Hijosa-Valsero *et al.*, 2012; Pedescoll *et al.*, 2013a; Webb *et al.*, 2013; Huang *et al.*, 2013; Seeger *et al.*, 2013; Lv *et al.*,2013; Chen *et al.*, 2014).

Carballeira *et al.* (2016) reported a linear correlation between ammonia nitrogen removal rate (g N/$m^2$·d) and biomass production (g volatile solids VS/$m^2$·d) in HSSF CWs treating domestic wastewater. According to that correlation, biomass production explains 91% of additional nitrogen removal in planted systems. Unplanted CW removed 0.34 g $NH_3$/$m^2$·d while planted systems removed an additional 13.4% per each g VS/$m^2$·d of above-ground biomass produced or the equivalent 36.6% additional per each kg VS/$m^2$ year.

The contribution of plants, in terms of nitrogen removal has been reported within the range 0.5-40.0% of the total nitrogen removal (Drizo *et al.*, 1997; Shamir *et al.*, 2001; Healy and Cawley, 2002; Meers *et al.*, 2008; Kantawanichkul *et al.*, 2009; Bialowiec *et al.*, 2011, Saeed and Sun, 2012b). Chen *et al.*, (2014) reported that N mass balance showed that denitrification, sedimentation burial and plant uptake respectively contributed 54%-94%, 1%-46% and 7.5%-14.3% to the N removal in CWs, mainly stored in aboveground biomass. Zheng *et al.*, (2015) reported that plants harvesting in the first year improved nutrients removal by plant uptake (41.9 g N/$m^2$ and 3.7 g P/$m^2$ versus 37.3 g N/$m^2$ and 3.2 g P/$m^2$) as well as in the substrate layer (216.9 g N/$m^2$ and 8.0 g P/$m^2$ versus 191.0 g N/$m^2$ and 5.7 g P/$m^2$) during the second year.

## Phosphorus Removal in CWs

In FWS, phosphorus is removed, due to three processes in wetlands: sorption, utilization to build a larger biomass compartment and storage as newly created, refractory residuals (burial). Sorption and biomass storage have a limited phosphorus retention capacity. Moreover, there are secondary processes like particulate settling that can rapidly remove high amounts of phosphorus from runoff water with a high level of suspended sediments (Kadlec & Wallace, 2009).

Similar phosphorus removal processes take place in subsurface flow CWs (HSSF and VF), biomass accumulation, sorption in the gravel media and sediment accretion. Sorption mechanisms are limited to the gravel media capacity, which can be high, but only when considered in a short term. In this sense, latest research in new materials with higher phosphorus sorption capacity has been done.

Like in HSSF systems, phosphorus removal in VF systems takes place mainly by sorption in the filter media. VF systems are designed principally to remove organic matter, solids and ammonia. As a result, phosphorus loading rate is bigger than the capacity of plants to accumulate phosphorus as vegetal biomass.

The biological removal of phosphorus by microorganisms in tidal flow CWs has never been considered as a significant factor in CW design. However, results from Li *et al.* (2015a) suggest that the cycling of phosphorus in CWs might depend significantly on the type of operations employed. Nevertheless, further investigations on the detailed mechanism of phosphorus removal are needed before any concrete conclusion can be made regarding phosphorus removal efficiency in tidal flow CWs.

## Influence of Filter Media of CWs on Phosphorus Removal

A review of different filter media used for phosphorus removal was elaborated by Vohla *et al.* (2011). These researchers reported that the most commonly used materials are described as natural materials, industrial byproducts and man-made products. Most of the studied materials had a pH >7 and high Ca (CaO) content. Several industrial byproducts achieved the highest P removal capacity, including some furnace slags (up to 420 g P/kg), followed by natural materials (maximum 40 g P/kg for heated opoka) and man-made filter media (maximum 12 g P/kg for Filtralite-P®). Some important factors such as saturation time, availability at a local level, the content of heavy metals, and the reuse of the saturated filter media as a fertilizer should be taken into consideration regarding the applicability of filter materials.

Moreover, a good hydraulic conductivity and the chemical composition of the adsorption media are key factors for CW design. Since phosphorus is removed via sorption and precipitation, Ca, Fe and Al content of the filter media is important for P removal. Even if a medium with high P binding capacity has been selected, it may be saturated after a few years (Arias *et al.*, 2001). An easy, practical and sustainable solution might be to install a separate well filled with replaceable high binding P capacity media after the treatment for P removal (Brix *et al.*, 2001).

## Recovery of Nutrients in Constructed Wetlands

Recovery of nutrients by CWs is generally not very effective, comparing with other wastewater treatment systems and will depend on the type of CW and the water quality. For example, harvesting of plants for uptaken nutrient recovery might be interesting, but it is only feasible for large surface areas. Another example for carbon and nutrient recovery can be sludge treatment wetlands (STW).

STW is a type of VF, and were developed to treat sludge produced as a result of activated sludge wastewater treatment. The required specific surface area can be roughly assumed as 0.25 to 0.50 $m^2$/PE (Kainz, 2006). These systems consist of several parallel beds filled with three or more layers with a bottom gravel drainage layer, an intermediate layer with coarse material and a soil/sand layer for favoring the vegetation establishment (commonly reeds, *Phragmites australis*). In STWs homogenized sludge is pumped intermittently and sequentially to the different beds by alternating dosing regimens and allowing resting periods for the different beds. The resting periods will depend on the design conditions, but the intervals between loadings must be sufficient to allow the drainage of trapped water from the sludge. As sludge is accumulated and built up, the rhizomes will develop in the sludge and penetrate, increasing dewatering by evapotranspiration process.

After the sludge storage capacity of the bed is filled, the sludge from the STW is stable. This process reaches a final product with a dry solid content higher than 25%, and is suitable for direct reuse. As an example, the sludge produced in the Helsinge STW after 10 years of operation resulted in total phosphorus content of 30-35 g/kg dry solids and total nitrogen content of 20-25 g/kg dry solids (Nielsen, 2015).

Nutrients concentration in biosolids depends on the wastewater composition and treatment, and on subsequent sludge management. Nitrogen comes from biomass present in the sludge and form wastewater residues. In a study carried out by Uggetti *et al.*, (2012), total Kjeldahl nitrogen (TKN) values in biosolids ranged from 0.03 to 0.25% TKN/TS (TS total solids), being significantly lower than in activated sludge (Andreoli *et al.*, 2007). TKN values were significantly higher (2.30-2.53% TKN/TS) in compost of sewage sludge (Ruggieri *et al.*, 2008; Sanchez *et al.*, 2010). Uggetti *et al.*, (2012) reported a certain nitrogen reduction in STW, due to sludge mineralization, ammonification and plant uptake.

The main sources of phosphorus in sludge are biomass formed during wastewater treatment, residues and phosphate-containing detergents and soaps. Uggetti *et al.*, (2012) reported a decrease in total phosphorus values from the influent (2.7-3.0% TP/TS) to treated sludge (0.07-0.39% TP/TS), probably due to phosphate immobilization in microbial cells (Elvira *et al.*, 1996).

# STATE OF THE ART OF CONSTRUCTED WETLANDS TREATING INDUSTRIAL EFFLUENTS

## Industrial Applications of Constructed Wetlands

CWs were initially used for treating domestic wastewater. But in the last two decades, industrial wastewater has been treated with hybrid CWs, both surface and subsurface flow. CWs started treating industrial effluents like petrochemical, dairy, meat processing, abattoir, and pulp and paper factory effluents. Brewery, tannery and olive mills wastewaters have been recently added to CWs applications. So, CWs can be applied to several and different kinds of industrial wastewaters, including acid mine wastewater with low organic matter content and landfill leachate. Vymazal (2013) reported industrial wastewater treatment with hybrid CWs systems with influent concentrations up to 10,000-24,000 mg COD/L and up to 496 mg $NH_4^+$/L. But there are not general rules to select the most suitable type of CW for a certain industrial wastewater or even urban wastewater. Every single case must be studied particularly due to several conditions: type of wastewater, land availability, the amount of flow and pollutant load, outlet discharge limits, etc. Some recommendations to be taken into account could be:

- Good pretreatments like anaerobic digesters (UASB, HUSB) to remove solids and avoid or at least delay clogging of media. A degreaser is necessary if wastewater contains huge grease concentration.
- If the industrial wastewater has a low volume to be treated and land availability, the simplest type of CW like HSSF can be a suitable solution.
- If high surface loading rate needs to be applied, VF or aerated wetlands are recommended.
- For the removal of nutrients, hybrid systems including VF and HSSF are specially indicated.

In this section, the authors will revise a few of the industrial effluents treated by CWs.

As shown in Table 1, the efficiencies can vary extensively, depending on the wastewater characteristics and the system design features and operational conditions. Regarding environmental conditions, CWs can work in extreme weather conditions ranging from the coldest temperatures to tropical conditions.

It seems hard that natural systems like CWs can work treating industrial wastewater, for instance, refinery, pulp and paper, tannery or textile effluents, where hardly biodegradable compounds are present. In wastewater with a high BOD/COD ratio, CWs can perform well as a secondary or tertiary treatment, with good pretreatments that can reduce the organic loads. For these types of wastewaters, the conventional primary treatment and activated sludge or SBR reactors as secondary treatment are the most common used systems. The variety of different industrial wastewater treated by CWs is high, and the latest research in the field is increasing this range of possibilities. The evolution of CWs has been enormous, from the simple FWS to the intensified systems. The treatment technology of CWs has evolved into a reliable technology which is nowadays successfully used for many types of industrial effluents (Vymazal, 2014).

### CWs for Dairy Parlor Wastewater Treatment

CWs have gained a place in dairy parlor and wastewater treatment from different farm activities. Table 2 summarizes main results for some applications.

*Table 1. Application of several types of CWs for the treatment of industrial wastewater*

| Type of Wastewater [a] | Type of CW (n) [c] | HLR (cm/d) | Treatment Efficiency (%) [d] | | | | | | Others Pollutants Assessed |
|---|---|---|---|---|---|---|---|---|---|
| | | | TSS | COD | BOD | $NH_4^+$ | TN | TP | |
| Refinery | FWS (8) | 0.1-128 | 44-95 | 54-93 | 50-98 | 84-94 | 19-22 | 43-68 | Phenol, HM, mineral oils, TKN |
| | HSSF (4) | 2-3 | | | 99 | 98 | | | Hydrocarbons |
| | VF (2) | 3.4-10 | 47 | 50 | | | 53 | | Hydrocarbons, HM |
| | Hybrid | 2-13 | | 30-50 | | | 48-80 | 50-60 | Hydrocarbons, HM |
| Pulp and paper | FWS (5) | 1.5-22 | 33-84 | | 6-77 | | 29 | 32-54 | $NO_3^-$, |
| | HSSF (6) | 2-15 | 55-68 | | 27-90 | 31-88 | | 14-31 | Phenol, Resin and fatty acids |
| Woodwaste | FWS (2) | 1.2-4.7 | | 6-50 | 60 | | | | Tannins, lignins |
| Tannery | FWS (1) | | | 30 | | 95 | | | HM |
| | HSSF (3) | 3-10 | 65-72 | 59-65 | 48-73 | 73 | | | TKN, HM, turbidity |
| | Hybrid (1) | 6 | 55 | 98 | 98 | 86 | | 87 | |
| Textile | HSSF (3) | 10-45 | 88 | 70 | | | | | Color, sulfate, HM |
| | VF (2) | | | 64 | | | | | Colorants |
| | Hybrid (2) | 1-6 | 93 | 84 | 66 | | 52 | | Color, colorants |
| Aquaculture | FWS (2) | 7.7-9.1 | 89-91 | | 65-82 | 47-75 | | 82-86 | $NO_3^-$ |
| | HSSF (8) | 1-790 | 81-95 | 12-94 | 72 | 7-61 | 21-69 | 43-89 | TKN, $NO_3^-$ |
| | VF (4) | 16-39 | 39-82 | 26-99 | 53-99 | 62-86 | 48-96 | 17-84 | $NO_3^-$ |
| | Hybrid (4) | 1-300 | 66-71 | 27-94 | 24-98 | 86-98 | 67 | 24-87 | $NO_3^-$, TKN |
| Winery | HSSF (7) | 2.5-48 | 30-97 | 49-99 | 92 | 54 | 58 | | $S_2^-$, TKN, phenols, tannin |
| | Hybrid (4) | 2-3 | 75-89 | 71-98 | 70-99 | 90 | 81-90 | 60-72 | $S_2^-$, TKN, $PO_3^{4-}$ |
| Destillery | HSSF (3) | 2.7-58 | | 64-80 | 84-85 | 2-10 | | 79 | TKN, $NO_3^-$, $SO_4^{2-}$ |
| Abbatoir | FWS (2) | 1.4-4.6 | 94-95 | 42 | 85 | 54 | 30 | 74 | NOx-N, TKN |
| | HSSF (6) | 1.7-7.2 | 44-99 | 65-98 | 77-98 | 20-82 | 18-75 | 37-79 | TKN, $PO_4^{3-}$ |
| | Hybrid (1) | 0.6 | 94 | 97 | 97 | 99 | 74 | | |
| Dairy/cheese | HSSF (8) | 0.1-4.4 | 62-96 | 92-97 | 34-98 | 9-92 | 44-80 | 45-81 | TKN |
| | Hybrid (4) | 0.7-10 | 40-90 | 72-96 | 55-90 | 62-82 | 40-90 | 30-88 | |
| | VF (3) | 10 | | 73-86 | | 43-72 | | | TKN, $PO_4^-$, color, phenol |
| Food industry [b] | FWS (4) | 1.4-39.5 | 52-93 | 68 | 89-99 | 22-99 | 72-92 | 21-77 | TKN |
| | HSSF (4) | 1.0-4.3 | 81 | 11-92 | 89-94 | 43-95 | | 18 | TKN,$PO_4^-$,$NO_3^-$ |
| | VF (1) | 2.5-7.4 | 88 | 87 | | 80 | | | |
| | Hybrid (2) | 0.3-3 | 83 | 93-94 | | 47-72 | 54-90 | 83 | |
| Laundry | HSSF (3) | 3.7 | 83 | | 61 | | 62 | 32 | LAS |
| Mixed industrial | FWS (2) | 10-19 | 81 | 61 | 89 | 53 | | 35 | HM |
| | Hybrid (2) | 36.5 | 89 | 67-81 | 66-69 | 24 | 30 | 62 | $PO_4^-$ |
| Steel industry | HSSF (2) | 1.-2.6 | | 50-77 | | 77 | | 6 | HM |
| | VF (1) | | | 55 | | 67 | | 93 | HM |

[a]Other industrial effluents with one reported case are brewery (HSSF), olive mill (FWS), chemical industry (VF), explosives (FWS), coke (HSSF), coal gasification (FWS), lignite pyrolysis (HSSF), tool industry (FWS and flower farm (hybrid).[b]Sugar, potato, fish/seafood and mixed food industry. [c]Number of cases. [d]Ranges of available values (note that not all the reported cases include data about all these parameters). Acronyms: HM, heavy metals, TKN, total Kjedahl nitrogen, LAS, linear alkylbenzene sulfonates.

(modified from Vymazal, 2014)

*Table 2. Results of dairy farms wastewater treatment with CWs*

| Reference | Country | Source of wastewater | Pre-treatment | Treatment[a] | | T (°C) | Efficiency (%removal) | | Surface loading rate SLR (g /m²·d) | | HRT (d) | Surface rate (m²/ cow) | Effluent concentration (mg/L) | |
|---|---|---|---|---|---|---|---|---|---|---|---|---|---|---|
| Tanner *et al.*, 1995 | Argentina | Dairy parlor (pre-treated) | | HSSF | 40 cm | 7-22 | 92-76 (BOD) 76-83 (TSS) | | 0.9-4.1 (BOD) | 1.9-8.5 (TSS) | 2 – 7 | - | 11-27 (BOD) 30-36 (TSS) | |
| Geary & Moore. 1999 | Australia | Farm (pre-treated) | Pond | FWS / HSSF | 24 cm/52cm | | 61 (BOD) 27 (TKN) | 43 (Norg.) 28 (TP) | 5.6 (BOD) 5.7 (TKN) | 3.2 ($NH_4^+$) 1.5 (TP) | oct-14 | 2.7 | 90 (BOD) 105 ($NH_3$) | 49 (TP) |
| Kern & Idler. 1999 | Germany | Dairy and domestic (1:7) | Septic tank | VF | 1.5 m | | 86 (COD) 35 (TN) | 67 ($NH_3$) 55 (TP) | 18.2 (COD) 2.8 (TN) | 2.1 ($NH_4^+$) 0.3 (TP) | 3 | - | 89 (COD) 63 (TN) 25 ($NH_4^+$) 32 ($NO_3^-$) 4.7 (TP) | |
| Newman *et al.* 2000 | USA | Dairy parlor | | FWS | | 3-20 | 94 (TSS) 85 (BOD) | 68 (TP) 53 (TKN) | 17.0 (BOD) 9.0 (TSS) 0.68 (TKN) | 0.19 (TP) | 41 (<12) | 4 | 692 (BOD) 130 (TSS) 52 ($NH_4^+$) 14 (TP) | |
| Knight *et al.* 2000 | USA: 38 cases | Farm | | FWS (/HSSF) | | | | | <10.3 (BOD) <10.3 (TSS) <2.8 ($NH_4^+$) | | | | Secondary and partial nitrification | |
| Mantovi *et al.*. 2003 | Italy | Dairy parlor and domestic[b] | Imhoff tank | 2xHSSF | 100 cm | | Global (IT +CW) > 90% (TSS, COD, BOD) 48.5 (TKN) 60.6 (TP) | | 12.7 (BOD) 27.0 (COD) 10.8 (TSS) | 2.0 (TKN) 0.42 (TP) | 10 | 1.9 | 28 (BOD) 60 (TSS) 98 (COD) 33 (TKN) 5 (TP) | |
| Dunne *et al.*. 2005 | Ireland | Farm | | FWS | | | 99 (TSS) 99 (BOD) | 96 ($NH_3$) 93 (SRP) | 3.6 (BOD) 1.0 (TSS) | 0.03 (SRP) 0.07 ($NH_4^+$) | | 114 | 20 (BOD) 1.6 ($NH_4^+$) 11 (TSS) | |
| Lansing & Martin. 2006 | USA | Farm | | CW Hybrid (11 units in series) | | | 99 (BOD) 99 ($NH_3$) | 79 ($PO_4^{3-}$) | | | 10 | - | 5.2 (BOD) 0.07 ($NH_3$) 0.53 (NOx) | 4.4 ($PO_4^{3-}$) |
| Healy *et al.*. 2007 | Ireland | Dairy parlor | Septic tank | ISF / VF with recirculation | | | 99 (BOD) 86 (TN) | | 30 (COD) | | | 5 | BOD removal and complete nitrification | |

[a]IT: Inhoff tank, ISF: Intermittent sand filter .[b]Domestic effluent means a dilution of dairy parlor effluent in this case, reducing 70% concentration. Effluent from holding room is separated to the slurry tank, not contributing to reported concentration. Pretreatment, IT (40 PE), includes a filter and reaches removal rates of 64% TSS, 48% COD and 35% BOD in the original mixed effluent.

Three cases are shown in this section. Newman *et al.* (2000) designed a FWS of 400 $m^2$ for dairy parlor wastewater treatment from a farm with 100 milk cows. Real $BOD_5$ concentration was much higher (2683 mg/L) than expected and the treatment was not sufficient due to effluent values of 692 mg/l with a 74% $BOD_5$ removal. After this experience, an estimation of pollutant loadings when designing the systems is not recommended by those authors.

Mantovi *et al.* (2003) described a two-stage HSSF system in series to treat a mixture of domestic and dairy parlor effluent, previously treated in an Inhoff tank and a well with plastic filter for the removal of sedimentable suspended solids. Removal efficiency was satisfactory, with values of more than 90% in organic matter (TSS, COD and BOD), 48.5% of TKN and 60.6% of TP.

The 16-years of expertise of van Oirschot (2015) treating dairy wastewater is described in this section. In rural areas from the Netherlands and Belgium the main sources of wastewater discharged into fresh water are untreated widespread domestic and dairy farms wastewater. Several projects were developed, testing different additions to the substrate for their efficiency in the removal of phosphorus. The ratio BOD/COD in wastewater coming from milking parlors, seems to be greater than domestic wastewater, and low N-values due to the presence of milk. Results of nitrogen and phosphorus removal of 13 CWs treating dairy farms were 84.2%, 78.7%, 47.9% and 72.3% for TKN, $NH_4^+$, TN and TP, respectively.

These results from van Oirschot (2015) show a high removal of organic matter and organic nitrogen, as well as a significant reduction of total phosphorus. Denitrification occurs to about 50%, good if considering that is single-stage VF being used. This denitrification rate is higher than the same CW treating domestic wastewater, due to the relatively high C/N ratio of the dairy wastewater. In VF systems, it seems that the lack of carbon in the lower anaerobic parts of the wetland is the main cause for the low denitrification rate. The total P concentrations of 6 CWs at dairy farms with $CaCO_3$ and iron additions in the substrate ranged from 8.8 to 29.1 mg P/L and 0.14 to 8.8 mg P/L in the inlet and the outlet respectively.

## Cheese Production Wastewater

Gorra *et al.*, (2014) reported the performance of a HSSF from a medium-sized cheese factory which processes 8 ± 4 $m^3$ of milk. The local climate conditions were a mean air temperature of 12°C and mean total precipitation of 430 mm. The system consists of a pretreatment by means of sedimentation tank (45 $m^3$) and a HSSF CW of 200 $m^2$ of total area and planted with common reed (*Phragmites australis*). The CW is divided into five sections, each one filled with a different material: (1) gravel from metamorphosed limestone, (2) ground ceramic wastes, (3) by-products from magnetite extraction, (4) zeolitite and (5) a horizon of a Dystrict Endoskeletic (Siltic) Cambisol. The input average flow was 10 ± 2 $m^3$/d, with a maximum during spring of 12 ± 1$m^3$/d and a minimum of 8 ± 2$m^3$/d during summer. Different volumes of influent treated have influenced the retention time, which for such reason varied from 15 to 5 days (average value about 7 days). Average SLRs were 40 ($BOD_5$), 10 (TN) and 0.5 (TP) g/$m^2$·d, while maximum SLR reached approximately 100 ($BOD_5$) and 25 (TN) g/$m^2$·d. The average mass removal was 55%, 55%, 59% and 58% for TN, organic N, $NH_4^+$ and TP, respectively.

Most of the influent nitrogen was in the form of organic N (>85%), while ammonium and nitrate were around 8-14% and 2-4% respectively. The average removal of biochemical oxygen demand ($BOD_5$) was higher than 90%.

The treated effluent had a final quality that would allow the use for direct irrigation in agriculture.

## Tannery Wastewater

CWs have been used for secondary or tertiary treatment for the treatment of tannery wastewater as a cost-effective and environmentally friendly technology. Calheiros *et al.*, (2014) reviewed the tannery sector in Portugal and the performance of CWs for treating this wastewater. Depending on the kind of treatment, tannery wastewater has different BOD/COD ratios, ranging from 0.23 and 0.66 for water that had primary treatment, and from 0.1 to 0.65 for water after secondary treatment. As shown by the ratio, these waters are not easy biodegradable, if compared to domestic wastewater or even with most industrial wastewater. When considering CW as part of a treatment system an efficient pretreatment and the primary treatment and precious treatments are of upmost importance before the wastewater is conducted to the CW (Calheiros *et al.* 2007). The reported treatments for tannery wastewater CWs are conventional primary treatments and an activated sludge as a secondary treatment.

Regarding the construction and establishment costs CWs correspond to circa 10–20% of the conventional secondary biological treatment and the running costs account for approximately 10% (Daniels, 2007), although it may vary according to the country where the system is implemented. Dias *et al.*, (2006) reported that in the Portuguese scenario the capital costs, with pretreatment included, are equivalent to a conventional treatment, however, concerning the operation and maintenance cost the same authors state they are at least 6–20 times lower.

Efficiencies of CWs treating tannery wastewater in Portugal ranged from 2.3 to 160 g $COD/m^2 \cdot d$ and 0.6 to 61 g $BOD/m^2 \cdot d$ removed.

Focusing on the phosphorus and nitrogen removal in tannery wastewater, Kadlec *et al.* 2000 reported that the performance is lower when compared to the organics removal, although in some way expected for this type of systems. For instance, Calheiros *et al.* (2008), reported that for an inlet of 0.21–0.43 mg P/L the outlet was 0.18–0.35 mg P/L and for an inlet of 63–87 mg $NH_3$/L the outlet was 40–56 mg $NH_3$/L. Removal of $PO_4$ reaching 61% (inlet 15 $\pm$ 7 mg/L, hydraulic retention time (HRT) of 12.5 d) has been reported for the treatment of tannery wastewater by CW in Bangladesh (Saeed *et al.* 2012a). High removal of $NH_4^+$ (99%) from tannery effluents has been reported by Kucuk *et al.* (2003) for an inlet of 20 mg/L at a HRT of 7 days.

## Greenhouse Gas Emissions (GHG) in CWs

Greenhouse gases namely $CO_2$, $CH_4$ and $N_2O$ are emitted from both natural and constructed wetlands. During the mineralization of organic matter, $CO_2$ production is desirable but, on the other hand, $CH_4$ and $N_2O$ emissions must be avoided due to their high greenhouse effect potential (they have a global warming potential of 34 and 298 relative to $CO_2$ over a 100 year time horizon (Myhre *et al.*, 2013).

HSSF CWs have the bottom as the principal anaerobic environment, and the zone close to the water surface and the rhizosphere as aerobic zones. Both organic and hydraulic loading rates are key operational variables that can influence the importance of anaerobic and aerobic roles, together with water level as a design variable. As a result, these factors determine the influence on methane generation and emission (Pedescoll *et al.*, 2011).

Methane emission rates from CWs treating different types of wastewaters can vary from -208 to 36,792 mg $CH_4/m^2 \cdot d$ (Pei-Dong *et al.*, 2002; Wang *et al.*, 2008) but more usual range is from 0 to 2000 mg $CH_4/m^2 \cdot d$. Some important factor affecting the measured gas emissions are the position on the length of the wetland in the direction of flow, and the season of the year (Tanner *et al.*, 1997; Mander *et al.*, 2003; Johansson *et al.*, 2004; de la Varga *et al.*, 2015).

According to the literature review reported by Mander *et al.*, (2014), $CO_2$ emissions are significantly lower in FWS than in HSSF CWs, with average values of 2208 and 4440 mg $CO_2/m^2 \cdot d$, respectively. Additionally, in VF systems, methane emissions were significantly lower than in HSSF and FWS (average values of 97, 178 and 142 mg $CH_4$ $m^2 \cdot d$, respectively). For $N_2O$ emissions, no significant differences were found for several types of CWs (average values for FWS, VF and HSSF of 2.2, 2.9 and 3.1 mg $N_2O/m^2 \cdot d$, respectively). As a conclusion, Mander *et al.*, (2014) reported that hybrid systems (VF, HSSF and/or FWS) perform better for treating wastewater and reducing direct greenhouse gas emissions. The emissions are affected by the loading, the operation mode as well as the local climatic conditions.

The emission factor for methane might be defined as the ratio of carbon emitted as methane to the total organic carbon influent to the CW ($CH_4/TOC_{in}$). Mander *et al.*, (2014) reported that the greatest emission factor for methane was found for FWS (average of 16.9%), followed by HSSF (4.5%) and VF (1.2%). These researchers reported that significant differences of emission factor for $N_2O$ (percentage ratio of $N_2O$ emitted to influent total nitrogen TN) were found for HSSF (0.79%), FWS (0.13%) and VF (0.023%). Moreover, according to Mander *et al.*, (2014), a significant correlation between the influent rate of total organic carbon (TOC) and the methane emission rate and between the influent rate or total nitrogen (TN) and $N_2O$ emission rate for all types of studied CWs was found as well.

Data about gas emissions from CW treating industrial effluents is still scarce. Tanner *et al.*, (1997) found an average methane emission rate of 225 mg $CH_4/m^2 \cdot d$ (standard deviation SD of 28-482) in a HSSF treating dairy wastewater, with a surface loading rate from 2.5 to 4.2 g $BOD/m^2 \cdot d$. Maltais-Landry *et al.*, (2009) reported an average methane emission rate of 20 to 120 mg $CH_4/m^2 \cdot d$ in an aerated HSSF pilot plant treating fish factory effluent. De la Varga *et al.* (2013b) reported methane emissions ranging from 143 to 1899 mg $CH_4/m^2.d$ for HSSF system treating winery wastewater while carbon dioxide emissions in this system ranged from 1783 to 6137 mg $CO_2/m^2 \cdot d$.

## Economics of CWs

CWs include mainly two types of costs: capital costs (construction, engineering and indirect costs) and operation and maintenance (O&M) costs. The main difference between FWS and subsurface flow system is the amount of gravel media which affects the construction costs. Although it is not possible to stablish universal cost guidelines, every system shares a similar set of construction components. Kadlec and Wallace, (2009) reported some equations for calculating costs of FWS and HSSF systems depending on surface needed or flow rate. As an example, for a 1 ha FWS system, total costs would be $480,000 (for 2006 USD), which means $48/m$^2$.

Area:

FWS: $C = 194A^{0.690}$ $R^2=0.79$ $0.03<A<10{,}000$

HSSF: $C = 652A^{0.704}$ $R^2=0.75$ $0.005<A<20$

Flow:

FWS: $C = 518A^{0.729}$ $R^2$=0.79 1<Q<5,000,000

HSSF: $C = 561A^{0.498}$ $R^2$=0.76 0.5<Q<10,000

where:

A= wetland area, ha

C=cost, 2006 dollars in thousands

Q=flow rate, $m^3$/d in thousands

DiMuro *et al.*, (2014) reported a study in which a comparison of economic and environmental costs of a constructed wetland vs. a sequential batch reactor (SBR) was assessed. The selected industry was the wastewater treatment facility of a large industrial complex containing several manufacturing units involved in the production of plastic resins and other organic chemicals. Wastewater and storm water captured in containment areas are routed through the wastewater treatment system. The input variables for the design of both systems were flow rate ($6.9 \cdot 10^6$ $m^3$/year), total organic carbon (TOC, 500 mg/L), COD (1500 mg/L), $BOD_5$ (875 mg/L) and TSS (918 kg/d). The financial results indicated that the total net present value savings calculated for implementing the CW instead of the SBR reactor was $282 million over the project's lifetime (30 years). The Life Cycle Assessment (LCA) demonstrated that the lower energy and material inputs to the CW resulted in lower potential impacts for fossil fuel use, acidification, smog formation, and ozone depletion and likely lead to lower potential impacts for global warming and marine eutrophication.

## CASE STUDIES

### Winery Wastewater Treatment

Winery wastewater is generated by the various processes and operations carried out in wine production, and wastewater comes mainly from the water used to wash equipment and bottles and from the cooling processes.

Winery wastewaters contain COD concentrations ranging from 500 to 45000 mg/L and total suspended solids (TSS) from 12 to 7300 mg/L, while $BOD_5$ is about 0.4-0.9 of the COD value (Shepher *et al.*, 2001; Petruccioli *et al.*, 2002; Masi *et al.*, 2002; Mosteo *et al.*, 2008). The concentrations of nitrogen and phosphorus compounds in winery wastewater are usually low and the pH varies from 3.5 to 7 (Serrano *et al.*, 2009).

Conventional biological treatment systems do not effectively treat wastewater generated by wineries (Petruccioli *et al.*, 2002; Mosteo *et al.*, 2006), as wastewater from these facilities are highly concentrated in terms of COD and show highly variable flows and loadings, both daily and seasonally. CWs have also been considered as a means to treat this wastewater (Masi *et al.*, 2014). Wetlands remove efficiently organic matter, excess nutrients and other pollutants. CWs are robust and can assimilate large and highly variable flows and organic loadings and neutralise low pH.

As a complement and to ensure the best performance, anaerobic pretreatment such as hydrolytic upflow sludge bed (HUSB) digesters are recommended. HUSB is an intensive, efficient and low-cost technology for domestic wastewater and industrial wastewaters (Álvarez *et al.*, 2008) that can also be applied to the pretreatment of winery wastewater (de la Varga *et al.*, 2013a).

## Winery Wastewater Treatment in Hybrid CW System in Spain

In this section the authors summarize the results obtained on a full-scale hybrid CW with an anaerobic digester as primary treatment, followed by CW which combined VF and HSSF units for treating winery and a touristic and restaurant wastewater component (which provided higher nutrients concentration than a normal winery). (Serrano *et al.*, 2011; de la Varga *et al.*, 2013b). The wastewater was previously pretreated in an HUSB digester (de la Varga *et al.*, 2013a). The system installed showed an average removal efficiency of 76, 26 and 21% of TSS, COD and BOD respectively. The anaerobic digester mainly initiated the ammonification of organic nitrogen compounds and the acidification of organic matter.

## Description of the Plant

The full-scale plant was designed to treat the wastewater coming from a winery with a production capacity of 315000 L of white wine. The touristic and restaurant component increased nutrient concentration of wastewater.

The constructed-wetland system consisted of a HUSB anaerobic digester 6 $m^3$ in volume, a VF CW 50 $m^2$ in area and three HSSF CW at 100 $m^2$ each. The water depth for HSSF1 was 0.3 m, while the depth for HSSF2 and HSSF3 was 0.6 m, respectively. CWs were planted with three to four units/$m^2$ of *Phragmites australis* (VF) and *Juncus effusus* (HF) and started up on April 14$^{th}$ 2008.

## CW Performance and Efficiency

Average influent temperature was 20.2±4.5 °C, while effluent temperature was in general 3 °C lower (17.2±4.6 °C in average, ranging from 12 to 28 °C). Average influent pH was 6.41±0.95, while effluent pH was 6.7±0.5 (ranging from 5.4 to 7.4), and then the wastewater increased pH in 0.5 units passing through the HSSF wetlands.

The average surface loading rate (SLR) to the VF ranged from 43 to 466 g COD/$m^2$d and from 22 to 296 g $BOD_5$/$m^2$d, while COD removal ranged from 29% to 70% and $BOD_5$ removal from 36% to 68%. Thus the VF unit reached high surface removal rates of up to 153 g COD/$m^2$d and 107 g $BOD_5$/$m^2$/d.

Influent TSS levels to HSSF were generally low as they had been efficiently removed in the preceding steps of the plant (mainly in the HUSB reactor). Therefore, SLR of suspended solids remained low (ranging from 0.5 to 5 g TSS/$m^2$·d) and solids have been efficiently removed in HSSF units, which reached TSS effluent concentrations below 44 mg/L (17 mg/L on average) and TSS removal efficiencies of about 74%.

Influent COD and $BOD_5$ were highly variable, ranging from 50 to 4000 mg COD/L and from 12 to 2400 mg $BOD_5$/L (n=68). Higher influent concentrations and flows were registered for vintage phases of years 2008 and 2009, respectively. This led to highly variable SLR to the HSSF units, ranging from

4-85 g TCOD/m$^2$·d and from 2 to 49 g $BOD_5$/m$^2$·d. TCOD and $BOD_5$ removals were generally between 60 and 80% but clearly decreased when the influent concentration or the SLR applied increased.

Regarding nutrient removal, variable concentrations of nitrogen and phosphorus compounds were present in the influent, due to touristic and restaurant component of the wastewater. Influent TKN of VF CW was between 19 and 109 mg N/L (average of 52.9 mg N/L) and the VF effluent ranged from 11 to 63 mg N/L (average of 26 mg N/L). Ammonium concentration ranged from 1 to 101 mg $NH_3$/L in the VF influent (average of 28 mg $NH_3$/L) and from 0.6 to 76 mg N/L (average of 19 mg $NH_3$/L) in the VF effluent with an average removal of 32% (made on average values). This relatively low nitrification might be due to high organic load in the VF CW which primary consumed the oxygen present and subsequently reducing nitrification. Phosphate in the VF influent ranged from 0.28 to 5 mg P/L (average of 2.3 mg P/L) and from 0.9 to 6.3 mg P/L (average of 2.4 mg P/L) in the VF effluent. HSSF effluent concentrations of TKN, ammonia and phosphate ranged from 12 to 52 mg N/L (average of 25.2 mg N/L), from 2 to 52 mg (average of 12.5 mg N/L) and from 0.3 to 6 (average of 1.9 mg P/L), respectively. Removal of nitrogen compounds and phosphate in HSSF units was reduced, ranging from 6 to 29% on average. Limited data indicates that polyphenols removal averaged 39% in HSSF units.

The full-scale hybrid CW system successfully treated winery wastewater. Surface removal rates up 41 g COD/m$^2$·d, 25 g$BOD_5$/m$^2$·d and 4.4 g TSS/m$^2$·d were reached by the overall (VF + HF) system. Overall percentage removals ranged from 54% to 93% of COD, 45% to 95% of $BOD_5$ and 75% to 94% of TSS. The overall system also removed up to 1.4 gTKN/m$^2$·d (52%), 0.7 g$NH_3$/m$^2$·d and 0.04 g $PO_4^{3-}$/m$^2$·d (17%).

## Upgrading of a CW System for Winery Effluents in Italy

The Cecchi system in Italy (Masi *et al.*, 2002; 2014) consisted of an Inhoff tank, a single stage of HSSF of 480 m$^2$ and a FWS of 850 m$^2$. Due to growing production of the winery, the treated flow increased from 35 m$^3$/d to 70 m$^3$/d. An upgrading of the treatment plant was developed in 2009, with a French reed bed (FRB), followed by four parallel HSSF, with the existing FWS and optional sand filter before discharge into fresh water (Gena River). Results from a 3 years period show a COD reduction of up to 96% even at peak loads while COD average removal efficiency was 75%. For the HSSF unit, average removal was 87%.

## Further Considerations about CW Design Criteria for Winery Wastewater Treatment

Hybrid systems of CWs can themselves treat winery wastewater, as well as a complement to other systems like sequential batch reactors (SBR) or activated sludge. For small wineries producing less than 100.000 L wine/year, a septic tank +HSSF CW with more than 5 days of HRT during peak flow period can ensure a proper treatment for fresh water discharge or for irrigation purposes (Masi *et al.*, 2014).

A good pretreatment or primary treatment to decrease the solids and the organic particulate matter is important. For small wineries, a septic tank could be enough if a sort of resting period of 10 months with lower organic and hydraulic loads takes place. For bigger wineries, this first stage (SBR, FRB, etc.) should remove 40-50% of pollutants from the raw wastewater.

Some of the wineries have additional nutrients in the wastewater, like black waters from employees, small households or restaurants. No supplementation of nutrients has been necessary, as well as neutralization of low pH. If we do so, performance of CWs might be increased, and this is a good point for future research.

## HIGHWET Project for Food Processing Industry

The research project "Performance and validation of HIGH-rate constructed WETlands" (HIGHWET), under the European call FP7-SME-2013-1, was developed during years 2013 to 2015. There were two pilot plants, one located in Spain to treat domestic wastewater, and the other "real" plant was implemented in Denmark, to treat the discharges generated from a food production factory, a house inhabited by the factory owners and a small dairy factory.

The final design includes a combination of the anaerobic treatment HUSB reactor, aerated *CWs* and conventional CWs. In addition, the system was fitted with reactive media to improve phosphorus removal in order to achieve discharge limits.

Figure 2 presents the layout and the different units of the treatment plant. During the project, inflow was limited to 1000 L/d to control the process. After a pumping well, water is conducted to the primary treatment, which is a HUSB reactor. Then, the system is split in two treatment trains. The first one (aerated train) is composed by an aerated VF, an aerated HSSF and a phosphorus removal well, filled with tobermorite. The second train (conventional train) has a conventional VF, an aerated HSSF and a second phosphorus removal well, with polonite. Both VF beds have an area of 16 $m^2$ while the HSSF beds are of 3 $m^3$ followed by wells of 1 $m^3$. The aeration systems installed in the beds supply air to the beds using individual blowers that provide atmospheric air to increase the oxygen availability and improved the aerobic depletion processes of pollutants. The aeration time to each wetland cell can be regulated depending on the operation planned. The aerated train treated 80% of the total flow, and conventional train treated 20% of the total flow.

*Figure 2. Sampling points along the treatment trains 1) inlet; 2) after the HUSB, 3) after the aerated VF; 4) after the aerated HF bed; 5) after the P removal filtered filled with tobermorite 6) after the non-aerated bed 7) after the aerated bed 8) after the P removal filter filled with Polonite 9) final effluent*

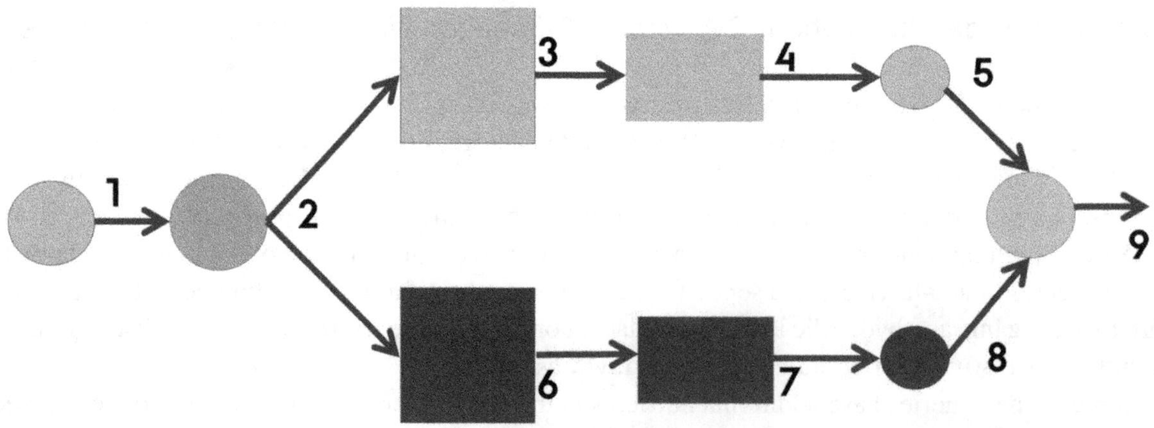

## Operation of the Plant

During the plant operation, pollutant loadings, aeration times and hydraulic loadings by recirculation of VF effluent to the pumping well after the HUSB reactor were changed to determine the capacity of the plant. In this sense, five campaigns were carried out under different operational conditions. As the wastewater produced at the site did not reach the projected organic and nutrient loadings, a system with a prepared solution was installed to increase the pollutant and hydraulic loads. The loading solution was prepared using a blend of pig manure, molasses, starch, urea and fertilizer.

After plant start-up (campaigns 1 and 2), several monitoring campaigns with high nutrient loadings were carried out (campaigns 3, 4 and 5). During campaigns 3 and 4, aeration schemes were changed in order to increase the total nitrogen removal, and during campaign 5, recirculation (increasing to around 80% the hydraulic load to each bed) from the VF effluent to the well after the HUSB was implemented. Following, the efficiency reached during campaign 3 is presented (Table 3). Similar results were obtained for campaigns 4 and 5.

The third campaign took place in May when temperatures increased and the local weather was mild. The vegetation growth was better in the aerated beds, due to higher water flow that allowed major nutrient availability. On the other hand, plant growth in the passively aerated beds was poor. The water temperature did not change too much, compared to the winter temperatures, due to low temperatures at night. As expected, anaerobic conditions were prevalent before and after the HUSB digester. After this primary treatment, dissolved oxygen was close to saturation generally, except in aerated beds when aeration regime was decreased to improve denitrification.

Total suspended solids were efficiently removed, with a concentration in the final effluent below 3 mg/L. A high COD percentage was removed at the HUSB reactor, and further COD removal was high in both VF beds which reached similar effluent concentrations. Along the system, biological oxygen demand ($BOD_5$) was removed to very low concentrations.

*Table 3. Pollutant concentrations at the different sampling points during campaign 3. Concentrations are presented in average. BDL: Below detection limit*

| Sampling point | TSS (mg/L) | COD (mg/L) | $BOD_5$ (mg/L) | $NH_4^+$ (mg/L) | TN (mg/L) | TP (mg/L) |
|---|---|---|---|---|---|---|
| Inlet | 117 | 5268 | 4167 | 250 | 382 | 25.1 |
| After HUSB | 25.2 | 2055 | 1383 | 197 | 201 | 18.2 |
| **Aerated train** | | | | | | |
| After aerated VF | 22.6 | 211 | 44 | 5.2 | 11 | 5.2 |
| After aerated HSSF | 33.6 | 156 | 6 | 0.12 | 9 | 4.0 |
| After tobermorite | 33.8 | 156 | 6 | 0.14 | 11 | 2.2 |
| **Conventional train** | | | | | | |
| After VF | 86.5 | 174 | 3 | 0.10 | 42 | 2.9 |
| After aerated HSSF | 101.7 | 82 | 2 | BDL | 40 | 1.2 |
| After polonite | 102.7 | 61 | 2 | 0.7 | 45 | 0.5 |

During the third campaign, the HUSB reactor removed around 20% of $NH_4^+$, so $NO_3^-$ was denitrified effectively in this stage. After the HUSB reactor and in both treatment trains, N-removal was effective producing nitrified effluent with a $NH_4^+$ concentration below 1 mg/L. The results show that the $NH_4^+$ nitrified in the aerated VF bed is also denitrified during the shut-off time after the oxygen supply. The conventional VF bed does not show the same denitrification rates as the aerated bed and the effluent from the bed has a $NO_3^-$ concentration of 40 mg/L. Even though $NO_3^-$ seems high in the outlet, concentration in the inlet was much higher and denitrification was considerable in the passive VF. Total nitrogen followed the same dynamics as the nitrogen species. Phosphorus removal was effective in both tested materials, but the P concentration in the effluent after tobermorite material was higher than the Danish limits.

In all campaigns, the percentage of total nitrogen removal was higher than 80%. The HUSB reactor removed between 30% and 40%, corresponding to $NO_3^-$ denitrified. The aerated VF bed performance of total nitrogen removal was better for periods with discontinuous aeration compared to continuous aeration, as well as with recirculation of nitrified VF effluent. The non-aerated VF nitrified quite well, and removed total nitrogen as well, which means that some parts of the bed have good denitrifying environmental conditions.

## CONCLUSION

Constructed wetlands are engineered systems that have been designed to utilize natural processes involving wetland vegetation, soils, and their associated microbial assemblages to assist in treating wastewater. In CWs, removal of pollutants takes place due to physical, chemical and microbial processes including, among others, sedimentation, sorption, precipitation, evapotranspiration, volatilization, photodegradation, diffusion, plant uptake, and microbial degradation (i.e. nitrification, denitrification, sulphate reduction, carbon metabolization, etc.). The most used types of CWs are the FWS, the HSSF and the VF. For improving the performance and the removal of pollutants and nutrients, a combination of these systems, the hybrid systems, may be used. CWs are also used to recover nutrients from the macrophyte biomass and particularly from sludge in the sludge treatment wetlands. Nowadays, CWs have been applied to the treatment of different kind of wastewaters as such as refinery and petrochemical industry effluents, food industry effluents including abattoir, dairy, meat, fruit and vegetables processing industries, distillery and winery effluents, pulp and paper, textile, tannery, aquaculture, steel and mixed industrial effluents.

The reached efficiencies can vary depending on the wastewater characteristics and the system design features and operational conditions.

Regarding environmental conditions, CWs can work in extreme weather conditions ranging from the coldest one to tropical conditions. TSS, COD and $BOD_5$ are potentially removed in any kind of CWs if proper design and loading rates are provided.

On the other hand, nitrogen removal requires the combination of adequate oxic and anoxic/anaerobic conditions which can be provided by hybrid systems and other intensified systems involving water recirculation or artificial aeration.

Phosphorus removal requires extensive application of CWs (large footprint) or specific filtering substrates to absorb phosphorus compounds.

As reviewed, CWs are particularly suitable for the treatment of dairy parlour wastewater, cheese production wastewater and other food industry wastewater, as well as for tannery and winery wastewater.

## REFERENCES

Álvarez, J. A., Ruíz, I., & Soto, M. (2008). Anaerobic digesters as a pretreatment for constructed wetlands. *Ecological Engineering*, *33*(1), 54–67. doi:10.1016/j.ecoleng.2008.02.001

Andreoli, C. V., Pegorini, E. S., Fernandes, F., & Santos, H. F. (2007). Land application of sewage sludge. In V. Cleverson, M. Von Sperling, & F. Fernandes (Eds.), *Sludge Treatment and Disposal*. London, UK: IWA Publishing.

Arheimer, B., & Wittgren, H. B. (1994). Modelling the effects of wetlands on regional nitrogen transport. *Ambio*, *23*(6), 378–386.

Arias, C.A., Brix, H., Kilian, R.M. (2011). *Beplantede filteranlæg til miljøeffektiv spildevandsrensning i det åbne land*. Article Teknik og Miljø.

Arias, C. A., Del Bubba, M., & Brix, H. (2001). Phosphorus removal by sands for use as media in subsurface flow constructed reed beds. *Water Research*, *35*(5), 1159–1168. doi:10.1016/S0043-1354(00)00368-7 PMID:11268836

Austin, D., Maciolek, D., Davis, B., & Wallace, S. (2006). Damköhler number design method to avoid plugging of tidal flow constructed wetlands by heterotrophic biofilms. *10th International Conference on Wetland Systems for Water Pollution Control*.

Bialowiec, A., Janczukowicz, W., & Randerson, P. F. (2011). Nitrogen removal from wastewater in vertical flow constructed wetlands containing LWA/gravel layers and reed vegetation. *Ecological Engineering*, *37*(6), 897–902. doi:10.1016/j.ecoleng.2011.01.013

Brix, H. (1997). Do macrophytes play a role in constructed treatment wetlands? *Water Science and Technology*, *35*(5), 11–17. doi:10.1016/S0273-1223(97)00047-4

Brix, H., & Arias, C. A. (2005). Danish guidelines for small-scale constructed wetland systems for onsite treatment of domestic sewage. *Water Science and Technology*, *51*(9), 1–9. PMID:16042237

Brix, H., Arias, C. A., & Del Bubba, M. (2001). Media selection for sustainable phosphorus removal in subsurface flow constructed wetlands. *Water Science and Technology*, *44*(11–12), 47–54. PMID:11804137

Brix, H., & Schierup, H. H. (1990). Soil oxygenation in constructed reed beds: The role of macrophyte and soil-atmosphere interface oxygen transport. *Water Research*, *29*(Part 2), 259–266. doi:10.1016/0043-1354(90)90112-J

Calheiros, C., Rangel, A., & Castro, P. (2014). Constructed Wetlands for Tannery Wastewater Treatment in Portugal: Ten Years of Experience. *International Journal of Phytoremediation*, *16*(9), 859–870. doi:10.1080/15226514.2013.798622 PMID:24933889

Calheiros, C. S. C., Rangel, A. O. S. S., & Castro, P. M. L. (2007). Constructed wetland systems vegetated with different plants applied to the treatment of tannery wastewater. *Water Research*, *41*(8), 1790–1798. doi:10.1016/j.watres.2007.01.012 PMID:17320926

Calheiros, C. S. C., Rangel, A. O. S. S., & Castro, P. M. L. (2008). Evaluation of different substrates to support the growth of *Typha latifolia* in constructed wetlands treating tannery wastewater over long-term operation. *Bioresource Technology*, *99*(15), 6866–6877. doi:10.1016/j.biortech.2008.01.043 PMID:18334290

Carballeira, T., Ruiz, I., & Soto, M. (2016). Effect of plants and surface loading rate on the treatment efficiency of shallow subsurface constructed wetlands. *Ecological Engineering*, *90*, 203–214. doi:10.1016/j.ecoleng.2016.01.038

Chang, Y., Wu, S., Zhang, T., Mazur, R., Pang, C., & Dong, R. (2014). Dynamics of nitrogen transformation depending on different operational strategies in laboratory-scale tidal flow constructed wetlands. *The Science of the Total Environment*, *487*, 49–56. doi:10.1016/j.scitotenv.2014.03.114 PMID:24768911

Chen, Y., Wen, Y., Zhou, Q., & Vymazal, J. (2014). Effects of plant biomass on nitrogen transformation in subsurface-batch constructed wetlands: A stable isotope and mass balance assessment. *Water Research*, *63*, 158–167. doi:10.1016/j.watres.2014.06.015 PMID:25000198

Cui, L., Feng, J., Ouyang, Y., & Deng, P. (2012). Removal of nutrients from septic effluent with re-circulated hybrid tidal flow constructed wetland. *Ecological Engineering*, *46*(0), 112–115. doi:10.1016/j.ecoleng.2012.06.003

Daniels, R. P. (2007). Tannery effluent and reed beds: Working with nature. *The Journal of American Leather Chemists Association*, *102*(8), 248–253.

de la Varga, D., Díaz, A., Ruiz, I., & Soto, M. (2013a). Avoiding clogging in constructed wetlands by using anaerobic digesters as pre-treatment. *Ecological Engineering*, *52*, 262–269. doi:10.1016/j.ecoleng.2012.11.005

de la Varga, D., Ruíz, I., Álvarez, J. A., & Soto, M. (2015). Methane and carbon dioxide emissions from constructed wetlands receiving anaerobically pretreated sewage. *The Science of the Total Environment*, *538*, 824–833. doi:10.1016/j.scitotenv.2015.08.090 PMID:26342902

de la Varga, D., Ruiz, I., & Soto, M. (2013b). Winery wastewater treatment in subsurface constructed wetlands with different bed depths. *Water, Air, and Soil Pollution*, *224*(4), 1–13. doi:10.1007/s11270-013-1485-5

Dias, V. N., Canseiro, C., Gomes, A. R., Correia, B., & Bicho, C. (2006). Constructed wetlands for wastewater treatment in Portugal: a global review.*10th International Conference on Wetland Systems for Water Pollution Control.*

DiMuro, J. L., Guertin, F. M., Helling, R. K., Perkins, J. L., & Romer, S. (2014). A financial and environmental analysis of constructed wetlands for industrial wastewater treatment. *Journal of Industrial Ecology*, *18*(5), 631–640. doi:10.1111/jiec.12129

Drizo, A., Frost, C. A., Smith, K. A., & Grace, J. (1997). Phosphate and ammonium removal by constructed wetlands with horizontal subsurface flow, using shale as a substrate. *Water Science and Technology*, *35*(5), 95–102. doi:10.1016/S0273-1223(97)00057-7

Dunne, E. J., Culleton, N., O'Donovan, G., Harrington, R., & Olsen, A. E. (2005). An integrated constructed wetland to treat contaminants and nutrients from dairy farmyard dirty water. *Ecological Engineering*, *24*(3), 221–234. doi:10.1016/j.ecoleng.2004.11.010

Elvira, C., Goicoechea, M., Sampedro, L., Mato, S., & Nogales, R. (1996). Bioconversion of solid paperpulp mill sludge by earthworms. *Bioresource Technology*, *57*(2), 173–177. doi:10.1016/0960-8524(96)00065-X

Geary, P. M., & Moore, J. A. (1999). Suitability of a Treatment Wetland for Dairy Wastewaters. *Water Science and Technology*, *40*(3), 179–185. doi:10.1016/S0273-1223(99)00464-3

Gorra, R., Freppaz, M., Zanini, E., & Scalenghe, R. (2014). Mountain dairy wastewater treatment with the use of a 'irregularly shaped' constructed wetland (Aosta Valley, Italy). *Ecological Engineering*, *73*, 176–183. doi:10.1016/j.ecoleng.2014.09.013

He, Y., Tao, W., Wang, Z., & Shayya, W. (2012). Effects of pH and seasonal temperature variation on simultaneous partial nitrification and anammox in free-water surface wetlands. *Journal of Environmental Management*, *110*, 103–109. doi:10.1016/j.jenvman.2012.06.009 PMID:22762912

Healy, M., & Cawley, A. M. (2002). Nutrient processing capacity of a constructed wetland in western Ireland. *Journal of Environmental Quality*, *31*(5), 1739–1747. doi:10.2134/jeq2002.1739 PMID:12371194

Healy, M. G., Rodgers, M., & Mulqueen, J. (2007). Treatment of dairy wastewater using constructed wetlands and intermittent sand filters. *Bioresource Technology*, *98*(12), 2268–2281. doi:10.1016/j.biortech.2006.07.036 PMID:16973357

Hijosa-Valsero, M., Sidrach-Cardona, R., & Bécares, E. (2012). Comparison of interannualremoval variation of various constructed wetland types. *The Science of the Total Environment*, *430*, 174–183. doi:10.1016/j.scitotenv.2012.04.072 PMID:22647241

Huang, J., Cai, W., Zhong, Q., & Wang, S. (2013). Influence of temperature on micro-environment, plant eco-physiology and nitrogen removal effect in subsurface flow constructed wetland. *Ecological Engineering*, *60*, 242–248. doi:10.1016/j.ecoleng.2013.07.023

Johansson, A. E., Gustavsson, A. M., Öquist, M. G., & Svensson, B. H. (2004). Methane emissions from a constructed wetland treating wastewater — seasonal and spatial distribution and dependence on edaphic factors. *Water Research*, *38*(18), 3960–3970. doi:10.1016/j.watres.2004.07.008 PMID:15380986

Ju, X., Wu, S., Zhang, Y., & Dong, R. (2014). Intensified nitrogen and phosphorus removal in a novel electrolysis-integrated tidal flow constructed wetland system. *Water Research*, *59*(0), 37–45. doi:10.1016/j.watres.2014.04.004 PMID:24784452

Kadlec, R. H., Knight, R. L., Vymazal, J., Brix, H., Cooper, P., & Haberl, R. (2000). *Constructed wetlands for pollution control – processes, performance, design and operation*. IWA Scientific and Technical Report No. 8. London, UK: IWA Publishing.

Kadlec, R. H., & Wallace, S. D. (2009). *Treatment Wetlands* (2nd ed.). Boca Raton, FL: CRC Press.

Kainz, H., & Gruber, G. (2006). *Wastewater treatment, sludge treatment. Lecture notes*. Austria: Technical University Graz.

Kantawanichkul, S., Kladprasert, S., & Brix, H. (2009). Treatment of high-strength wastewater in tropical vertical flow constructed wetlands planted with *Typha angustifolia* and *Cyperus involucratus*. *Ecological Engineering*, *35*(2), 238–247. doi:10.1016/j.ecoleng.2008.06.002

Kartal, B., Kuenen, J. G., & van Loosdrecht, M. C. M. (2010). Sewage treatment with anammox. *Science*, *328*(5979), 702–703. doi:10.1126/science.1185941 PMID:20448175

Kern, J., & Idler, C. (1999). Treatment of domestic and agricultural wastewater by reed bed systems. *Ecological Engineering*, *12*(1), 13–25. doi:10.1016/S0925-8574(98)00051-2

Knight, R. L., Payne, V. W. E. Jr, Borer, R. E., Clarke, R. A. Jr, & Pries, J. H. (2000). Constructed wetlands for livestock wastewater management. *Ecological Engineering*, *15*(1-2), 41–55. doi:10.1016/S0925-8574(99)00034-8

Kucuk, O. S., Sengul, F., & Kapdan, I. K. (2003). Removal of ammonia from tannery effluents in a reed bed constructed wetland. *Water Science and Technology*, *48*(11–12), 179–186. PMID:14753534

Lansing, S., & Martin, J. F. (2006). Use of an ecological treatment system (ETS) for removal of nutrients from dairy wastewater. *Ecological Engineering*, *28*(3), 235–245. doi:10.1016/j.ecoleng.2006.04.006

Li, C., Wu, S., & Dong, R. (2015a). Dynamics of organic matter, nitrogen and phosphorus removal and their interactions in a tidal operated constructed wetland. *Journal of Environmental Management*, *151*, 310–316. doi:10.1016/j.jenvman.2015.01.011 PMID:25585144

Li, L., He, C., Ji, G., Zhi, W., & Sheng, L. (2015b). Nitrogen removal pathways in a tidal flow constructed wetland under flooded time constraints. *Ecological Engineering*, *81*, 266–271. doi:10.1016/j.ecoleng.2015.04.073

Lv, T., Wu, S., Hong, H., Chen, L., & Dong, R. (2013). Dynamics of nitrobenzene degradation and interactions with nitrogen transformations in laboratory-scale constructed wetlands. *Bioresource Technology*, *133*, 529–536. doi:10.1016/j.biortech.2013.02.003 PMID:23455225

Maltais-Landry, G., Maranger, R., Brisson, J., & Chazarenc, F. (2009). Greenhouse gas production and efficiency of planted and artificially aerated constructed wetlands. *Environmental Pollution*, *157*(3), 748–754. doi:10.1016/j.envpol.2008.11.019 PMID:19110349

Mander, Ü., Dotro, G., Ebie, Y., Towprayoon, S., Chiemchaisri, C., Nogueira, S. F., & Mitsch, W. J. et al. (2014). Greenhouse gas emission in constructed wetlands for wastewater treatment: A review. *Ecological Engineering*, *66*, 19–35. doi:10.1016/j.ecoleng.2013.12.006

Mander, Ü., Kuusemets, V., Lõhmus, K., Mauring, T., Teiter, S., & Augustin, J. (2003). Nitrous oxide, dinitrogen and methane emission in a subsurface flow constructed wetland. *Water Science and Technology*, *48*(5), 135–142. PMID:14621157

Mantovi, P., Marmiroli, M., Maestri, E., Tagliavini, S., Piccinni, S., & Marmiroli, N. (2003). Application of a horizontal subsurface flow constructed wetland on treatment of dairy parlor wastewater. *Bioresource Technology*, *88*(2), 85–94. doi:10.1016/S0960-8524(02)00291-2 PMID:12576000

Masi, F., Conte, G., Martinuzzi, N., & Pucci, B. (2002). *Winery high organic content wastewater treated by constructed wetland in mediterranean climate.* IWA 8th International Conference on Wetland Systems for Water Pollution Control, Arusha, Tanzania.

Masi, F., Rochereau, J., Troesch, S., Ruiz, I., & Soto, M. (2014). Wineries wastewater treatment by constructed wetlands: A review. *Water Science and Technology*, *71*(8), 1113–1127. doi:10.2166/wst.2015.061 PMID:25909720

Meers, E., Tack, F. M. G., Tolpe, I., & Michels, E. (2008). Application of full-scale constructed wetland for tertiary treatment of piggery manure: Monitoring results. *Water, Air, and Soil Pollution*, *193*(1-4), 15–24. doi:10.1007/s11270-008-9664-5

Mosteo, R., Ormad, P., Mozas, E., Sarasa, J., & Ovelleiro, J. L. (2006). Factorial experimental design of winery wastewater treatment by heterogeneous photo-Fenton process. *Water Research*, *40*(8), 1561–1568. doi:10.1016/j.watres.2006.02.008 PMID:16574187

Mosteo, R., Sarasa, J., Ormad, M., & Ovelleiro, J. (2008). Sequential Solar Photo-Fenton-Biological System for the Treatment of Winery Wastewater. *Journal of Agricultural and Food Chemistry*, *56*(16), 7333–7338. doi:10.1021/jf8005678 PMID:18642841

Myhre, G., Shindell, D., Bréon, F.-M., Collins, W., Fuglestvedt, J., Huang, J., . . . Zhang, H. (2013). Anthropogenic and Natural Radiative Forcing. In T. F. Stocker, D. Qin, G.-K. Plattner, M. Tignor, S. K. Allen, J. Boschung, . . . Midgley, P. M. (Eds.), Climate Change 2013: The Physical Science Basis. Contribution of Working Group I to the Fifth Assessment Report of the Intergovernmental Panel on Climate Change. Cambridge, UK: Cambridge University Press. Available from http://www.ipcc.ch/pdf/assessment-report/ar5/wg1/WG1AR5_Chapter08_FINAL.pdf

Newman, J. M., & Clausen, J. C. (2000). Seasonal performance of a wetland constructed to process milkhouse wastewater in Connecticut. *Ecological Engineering*, *14*(1), 181–198.

Nielsen, S. (2015). *Sludge Treatment in Reed Beds System Technology: Theory, design and application.* IRSTEA 2015 6-8th July Workshop program, Lyon, France.

Pei-Dong, . (2002). Greenhouse gas emissions from a constructed wetland for municipal sewage treatment. *Journal of Environmental Sciences (China)*, *14*(1), 27–33. PMID:11887314

Petruccioli, M., Duarte, J. C., Eusebio, A., & Federici, F. (2002). Aerobic treatment of winery wastewater using a jet-loop activated sludge reactor. *Process Biochemistry*, *37*(8), 821–829. doi:10.1016/S0032-9592(01)00280-1

Ruggieri, L., Gea, T., Artola, A., & Sanchez, A. (2008). Influence of different co-substances biochemical composition on raw sludge co-composting. *Biodegradation*, *19*(3), 403–415. doi:10.1007/s10532-007-9146-2 PMID:17674125

Saeed, T., Afrin, R., Muyeed, A. A., & Sun, G. (2012a). Treatment of tannery wastewater in a pilot-scale hybrid constructed wetland system in Bangladesh. *Chemosphere*, *88*(9), 1065–1073. doi:10.1016/j.chemosphere.2012.04.055 PMID:22673399

Saeed, T., & Sun, G. (2012b). A review on nitrogen and organics removal mechanisms in subsurface flow constructed wetlands: Dependency on environmental parameters, operating conditions and supporting media. *Journal of Environmental Management*, *112*, 429–448. doi:10.1016/j.jenvman.2012.08.011 PMID:23032989

Sanchez, M., Hernadez, S., Bravo, C., Antolin, J. M., & Gallardo, T. (2010). *Sustainable use of biodegradables products in agricultural fertilization, reforestation, plant production and recovery of degraded areas.* Special Orbit. International Congress about Organic Soils and Organic Matter from Waste, Valladolid, Spain. (in Spanish)

Seeger, E. M., Maier, U., Grathwohl, P., Kuschk, P., & Kaestner, M. (2013). Performance evaluation of different horizontal subsurface flow wetland types by characterization of flow behavior, mass removal and depth-dependent contaminant load. *Water Research*, *47*(2), 769–780. doi:10.1016/j.watres.2012.10.051 PMID:23200508

Seidel, K. (1961). Zur Problematik der Keim- und Pflanzengewasser. Verh Internat. Verein. *Limnology*, *14*, 1035–1039.

Serrano, L., de la Varga, D., Díaz, M. A., Ruiz, I., Bondelle, R., & Soto, M. (2009). HUSB-CW treatment system for winery wastewater.*Proceedings of the 5th International Specialized Conference on Sustainable Viticulture: Winery Waste and Ecologic Impacts Management.* (pp.317-324).

Serrano, L., de la Varga, D., Ruiz, I., & Soto, M. (2011). Winery wastewater treatment in a hybrid constructed wetland. *Ecological Engineering*, *37*(5), 744–753. doi:10.1016/j.ecoleng.2010.06.038

Shamir, E., Thompson, T. L., Karpiscak, M. M., Freitas, R. J., & Zauderer, J. (2001). Nitrogen accumulation in a constructed wetland for dairy wastewater treatment. *Journal of the American Water Resources Association*, *37*(2), 315–325. doi:10.1111/j.1752-1688.2001.tb00971.x

Shepher, H. L., Grismer, M. E., & Tchobanoglous, G. (2001). Treatment of high-strength winery wastewater using a subsurface-flow constructed wetland. *Water Environment Research*, *73*(4), 394–403. doi:10.2175/106143001X139434

Singer, P. C., & Stumm, W. (1970). Acid mine drainage—the rate limiting step. *Science*, *167*(3921), 1121–1123. doi:10.1126/science.167.3921.1121 PMID:17829406

Smith, L. K., Sartoris, J., Thullen, J., & Anderson, D. C. (2000). Investigation of denitrification rates in an ammonia-dominated treatment wetland. *Wetlands*, *20*(4), 684–696. doi:10.1672/0277-5212(2000)020[0684:IODRIA]2.0.CO;2

Stumm, W., & Morgan, J. J. (1981). *Aquatic Chemistry. An Introduction Emphasizing Chemical Equilibria in Natural Waters* (2nd ed.). New York: Wiley-Interscience.

Sun, G., Zhao, Y., & Allen, S. (2005). Enhanced removal of organic matter and ammoniacal-nitrogen in a column experiment of tidal flow constructed wetland system. *Journal of Biotechnology*, *115*(2), 189–197. doi:10.1016/j.jbiotec.2004.08.009 PMID:15607237

Sun, G. J., Zhao, Y., Allen, S., & Cooper, D. (2006). Generating "tide" in pilot-scale constructed wetlands to enhance agricultural wastewater treatment. *Engineering in Life Sciences*, *6*(6), 560–565. doi:10.1002/elsc.200620156

Tanner, C. (2001). Plants as ecosystem engineers in subsurface-flow treatment wet-lands. *Water Science and Technology*, *44*(11–12), 9–17. PMID:11804163

Tanner, C. C., Adams, D. D., & Downes, M. T. (1997). Methane emissions from constructed wetlands treating agricultural wastewaters. *Journal of Environmental Quality*, *26*(4), 1056–1062. doi:10.2134/jeq1997.00472425002600040017x

Tanner, C. C., Clayton, J. S., & Upsdell, M. P. (1995). Effect of loading rate and planting on treatment of dairy farm wastewaters in constructed wetlands I. Removal of oxygen demand, suspended solids and faecal coliforms. *Water Research*, *29*(1), 17–26. doi:10.1016/0043-1354(94)00139-X

Torrijos, V., Trueba, A., Ruiz, I., & Soto, M. (2015). Investigating different hybrid configurations of constructed wetlands for nitrogen removal from domestic and industrial wastewaters.*Proceedings of the WETPOL 2015 International Symposium on Wetland Pollutant Dynamics and Control and the Annual Conference of the Constructed Wetland Association* (pp. 294-295).

Uggetti, E., Ferrer, I., Nielsen, S., Arias, C., Brix, H., & García, J. (2012). Characteristics of Biosolids from Sludge Treatment Wetlands for Agricultural Reuse. *Ecological Engineering*, *40*, 210–216. doi:10.1016/j.ecoleng.2011.12.030

van der Star, W. R. L., Abma, W., Blommers, D., Mulder, J. W., Tokutomi, T., Strous, M., & van Loosdrecht, M. C. M. et al. (2007). Startup of reactors for anoxic ammonium oxidation: Experiences from the first full-scale anammox reactor in Rotterdam. *Water Research*, *41*(18), 4149–4163. doi:10.1016/j.watres.2007.03.044 PMID:17583763

Van Oirschot, D. (2015). RietLand vertical flow wetlands: 16 years of experience in the Netherlands and Belgium. *Proceedings 12th IWA International Conference on Wetland Systems for Water Pollution Control.*

Vohla, C., Koiv, M., Bavor, H. J., Chazarenc, F., & Mander, U. (2011). Filter materials for phosphorus removal from wastewater in treatment wetlands—A review. *Ecological Engineering*, *37*(1), 70–89. doi:10.1016/j.ecoleng.2009.08.003

Vymazal, J. (2011a). Constructed wetlands for wastewater treatment: Five decades of experience. *Environmental Science & Technology*, *45*(1), 61–69. doi:10.1021/es101403q PMID:20795704

Vymazal, J. (2011b). Plants used in constructed wetlands with horizontal subsurface flow: A review. *Hydrobiologia*, *674*(1), 133–156. doi:10.1007/s10750-011-0738-9

Vymazal, J. (2013). The use of hybrid constructed wetlands for wastewater treatment with special attention to nitrogen removal: A review of a recent development. *Water Research*, *47*(14), 4795–4811. doi:10.1016/j.watres.2013.05.029 PMID:23866134

Vymazal, J. (2014). Constructed wetlands for treatment of industrial wastewater: A review. *Ecological Engineering*, *73*, 724–751. doi:10.1016/j.ecoleng.2014.09.034

Wang, Y., Inamori, R., Kong, H., Xu, K., Inamori, Y., Kondo, T., & Zhang, J. (2008). Influence of plant species and wastewater strength on constructed wetland methane emissions and associated microbial populations. *Ecological Engineering*, *32*(1), 22–29. doi:10.1016/j.ecoleng.2007.08.003

Webb, J. M., Quintã, R., Papadimitriou, S., Norman, L., Rigby, M., Thomas, D. N., & Le Vay, L. (2013). The effect of halophyte planting density on the efficiency of constructed wetlands for the treatment of wastewater from marine aquaculture. *Ecological Engineering*, *61*, 145–153. doi:10.1016/j.ecoleng.2013.09.058

Wu, S., Kuschk, P., Brix, H., Vymazal, J., & Dong, R. (2014). Development of constructed wetlands in performance intensifications for wastewater treatment: A nitrogen and organic matter targeted review. *Water Research*, *57*(15), 40–55. doi:10.1016/j.watres.2014.03.020 PMID:24704903

Wu, S., Zhang, D., Austin, D., Dong, R., & Pang, C. (2011). Evaluation of a lab-scale tidal flow constructed wetland performance: Oxygen transfer capacity, organic matter and ammonium removal. *Ecological Engineering*, *37*(11), 1789–1795. doi:10.1016/j.ecoleng.2011.06.026

Zheng, Y., Wang, C. X., Ge, Y., Dzakpasu, M., Zhao, Y., & Xiong, J. (2015). Effects of annual harvesting on plants growth and nutrients removal in surface-flow constructed wetlands in northwestern China. *Ecological Engineering*, *83*, 268–275. doi:10.1016/j.ecoleng.2015.06.035

## KEY TERMS AND DEFINITIONS

**Hydraulic Loading Rate (HLR):** Amount of influent wastewater per unit of wetland surface. It is often expressed as mm/d, which means $L/m^2 \cdot d$, or cm/d.

**Hydrolytic Up-Flow Sludge Blanket Reactor (HUSB):** Up-flow anaerobic reactor in which only hydrolysis and sedimentation of particulate organic matter and solids are aimed, avoiding methanogenic reactions.

**Surface Loading Rate (SLR):** Amount of a certain pollutant applied to a wastewater treatment system per surface and time units.

**Surface Removal Rate (SRR):** Amount of a certain pollutant removed in a wastewater treatment system per surface and time units.

## ENDNOTE

1 PE: People equivalents.

# Chapter 9
# Aerobic Granular Sludge:
## Treatment of Wastewaters Containing Toxic Compounds

**Catarina L. Amorim**
*Universidade Católica Portuguesa, Portugal & University of Aveiro, Portugal*

**Anouk F. Duque**
*Universidade Católica Portuguesa, Portugal & Universidade Nova de Lisboa, Portugal*

**Irina S. Moreira**
*Universidade Católica Portuguesa, Portugal*

**Mark C. M. van Loosdrecht**
*Delft University of Technology, The Netherlands*

**Paula M. L. Castro**
*Universidade Católica Portuguesa, Portugal*

## ABSTRACT

*Aerobic Granular Sludge (AGS) has been successfully applied for carbon, nitrogen and phosphorous removal from wastewaters, in a single tank, reducing the space and energy requirements. This is especially beneficial for, often space restricted, industrial facilities. Moreover, AGS holds a promise for the toxic pollutants removal, due to its layered and compact structure and the bacteria embedding in a protective extracellular polymeric matrix. These outstanding features contribute to AGS tolerance to toxicity and stability. Strategies available to deal with toxic compounds, namely granulation with effluents containing toxics and bioaugmentation, are addressed here. Different applications for the toxics/micropollutants removal through biosorption and/or biodegradation are presented, illustrating the technology versatility. The anthropogenic substances effects on system performance and bacterial populations established within AGS are also addressed. Combination of contaminants removal to allow water discharge, and simultaneous valuable products recovery are presented as final remark.*

## INTRODUCTION

Aerobic granular sludge (AGS) has emerged as a promising biotechnology for the treatment of industrial wastewaters. AGS consists of near spherical sludge aggregates formed by self-immobilization of microorganisms commonly found in activated sludge systems, when specific operational conditions are applied

DOI: 10.4018/978-1-5225-1037-6.ch009

(D. Gao, Liu, Liang, & Wu, 2011; Q. Zhang, Hu, & Lee, 2016). Compared with conventional activated sludge flocs, AGS has a compact structure with excellent settling capability, reducing the settling time applied in each treatment cycle and avoiding the need for separate settling tanks, thereby reducing the space needed for its implementation. However, in general, AGS system start-up is usually longer than that of conventional activated sludge system, which is a disadvantage of this kind of treatment systems.

Nevertheless, these microbial aggregates possess higher biomass retention and tolerance to toxic substrates than activated sludge flocs (Adav, Lee, Show, & Tay, 2008), which contribute to the great potential of the AGS technology to operate with industrial wastewater, including those effluents where toxic substrates are commonly present. Moreover, the aerobic granules contain diverse microbial communities that allow the establishment of efficient microbial communities and metabolic networks (Winkler, 2012). With a proper system operation mode, characterized by cyclic, fill-an-draw operation, the simultaneous removal of carbon, nitrogen and phosphorus from wastewater can be accomplished in a single tank. Also, as no recycle pumps and mixers are needed the energy requirements can be reduced (de Kreuk, Heijnen, & van Loosdrecht, 2005). These features could be beneficial for the implementation of the AGS technology, especially in industrial facilities.

The requirement of a minimal amount of mechanical equipment and more efficient aeration results in energy saving as an extra attractive aspect of the AGS technology. For the industrial sector, generating large quantities of wastewater, the application of this technology reduces the economic costs to treat their effluents and make it available for reuse. Nevertheless, generally for large-scale industrial operation, continuous-flow reactors are preferred over batch-reactors, such as SBR. Up to date, AGS technology at full-scale was implemented using SBR, which is usually more suitable for the treatment of smaller amounts of wastewater. Therefore its implementation on industrial facilities can become more difficult. Much effort has been made in the development of AGS continuous-flow reactors which can lower plant installation cost, maintenance and control. Recently, the possibility of operating an AGS reactor in continuous mode for the treatment of an industrial wastewater from a milk-processing factory was demonstrated with satisfactory treatment performances (Bumbac, Ionescu, Tiron, & Badescu, 2015). However, in the long-run AGS lost its structural stability faster when compared with granules from a SBR. AGS continuous-flow reactors could be more feasible in large scale application; however further knowledge on the reactor design optimisation and operation conditions is needed for its scale-up.

In spite of some limitations, the AGS system is an upcoming technology for the biological wastewater treatment and research on this technology is a hot topic in the field of wastewater treatment, opening new perspectives for its upgrading. At the present, Netherlands, Portugal, South Africa, Brasil, Poland, Ireland, Australia and China have full-scale AGS systems in operation, and it is expected that the number of installations will increase rapidly worldwide (http://www.royalhaskoningdhv.com/en-gb/nereda). Recent reports revealed that there are projects in the pipeline to apply the AGS systems for the treatment of effluents from food, beverage and chemical industries (Giesen, de Bruin, Niermans, & van der Roest, 2013). So far, AGS full-scale operation processes available describe the system performance treating domestic wastewater (Pronk, de Kreuk, et al., 2015a) or a mixture of domestic and industrial wastewater (Giesen et al., 2013; J. Li, Ding, Cai, Huang, & Horn, 2014a). Currently, data on AGS full-scale performance when treating industrial wastewater containing toxic compounds is less available.

## GENERAL OVERVIEW ON THE AGS TECHNOLOGY

AGS technology represents a good alternative to conventional activated sludge systems, in particular because the former allows the combined carbon, nitrogen and phosphorous removal. This is an interesting aspect of this technology since it allows that the removal of nutrients could be accomplished in a single reactor, because the nitrification, denitrification and phosphorus removal processes are carried-out within the granules (Figure 1) (de Kreuk et al., 2005). In fact, the morphology of the granules leads to the formation of an $O_2$ diffusion gradient within these aggregates thus creating a layered structure with aerobic and anoxic zones (Chiu, Chen, Lee, Wang, & Lai, 2007). The existence of these different zones within the granules allows the coexistence of heterotrophic, nitrifying, and denitrifying populations. In this way, in the aerobic outer layer of the granule takes place nitrification carried out by ammonia-oxidizing bacteria (AOB) and nitrite-oxidising bacteria (NOB) and aerobic P-uptake by polyphosphate-accumulating bacteria (PAO). Denitrification coupled to anoxic phosphate removal occurs in the anoxic zone of the granule (de Kreuk et al., 2005). One of the challenges of the AGS technology is the simultaneous nitrification and denitrification (SND). The main parameters governing SND are the dissolved oxygen concentration in the bulk liquid and the granule sizes. The thickness of anoxic/aerobic zones within the granules is correlated with the granules size, which seemed to have a more important role in the SND process (Di Bella & Torregrossa, 2013). In fact, when the anoxic layer was quite thin, especially during the famine period when the nitrification process occurs, SND can be compromised. It is also interesting to notice that, apart from occurring in different zones within granules, SND takes place at different times of the treatment cycle. For SND to occur requires alternating anaerobic and aerobic cyclic conditions and the feast and famine period coincides with the dominant process of denitrification and nitrification, respectively (Di Bella & Torregrossa, 2013). During the feast period, heterotrophic organisms in the outer layer consume the available dissolved oxygen, avoiding their penetration; thus nitrification hardly occurs.

*Figure 1. Aerobic granular sludge (AGS) and the typical structure of a heterotrophic granule with nitrifiers (green) in the aerobic zone followed by PAOs (Blue), GAOs (yellow) and denitrifies (pink) in the up following anoxic zone*
*Adapted from Winkler, 2012*

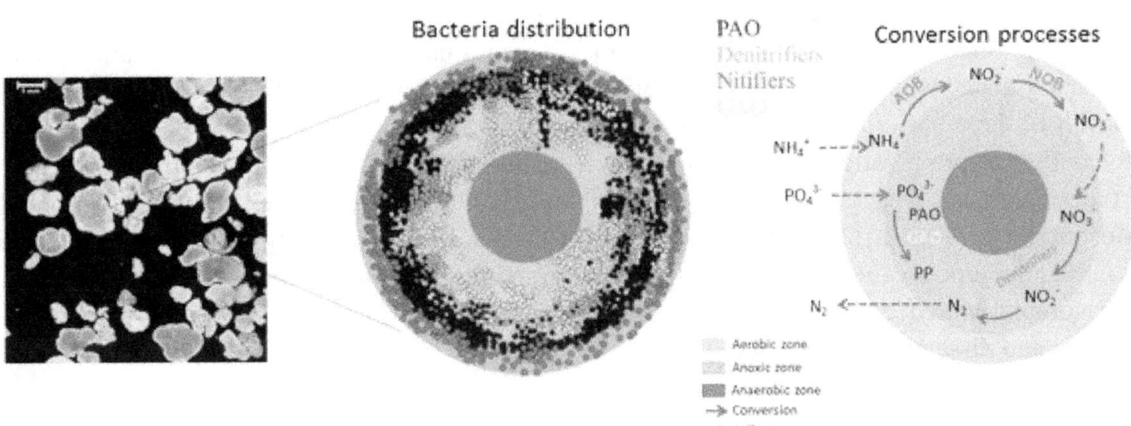

The well-functioning of the AGS technology relies on the successful cultivation of granules. The granulation process consists of bacterial cells self-immobilization during which activate sludge flocs assemble and agglomerate as dense and compact granular sludge under controlled loading and operating conditions. Recent studies point out that granulation is a dynamic process which can be divided into sequential steps:

1. Formation of young granules based on the initial aggregation of bioflocs;
2. Granules growth due to biomass attachment;
3. Aggregation of the remaining flocculent biomass and newborn cells to granules;
4. Re-granulation of the detached bioflocs, newborn cells and crushed aged granules (Zhou, Niu, Xiong, Yang, & Dong, 2014).

The complex process of granulation is influenced by a variety of factors (Show, Lee, & Tay, 2012). In most of the reported studies, the granulation process is usually conducted using sequencing batch reactor (SBR) configurations (Beun et al., 1999) whose cyclic operation allows the granulation process to occur. Even if more scarcely, the formation of AGS was successfully achieved with other reactor configurations. Zhou et al., (2013) reported AGS formation using a continuous flow airlift fluidized bed reactor (CAFB). In comparison with SBR, it has a simple structure, is easier to operate and allows a continuous flow of wastewater. AGS was observed after 12 days of cultivation which was possibly due to the strong shear force exerted on the activated flocs. However, filamentous bulking occurred at the end of 35 days of operation, pointing out the instability of such configuration for long-term running. The granulation process was also reported to be attained using a sequencing batch airlift reactor (SBAR) with a box structure (Rezaei, Ayati, & Ganjidoust, 2012). Airlift reactors could provide higher hydrodynamic shear forces, better mixing and mass transfer than SBR. However, the granulation process is time consuming since only after 120 days, yellow granules with a diameter of 0.5–2 mm were dominant in the reactor. The different reactor configurations presented above were indeed able to form AGS but, till date aerobic granulation on full scale was only successfully reported using SBR (Pronk, de Kreuk, et al., 2015b). In fact, the formation of AGS in the full scale plant was achieved after a start-up period of approximately 5 months, and the stable granule bed formed could be maintained thereafter.

Maintaining granule structural stability poses a major challenge for the AGS operation and its implementation. Generally, the deterioration of granule stability was observed in the long-run bioreactor operation (Show et al., 2012). Overgrowth of filamentous bacteria in the surface of mature granules has been correlated with the loss of AGS stability. Indeed these bacteria are crucial for granulation, and low levels of these bacteria, in general, do not affect the system.

Recently, the proliferation of filamentous bacteria in mature AGS was investigated demonstrating that the source of such bacteria could be the initial inoculum (Figueroa, Val del Rio, Campos, Méndez, & Mosquera-Corral, 2015). In this case, filamentous bacteria were not completely washed-out during the granulation process, remaining within the biomass and proliferating when environmental conditions were favorable. Moreover, when real industrial wastewater was used to feed, filamentous bacteria could be present and disseminate within the system.

## AGS APPLICATION FOR THE TREATMENT OF INDUSTRIAL EFFLUENTS

The treatment of industrial effluents can be challenging due to the wide variability within these effluents. Biological wastewater treatment, including AGS processes, can face several problems when treating industrial wastewater due to:

1. The variable chemical composition of the wastewater;
2. The presence of recalcitrant compounds;
3. The problem of sequential alternating presence of pollutant; and
4. The starvation periods occurring due to temporary stops of the production processes.

Several papers highlight the great potential of AGS systems to deal with industrial effluents, especially those containing toxic compounds or facing high salinity, compared to conventional activated sludge systems and other biofilm technologies. The layered and compact structure of AGS is a feature that contributes to the capacity of these aggregates to withstand toxicity since a concentration gradient is created within these structures. As such, mass transfer limitation can result in limited penetration of toxic compounds in the granules, thereby protecting the microbial populations in the core of the granules. Other attributes also contribute to the AGS tolerance towards toxic compounds, such as the presence of extracellular polymeric substances (EPS), which is usually the first barrier for microbial cells that directly contact with the toxic compounds.

EPS are a complex high-molecular-weight mixture of polymers that determine the physicochemical and biological properties of the biomass. EPS molecules are mainly constituted by polysaccharides and proteins but humic-like substances are often also present in EPS from sludge (Sheng, Yu, & Li, 2010). The high AGS tolerance to toxics is in part due to its capacity to produce EPS when in contact with such compounds. This self-protection strategy against toxic compounds was observed after the gradual addition of 4-chlorophenol to AGS bioreactors, during which the production of EPS was promoted as revealed by the increase in the polysaccharides and proteins levels (Wei, Wang, Wang, et al., 2015). A similar stress response was reported due to the exposure of AGS to tetracycline, but a higher production of proteins over polysaccharides was observed (Y. Shi, Xing, Wang, & Wang, 2013). Therefore, the AGS capability to produce more EPS to protect itself from the toxicity of those compounds seems to be an important defense strategy. In addition, the presence of functional groups in EPS, such as carboxyl, phosphoric, sulfhydryl, phenolic and hydroxyl groups, can also promote the adsorption of toxics to the EPS matrix (Sheng et al., 2010).

Several studies have addressed the feasibility of the AGS technology to deal with real industrial effluents, demonstrating the applicability and effectiveness of these systems when facing complex wastewaters. The capacity to treat high salt fish canning wastewater using AGS technology was demonstrated recently (Corsino, Capodici, Morici, Torregrossa, & Viviani, 2016). These wastewaters contained amounts of salt up to 75 $g_{NaCl}$ $L^{-1}$, which could be problematic due to inhibitory effects exerted in the biological removal processes. Salinity did not affect the organic carbon removal. Fully nitrogen removal by the nitritation-denitritation process was achieved, with a maximum efficiency of 98% at 50 $g_{NaCl}$ $L^{-1}$, but above this salinity nitritation collapsed. AGS was successfully cultivated using papermaking wastewater exhibiting good treatment performance (Hailei, Guosheng, Ping, & Feng, 2006). The incorporation of specific strains on those granules allowed improving the ability for biodegrading lignin, which is difficult to be degraded by conventional activated sludge. AGS cultivated with a raw chemical industrial wastewater composed by a mixture of effluents from a pesticide production plant, a chlor-alkali industrial plant and several other chemical plants, led to an effective treatment demonstrating the ability of AGS to withstand toxic substances (L. Liu et al., 2011). The core structure exhibited by the granules might allow microorganisms to survive and persist in such toxic environment and thus to achieve an effective treatment. Also, other studies conducted in lab-scale AGS-SBRs using real wastewater from different industrial sectors such as from a rubber factory industry (Rosman et al., 2013), from a large-scale malt factory (Schwarzenbeck, Erley, Mc Swain, Wilderer, & Irvine, 2004), from a brewery plant (S.-G. Wang, Liu, Gong, et al., 2007), from a palm oil mill industry (Abdullah, Yuzir, Curtis, Yahya, & Ujang, 2013), from a pig farm (Figueroa, Val del Rio, Campos, Mosquera-Corral, & Mendez, 2011), raw undiluted livestock wastewater from a cattle farm, containing urine, washing water, blood and feces (Othman et al., 2013), and from a laboratory for analysis of dairy products (Arrojo, Mosquera-Corral, Garrido, & Méndez, 2004), demonstrated the robustness and stability of the AGS systems for the treatment of complex wastewater. Moreover, the successful full-scale AGS-SBR operation for the treatment of raw wastewater composed by approximately 30% domestic sewage and 70% industrial wastewater, originated from printing, dyeing, chemical, textile and beverage industries, was recently demonstrated (J. Li et al., 2014). All these studies contribute to further the scale-up and implementation of the technology to a wide range of industrial effluents.

## AGS TREATING WASTEWATER CONTAINING TOXIC POLLUTANTS

### Removal of Contaminants Through Biodegradation

The properties of AGS, such as excellent settleability, high and stable rates of metabolism, resilience to shocks and toxins due to the protection by a matrix of EPS, long biomass residence time, biomass immobilization inside the aggregates and the possibility for bioaugmentation, hold a promise for the removal of toxic pollutants and/or micro-pollutants through biosorption and/or biodegradation. Another important feature of AGS is the stratification of conversion processes and redox zones within the granules which provide aerobic and anaerobic/anoxic layers. This is an advantage for the treatment of effluents containing toxics which require the combination of anaerobic and aerobic processes. This is the case of the color removal from textile dyeing wastewater. The cleavage of the N≡N bond of the azo dyes occurs during anaerobic stage along with the generation of toxic aromatic amines, while mineralization of the amines only takes place under aerobic condition (Işik & Sponza, 2008). AGS allows simplifying the system since both anaerobic and aerobic conditions take place in only one reactor. Strict anaerobic microorganisms can survive since oxygen only penetrates partially into granules during the aerated phase of the process. Several studies highlighted the AGS technology capacity to deal with toxic pollutants often found in industrial wastewaters (Table 1).

AGS technology has been applied to degrade aromatic toxic compounds successfully, making use of different strategies (Anouk F Duque et al., 2011; S. T.-L. Tay, Moy, Jiang, & Tay, 2005; Lan Wang et al., 2014; S.-G. Wang, Liu, Zhang, et al., 2007; Yi, Zhuang, Wu, Tay, & Tay, 2006). Aerobic microbial granules grown in acetate were successfully used to degrade phenol (Tay, Moy, Jiang, & Tay, 2005), 4-chlorophenol (Alessandra Carucci, Milia, De Gioannis, & Piredda, 2009) or even the highly chlorinated

*Table 1. Removal of toxic compounds through biodegradation by AGS in SBR systems*

| Organic Pollutants | | Concentration Range | Removal Efficiency (%)/ Degradation Rate | Reference |
|---|---|---|---|---|
| **Aromatics** | 2-Fluorophenol | 25-50 mg $L^{-1}$ | up to 100% | Duque et al., 2011 |
| | 2-; 3-; 4-chloroaniline and 3,4-dichloroaniline (in a mixture) | 20-450 mg $L^{-1}$ (total inlet concentration; equal amounts of each toxic) | 50-100, 40-100, 10-100 and 10-100, respectively | L Zhu, Xu, Luo, Cao, & Yang, 2008 |
| | 2,4-Dichlorophenol | 50-100 mg $L^{-1}$ | up to 94% | S.-G. Wang, Liu, Zhang, et al., 2007 |
| | 2,4,6-trichlorophenol | 0-15 mg $L^{-1}$ | 100%/ 0.45-2.36 mg $g_{VSS}^{-1}$ $h^{-1}$ | A. Carucci, S. Milia, G. De Gioannis, 2008 |
| | 4-Chloroaniline | 20-200 mg $L^{-1}$ | 40-98% | Liang Zhu, Lv, Dai, Zhou, & Xu, 2013 |
| | 4-Chlorophenol | < 50 mg $L^{-1}$ | 18-100% / 3.40 mg $g_{VSS}^{-1}$ $h^{-1}$ | Carucci, Milia, De Gioannis, & Piredda, 2009 |
| | 4-Nitrophenol | < 40.1 mg $L^{-1}$ | 19.3 mg $g_{VSS}^{-1}$ $h^{-1}$ | Yi, Zhuang, Wu, Tay, & Tay, 2006 |
| | | 5-15 mg $L^{-1}$ | up to 95% | Jemaat, Suárez-Ojeda, Pérez, & Carrera, 2013 |
| | Nitrobenzene | 50-1000 mg $L^{-1}$ | 30-100% | D. Zhao, Liu, Zhang, & Liu, 2011 |
| | *O*-cresol | 100-1000 mg $L^{-1}$ | 100% | Jemaat, Suárez-Ojeda, Pérez, & Carrera, 2014a |
| | Phenol | 50-500 mg $L^{1}$ | 1.4 g $g_{VSS}^{-1}$ $d^{-1}$ | Jiang, Tay, & Tay, 2002 |
| | | 1000 mg $L^{-1}$ | 100% | Tay, Moy, Jiang, & Tay, 2005 |
| | *p*-nitrophenol, phenol and *o*-cresol (in a mixture) | Up to 371, 598 and 103 mg $L^{-1}$, respecively | 100% | Ramos, Suárez-Ojeda, & Carrera, 2016b |
| **Other** | Methyl tert-butyl ether | 25-500 mg $L^{-1}$ | 0.544 $g_{MTBE}$ $g_{VSS}^{-1}$ $d^{-1}$ | Zhang, Chen, & Fang, 2008 |
| | Pyridine | 250-2500 mg $L^{-1}$ | 73.0 mg $g_{VSS}^{-1}$ $h^{-1}$ | Adav, Lee, & Ren, 2007 |
| | tert-Butyl Alcohol | 100-600 mg $L^{-1}$ | 7.4 $mg_{TBA}$ $g_{VSS}^{-1}$ $h^{-1}$ | S. T. Tay, Zhuang, & Tay, 2005 |
| | Tetracycline | 10-30 mg $L^{-1}$ | 34.14-44.38% | Y.-J. Shi et al., 2011 |

2,4,6-tri-chlorophenol (A. Carucci, Milia, De Gioannis, & Piredda, 2008). In the study conducted by Duque et al. (2011), the acetate-fed microbial granules failed to degrade a compound such as 2-fluorophenol and the degradation was only achieved after bioaugmentation of the AGS with a 2-fluorophenol degrading strain. On the other hand, an alternative strategy based on granulation using activated sludge acclimated to the respective toxic compound, with addition of an easily degradable carbon source (e.g. glucose, sucrose or acetate) as a co-substrate and step increase in the influent toxic concentration, proved to be successful in the development of aerobic toxic-degrading granules capable to remove toxics such as 2,4-dichlorophenol (S.-G. Wang, Liu, Zhang, et al., 2007), 4-nitrophenol (Yi et al., 2006), 4-chloroaniline (Liang Zhu et al., 2013) or even a mixture of 2-, 3-, 4-chloroaniline and 3,4-dichloroaniline (L Zhu, Xu, Luo, Cao, & Yang, 2008). Moreover, AGS cultivated in a medium containing phenol was also able to degrade phenol at high concentrations (Jiang, Tay, & Tay, 2002). In other study AGS cultivated in synthetic wastewater containing phenol was also able to degrade pyridine and the presence of this compound had minimal effect on phenol degradation (Adav, Lee, & Ren, 2007). However, the presence of phenol resulted in a competitive type of inhibition, which limited pyridine degradation rates by the granules. Degradation of other toxic compounds with AGS has been achieved. Zhang et al. (2008) reported efficient degradation of methyl t-butyl ether by AGS with ethanol as a co-substrate; *o*-cresol was totally biodegraded by AGS in a continuous airlift reactor (Jemaat et al., 2014a) and nitrobenzene was degraded as sole carbon and nitrogen source by AGS cultivated in SBR (D. Zhao et al., 2011). Moreover, AGS applied to treat a simulated textile wastewater in a SBR system demonstrated good removal of COD and ammonia accompanied by 62% of color removal due to dye degradation by the biomass (Muda et al., 2010).

## Removal of Contaminants Through Biosorption

The high biosorption capacity is an attractive AGS property that also contributes to its capacity for removal of toxic compounds. This feature results in part from the high surface area and porosity of these compact biological aggregates. In addition, the presence of EPS in the AGS structure also contributes for the sorption of toxic compounds. The EPS constituents contribute with many functional groups on the AGS surface, including carboxyl, phosphoric, amine and hydroxyl groups, thus providing effective sites to trap toxics (Fomina & Gadd, 2014). The AGS binding ability to heavy metals has been largely reported but this ability also extends to organic pollutants, often present in industrial effluents, such as pharmaceuticals, and dyes. Several studies reporting the removal of toxic compounds demonstrated that AGS biosorption could be an efficient and cost- effective technology for the removal of toxic compound (Table 2). Moreover, the settling velocity and density of AGS is an advantage over other biosorbents currently used, facilitating the post-separation of the biomass from the treated effluent and the regeneration of biosorbent. Most of the works reporting the AGS biosorption of toxics were conducted in batch mode and only evaluate the capability of the biosorbent to trap the toxics. Nevertheless, more recent studies conducted in reactor systems demonstrated that AGS could be used as treatment process and furthermore act as biosorbent. This synergy could be beneficial for the implementation the AGS technology however, more research is necessary.

The highly toxic heavy metals removal by AGS through biosorption has been largely reported in literature. Some of the works demonstrating the applicability of AGS to act as biosorbent for the removal of metals in batch or reactor systems are summarized in Table 2 (Y Liu, Yang, Tan, Lin, & Tay, 2002; Y Liu, 2003; Yu Liu & Xu, 2007; Yu Liu et al., 2003; Marques et al., 2013; Wei et al., 2016; Xu, Liu, & Tay, 2006). Initial studies conducted in batch mode revealed that, within the concentration range tested, an increase in the initial concentration of $Ce^{3+}$ lead to an increase in the AGS sorption ability (L.-L. Zhang, Feng, Xu, Xu, & Cai, 2005). The microstructure of granules seems to be beneficial for the diffusion of ions into granules.

In relation to the biosorption mechanism, in a study for $Ni^{2+}$ removal, it was observed that most of $Ni^{2+}$ biosorption occurred through an ion exchange mechanism, while the EPS-associated $Ni^{2+}$ biosorption only accounted for about 14.2% of the total $Ni^{2+}$ removed. Moreover, it was also observed that the adsorbed $Ni^{2+}$ ions were uniformly distributed from the surface to the center of aerobic granules (Yu Liu & Xu, 2007). In a study where AGS was applied as a biosorbent for removing $Sr^{2+}$ from waters, the main mechanism of adsorption was suggested to be complexation with functional groups on the polysaccharides of EPS (Li Wang et al., 2013). More recently, Wei and co-workers (2016) conducted batch experiments to evaluate the interaction between EPS and $Zn^{2+}$ during the sorption process of this heavy

*Table 2. Removal of toxic compounds through biosorption onto AGS*

| Organic Pollutants | | Concentration Range | Removal Efficiency (%)/ Equilibrium Biosorption Capacity | Reference |
|---|---|---|---|---|
| **Metals** | $Zn^{2+}$ | 50-100 mg $L^{-1}$ | up to 90% | Marques et al., 2013 [R] |
| | | 25-1000 mg $L^{-1}$ | 10-60 mg $g^{-1}$ | Wei et al., 2016 |
| | $Cd^{2+}$ | 10-200 mg $L^{-1}$ | 40-200 mg $g^{-1}$ | Yu Liu et al., 2003 |
| | $Sr^{2+}$ | 10-160 mg $L^{-1}$ | 65-28% / 4-27 mg $g^{-1}$ | Li Wang et al., 2013 |
| | $Sb^{5+}$ | 20-100 mg $L^{-1}$ | up to 99% / 36.6-146 mg $g^{-1}$ | Li Wang et al., 2014 |
| | $Ce^{3+}$ | 25-150 mg $L^{-1}$ | 58-357 mg $g^{-1}$ | L.-L. Zhang, Feng, Xu, Xu, & Cai, 2005 |
| **Pharmaceuticals** | 17β-estradiol | 400-1000 µg $L^{-1}$ | 141-300 µg $g^{-1}$ $_{MLSS}$ | Zheng et al., 2016 |
| | Cefalexin | 20 mg $L^{-1}$ | 70-90% | Kong, Zhi-Bin, Shu, & Miao, 2015 [R] |
| | Ofloxacin, norfloxacin and ciprofloxacin | 9 µM | 10.1, 12.1 and 17.9 mg FQ/ kgVSS, respectively | Amorim et al., 2014 [R] |
| | Tetracycline | 10-30 mg $L^{-1}$ | 2-4 mg $g^{-1}$ | Y.-J. Shi et al., 2011« |
| | Prednisolone, naproxen, norfloxacin, sulfamethoxazole and ibuprofen | 50–55 µg $L^{-1}$ | 98.56, 84.02, 87.85,77.83 and 63.32%, respectively | Xia, Xiao-chun, Zhong-lin, Hao, & Qing-fang, 2014 [R] |
| | Fluoxetine | 3-4 µM | 0.20-0.50 mol FLX/g VSS | Moreira et al., 2015 [R] |
| **Dyes** | Malachite green | 50-80 mg $L^{-1}$ | 47.95-85.81 mg $g^{-1}$ MLSS (eq) | Sun et al., 2008 |
| | Methylene blue | 25–1500 mg $L^{-1}$ | 18-90% / 18-350 mg $g^{-1}$ | Wei, Wang, Ngo, et al., 2015 |
| | Acid Yellow 17 | 20-100 mg $L^{-1}$ | 5.86-32.26 mg $g^{-1}$ | J. Gao, Zhang, Su, Chen, & Peng, 2010 |
| | Yellow 2G and Brilliant Red K-2G (in single or binary solutions) | 39-160 mg $L^{-1}$ (in single or in mixture) | 17.05-66.43 and 17.20-69.43 mg $g^{-1}$ (in single solutions, respectively); 14.30-56.74 and 15.52-69.72 (in binary solutions, respectively) | J. F. Gao, Zhang, Su, & Wang, 2010 |

Source of data. [R]-experience performed in reactor

metal onto AGS. Results showed that in this case EPS, in particular its PN-like substances, played a primary role in the sorption process.

The literature available seems to indicate that, when AGS is exposed to the class of pharmaceuticals compounds, sorption is the main mechanism for the removal of those micropollutants. This was the case in a study on the removal of antibiotics, such as fluoroquinolones (Amorim et al., 2014) and cefalexin (Kong et al., 2015). It is important to notice that previously cited works were conducted in reactor systems, and AGS was used as treatment process and furthermore acted as biosorbent. Similarly, the removal of fluoxetine on an AGS reactor was due to sorption onto biomass (Moreira et al., 2015). On a first phase, fluoxetine was adsorbed onto AGS until their sorption capacity was exhausted and from that point onwards desorption occurred. Furthermore, desorption occurred also when contaminant was no more present in the inlet stream. In a study also conducted in a reactor for the removal of pharmaceuticals and personal care products (PPCPs), AGS was able to remove sulfamethoxazole, norfloxacin, prednisolone, naproxen and to a less extent ibuprofen by adsorption and/or degradation mechanisms (Xia et al., 2014). Removal of tetracycline was also characterized by a quick sorption and a slow process of biodegradation during the operation of a nitrifying granular sludge reactor (Shi et al., 2011). The adsorption process occurred through a complex mechanism of surface adsorption and intra-particle diffusion, while biodegradation was likely attributed to the synergistic effects of multiple microbes. More recently, Zheng et al. (2016) compared the adsorption of the hydrophobic 17β-estradiol onto AGS and activated sludge in batch mode, both pre-treated by a modified non-chemical wet-heat method to inhibit the biological activity of microorganisms in order to evaluate the physical adsorption effect. AGS had greater biosorption capacity than activated sludge for the same concentration of the compound and at low pH the adsorption of the target compound onto the sludge was promoted.

The biosorption of dyes such as Acid Yellow 17, Yellow 2G and Reactive Brilliant Red K-2G from aqueous solution was also studied using non-living AGS, to evaluate the effective sorption capacity of these microbial aggregates without any contribution from biological processes (J. F. Gao et al., 2010; J. Gao et al., 2010). Through Fourier Transmission Infrared Spectroscopy (FTIR) analysis, both studies revealed that functional groups such as amine, hydroxyl and carboxyl would be the active binding sites for the biosorption in non-living granules. Sun et al., (2008) reported the biosorption of the dye malaquite green onto AGS through boundary layer and pore diffusion. The role of EPS in the biosorption of the methylene blue dye from wastewater using AGS was recently explored (Wei, Wang, Ngo, et al., 2015). The results point out that EPS sorption played an important role in the dye removal, and three-dimensional excitation-emission matrix revealed that proteins and humic acid-like substances were the two EPS components that were gradually quenched in the presence of the contaminant.

Due to the AGS biosorption capacity and its importance for developing an efficient removal technology, optimization of the adsorption efficacy by AGS has revealed an increasing interest. Different strategies have been applied in order to improve the AGS removal capacity. However, biosorption onto biological material such as AGS is complex due to the presence of many functional groups (e.g. carboxyl, phosphate, hydroxyl, amino, thiol, etc) able to interact with sorbate molecules. In addition, the sorption capacity of these functional groups depends on the physical-chemical factors existent on the system (Fomina & Gadd, 2014). Therefore, strategies to improve the AGS biosorption should take into account the characteristics of the sorbent and the sorbate in a given environment. Li Wang et al. (2014) found that $Fe^{3+}$ treatment significantly increased the adsorption capacity of $Sb^{5+}$ onto AGS. The influence of different parameters namely pH, biosorbent dosage, initial dye concentration, salinity and temperature, on the biosorption of the Acid Yellow to AGS has been a subject of investigation (J. Gao et al., 2010). It was found that the biosorption process was promoted under acidic pH (pH 2.0) because the amino groups ($R{-}NH_2$) of non-living AGS were protonated at this pH thus promoting its interaction with the anionic ions ($SO_3^-$) that the dye produced in aqueous solutions. Moreover, in this case, the biosorption capacity decreased with increasing the biosorbent dosage and salinity. Acidic conditions were also found to be preferable for the biosorption of 17β-estradiol by AGS since it favors the occurrence of hydrogen bonding between carboxylic groups on the AGS surface and hydroxyl groups of the pharmaceutical (Zheng et al., 2016). In the same work, authors also examined the effects of ionic strength on adsorption behavior, demonstrating that ionic strength up to 0.4 mol $L^{-1}$ promoted the biosorption but beyond this limit the adsorption decreased sharply.

Another important aspect to take into account to develop biosorption-based technologies is the desorption capacity of the previously sorbed compound. This step will enable the reuse of the AGS and possibly the recovery of the sorbed compound. Different strategies have been reported to promote the desorption process from AGS. Li Wang et al. (2013) used $Na^{1+}$ and $K^{1+}$ as desorbent to regenerated AGS and concentrate $Sr^{2+}$ from polluted waters. Recently, W. Liu and co-workers (2015) reported that the metal ions $Pb^{2+}$, $Cd^{2+}$ and $Zn^{2+}$could be efficiently desorbed from EPS under acidic conditions and the biosorbent could be reused. In this case, at low pH, the EPS functional groups were protonated and thus positively-charged, therefore the adsorption of metal cations was unfavorable due to electrostatic repulsion.

A wide range of target pollutants can be removed from aqueous solution through AGS biosorption. However, so far the work on the AGS biosorption capacity was mostly conducted in batch mode and using synthetic wastewater. Studies performed with real industrial effluents containing multiple pollutants are needed in order to prove the feasibility of the process and its further application. To progress the application of AGS for the biosorption of toxics from industrial wastewaters, studies focusing the long-term application of the biosorbent and exploring its stability and regeneration capacity are required. Also, it would be interesting to develop processes coupling the wastewater treatment and the biosorption, offering a potentially more economical treatment approach.

## Bioaugmentation with Specialized Strains

The biological treatment of industrial wastewaters is often disrupted by organic (e.g. chlorinated compounds, phenolic compounds, surfactants and herbicides) and inorganic (e.g. heavy metals, sulfides and ammonia) chemicals present in wastewater streams (Bitton, 1994). The need of specialized microorganisms able to degrade specific pollutants and the transient state of industrial wastewater (highly variable xenobiotics content) might be the main limitations of the biotreatment of wastewater contaminated with such pollutants. As microorganisms present in the native sludge do not always harbor the capacity to degrade pollutants, an exploitable strategy that could be adopted to deal better with toxic compounds in effluents is bioaugmentation.

Bioaugmentation is commonly defined as the introduction of specialized strains, which may include bacterial augmentation or gene-augmentation, aiming at enhancing the degradation efficiency of toxic compounds and protecting the essential metabolic processes (COD, N and P removal) (Boon, Top, Verstraete, Siciliano, & Icrobiol, 2003; Gentry, Rensing, & Pepper, 2004; Nancharaiah, Joshi, Hausner, & Venugopalan, 2008).

The degradation of the toxic compound 2-fluoropenol in AGS reactor achieved through the addition of a specialized degrading strain was an example of the robustness of the technology for long term operation, since that was shown to be stable for more than a year that the reactor was operated, with the degrading strain surviving till the end of operation (Anouk F Duque et al., 2011). Another successful example was the bioaugmentation of half-matured sludge granules pre-grown on glucose with a strain able to degrade the herbicide 2,4-dichlorophenoxyacetic acid, which allowed the successful utilization of this substrate by the granules. The inoculated strain survived incorporated on the granules till the end of the operation (220 days) (Xiangchun Quan, Ma, Xiong, & Wang, 2015). AGS was successfully in using papermaking wastewater by bioaugmentation with strains possessing the ability to degrade lignin (Hailei, Guosheng, Ping, & Feng, 2006). Kee and co-workers (2015) successfully developed granules using four dye-degrading bacteria (*Bacillus pumilus* strain ZK1, *Bacillus cereus* strain ZK2, *Brevibacillus panacihumi* strain ZB1, and *Lysinibacillus fusiformis* strain ZB2) and sterile sludge as seeding agent. The SBR was operated under facultative anaerobic–aerobic cycles, meaning that the reaction phase was divided into two steps: a facultative anaerobic step that would favor azo bond cleavage and an aerobic step where mineralization of aromatic amines occurred. The results showed that the combination of mixed culture of bacteria and the bioreactor operation mode could improve textile wastewater treatment efficiency. Bacterial community structure revealed that dye-degrading bacteria distribution in the mature granules was different from that in the initial inoculum. These granules were further used to treat real textile wastewater showing 61% of decolorization and 46% of COD removal at an HRT of 24 h.

Successful cultivation of AGS was achieved in a SBR inoculated with *Rhizobium* sp. NJUST18, that is both an effective pyridine degrader and a good auto-aggregator (X. Liu et al., 2015). The star-up with an auto aggregative pyridine-degrading bacterium allowed the formation of stable granules which also exhibited excellent pyridine degradation performance. However, other type of bioaugmentation aiming at granulation rather than degradation of the contaminants could also be beneficial for the system. Ivanov and co-workers (2006) reported that bioaugmentation of activated sludge with self-aggregating strains improved the formation of aerobic granules, which in turn could be further applied for the treatment of wastewater. The success of the granulation process relies on the auto-aggregation capacity of the bacterial inoculum since that helps the formation of the granules, even if they do not participate critically in the degradation steps. This could be especially interesting for the scale-up of the technology, shortening the start-up of AGS reactors.

Bioaugmentation with strains carrying catabolic plasmid could also be employed as a viable strategy to enhance the ability of AGS to degrade toxic compounds. Nancharaiah and co-workers (2008) demonstrated the feasibility of such process when augmented an AGS-SBR with a strain of *Pseudomonas putida* strain carrying the plasmid pWWO, for toluene degradation. 2,4-Dichlorophenoxyacetic acid degrading AGS was cultivated through bioaugmentation of glucose-fed granules with plasmid pJP4, carried by the donor strain *Pseudomonas putida* SM1443 (Xiang-chun Quan, Tang, Xiong, & Yang, 2010). These granules exhibited good performance in degrading a mixture of 2,4-dichlorophenoxyacetic and chlorophenols (Ma, Quan, Yang, & Li, 2012).

Gene-augmentation is a promising strategy for the fast establishment of granular sludge possessing specific functions, such as degradation of toxic pollutants. Compared with the traditional cell augmentation, it shows the advantage of less requirements for the survival of the added strains. However, successful application of gene augmentation in real wastewater treatment system depends on many factors, including influent components, reactor configuration, operation mode and selective pressure. In fact, selective pressure is an important factor influencing the effectiveness of bioaugmentation. It has been reported that a strong selective pressure of target recalcitrant compounds existing alone and persistently was more beneficial for the effectiveness of gene augmentation (Xiangchun Quan, Tang, & Ma, 2011).

Besides the benefits of bioaugmentation related to toxic (micro)pollutants removal, whether bioaugmentation affects nutrients removal is still an important issue. Bioaugmentation of an AGS reactor for the degradation of 2-FP with a specialized strain demonstrated that bioaugmentation may not improve nutrients removal, but rather enable the system to continue removing nutrients efficiently by removing toxicity caused by pollutants (A. F. Duque, 2011). Studies on the effect of bioaugmentation with specialized strains able to degrade toxic pollutants on nutrients removal are still scarce, even though this is an emerging discussion topic.

## AGS Granulation and Granule Stability Treating Effluent with Toxic Compounds

The treatment of industrial wastewater containing toxic pollutants is often complex, and the application of AGS technology has demonstrated a great potential. However, its formation and structural stability can be susceptible to toxic shock. A key strategy to better deal with wastewater containing toxics is the development of AGS using such compounds as substrate. Several studies reported a successful development of granular sludge using toxic compounds as substrates; this can potentially be applied to effectively treat contaminated industrial water streams. When dealing with toxics such as phenol, AGS cultivated with acetate as sole carbon source was able to quickly acclimate to the phenol loading, and stabilized only one week after phenol was introduced in the feeding (S. T.-L. Tay et al., 2005). However, when dealing with more recalcitrant toxic compounds such 2,4-dicholorophenol (S.-G. Wang, Liu, Zhang, et al., 2007), *p*-nitrophenol (Yi et al., 2006) and methyl *tert*-butyl ether (L. L. Zhang et al., 2008), a different strategy involving the addition of an easy degradable compound as a co-substrate and the stepwise increase of the toxic content in influent was demonstrate to be a feasible option to cultivate AGS able to effectively degrade the toxic substrate. Successful microbial granulation was obtained in the presence of a synthetic chelating agent, nitrilotriacetic acid, fed with acetate as a co-substrate in the inlet of a SBR (Nancharaiah, Joshi, Krishna Mohan, Venugopalan, & Narasimhan, 2008). The presence of the toxics during cultivation of the granules acted as a selective pressure for strain enrichment allowing the establishment of granules with degrading abilities.

It is a fact that when dealing with industrial wastewater different easy biodegradable substrates are also often present and can challenge the granulation process. The impact of a set of carbon compounds frequently encountered in industrial wastewaters, namely acetate, methanol, butanol, propanol, propionaldehyde and valeraldehyde, on the granular sludge formation was recently investigated showing that all these compounds lead to stable granular sludge (Pronk, Abbas, et al., 2015). Easy biodegradable substrates that are not converted into storage polymers during the anaerobic period of the AGS process would, however, lead to unstable granular sludge formation (Pronk, Abbas, et al., 2015). Particulate and slowly biodegradable substrates often form an important fraction of industrial wastewater. In a study conducted to evaluate the effect of such substrates on the system performance and AGS morphology, starch was used as a model substrate (de Kreuk, Kishida, Tsuneda, & van Loosdrecht, 2010). In fact, AGS could be maintained on starch, but granule stability, COD and nutrient removal efficiencies decreased, which was reported to be due to irregular filamentous outgrowth on the granules. Starch was removed by adsorption at the AGS surface where it was hydrolyzed instead of being accumulated inside AGS as usually occurs for soluble substrates. Thus, the substrate will be available during the aeration period, stimulating the outgrowth of filamentous structures which in turn affected granules stability.

Successful cases of AGS granulation processes with real industrial wastewater have been recently reported. AGS was successfully cultivated in a sequencing batch reactor fed with a mixture of chemical industrial wastewater rich in toxic organics and the effluent from an anaerobic acidogenic reactor (L. Liu et al., 2011). AGS was successfully cultivated using high-strength agro-based industry effluents such as rubber wastewater from a rubber factory in Malaysia (Rosman et al., 2013), barley dust wastewater supplied by a large-scale malt factory (Norbert Schwarzenbeck, Erley, Mc Swain, Wilderer, & Irvine, 2004), palm oil mill wastewater (Abdullah, Yuzir, Curtis, Yahya, & Ujang, 2013), livestock wastewater (Othman et al., 2013), swine slurry (Figueroa, Val Del Río, Campos, Mosquera-Corral, & Méndez, 2011), landfill leachate (Di Bella & Torregrossa, 2014) and brewery wastewater (S.-G. Wang, Liu, Gong, et al.,

2007). The formation of granules was also achieved by using industrial wastewater coming from a dairy analysis laboratory as influent (Arrojo, Mosquera-Corral, Garrido, & Méndez, 2004).

One of the problems when dealing with industrial wastewater is the possible combination of different substrates (e.g., solvents, surfactants, metal salts, oils) in the same effluent. Oily wastewater generated by ships in the engine-rooms (bilgewater) and by the washing of oil tanks (slops) combined with high salinity levels (up to 25.000 ppm) is an example of a complex mixture. Nevertheless, granulation was achieved using a mixture of real and simulated slop (Corsino, Campo, Di Bella, Torregrossa, & Viviani, 2015). In fact, AGS was quickly developed but granules were slightly unstable and susceptible to breaking.

In addition, the feasibility of cultivating AGS in a full-scale SBR was successfully achieved with wastewater that was constituted by approximately 30% of domestic sewage and 70% of industrial wastewater from printing and dyeing, chemical, textile and beverage (J. Li et al., 2014a). To date, AGS formation in full-scale installations exclusively treating industrial wastewater containing toxic compounds was not reported in the scientific literature.

## AGS NUTRIENT REMOVAL IN THE PRESENCE OF POLLUTANTS

### General Performance on Nitrogen and Phosphorous Removal

The main purpose of the biological wastewater treatment is to convert the organics present in an upcoming wastewater stream to $N_2$ and $CO_2$, as well as to reduce the amount of suspended solids entering the environment (Boon et al., 2003). The biological processes responsible for these conversions are designated as organic removal, nitrification, denitrification and phosphate removal. Briefly, the organic carbon is, usually, directly oxidized to $CO_2$ by heterotrophic organisms; N removal is performed in two steps, nitrification, which is the oxidation of ammonia to nitrite by ammonia oxidizing bacteria (AOB) and nitrite to nitrate by nitrite oxidizing bacteria (NOB), and denitrification, which consists in the reduction of nitrites and nitrates by means of heterotrophic denitrifying bacteria that utilize organic carbon as electron donor and the nitrites and nitrates as electron acceptors; biological phosphate removal is carried out by phosphate accumulating organisms (PAO), which during the anaerobic period consume easily degradable substrates, such as acetate, and store them as polyhydroxyalkanoates (PHA) and in the aerobic phase, phosphate is taken up in the cell, using PHA as substrate (de Kreuk, 2006).

In spite of the ability of AGS to remove pollutants by sorption and/or biodegradation, the main biological processes, such as the removal of nutrients, can be affected by the presence of these compounds. In a study conducted to evaluate the impact of different fluorinated drugs (ciprofloxacin - CPF, norfloxacin - NOR, ofloxacin - OFL and fluoxetine - FLX) on the performance of an AGS-SBR system (Amorim et al., 2014; Moreira et al., 2015), a sequence of intermittent, alternating and/or continuous shock loads of pharmaceuticals, at concentrations within the higher ranges reported for effluents, were applied during 468 days. The operation of the bioreactor was split into seventeen phases during which different conditions were tested. The reactor was first exposed to fluoroquinolones (at 32 μM), following the order CPF, NOR, OFL (single cycle shock), in phases II, III and IV, respectively. After a starvation phase, feed with 9 μM of OFL, NOR, CPF and mix of the three fluoroquinolones (continuous shock) was performed in phases VI, VII, VIII and IX. After having resumed performance, the behavior towards the antidepressant pharmaceutical FLX was evaluated. The reactor was fed with 4 μM of FLX in phase XI, followed by 3 μM in continuous mode, except in the phase XIII, in which the shock load was applied 2 days per week.

In the phases I, V, XV and XVII, there was no pharmaceutical feed. The effects on the main biological processes were evaluated. It was observed that the organic removal, measured by COD, was not markedly affected by the pharmaceuticals shock loads. The COD concentration observed in the effluent was consistently low (0-60 mg $L^{-1}$) (Figure 2). Ammonium and nitrite were practically not detected in the AGS bioreactor effluent, maximum 0.03 and 0.01 mM for ammonium and nitrite, respectively, indicating that the presence of these pharmaceuticals did not inhibit nitrification, whereas accumulation of nitrate (up to 5.43 mM) in the effluent was observed, indicating that denitrification was affected (Figure 2b and 2d). In the same studies, phosphate removal was affected to some extent during the period of feeding with fluoroquinolone antibiotics and at the beginning of FLX feeding, showing, however, a gradual adaptation, being practically not detected in the treated effluent at the end of bioreactor operation, maximum 0.04 mM (Figure 2c). The results suggested that PAO activity was not to severely affected by the load of pharmaceuticals. In another study, COD and $NH_4^+$-N removal efficiencies were slightly decreased and the denitrification process was inhibited in the presence of cefalexin (Kong et al., 2015).

The short term effect of several 2-fluorophenol (2-FP) concentrations (0-200 mg $L^{-1}$) on the activity of nitrite oxidizing bacteria (NOB) in granular sludge was evaluated in batch cultures (A.F. Duque, 2011). The same experiments were also conducted with crushed granular sludge and with conventional activated sludge. By crushing the granules, the possible protective structure of the granules, which might shield the interior biomass from toxic effects, is destroyed, leading to a morphology similar to activated sludge (small flocs). The effect of 2-FP on NOB activity is shown in Table 3 as an example.

*Figure 2. COD (a), $PO_4^{3-}$(b) and $NH_4^+$(c) concentration profiles in AGS-SBR along its operation. Concentrations at inlet ( ♦ ), effluent (▲) and removal efficiency ( – ) are shown. N profile (d). Effluent concentration of $NH_4^+$ ( ▲ ), $NO_3^-$ (●) and $NO_2^-$ (x) are shown. Numbers are indicative of the different phases listed in the text*
*Adapted from Amorim et al., 2014 and Moreira et al., 2015*

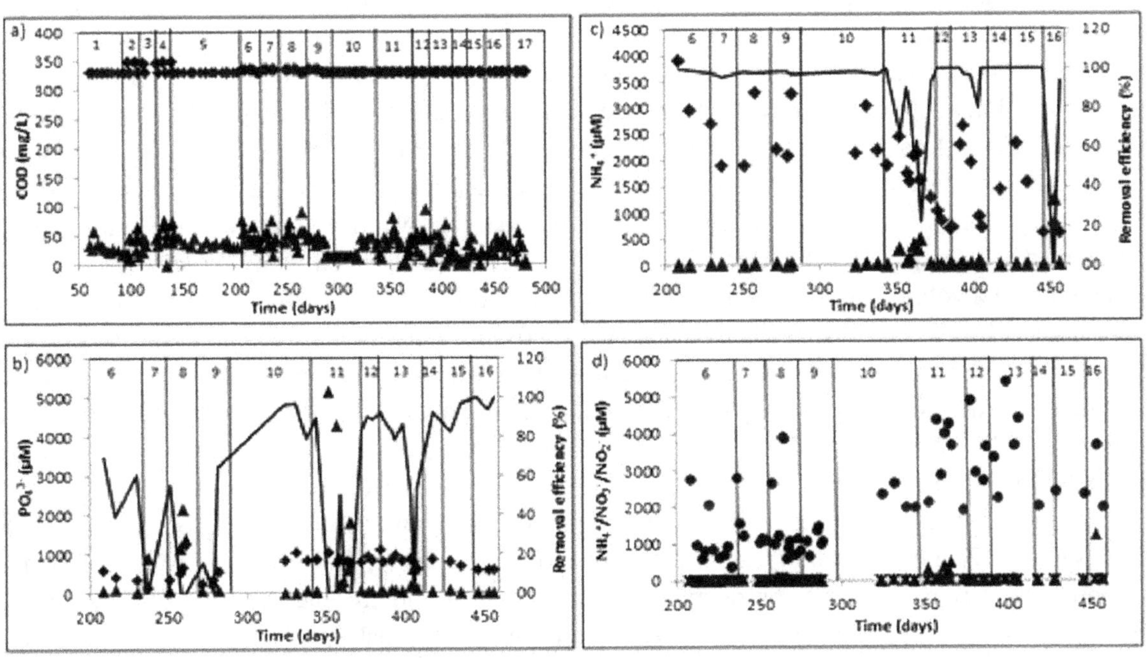

*Table 3. NOB activity in the presence of different 2-FP concentrations*

| 2-Fluorophenol concentration (mg $L^{-1}$) | NOB activity[a] (%) | | |
|---|---|---|---|
| | Activated sludge | Granular sludge | Crushed granular sludge |
| 0 | 100 | 100 | 100 |
| 5 | 15.4 | 31.9 | 4.67 |
| 10 | 4.35 | 3.59 | 1.65 |
| 15 | 2.61 | 3.54 | 0 |

[a]100% NOB activity is 1.57, 0.43 and 1.98 mg N/(h.gVSS) for AS, GS and crushed GS, respectively

At the end of the experiments, 2-FP concentration in the bulk liquid was the same as the initial concentration, indicating that there was neither consumption nor adsorption by the various types of sludge used. Generally, NOB were significantly affected by the presence of 2-FP for the range of concentrations tested from 5 to 200 mg $L^{-1}$, being completely inhibited at concentrations higher than 15 mg $L^{-1}$. In the presence of 10 and 15 mg $L^{-1}$, NOB were severely inhibited (> 95% inhibition), as detected by nitrite accumulation, independently of the sludge type. However, some differences were obtained for the different types of sludge tested in the presence of 5 mg $L^{-1}$ of 2-FP. NOB flocs were highly affected by the presence of 2-FP (inhibition of 85% (activated sludge) and 95% (crushed granular sludge) at 5 mg $L^{-1}$ of 2-FP). Amor, Eiroa, Kennes, and Veiga (2005), who have studied the effects of phenol on the nitrification process with activated sludge, have shown that phenol concentrations above 1000 mg $L^{-1}$ negatively affected the nitrification process, leading to nitrite and ammonium accumulation. Clearly, the fluoride addition to the phenol makes the compound much more toxic. Granular sludge NOB were less inhibited, ca. 68% inhibition at 2-FP concentrations of 5 mg $L^{-1}$. This suggests that the granular structure protected NOB against toxic effects. EPS have been detected in significant amounts in the aerobic granule structure (Y.-Q. Liu, Liu, & Tay, 2004; McSwain, Irvine, Hausner, & Wilderer, 2005) and can play an important role in the AGS structure, protecting the bacterial cells and reducing the impact of severe conditions towards the cells (Z. Wang, Liu, Yao, & Cai, 2006) which can explain the lower NOB activity inhibition obtained for the granular sludge. The long-term effects of 2-FP on granular sludge were evaluated in a laboratory scale AGS-SBR which was inoculated with a specialized strain capable to degrade 2-FP as the sole source of carbon and energy (Anouk F Duque et al., 2011). The investigation on the long-term effect of 2-fluorophenol on nitrification and phosphate removal by AGS revealed that it might be very promising for the simultaneous COD, N and P removal from wastewaters containing toxic compounds (A.F. Duque, 2011). During 5 months, the AGS-SBR was fed with 50 mg $L^{-1}$ of 2-FP but it was operated at a volume exchange ratio of 40%, whereby the 2-FP concentration inside the bioreactor was 19 mg $L^{-1}$. During the long-run, PAOs and AOB activities remained stable, 2.66 mg P $(h.g_{VSS})^{-1}$ and 1.38 mg N $(h.gVSS)^{-1}$, respectively, without accumulation of either ammonium, phosphate or nitrate in the effluent. In the short term toxicity tests for 2-FP on granular sludge, this concentration was strongly inhibitory (>97%) to NOB, but during reactor operation it didn't affect its activity, which is most probably due to the protection provided by the granular structure (A.F. Duque, 2011).

Good nutrient removal efficiencies were also obtained by mature aerobic 4-chloroaniline-degrading granules (Liang Zhu et al., 2008). The effect of phenol on the nitrifying activity of AGS revealed that ammonium oxidation and nitrate production rates decreased in the presence of phenol. However, after phenol degradation, nitrifying activity was recovered, indicating that nitrifying bacteria embedded in the microbial granules were relatively resilient to the inhibitory effects exerted by phenol (Y.-Q. Liu et al., 2005). In the case of reactors operated for the removal of heavy metals, the reactor performance generally complied with the mandatory levels for COD and total N. The reactor's performance was able to return to the initial condition, resisting the metal shock loading on the long run (Marques et al., 2013). A study conducted by Wei et al. (2014) demonstrated the possibility for the simultaneous nitrogen and phosphorus removal when treating ammonia-rich wastewater.

When dealing with industrial wastewater, one of the problems that biological systems have to face is to deal with the variable influent composition, with wastewaters characterized by sequentially alternating pollutants (SAP) scenarios. Jemaat and co-workers (2014b) demonstrated that the feasibility of the AGS technology to deal with SAP scenarios depended on the nature of the pollutant. In their study, partial nitrification and *o*-cresol biodegradation were successfully achieved when phenol or *p*-nitrophenol were present in the feeding wastewater but the presence of a more recalcitrant compound (e.g. 2-chlorophenol) strongly inhibited both processes. An AGS cultivated with modified piggery wastewater, showed to be capable of simultaneous removal of COD and N. The study revealed that the heterotrophic nitrification and autotrophic nitrification coexisted in the granules; with the contribution of the heterotrophic nitrification more important to the N removal at high loading rates (Y. Zhao, Huang, Zhao, & Yang, 2013). In AGS cultivated with a mixed wastewater rich in toxic organics, when the industrial wastewater fraction in the influent of the SBR was increased, the COD and ammonia removal efficiency decreased but then increased when granules were formed (L. Liu et al., 2011).

Other factor that could affect the performance of the biological treatment when dealing with industrial wastewater is the high salinity content of wastewater. During the treatment of a mixture of real and simulated wastewater containing high salt and hydrocarbon content, the efficiency of phosphorous and carbon removal was satisfactory (Corsino et al., 2015). However, nitrification was absent probably because of the inhibition of the autotrophic biomass.

Generally, AGS showed a great capacity to treat real wastewater, namely rubber, brewery, livestock, dairy, chemical, swine slurry, textile wastewater, with simultaneous carbon and nitrogen removal (Table 4). Reported removal efficiencies usually are lower in the startup of the reactor system when feeding with industrial wastewater but during the gradual transformation of flocculent sludge into granular sludge, removal efficiencies improve and become high and stable after the granulation process (Rosman et al., 2013; Liu et al., 2011; Li et al., 2014; Schwarzenbeck et al., 2005). The formation of granules significantly improved the ability of sludge to withstand the toxic substances and high loading rates.

Kocaturk and Erguder (2015) compared the efficiency of AGS to treat raw sugar beet processing wastewater and the same wastewater after being anaerobically digested. AGS proved to be efficient to treat both types of wastewater however when bioreactor was fed with raw wastewater, AGS properties slightly changed but any drastic and negative effect on treatment performance was observed.

The feasibility of treating real petrochemical wastewater using AGS technology has been investigated (H. Zhang, He, Jiang, & Yang, 2011). By gradually increasing the proportion of petrochemical wastewater, granule settleability became poor, and the removal efficiency of COD and $NH_4^+$ decreased. When feed was performed with 100% of the petrochemical wastewater without any external carbon, both removal efficiencies of COD and $NH_4^+$ deteriorated sharply but addition of sodium propionate as a co-substrate for co-metabolism in the inlet stream significantly improved the removal performance of carbon and nitrogen, indicating that co-metabolism played an important role in the biological conversions and could be a way to overcome some limitations of the technology.

*Table 4. N and P removal in aerobic granular sludge application to industrial wastewaters*

| Wastewater Type | Organic Loading Rate | N-Removal Efficiency | P-Removal Efficiency | Reference |
|---|---|---|---|---|
| Rubber wastewater | 3.7 $kg_{COD}$ $m^{-3}$ $d^{-1}$ | 94.7% ($NH_4^+$-N)<br>89.4% (TN) | - | Rosman et al., 2013 |
| Livestock wastewater from a cattle farm | 9 $kg_{COD}$ $m^{-3}$ $d^{-1}$ | 60-73% (TN) | 59-70% (TP) | Othman et al., 2013 |
| Chemical industrial wastewater | 500-1000 $mg_{COD}$ $L^{-1}$ | 14-90% ($NH_4^+$)<br>40% (TN) | - | L. Liu et al., 2011 |
| Dyeing textile wastewater | 0.4-3.4 $kg_{COD}$ $m^{-3}$ $d^{-1}$ | 28.6-90.4% (TKN) | - | Lotito, De Sanctis, Di Iaconi, & Bergna, 2014 |
| Industrial wastewater from analysis of dairy products | 7 $g_{COD}$ $L^{-1}$ $d^{-1}$ | 70% (TN) | - | Arrojo et al., 2004 |
| Swine slurry wastewater | 2.2-7.3 $kg_{COD}$ $m^{-3}$ $d^{-1}$ | 64-70% (TN) | - | Figueroa et al., 2011 |
| Wastewater from Yancang WWTP (approx. 30% domestic sewage and 70% industrial wastewater from printing and dyeing, chemical, textile and beverage) | 3.9-4.5 $kg_{COD}$ $m^{-3}$ $d^{-1}$ | 94-95% ($NH_4^+$-N)<br>23-60% (TN) | - | J. Li, Ding, Cai, Huang, & Horn, 2014b |
| Brewery wastewater | 3.5 $kg_{COD}$ $m^{-3}$ $d^{-1}$ | 88.9% ($NH_4^+$) | - | S.-G. Wang, Liu, Gong, et al., 2007 |
| Wastewater including 40% domestic sewage and 60% industrial wastewater | 1-4.6 $g_{COD}$ $g^{-1}$ $d^{-1}$ (F/M) | 90% ($NH_4^+$-N) | - | L. Liu et al., 2011 |
| Dairy wastewater | 4.5-5.9 $kg_{COD}$ $m^{-3}$ $d^{-1}$ | 80% (TN) | 67% (TP) | N Schwarzenbeck, Borges, & Wilderer, 2005 |

TN – Total nitrogen; TP – Total phosphorous; TKN – Total Kjeldahl Nitrogen; F/M – Food to microorganism ratio; - - not reported

However, data on P removal from real wastewaters are scarce. Removal efficiencies around 70% for TN and TP were achieved in studies with high OLR and high levels of phosphorus and nitrogen, like wastewater from livestock industry (Othman et al., 2013). Similar nitrogen removal efficiency was observed for industrial wastewater produced in a laboratory for analysis of dairy products, in which the ammonium was nitrified even at high OLR (Arrojo et al., 2004; Wang et al., 2007). However, when NRL of swine raw slurry reached values of 1.26 kg N $m^{-3}$ $d^{-1}$, the nitrogen removal via nitrification–denitrification vanished (Figueroa et al., 2011).

## Impact on the Microbial Community

In spite of the compact and strong microbial structure of AGS, the presence of toxic compounds can affect the bacterial populations and hence disturb the accomplishment of their functions within the community. Therefore, the variation analysis in microbial communities could enhance our understanding for the industrial wastewater treatment process and might be helpful for the future design of a biological process with better efficiency. Studies reporting the effect of high and variable loads of anthropogenic substances on the AGS bacterial community are well documented in the literature. Dai and co-workers (2015) observed that the load of aniline, a widely used compound in the pharmaceutical, pesticide and textile dye industry, changed the community structure of the bioreactor. On one hand, the aniline supplementation enhanced the dominance of microbes with aniline-biodegradation ability such as *Burkholderia thailandensis*, *Cohaesibacter marisflavi* and *Paracoccus yeei*. On the other hand, 4-chloroaniline shock load resulted in the disappearance of dominant bacteria like *Kineosphaera limosa*, related to EPS production, and appearance of *Acinetobacter*, involved in proteolysis, which might have contributed to the verified disintegration of AGS (Liang Zhu et al., 2013).

The main nitrobenzene-degrading bacteria found in the granular sludge able to remove nitrobenzene as sole carbon and nitrogen were *Micrococcus luteus*, *Bacillus* sp., *Pseudomonas* sp., *Comamonas* sp., *Acinetobacter* sp. (D. Zhao et al., 2011). Aerobic granules cultivated with wastewater containing *p*-nitrophenol presented diverse microbial morphotypes, with degrading bacteria accounting for 49% of the total culturable heterotrophic bacteria. Denaturing gradient gel electrophoresis (DGGE) analysis of 16S rRNA gene fragments showed a gradual temporal shift in microbial community succession as the granules developed from the activated sludge seed (Yi et al., 2006). DGGE fingerprint profiles of the microbial community of the bacteria inhabiting the AGS responsible for the degradation of methyl *tert*-butyl ether revealed that the reactor was inhabited by microorganisms related to the genera *Sphingomonas*, *Methylobacterium*, and *Hyphomicrobium vulgare*, strains previously reported to be associated with methyl tert-butyl ether biodegradation (L. L. Zhang et al., 2008). The DGGE fingerprint pattern of cultivated phenol-fed aerobic granules revealed a shift in the microbial community when pyridine was present. Moreover, *Acinetobacter* sp. dominated the pyridine–phenol aerobic granule system (Adav et al., 2007). In relation to the impact of heavy metals on AGS microbial community, it was found that the addition of $Cu^{2+}$ drastically reduced the biodiversity, while the toxic effect of $Ni^{2+}$ addition was milder (X.-H. Wang et al., 2010).

Recently different approaches were used to elucidate the impact of a load of pharmaceutical compounds in the bacterial population. Zhao et al. (2015) used solexa sequencing with microbial taxonomic profiling analyses to characterize the effect of pharmaceuticals and personal care products dosing, which revealed that these contaminated water streams promoted the dominance of the *Proteobacteria* and *Bacteroidetes* phyla and the *Betaproteobacteria* class. PCR-DGGE of bacterial 16S rRNA gene also revealed that the load of fluorinated pharmaceuticals namely fluoroquinolones and fluoxetine, induced shifts in the AGS bacterial population (Amorim et al., 2014; Moreira et al., 2015). High-throughput sequencing analysis showed that aerobic granules exposed to cefalexin had a superior diversity of microbial species. The abundance of *Proteobacteria* significantly decreased, while the abundance of *Bacteroidetes* and *Chloroflexi* underwent a highly significant increase (Kong et al., 2015).

The levels of carbon, phosphorus and nitrogen in the effluent are of primary concern since it must meet the standards for water discharge. Therefore, studies on the elucidation of the impact of the load of toxics or micro-pollutants on the structural diversity and evolutionary changes of the microbial populations responsible for those biological processes within AGS are increasing. Franca et al. (2015) showed that the AGS technology was indeed a robust system to deal with shocks of textile dye wastewater. Even though the overall COD removal efficiency was not affected, FISH analysis demonstrated that the number of PAOs was significantly reduced whereas the number of *Competibacter* GAOs doubled due to the presence of the dye in the feeding. A study conducted with wastewaters with a high salt content, like those generated in seafood canning, pickling, and cheese-processing industries, demonstrated that changes in the main biological processes and the dynamics of microbial populations occurred due to the increasing salt concentrations on the feeding wastewater (Bassin et al., 2011). Using FISH analysis, these authors found that high salt concentration (33 g $L^{-1}$) lead to the disappearance of PAOs, and in parallel

GAOs became more dominant, severely affecting the phosphorus removal. Also, PCR-DGGE performed using amoA gene-based primers as molecular marker revealed that *Nitrosomonas* sp. was present in the operational period which is in line with the bioreactor performance for ammonium removal. Actually, the authors find out that AOB in aerobic granules are less sensitive to salt concentrations than NOB, namely *Nitrospira* sp.. The effect of salinity on AGS system treating a simulated industrial wastewater containing a mixture of aromatic compounds (phenol, *o*-cresol and *p*-nitrophenol) was evaluated by Ramos, Suárez-Ojeda, & Carrera (2015). Complete biodegradation of the target aromatic compounds was achieved, but the increase of salinity caused the loss of the aerobic granules integrity and also changed the microbial populations at class and genus level. The pyrosequencing analysis revealed that *Sphingobium, Cythophaga* and *Comamona* were the most abundant genera at low salinity while *Devosia*, *Sphingomonas* and *Muricauda* the most abundant genera at high salinity. Real-time PCR performed by Liu and co-workers (H. Liu, Li, Li, & Chen, 2008) demonstrated that the AOB quantity did not decrease proportionally with the increase in pentachlorophenol concentration. The observed decrease in the nitrogen removal capacity was probably due to the inhibition of the AOB activity by pentachlorophenol and not to the decrease in AOB cells quantity. More recently, simultaneous nitritation and biodegradation of a mixture of aromatic compounds (*o*-cresol, phenol and quinoline) by AGS at high salinity levels were achieved (Ramos, Suárez-Ojeda, & Carrera, 2016a). *Nitrosomonas* genus was the only AOB detected within AGS both by pyrosequencing and FISH whereas no NOB genus was detected by each molecular technique.

J. Li et al. (2014b), who have studied aerobic granulation using a 30% domestic sewage and 70% industrial wastewater mixture, reported the presence of *Nitrospira* sp. and *Thauera* sp., through DGGE analysis. These observations are in accordance with the nitrogen removal results (Table 4). Interestingly, *Thauera* genus was considered as the main player in nitrogen removal, due to its dominance in the reactors. Moreover, fluctuations on the COD/N ratio, usual in industrial wastewaters, can be a destabilizing factor of reactor performance associated with shifts in the microbial community. Luo and co-workers (2014) verified that when the COD/N ratio decreased from 4 to 1, abundance of genera associated with incomplete denitrification, such as *Roseomonas*, *Shinella*, *Devosia*, *Diaphorobacter*, *Legionella* and *Lysobacter* shifted to *Thauera*, a genus capable of complete denitrification. Enhanced nitrification was verified and supported by the greater abundance of the AOB genus *Nitrosomonas*. However NOB genera, *Nitrospira* and *Nitrobacter*, decreased. Moreover, at a COD/N ratio of 4 *Flavihumibacter* was a dominant genus and *Filimonas* was also present. Both are filamentous bacteria which might promote granular sludge development. The overall reduction in filamentous genera under COD/N of 1 could have contributed to granule disintegration as they provide a structural network to the granule. Ammonia load in wastewater is one of the major factors influencing the composition of AOB species in aerobic granules. While low ammonia environments usually harbor *Nitrosospira* ssp., the predominant AOB in ammonium-rich systems are members of the genus *Nitrosomonas* (Mobarry, Wagner, Urbain, Rittmann, & Stahl, 1996). Cydzik-Kwiatkowska and Wojnowska-Baryła (2011) observed that with an increase in the ammonia load in wastewater, members of the *Nitrosospira* and *Nitrosomonas* genera coexisted in the granules, representing high AOB species diversity, advantageous for the stability of wastewater treatment. Many investigations focused on micropollutants removal using AGS and its impact mainly on N removal. However, N and P microbial communities associated to real industrial wastewater treatment were scarcely studied.

## COMBINED PROCESSES FOR CONTAMINANT REMOVAL AND RESOURCE RECOVERY

The high phosphorus and nitrogen content in wastewater leads to serious problems of eutrophication into receiving waters and thus the resulting effluent should comply with the stringent discharge limits. Moreover, these nutrients are non-renewable resources which abundance is decreasing in the world, and presently there is an increasing demand for these natural resources. Through nutrient recovery, the scarcity of usable nutrients could be mitigated while improving water quality to be discharged into the environment. Also, the recovery processes could represent an opportunity to generate revenue while producing nutrients that can be used for different purposes such as fertilizers in agriculture. Therefore, combined processes that remove contaminants from wastewaters and simultaneously recover valuable products, strongly increase the sustainability and the economic value of the wastewater treatment.

Recovery of phosphorous precipitates during wastewater treatment with activated sludge has been largely reported and recently, the applicability of the AGS systems to P-recovery was shown but it still is a relatively unexplored field. The first indication of biologically induced phosphate precipitation in a granular sludge system was reported by de Kreuk et al. (2005). The authors suggested that part of the phosphate removal was due to precipitation inside the granules. Later, Yilmaz and co-workers (2008) reported the transient formation of struvite during the treatment of a nutrient-rich abattoir wastewater with AGS. However, the overall mineral precipitation was low since most of the mineral complexes formed during the anaerobic feeding period were subsequently dissolved during the aerobic period. Both studies reported the formation of phosphorous precipitates, although the characterization of those complexes was not performed.

Recently, Mañas et al. (2011) for the first time have revealed the nature of phosphorus precipitates formed in the AGS process. Biologically induced precipitation of phosphorus in the core of granules occurred in the form of hydroxyapatite. The formation of such mineral represents about 45% of the total P removal which was much more significant than those estimated for flocculated sludge. The accumulation of phosphorus precipitated as calcium phosphate was observed at high influent calcium concentration (Barat, Montoya, Borrás, Ferrer, & Seco, 2008). The formation of such mineral was initiated by precipitation of the amorphous calcium phosphate, and later crystallization of hydroxyapatite. Also, struvite (potassium/ammonium magnesium phosphate) accumulation in AGS induced by alginate-like EPS was reported by Lin et al. (2012). These authors also observed that this mineral was a source for ammonium adsorption. More recently, the accumulation of mineral phosphorus in the forms of Ca–Mg phosphate ($Ca_7Mg_2(PO_4)_6$) and whitlockite ($Ca_3(PO_4)_2$) - a calcium orthophosphate crystal in which magnesium may be partly substituted for calcium- instead of hydroxyapatite, was reported (Y. Li, Zou, Zhang, & Sun, 2014). Also, the authors reported that the richness in filamentous bacteria belonging to the genus *Thiothrix* within P-rich granules might have contributed to enhancing the P content in AGS since these bacteria were able to trap minerals.

Presently, the extraction of other valuable compounds, especially carbon rich compounds, has been reported in the literature. Alginate-like exopolysaccharide molecules were successfully extracted from AGS feed with synthetic wastewater in lab-scale reactors (Y. M. Lin, Wang, Chi, & Liu, 2008). Also, the extraction from AGS cultivated in a pilot plant treating municipal sewage was proven to be feasible (Y. Lin, de Kreuk, van Loosdrecht, & Adin, 2010). The alginate-like exopolysaccharides derived from AGS possess stronger mechanical properties, comparing to that derived from aerobic flocculent sludge (Y. M. Lin, Sharma, & van Loosdrecht, 2013). Gobi and Vadivelu (2014) demonstrated the capability of AGS to accumulate polyhydroxyalkanoates (PHA) polymers while treating palm oil mill effluent. Fourier Transmission Infrared Spectroscopy (FTIR) analysis revealed that the PHA produced was a poly (hydroxybutyrate-co-hydroxyvalerate) (PHB-co-PHV) copolymer that can be used for plastic production due to its stability and robustness. Later the same authors demonstrated for the first time the feasibility to extract the PHA accumulated within AGS (Gobi & Vadivelu, 2015).

The recovery of biomaterials from waste AGS is other possible way to increase the sustainability and economics of wastewater treatment (Y. M. Lin, Nierop, Girbal-Neuhauser, Adriaanse, & van Loosdrecht, 2015). The polysaccharide-based biomaterial recovered had amphiphilic and film-forming properties, and thus the biomaterial has the potential to be used as a water-barrier coating material. In fact, the authors demonstrated the applicability of the polysaccharide-based biomaterial as a surface coating on paper packaging, proving that the biomaterial confers to paper interesting functionalities while maintaining its environment-friendly characteristics.

## CONCLUSION

Microbes exhibit unique capabilities for contaminant degradation and thus different strategies may be adopted for the treatment of industrial wastewater containing toxic pollutants.

The high number of laboratory scale, and some pilot scale, AGS reactors which point to successful operation in larger scale, stress the potential of the technology. The energy and space saving features of the AGS systems is a major driver for the water challenges that industry faces. However, there is a need for data on the large scale application of the AGS technology to industrial effluents in order to push the technology forward. Also, most of the studies focusing on pollutant removal using AGS reported the impact of such compounds on the C-, N- and P-removal performance but scarcely revealed their effects on N and P microbial communities that play a central role in the regulation of the biological processes occurring during the wastewater treatment. Presently, there is an increasing interest to understand microbial community structure and function and new insights into this matter will improve optimization and modeling strategies for AGS biological wastewater treatment. Furthermore, concerning resource recovery, the potential to use AGS for treating wastewater while recovering valuable products not only promotes the sustainability of the process but also could increase its economic value, thus benefiting its implementation in industrial facilities.

## ACKNOWLEDGMENT

C.L. Amorim, I.S. Moreira and A.F. Duque acknowledge research grants from Fundação para a Ciência e Tecnologia (FCT), Portugal (SFRH/BPD/96481/2013, SFRH/BPD/87251/2012 and SFRH/BPD/88705/2012) and Fundo Social Europeu (Programa Operacional Potencial Humano (POPH), Quadro de Referência Estratégico Nacional (QREN)). Authors also acknowledge COST Action ES1202: Conceiving Wastewater Treatment in 2020 - Energetic, environmental and economic challenges.

## REFERENCES


Abdullah, N., Yuzir, A., Curtis, T. P., Yahya, A., & Ujang, Z. (2013). Characterization of aerobic granular sludge treating high strength agro-based wastewater at different volumetric loadings. *Bioresource Technology*, *127*, 181–187. doi:10.1016/j.biortech.2012.09.047 PMID:23131639

Adav, S. S., Lee, D.-J., & Ren, N. Q. (2007). Biodegradation of pyridine using aerobic granules in the presence of phenol. *Water Research*, *41*(13), 2903–2910. doi:10.1016/j.watres.2007.03.038 PMID:17499839

Adav, S. S., Lee, D.-J., Show, K.-Y., & Tay, J.-H. (2008). Aerobic granular sludge: Recent advances. *Biotechnology Advances*, *26*(5), 411–423. doi:10.1016/j.biotechadv.2008.05.002 PMID:18573633

Amor, L., Eiroa, M., Kennes, C., & Veiga, M. C. (2005). Phenol biodegradation and its effect on the nitrification process. *Water Research*, *39*(13), 2915–2920. doi:10.1016/j.watres.2005.05.019 PMID:15998531

Amorim, C. L., Maia, A. S., Mesquita, R. B. R., Rangel, A. O. S. S., van Loosdrecht, M. C. M., Tiritan, M. E., & Castro, P. M. L. (2014). Performance of aerobic granular sludge in a sequencing batch bioreactor exposed to ofloxacin, norfloxacin and ciprofloxacin. *Water Research*, *50*, 101–113. doi:10.1016/j.watres.2013.10.043 PMID:24361707

Arrojo, B., Mosquera-Corral, A., Garrido, J. M., & Méndez, R. (2004). Aerobic granulation with industrial wastewater in sequencing batch reactors. *Water Research*, *38*(14-15), 3389–3399. doi:10.1016/j.watres.2004.05.002 PMID:15276756

Barat, R., Montoya, T., Borrás, L., Ferrer, J., & Seco, A. (2008). Interactions between calcium precipitation and the polyphosphate- accumulating bacteria metabolism. *Water Research*, *42*(13), 3415–3424. doi:10.1016/j.watres.2008.05.003 PMID:18538819

Bassin, J. P., Pronk, M., Muyzer, G., Kleerebezem, R., Dezotti, M., & van Loosdrecht, M. C. M. (2011). Effect of elevated salt concentrations on the aerobic granular sludge process: Linking microbial activity with microbial community structure. *Applied and Environmental Microbiology*, *77*(22), 7942–7953. doi:10.1128/AEM.05016-11 PMID:21926194

Beun, J. J., Hendriks, A., van Loosdrecht, M. C. M., Morgenroth, E., Wilderer, P. A., & Heijnen, J. J. (1999). Aerobic granulation in a sequencing batch reactor. *Water Research*, *33*(10), 2283–2290. doi:10.1016/S0043-1354(98)00463-1 PMID:11831218

Bitton, G. (1994). Toxicity testing in wastewater treatment plants using microorganisms: wastewater microbiology. New York: Wiley-Liss.

Boon, N., Top, E. M., Verstraete, W., Siciliano, S. D., & Icrobiol, A. P. P. L. E. N. M. (2003). *Bioaugmentation as a Tool To Protect the Structure and Function of an Activated-Sludge Microbial Community against a 3-Chloroaniline Shock Load*. doi:10.1128/AEM.69.3.1511

Bumbac, C., Ionescu, I. A., Tiron, O., & Badescu, V. R. (2015). Continuous flow aerobic granular sludge reactor for dairy wastewater treatment. *Water Science and Technology*, *71*(3), 440–445. doi:10.2166/wst.2015.007 PMID:25714645

Carucci, A., Milia, S., De Gioannis, G., & Piredda, M. (2008). Acetate-fed aerobic granular sludge for the degradation of chlorinated phenols. *Water Science and Technology*, *58*(2), 309–315. doi:10.2166/wst.2008.394 PMID:18701779

Carucci, A., Milia, S., De Gioannis, G., & Piredda, M. (2009). Acetate-fed aerobic granular sludge for the degradation of 4-chlorophenol. *Journal of Hazardous Materials*, *166*(1), 483–490. doi:10.1016/j.jhazmat.2008.11.049 PMID:19111986

Chiu, Z. C., Chen, M. Y., Lee, D. J., Wang, C. H., & Lai, J. Y. (2007). Oxygen diffusion in active layer of aerobic granule with step change in surrounding oxygen levels. *Water Research*, *41*(4), 884–892. doi:10.1016/j.watres.2006.11.035 PMID:17224172

Corsino, S. F., Campo, R., Di Bella, G., Torregrossa, M., & Viviani, G. (2015). Cultivation of granular sludge with hypersaline oily wastewater. *International Biodeterioration & Biodegradation*, *105*, 192–202. doi:10.1016/j.ibiod.2015.09.009

Corsino, S. F., Capodici, M., Morici, C., Torregrossa, M., & Viviani, G. (2016). Simultaneous nitritation-denitritation for the treatment of high-strength nitrogen in hypersaline wastewater by aerobic granular sludge. *Water Research*, *88*, 329–336. doi:10.1016/j.watres.2015.10.041 PMID:26512811

Cydzik-Kwiatkowska, A., & Wojnowska-Baryła, I. (2011). Nitrifying granules cultivation in a sequencing batch reactor at a low organics-to-total nitrogen ratio in wastewater. *Folia Microbiologica*, *56*(3), 201–208. doi:10.1007/s12223-011-0037-x PMID:21618047

Dai, Y., Jiang, Y., & Su, H. (2015). Influence of an aniline supplement on the stability of aerobic granular sludge. *Journal of Environmental Management*, *162*, 115–122. doi:10.1016/j.jenvman.2015.05.017 PMID:26233584

de Kreuk, M. K. (2006). *Aerobic granular sludge: scaling up a new technology*. Delft, The Netherlands: Technical University Delft.

de Kreuk, M. K., Heijnen, J. J., & van Loosdrecht, M. C. M. (2005). Simultaneous COD, nitrogen and phosphate removal by aerobic granular sludge. *Biotechnology and Bioengineering*, *90*(6), 761–769. doi:10.1002/bit.20470 PMID:15849693

de Kreuk, M. K., Kishida, N., Tsuneda, S., & van Loosdrecht, M. C. M. (2010). Behavior of polymeric substrates in an aerobic granular sludge system. *Water Research*, *44*(20), 5929–5938. doi:10.1016/j.watres.2010.07.033 PMID:20817210

Di Bella, G., & Torregrossa, M. (2013). Simultaneous nitrogen and organic carbon removal in aerobic granular sludge reactors operated with high dissolved oxygen concentration. *Bioresource Technology*, *142*, 706–713. doi:10.1016/j.biortech.2013.05.060 PMID:23751809

Di Bella, G., & Torregrossa, M. (2014). Aerobic Granular Sludge for Leachate Treatment. *Chemical Engineering Transactions*, *38*, 493–498. doi:10.3303/CET1438083

Duque, A. F. (2011).*Fluorophenol: Performance of biofilm and aerobic granular sludge reactors*. Universidade Católica Portuguesa.

Duque, A. F., Bessa, V. S., Carvalho, M. F., de Kreuk, M. K., van Loosdrecht, M. C. M., & Castro, P. M. L. (2011). 2-Fluorophenol degradation by aerobic granular sludge in a sequencing batch reactor. *Water Research*, *45*(20), 6745–6752. doi:10.1016/j.watres.2011.10.033 PMID:22060965

Figueroa, M., Val del Rio, A., Campos, J. L., Méndez, R., & Mosquera-Corral, A. (2015). Filamentous bacteria existence in aerobic granular reactors. *Bioprocess and Biosystems Engineering*, *38*(5), 841–851. doi:10.1007/s00449-014-1327-x PMID:25533039

Figueroa, M., Val Del Río, A., Campos, J. L., Mosquera-Corral, A., & Méndez, R. (2011). Treatment of high loaded swine slurry in an aerobic granular reactor. *Water Science and Technology*, *63*(9), 1808–1814. doi:10.2166/wst.2011.381 PMID:21902017

Fomina, M., & Gadd, G. M. (2014). Biosorption: Current perspectives on concept, definition and application. *Bioresource Technology*, *160*, 3–14. doi:10.1016/j.biortech.2013.12.102 PMID:24468322

Franca, R. D. G., Vieira, A., Mata, A. M. T., Carvalho, G. S., Pinheiro, H. M., & Lourenço, N. D. (2015). Effect of an azo dye on the performance of an aerobic granular sludge sequencing batch reactor treating a simulated textile wastewater. *Water Research*, *85*, 327–336. doi:10.1016/j.watres.2015.08.043 PMID:26343991

Gao, D., Liu, L., Liang, H., & Wu, W.-M. (2011). Aerobic granular sludge: Characterization, mechanism of granulation and application to wastewater treatment. *Critical Reviews in Biotechnology*, *31*(2), 137–152. doi:10.3109/07388551.2010.497961 PMID:20919817

Gao, J., Zhang, Q., Su, K., Chen, R., & Peng, Y. (2010). Biosorption of Acid Yellow 17 from aqueous solution by non-living aerobic granular sludge. *Journal of Hazardous Materials*, *174*(1-3), 215–225. doi:10.1016/j.jhazmat.2009.09.039 PMID:19783368

Gao, J. F., Zhang, Q., Su, K., & Wang, J. H. (2010). Competitive biosorption of Yellow 2G and Reactive Brilliant Red K-2G onto inactive aerobic granules: Simultaneous determination of two dyes by first-order derivative spectrophotometry and isotherm studies. *Bioresource Technology*, *101*(15), 5793–5801. doi:10.1016/j.biortech.2010.02.091 PMID:20303747

Gentry, T., Rensing, C., & Pepper, I. (2004). New Approaches for Bioaugmentation as a Remediation Technology. *Critical Reviews in Environmental Science and Technology*, *34*(5), 447–494. doi:10.1080/10643380490452362

Giesen, A., de Bruin, L. M. M., Niermans, R. P., & van der Roest, H. F. (2013). Advancements in the application of aerobic granular biomass technology for sustainable treatment of wastewater. *Water Practice and Technology*, *8*(1), 47–54. doi:10.2166/wpt.2013.007

Gobi, K., & Vadivelu, V. M. (2014). Aerobic dynamic feeding as a strategy for in situ accumulation of polyhydroxyalkanoate in aerobic granules. *Bioresource Technology*, *161*, 441–445. doi:10.1016/j.biortech.2014.03.104 PMID:24725384

Gobi, K., & Vadivelu, V. M. (2015). Polyhydroxyalkanoate recovery and effect of in situ extracellular polymeric substances removal from aerobic granules. *Bioresource Technology*, *189*, 169–176. doi:10.1016/j.biortech.2015.04.023 PMID:25889804

Hailei, W., Guosheng, L., Ping, L., & Feng, P. (2006). The effect of bioaugmentation on the performance of sequencing batch reactor and sludge characteristics in the treatment process of papermaking wastewater. *Bioprocess and Biosystems Engineering*, *29*(5-6), 283–289. doi:10.1007/s00449-006-0077-9 PMID:16983527

Işik, M., & Sponza, D. T. (2008). Anaerobic/aerobic treatment of a simulated textile wastewater. *Separation and Purification Technology*, *60*(1), 64–72. doi:10.1016/j.seppur.2007.07.043

Ivanov, V., Wang, X. H., Tay, S. T. L., & Tay, J. H. (2006). Bioaugmentation and enhanced formation of microbial granules used in aerobic wastewater treatment. *Applied Microbiology and Biotechnology*, *70*(3), 374–381. doi:10.1007/s00253-005-0088-5 PMID:16091930

Jemaat, Z., Suárez-Ojeda, M. E., Pérez, J., & Carrera, J. (2013). Simultaneous nitritation and p-nitrophenol removal using aerobic granular biomass in a continuous airlift reactor. *Bioresource Technology*, *150*, 307–313. doi:10.1016/j.biortech.2013.10.005 PMID:24177164

Jemaat, Z., Suárez-Ojeda, M. E., Pérez, J., & Carrera, J. (2014a). Partial nitritation and o-cresol removal with aerobic granular biomass in a continuous airlift reactor. *Water Research*, *48*(1), 354–362. doi:10.1016/j.watres.2013.09.048 PMID:24140352

Jemaat, Z., Suárez-Ojeda, M. E., Pérez, J., & Carrera, J. (2014b). Sequentially alternating pollutant scenarios of phenolic compounds in a continuous aerobic granular sludge reactor performing simultaneous partial nitritation and o-cresol biodegradation. *Bioresource Technology*, *161*, 354–361. doi:10.1016/j.biortech.2014.03.071 PMID:24727695

Jiang, H. L., Tay, J. H., & Tay, S. T. L. (2002). Aggregation of immobilized activated sludge cells into aerobically grown microbial granules for the aerobic biodegradation of phenol. *Letters in Applied Microbiology*, *35*(5), 439–445. doi:10.1046/j.1472-765X.2002.01217.x PMID:12390497

Kee, T. C., Bay, H. H., Lim, C. K., Muda, K., & Ibrahim, Z. (2015). Development of bio-granules using selected mixed culture of decolorizing bacteria for the treatment of textile wastewater. *Desalination and Water Treatment*, *54*(1), 132–139. doi:10.1080/19443994.2013.877853

Kocaturk, I., & Erguder, T. H. (2015). Investigation of the use of aerobic granules for the treatment of sugar beet processing wastewater. *Environmental Technology*, *36*(20), 2577–2587. doi:10.1080/09593330.2015.1039070 PMID:25851439

Kong, Q., Zhi-Bin, W., Shu, L., & Miao, M.-S. (2015). Characterization of the extracellular polymeric substances and microbial community of aerobic granulation sludge exposed to cefalexin. *International Biodeterioration & Biodegradation*, *102*, 375–382. doi:10.1016/j.ibiod.2015.04.020

Li, J., Ding, L.-B., Cai, A., Huang, G.-X., & Horn, H. (2014). Aerobic sludge granulation in a full-scale sequencing batch reactor. *BioMed Research International*. doi:10.1155/2014/268789 PMID:24822190

Li, Y., Zou, J., Zhang, L., & Sun, J. (2014). Aerobic granular sludge for simultaneous accumulation of mineral phosphorus and removal of nitrogen via nitrite in wastewater. *Bioresource Technology*, *154*, 178–184. doi:10.1016/j.biortech.2013.12.033 PMID:24388958

Lin, Y., de Kreuk, M., van Loosdrecht, M. C. M., & Adin, A. (2010). Characterization of alginate-like exopolysaccharides isolated from aerobic granular sludge in pilot-plant. *Water Research*, *44*(11), 3355–3364. doi:10.1016/j.watres.2010.03.019 PMID:20417539

Lin, Y. M., Bassin, J. P., & van Loosdrecht, M. C. M. (2012). The contribution of exopolysaccharides induced struvites accumulation to ammonium adsorption in aerobic granular sludge. *Water Research*, *46*(4), 986–992. doi:10.1016/j.watres.2011.11.072 PMID:22209260

Lin, Y. M., Nierop, K. G. J., Girbal-Neuhauser, E., Adriaanse, M., & van Loosdrecht, M. C. M. (2015). Sustainable polysaccharide-based biomaterial recovered from waste aerobic granular sludge as a surface coating material. *Sustainable Materials and Technologies*, *4*, 24–29. doi:10.1016/j.susmat.2015.06.002

Lin, Y. M., Sharma, P. K., & van Loosdrecht, M. C. M. (2013). The chemical and mechanical differences between alginate-like exopolysaccharides isolated from aerobic flocculent sludge and aerobic granular sludge. *Water Research*, *47*(1), 57–65. doi:10.1016/j.watres.2012.09.017 PMID:23084341

Lin, Y. M., Wang, L., Chi, Z. M., & Liu, X. Y. (2008). Bacterial Alginate Role in Aerobic Granular Bio-particles Formation and Settleability Improvement. *Separation Science and Technology*, *43*(7), 1642–1652. doi:10.1080/01496390801973805

Liu, H., Li, G., Li, X., & Chen, J. (2008). Molecular characterization of bacterial community in aerobic granular sludge stressed by pentachlorophenol. *Journal of Environmental Sciences (China)*, *20*(10), 1243–1249. doi:10.1016/S1001-0742(08)62216-0 PMID:19143350

Liu, L., Sheng, G.-P., Li, W.-W., Tong, Z.-H., Zeng, R. J., & Liu, J.-X. et al. (2011). Cultivation of aerobic granular sludge with a mixed wastewater rich in toxic organics. *Biochemical Engineering Journal*, *57*, 7–12. doi:10.1016/j.bej.2011.07.005

Liu, W., Zhang, J., Jin, Y., Zhao, X., & Cai, Z. (2015). Adsorption of Pb (II), Cd (II) and Zn (II) by extracellular polymeric substances extracted from aerobic granular sludge : Efficiency of protein. *Journal of Environmental Chemical Engineering*, *3*(2), 1223–1232. doi:10.1016/j.jece.2015.04.009

Liu, X., Chen, Y., Zhang, X., Jiang, X., Wu, S., & Shen, J. et al. (2015). Aerobic granulation strategy for bioaugmentation of a sequencing batch reactor (SBR) treating high strength pyridine wastewater. *Journal of Hazardous Materials*, *295*, 153–160. doi:10.1016/j.jhazmat.2015.04.025 PMID:25897697

Liu, Y. (2003). A general model for biosorption of Cd2+, Cu2+ and Zn2+ by aerobic granules. *Journal of Biotechnology*, *102*(3), 233–239. doi:10.1016/S0168-1656(03)00030-0 PMID:12730006

Liu, Y., & Xu, H. (2007). Equilibrium, thermodynamics and mechanisms of Ni2+ biosorption by aerobic granules. *Biochemical Engineering Journal*, *35*(2), 174–182. doi:10.1016/j.bej.2007.01.020

Liu, Y., Yang, S.-F., Tan, S.-F., Lin, Y.-M., & Tay, J.-H. (2002). Aerobic granules: A novel zinc biosorbent. *Letters in Applied Microbiology*, *35*(6), 548–551. doi:10.1046/j.1472-765X.2002.01227.x PMID:12460442

Liu, Y., Yang, S.-F., Xu, H., Woon, K.-H., Lin, Y.-M., & Tay, J.-H. (2003). Biosorption kinetics of cadmium(II) on aerobic granular sludge. *Process Biochemistry*, *38*(7), 997–1001. doi:10.1016/S0032-9592(02)00225-X

Liu, Y.-Q., Liu, Y., & Tay, J.-H. (2004). The effects of extracellular polymeric substances on the formation and stability of biogranules. *Applied Microbiology and Biotechnology*, *65*(2), 143–148. doi:10.1007/s00253-004-1657-8 PMID:15197510

Liu, Y.-Q., Tay, J.-H., Ivanov, V., Moy, B. Y.-P., Yu, L., & Tay, S. T.-L. (2005). Influence of phenol on nitrification by microbial granules. *Process Biochemistry*, *40*(10), 3285–3289. doi:10.1016/j.procbio.2005.03.018

Lotito, A. M., De Sanctis, M., Di Iaconi, C., & Bergna, G. (2014). Textile wastewater treatment: Aerobic granular sludge vs activated sludge systems. *Water Research*, *54*, 337–346. doi:10.1016/j.watres.2014.01.055 PMID:24583525

Luo, J., Hao, T., Wei, L., Mackey, H. R., Lin, Z., & Chen, G. H. (2014). Impact of influent COD/N ratio on disintegration of aerobic granular sludge. *Water Research*, *62*, 127–135. doi:10.1016/j.watres.2014.05.037 PMID:24950459

Ma, J., Quan, X., Yang, Z., & Li, A. (2012). Biodegradation of a mixture of 2,4-dichlorophenoxyacetic acid and multiple chlorophenols by aerobic granules cultivated through plasmid pJP4 mediated bioaugmentation. *Chemical Engineering Journal*, *181-182*, 144–151. doi:10.1016/j.cej.2011.11.041

Mañas, A., Biscans, B., Spérandio, M., Angela, M., Béatrice, B., & Mathieu, S. (2011). Biologically induced phosphorus precipitation in aerobic granular sludge process. *Water Research*, *45*(12), 3776–3786. doi:10.1016/j.watres.2011.04.031 PMID:21616518

Marques, A. P. G. C., Duque, A. F., Bessa, V. S., Mesquita, R. B. R., Rangel, A. O. S. S., & Castro, P. M. L. (2013). Performance of an aerobic granular sequencing batch reactor fed with wastewaters contaminated with Zn2+. *Journal of Environmental Management*, *128*, 877–882. doi:10.1016/j.jenvman.2013.06.052 PMID:23880431

McSwain, B. S., Irvine, R., Hausner, M., & Wilderer, P. A. (2005). Composition and Distribution of Extracellular Polymeric Substances in Aerobic Flocs and Granular Sludge. *Applied and Environmental Microbiology*, *71*(2), 1051–1057. doi:10.1128/AEM.71.2.1051-1057.2005 PMID:15691965

Mobarry, B., Wagner, M., Urbain, V., Rittmann, B., & Stahl, D. (1996). Phylogenetic probes for analyzing abundance and spatial organization of nitrifying bacteria. *Applied and Environmental Microbiology*, *62*(6), 2156–2162. PMID:8787412

Moreira, I. S., Amorim, C. L., Ribeiro, A. R., Mesquita, R. B. R., Rangel, A. O. S. S., & van Loosdrecht, M. C. M. et al. (2015). Removal of fluoxetine and its effects in the performance of an aerobic granular sludge sequential batch reactor. *Journal of Hazardous Materials*, *287*, 93–101. doi:10.1016/j.jhazmat.2015.01.020 PMID:25625631

Muda, K., Aris, A., Salim, M. R., Ibrahim, Z., Yahya, A., & van Loosdrecht, M. C. M. et al. (2010). Development of granular sludge for textile wastewater treatment. *Water Research*, *44*(15), 4341–4350. doi:10.1016/j.watres.2010.05.023 PMID:20580402

Nancharaiah, Y. V., Joshi, H. M., Hausner, M., & Venugopalan, V. P. (2008). Bioaugmentation of aerobic microbial granules with Pseudomonas putida carrying TOL plasmid. *Chemosphere*, *71*(1), 30–35. doi:10.1016/j.chemosphere.2007.10.062 PMID:18076969

Nancharaiah, Y. V., Joshi, H. M., Krishna Mohan, T. V., Venugopalan, V. P., & Narasimhan, S. V. (2008). Formation of aerobic granules in the presence of a synthetic chelating agent. *Environmental Pollution*, *153*(1), 37–43. doi:10.1016/j.envpol.2007.11.017 PMID:18166252

Othman, I., Anuar, A. N., Ujang, Z., Rosman, N. H., Harun, H., & Chelliapan, S. (2013). Livestock wastewater treatment using aerobic granular sludge. *Bioresource Technology*, *133*, 630–634. doi:10.1016/j.biortech.2013.01.149 PMID:23453799

Pronk, M., Abbas, B., Al-zuhairy, S. H. K., Kraan, R., Kleerebezem, R., & van Loosdrecht, M. C. M. (2015). Effect and behaviour of different substrates in relation to the formation of aerobic granular sludge. *Applied Microbiology and Biotechnology*, *99*(12), 5257–5268. doi:10.1007/s00253-014-6358-3 PMID:25616527

Pronk, M., de Kreuk, M. K., de Bruin, B., Kamminga, P., Kleerebezem, R., & van Loosdrecht, M. C. M. (2015a). Full scale performance of the aerobic granular sludge process for sewage treatment. *Water Research*, *84*, 207–217. doi:10.1016/j.watres.2015.07.011 PMID:26233660

Pronk, M., de Kreuk, M. K., de Bruin, B., Kamminga, P., Kleerebezem, R., & van Loosdrecht, M. C. M. (2015b). Full scale performance of the aerobic granular sludge process for sewage treatment. *Water Research*, *84*, 207–217. doi:10.1016/j.watres.2015.07.011 PMID:26233660

Quan, X., Ma, J., Xiong, W., & Wang, X. (2015). Bioaugmentation of half-matured granular sludge with special microbial culture promoted establishment of 2,4-dichlorophenoxyacetic acid degrading aerobic granules. *Bioprocess and Biosystems Engineering*, *38*(19), 1081–1090. doi:10.1007/s00449-014-1350-y PMID:25567414

Quan, X., Tang, H., & Ma, J. (2011). Effects of gene augmentation on the removal of 2,4-dichlorophenoxyacetic acid in a biofilm reactor under different scales and substrate conditions. *Journal of Hazardous Materials*, *185*(2-3), 689–695. doi:10.1016/j.jhazmat.2010.09.073 PMID:20951494

Quan, X., Tang, H., Xiong, W., & Yang, Z. (2010). Bioaugmentation of aerobic sludge granules with a plasmid donor strain for enhanced degradation of 2, 4-dichlorophenoxyacetic acid. *Journal of Hazardous Materials*, *179*(1-3), 1136–1142. doi:10.1016/j.jhazmat.2010.04.002 PMID:20430519

Ramos, C., Suárez-Ojeda, M. E., & Carrera, J. (2015). Long-term impact of salinity on the performance and microbial population of an aerobic granular reactor treating a high-strength aromatic wastewater. *Bioresource Technology*, *198*, 844–851. doi:10.1016/j.biortech.2015.09.084 PMID:26457833

Ramos, C., Suárez-Ojeda, M. E., & Carrera, J. (2016a). Biodegradation of a high-strength wastewater containing a mixture of ammonium, aromatic compounds and salts with simultaneous nitritation in an aerobic granular reactor. *Process Biochemistry*, *51*(3), 399–407. doi:10.1016/j.procbio.2015.12.020

Ramos, C., Suárez-Ojeda, M. E., & Carrera, J. (2016b). Long-term performance and stability of a continuous granular airlift reactor treating a high-strength wastewater containing a mixture of aromatic compounds. *Journal of Hazardous Materials*, *303*, 154–161. doi:10.1016/j.jhazmat.2015.10.031 PMID:26530892

Rezaei, L. S., Ayati, B., & Ganjidoust, H. (2012). Cultivation of aerobic granules in a novel configuration of sequencing batch airlift reactor. *Environmental Technology*, *33*(19-21), 2273–2280. doi:10.1080/09593330.2012.665490 PMID:23393968

Rosman, N. H., Nor Anuar, A., Othman, I., Harun, H., Sulong Abdul Razak, M. Z., & Elias, S. H. et al. (2013). Cultivation of aerobic granular sludge for rubber wastewater treatment. *Bioresource Technology, 129*, 620–623. doi:10.1016/j.biortech.2012.12.113 PMID:23317554

Schwarzenbeck, N., Borges, J. M., & Wilderer, P. (2005). Treatment of dairy effluents in an aerobic granular sludge sequencing batch reactor. *Applied Microbiology and Biotechnology, 66*(6), 711–718. doi:10.1007/s00253-004-1748-6 PMID:15558277

Schwarzenbeck, N., Erley, R., Mc Swain, B. S., Wilderer, P., & Irvine, R. L. (2004). Treatment of malting wastewater in a granular sludge sequencing batch reactor (SBR). *Acta Hydrochimica et Hydrobiologica, 32*(1), 16–24. doi:10.1002/aheh.200300517

Sheng, G.-P., Yu, H.-Q., & Li, X.-Y. (2010). Extracellular polymeric substances (EPS) of microbial aggregates in biological wastewater treatment systems: A review. *Biotechnology Advances, 28*(6), 882–894. doi:10.1016/j.biotechadv.2010.08.001 PMID:20705128

Shi, Y., Xing, S., Wang, X., & Wang, S. (2013). Changes of the reactor performance and the properties of granular sludge under tetracycline (TC) stress. *Bioresource Technology, 139*, 170–175. doi:10.1016/j.biortech.2013.03.037 PMID:23660380

Shi, Y.-J., Wang, X.-H., Qi, Z., Diao, M.-H., Gao, M.-M., & Xing, S.-F. et al. (2011). Sorption and biodegradation of tetracycline by nitrifying granules and the toxicity of tetracycline on granules. *Journal of Hazardous Materials, 191*(1-3), 103–109. doi:10.1016/j.jhazmat.2011.04.048 PMID:21570181

Show, K.-Y., Lee, D.-J., & Tay, J.-H. (2012). Aerobic granulation: Advances and challenges. *Applied Biochemistry and Biotechnology, 167*(6), 1622–1640. doi:10.1007/s12010-012-9609-8 PMID:22383048

Sun, X.-F., Wang, S.-G., Liu, X.-W., Gong, W.-X., Bao, N., Gao, B.-Y., & Zhang, H.-Y. (2008). Biosorption of Malachite Green from aqueous solutions onto aerobic granules: Kinetic and equilibrium studies. *Bioresource Technology, 99*(9), 3475–3483. doi:10.1016/j.biortech.2007.07.055 PMID:17855080

Tay, S. T., Zhuang, W.-Q., & Tay, J.-H. (2005). Start-up, microbial community analysis and formation of aerobic granules in a tert-butyl alcohol degrading sequencing batch reactor. *Environmental Science & Technology, 39*(15), 5774–5780. doi:10.1021/es050278x PMID:16124314

Tay, S. T.-L., Moy, B. Y.-P., Jiang, H.-L., & Tay, J.-H. (2005). Rapid cultivation of stable aerobic phenol-degrading granules using acetate-fed granules as microbial seed. *Journal of Biotechnology, 115*(4), 387–395. doi:10.1016/j.jbiotec.2004.09.008 PMID:15639100

Wang, L., Wan, C., Lee, D.-J., Liu, X., Zhang, Y., Chen, X. F., & Tay, J.-H. (2014). Biosorption of antimony(V) onto Fe(III)-treated aerobic granules. *Bioresource Technology, 158*(V), 351–354. doi:10.1016/j.biortech.2014.02.046 PMID:24630496

Wang, L., Wan, C., Lee, D.-J., Tay, J.-H., Chen, X. F., Liu, X., & Zhang, Y. (2013). Adsorption–desorption of strontium from waters using granules. *Journal of the Taiwan Institute of Chemical Engineers, 44*(3), 454–457. doi:10.1016/j.jtice.2012.12.005

Wang, L., Zheng, P., Xing, Y., Li, W., Yang, J., & Abbas, G. et al. (2014). Effect of particle size on the performance of autotrophic nitrogen removal in the granular sludge bed reactor and microbiological mechanisms. *Bioresource Technology*, *157*, 240–246. doi:10.1016/j.biortech.2014.01.116 PMID:24561629

Wang, S.-G., Liu, X.-W., Gong, W.-X., Gao, B.-Y., Zhang, D.-H., & Yu, H.-Q. (2007). Aerobic granulation with brewery wastewater in a sequencing batch reactor. *Bioresource Technology*, *98*(11), 2142–2147. doi:10.1016/j.biortech.2006.08.018 PMID:17071083

Wang, S.-G., Liu, X.-W., Zhang, H.-Y., Gong, W.-X., Sun, X.-F., & Gao, B.-Y. (2007). Aerobic granulation for 2,4-dichlorophenol biodegradation in a sequencing batch reactor. *Chemosphere*, *69*(5), 769–775. doi:10.1016/j.chemosphere.2007.05.026 PMID:17617438

Wang, X.-H., Gai, L.-H., Sun, X.-F., Xie, H.-J., Gao, M.-M., & Wang, S.-G. (2010). Effects of long-term addition of Cu(II) and Ni(II) on the biochemical properties of aerobic granules in sequencing batch reactors. *Applied Microbiology and Biotechnology*, *86*(6), 1967–1975. doi:10.1007/s00253-010-2467-9 PMID:20169340

Wang, Z., Liu, L., Yao, J., & Cai, W. (2006). Effects of extracellular polymeric substances on aerobic granulation in sequencing batch reactors. *Chemosphere*, *63*(10), 1728–1735. doi:10.1016/j.chemosphere.2005.09.018 PMID:16288800

Wei, D., Li, M., Wang, X., Han, F., Li, L., & Guo, J. et al. (2016). Extracellular polymeric substances for Zn (II) binding during its sorption process onto aerobic granular sludge. *Journal of Hazardous Materials*, *301*, 407–415. doi:10.1016/j.jhazmat.2015.09.018 PMID:26410269

Wei, D., Shi, L., Yan, T., Zhang, G., Wang, Y., & Du, B. (2014). Aerobic granules formation and simultaneous nitrogen and phosphorus removal treating high strength ammonia wastewater in sequencing batch reactor. *Bioresource Technology*, *171*, 211–216. doi:10.1016/j.biortech.2014.08.001 PMID:25203228

Wei, D., Wang, B., Ngo, H. H., Guo, W., Han, F., & Wang, X. et al. (2015). Role of extracellular polymeric substances in biosorption of dye wastewater using aerobic granular sludge. *Bioresource Technology*, *185*, 14–20. doi:10.1016/j.biortech.2015.02.084 PMID:25746473

Wei, D., Wang, Y., Wang, X., Li, M., Han, F., & Ju, L. et al. (2015). Toxicity assessment of 4-chlorophenol to aerobic granular sludge and its interaction with extracellular polymeric substances. *Journal of Hazardous Materials*, *289*, 101–107. doi:10.1016/j.jhazmat.2015.02.047 PMID:25710820

Winkler, M.-K. H. (2012). *Segregation of biomass in aerobic granular sludge*. Delft, The Netherlands: Technical University Delft.

Xia, Z., Xiao-chun, W., Zhong-lin, C., Hao, X., & Qing-fang, Z. (2014). Microbial community structure and pharmaceuticals and personal care products removal in a membrane bioreactor seeded with aerobic granular sludge. *Applied Microbiology and Biotechnology*, *99*(1), 425–433. doi:10.1007/s00253-014-5984-0 PMID:25099174

Xu, H., Liu, Y., & Tay, J.-H. (2006). Effect of pH on nickel biosorption by aerobic granular sludge. *Bioresource Technology*, *97*(3), 359–363. doi:10.1016/j.biortech.2005.03.011 PMID:15905090

Yi, S., Zhuang, W.-Q., Wu, B., Tay, S. T.-L., & Tay, J.-H. (2006). Biodegradation of p-nitrophenol by aerobic granules in a sequencing batch reactor. *Environmental Science & Technology*, *40*(7), 2396–2401. doi:10.1021/es0517771 PMID:16646480

Yilmaz, G., Lemaire, R., Keller, J., & Yuan, Z. (2008). Simultaneous nitrification, denitrification, and phosphorus removal from nutrient-rich industrial wastewater using granular sludge. *Biotechnology and Bioengineering*, *100*(3), 529–541. doi:10.1002/bit.21774 PMID:18098318

Zhang, H., He, Y., Jiang, T., & Yang, F. (2011). Research on characteristics of aerobic granules treating petrochemical wastewater by acclimation and co-metabolism methods. *Desalination*, *279*(1-3), 69–74. doi:10.1016/j.desal.2011.05.060

Zhang, L. L., Chen, J. M., & Fang, F. (2008). Biodegradation of methyl t-butyl ether by aerobic granules under a cosubstrate condition. *Applied Microbiology and Biotechnology*, *78*(3), 543–550. doi:10.1007/s00253-007-1321-1 PMID:18183384

Zhang, L.-L., Feng, X. X., Xu, F., Xu, S., & Cai, W. M. (2005). Biosorption of Rare Earth Metal Ion on Aerobic Granules. *Journal of Environmental Science and Health. Part A*, *40*(4), 857–867. doi:10.1081/ESE-200048290

Zhang, Q., Hu, J., & Lee, D. (2016). Aerobic granular processes : Current research trends. *Bioresource Technology*, *210*, 74–80. doi:10.1016/j.biortech.2016.01.098 PMID:26873285

Zhao, D., Liu, C., Zhang, Y., & Liu, Q. (2011). Biodegradation of nitrobenzene by aerobic granular sludge in a sequencing batch reactor (SBR). *Desalination*, *281*, 17–22. doi:10.1016/j.desal.2011.07.037

Zhao, X., Chen, Z., Wang, X., Li, J., Shen, J., & Xu, H. (2015). Remediation of pharmaceuticals and personal care products using an aerobic granular sludge sequencing bioreactor and microbial community profiling using Solexa sequencing technology analysis. *Bioresource Technology*, *179*, 104–112. doi:10.1016/j.biortech.2014.12.002 PMID:25531682

Zhao, Y., Huang, J., Zhao, H., & Yang, H. (2013). Microbial community and N removal of aerobic granular sludge at high COD and N loading rates. *Bioresource Technology*, *143*, 439–446. doi:10.1016/j.biortech.2013.06.020 PMID:23827440

Zheng, X., He, Y., Chen, W., Wang, M., Cao, S., Ni, M., & Chen, Y. (2016). A comparative adsorption study : 17β-estradiol onto aerobic granular sludge and activated sludge. *Environmental Technology*, *37*(1), 136–144. doi:10.1080/09593330.2015.1065004 PMID:26209151

Zhou, D., Liu, M., Wang, J., Dong, S., Cui, N., & Gao, L. (2013). Granulation of activated sludge in a continuous flow airlift reactor by strong drag force. *Biotechnology and Bioprocess Engineering*, *18*(2), 289–299. doi:10.1007/s12257-012-0513-4

Zhou, D., Niu, S., Xiong, Y., Yang, Y., & Dong, S. (2014). Microbial selection pressure is not a prerequisite for granulation: Dynamic granulation and microbial community study in a complete mixing bioreactor. *Bioresource Technology*, *161*, 102–108. doi:10.1016/j.biortech.2014.03.001 PMID:24690580

Zhu, L., Lv, M., Dai, X., Zhou, J.-H., & Xu, X.-Y. (2013). The stability of aerobic granular sludge under 4-chloroaniline shock in a sequential air-lift bioreactor (SABR). *Bioresource Technology*, *140*, 126–130. doi:10.1016/j.biortech.2013.04.017 PMID:23685649

Zhu, L., Xu, X., Luo, W., Cao, D., & Yang, Y. (2008). Formation and microbial community analysis of chloroanilines-degrading aerobic granules in the sequencing airlift bioreactor. *Journal of Applied Microbiology*, *104*(1), 152–160. doi:10.1111/j.1365-2672.2007.03548.x PMID:17887986

Zhu, L., Xu, X., Luo, W., Tian, Z., Lin, H., & Zhang, N. (2008). A comparative study on the formation and characterization of aerobic 4-chloroaniline-degrading granules in SBR and SABR. *Applied Microbiology and Biotechnology*, *79*(5), 867–874. doi:10.1007/s00253-008-1476-4 PMID:18449538

# Chapter 10
# Anammox Process: Technologies and Application to Industrial Effluents

**Ángeles Val del Río**
*Universidade de Santiago de Compostela, Spain*

**Elisa Amanda Giustinianovich**
*University of Concepción, Chile*

**Alba Pedrouso Fuentes**
*Universidade de Santiago de Compostela, Spain*

**José Luis Campos Gomez**
*University Adolfo Ibáñez, Chile*

**Anuska Mosquera-Corral**
*Universidade de Santiago de Compostela, Spain*

## ABSTRACT

*Application of anammox based processes is nowadays an efficient way to remove nitrogen from wastewaters, being good alternative to the conventional nitrification-denitrification process. This chapter reviews the possible configurations to apply the anammox process, being special attention to the previous partial nitritation, necessary to obtain the adequate substrates for anammox bacteria. Furthermore a description of the main technologies developed and patented by different companies was performed, with focus on the advantages and bottlenecks of them. These technologies are classified in the chapter based on the type of biomass: suspended, granular and biofilm. Also a review is presented for the industrial applications (food industry, agricultural wastes, landfill leachates, electronic industry, etc.), taking into account full scale experiences and laboratory results, as well as microbiology aspects respect to the anammox bacteria genera involved. Finally the possibility to couple nitrogen removal, by anammox, with phosphorus recovery, by struvite precipitation, is also evaluated.*

## INTRODUCTION

The anaerobic ammonium oxidation (anammox) process is the anaerobic oxidation of ammonium (van de Graaf, de Bruijn, Robertson, Jetten, & Kuenen, 1996) using nitrite as electron acceptor and carbonate for cell biosynthesis, according to the stoichiometry described by (Strous, Heijnen, Kuenen, & Jetten,

DOI: 10.4018/978-1-5225-1037-6.ch010

1998) (Equation 1) and recalculated latter by a kinetic characterization experiment with a suspended anammox culture by (Lotti, Kleerebezem, Lubello, & van Loosdrecht, 2014) (Equation 2):

$$NH_4^+ + 1.32\ NO_2^- + 0.066\ HCO_3^- + 0.13H^+ \rightarrow \\ 1.02\ N_2 + 0.26NO_3^- + 0.066\ CH_2O_{0.5}N_{0.15} + 2.03\ H_2O \tag{1}$$

$$NH_4^+ + 1.146\ NO_2^- + 0.071\ HCO_3^- + 0.057H^+ \rightarrow \\ 0.986\ N_2 + 0.161\ NO_3^- + 0.071\ CH_{1.74}O_{0.31}N_{0.20} + 2.002\ H_2O \tag{2}$$

This reaction is carried out by anammox bacteria, belonging to Planctomycetes phylum, including the genera *Brocadia*, *Kuenenia*, *Anammoxoglobus*, *Jettenia* and *Scalindua* (Jetten et al., 2009). According to Equation 1 and Equation 2, in order to treat ammonia rich effluents with the anammox process, half of the incoming nitrogen needs to be oxidized to nitrite. The latter process is called nitritation (Equation 3), and it is performed by the ammonia oxidizing bacteria (AOB).

$$NH_4^+ + 0.0743\ HCO_3^- + 1.404\ O_2 \rightarrow \\ 0.985\ NO_2^- + 0.0149\ C_5H_7O_2N + 1.911\ H^+ + 1.03\ H_2O \tag{3}$$

The fully autotrophic nitrogen removal from wastewater can be achieved by combining the appropriate ratio of nitritation (partial nitritation) with the anammox process, given the overall reaction as (Equation 4):

$$NH_4^+ + 0.804\ O_2 + 0.071\ HCO_3^- \rightarrow \\ 0.111\ NO_3^- + 0.436\ N_2 + 0.009\ C_5H_7O_2N + 0.028\ CH_2O_{0.5}N_{0.15} + 1.038\ H^+ + 1.46\ H_2O \tag{4}$$

Compared to conventional nitrification-denitrification processes, the fully autotrophic nitrogen removal allows for the saving of 100% organic matter, for a potential biogas production by anaerobic digestion (Wett, 2007), and the reduction of operational costs if the effluent COD/N ratio is lower than 5 g/g (Fux & Siegrist, 2004). Also, oxygen requirements are reduced around 60% (Ahn, 2006), sludge production around 90% (Vazquez-Padin et al., 2014) and $N_2O$ production around 83% (Morales et al., 2015b) of the values required for the conventional processes. This costs reduction makes the technologies based on the anammox process attractive for the industrial sector.

However, some questions have to be taken into account like the complexity of wastewater from industrial origin which makes its treatment more difficult. Furthermore, the presence of toxic compounds and the potential lack of nutrients and trace elements necessary for the growth of microorganisms are further factors to be considered.

Up to date, the anammox based process has been successfully applied for the treatment of different types of wastewater with high ammonium content such as landfill leachate, manure or food industry effluents. However, many other industrial sectors may benefit from the application of this technology and research is ongoing to broaden the field of applicability especially focusing on those industrial effluents whose production is increasing (e.g. the electronic industry).

The application of different configurations and technologies that utilize partial nitritation and anammox to achieve an efficient nitrogen removal from wastewater is under study. The purpose of the present chapter is to review existing technologies, paying special attention to the treatment of industrial effluents, currently applied at full scale and to study their future evolution.

## SYSTEMS CONFIGURATIONS

In order to autotrophic remove the nitrogen via anammox based processes, since nitrogen is present in the wastewater mainly as ammonium, half of the influent ammonia needs to be oxidized in a partial nitration step. The fully autotrophic nitrogen removal can be performed in two different configurations:

1. Two reactor systems, the first for partial nitritation (PN) (Equation 3) and the second one for the anammox process (Equation 1 and 2) or
2. In a single reactor by coupling both reactions (Equation 4) by means of controlling operational conditions, especially the dissolved oxygen (DO) concentration.

An overview of the main operational advantages of each configuration is presented in Table 1.

In the two-reactor configuration the anammox process strongly depends on the previous PN process. The first and more spread technology based on two steps partial nitritation-anammox, available in full-scale, is called SHARON (Single reactor system for High activity Ammonium Removal Over Nitrite) (Hellinga, Schellen, Mulder, van Loosdrecht, & Heijnen, 1998). The SHARON process bases the production of a suitable substrate composition (ammonia/nitrite ratio close to 1 g N/g N), to feed the next anammox unit, by operating at temperatures around 30°C, equal hydraulic and solid retention times (HRT and SRT), around 1 day, controlled DO concentration and ammonia and free nitrous acid inhibition level by means of the pH value. The process takes advantage of the faster growth rate of AOB over NOB (Nitrite Oxidizing Bacteria) at the working temperature to wash out NOB while AOB remain inside the system (van Dongen, Jetten, & van Loosdrecht, 2001; Van Loosdrecht, & Jetten, 1998).

The single reactor configuration is the most frequently applied at industrial scale basically because it involves less complex control systems and lower investment costs due to the use of only one reaction vessel (Vlaeminck, De Clippeleir, & Verstraete, 2012) as mentioned in Table 1. Processes performed at single stage were initially called as: CANON (Completely Autotrophic Nitrogen removal Over Ni-

*Table 1. Overview of the main operational advantages of the two and single stages configurations for the fully autotrophic nitrogen removal*

| Two Stages Configuration | Single Stage Configuration |
|---|---|
| Each stage operation can be optimized individually | Simplification of reactor control and operation, lower investment costs |
| More effective nitratation suppression in the first stage | Continuous depletion of nitrite, avoiding toxic concentrations for anammox |
| Influent COD oxidation is performed in the first stage, avoiding anammox to be overgrown by heterotrophs | Lower $N_2O$ emissions compared to two-stage (Vlaeminck et al., 2012) |
| Lower risk of dissolved oxygen inhibition over anammox | |

trite process), OLAND (Oxygen-Limited Autotrophic Nitrification–Denitrification), aerobic/anoxic deammonification and the SNAP process (Single-stage Nitrogen removal using Anammox and Partial nitritation), differentiated according to the organisms that were originally assumed to be responsible for anaerobic ammonium oxidation, now confirmed as anammox bacteria by molecular analysis (Van Hulle et al., 2010). This process performs two sequential reactions in a single reactor by establishing a cooperative bacterial consortium between aerobic and anaerobic ammonium oxidizing bacteria (Ahn, 2006). In the two-reactors configuration the separation between both processes is spatial, while in the one stage configuration can be applied a separation by:

1. Time, with alternation of aerobic and anoxic periods; or
2. Space by the use of biomass that grows as biofilm, where aerobic nitrifiers oxidize ammonia to nitrite and consume oxygen, creating anoxic conditions in the deeper zones of the biofilm.

Furthermore, the two strategies can be combined to take advantage of both mechanisms, applying alternating aerobic/anoxic periods with the use of biofilms. A general scheme of both configurations is presented in Figure 1.

Anammox bacteria are slow growing microorganisms, with reported doubling time of 3.3 to11 days (Lotti et al., 2014; Strous et al., 1998). Therefore, efficient biomass retention methods must be considered in the anammox reactor design process. The retention can be achieved by the use of carrier materials where biomass grows attached in the form of biofilm (Fernandez, Vazquez-Padin, Mosquera-Corral, Campos, & Mendez, 2008) or the promotion of the microbial self-aggregation in the form of granular biomass (Vázquez-Padín et al., 2009a). In the case of the latter, AOB grow located in the surface and anammox bacteria in the internal zones of the granules. In this way, AOB protect anammox bacteria from high DO levels, which are inhibitory (Varas et al., 2015). In these systems the control of the aera-

*Figure 1. Most applied configurations for autotrophic nitrogen removal via Anammox*

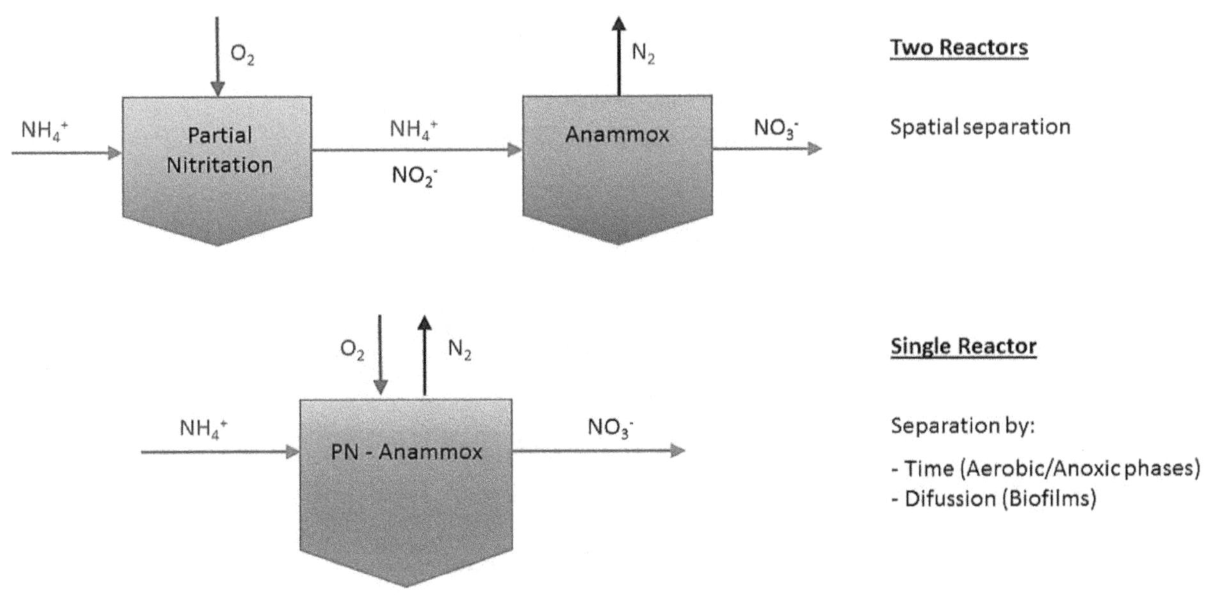

tion intensity or the DO concentration allows limiting the production of nitrite to avoid its accumulation inside the system up to inhibitory concentrations for the anammox bacteria (Dapena-Mora et al., 2007).

In these situations the process is quite sensitive to applied nitrogen-surface load, biofilm thickness, temperature, being the dissolved oxygen concentration the most important parameter (Ahn, 2006). DO mass transfer resistance is normally limiting the efficiency of the systems, where the oxygen transfer inside the biofilm/granule and the gas–liquid transfer control the overall nitrogen removal rate (Van Hulle et al., 2010). A co-culture of aerobic and anaerobic ammonium oxidizing bacteria in biofilm systems can be established under controlled oxygen-limited conditions (Khin & Annachhatre, 2004).

Industrial effluents characterized by high ammonia concentrations contain in many cases organic matter concentrations that can negatively affect the activity of the anammox bacteria (Chen, Liu, Yang, Xue, & Wang, 2009; Liang et al., 2014; Molinuevo, García, Karakashev, & Angelidaki, 2009; Mosquera-Corral, González, Campos, & Méndez, 2005; Vazquez-Padin et al., 2009b). The heterotrophic bacteria, which use this organic matter as substrate, compete with autotrophic bacteria, since they have lower duplication time overgrow the autotrophs in the bulk or in the surface of granules and biofilms, and have higher affinity for substrates like nitrite (Liang et al., 2014; Zhang, Li, Chen, Wang, & Bao, 2012). Recently, a process was developed that takes advantage of heterotrophic denitrification to improve nitrogen removal efficiency in anammox based reactors. In the SNAD process (Simultaneous Nitrification, Anammox and Denitrification) (Chen et al., 2009) the nitrate generated by anammox serves as an electron acceptor to oxidize organic carbon (Figure 2). Therefore, this system requires an operation enabling the balance among the different bacterial population activities. The operational parameter requirements are

*Figure 2. Diagram of coupled reactions of partial nitritation-anammox-denitrification in a granular system. Numbers indicate molar quantities.*

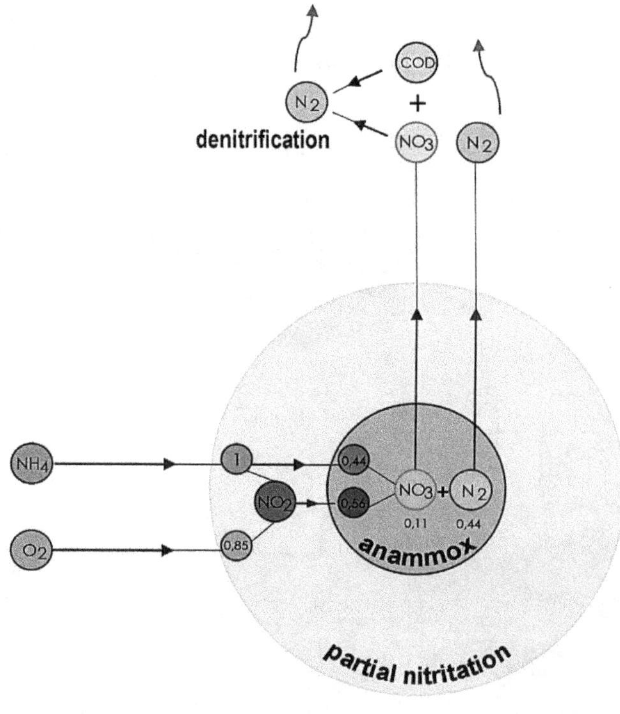

quite similar to those for other partial nitritation/anammox based processes, but its operational limits to maintain high nitrogen removal efficiency have not yet been defined.

Based on the different possible configurations to perform the autotrophic nitrogen removal via anammox process, a significant amount of technologies have been or are under development by companies around the world, and many of them have already been implemented at full scale. A description of the main characteristics of these technologies is presented in the following section.

## DEVELOPED TECHNOLOGIES

Diverse technologies have been already implemented at full scale to carry out the partial nitrification and anammox processes on effluents rich in ammoniacal nitrogen, either in municipal or industrial facilities (Abma, Schultz, Mulder, & van Loosdrecht, 2007; Lackner et al., 2014). They differ mainly on the procedures followed to develop and retain ammonia oxidizing and anammox bacteria: number of stages, strategy of feeding and/or aeration, implemented control strategies, etc. Existing technologies include reactors with:

1. Biomass in suspension,
2. Granular biomass and
3. Biofilm biomass.

According to a recent review of these technologies (Lackner et al., 2014) nowadays there are around 100 full scale plants of partial nitritation-anammox (PN/AA) in operation around the world. In all the available technologies the ammonia and total nitrogen removal efficiencies are around 90-95% and 80-85%, respectively.

### Technologies Based on Suspended Biomass

#### DEMON® Process

The DEMON® process, acronym of DE-amMONnification (World Water Works, Inc.) is based on the use of one reactor configuration for the partial nitritation-anammox processes. The type of reactor used is a sequencing batch reactor (SBR) that includes a hydrocyclone. In the hydrocyclone the heavier biomass fraction (granules composed mainly of anammox bacteria) is separated and reintroduced inside the reactor, whereas the small flocs (heterotrophs, AOB and NOB) are purged. With this strategy different sludge retention times (SRT) are achieved avoiding the NOB growth while enriching the anammox biomass.

The control and stability of the DEMON® process, to balance the activities of the AOB and anammox bacteria, has been defined based on three elements: the time for the reactor filling and withdrawal, the DO concentration and the pH value (O'Shaughnessy et al., 2008). The DO concentration is maintained at low values (around 0.3 mg $O_2$/L) to avoid a fast production of nitrite by the AOB and to repress the growth of NOB and the anammox inhibition by oxygen. The SBR operation comprises mini cycles of feeding, aeration and mixing without aeration. The mini cycles are repeated multiple times until the complete reaction and then the settling and withdrawal take place. The total length of the cycle can range

*Table 2. Number of full scale installations for the different technologies in operation in municipal and industrial WWTPs*

| Name of the Technology | Type of Biomass | Type of Reactor [a] | Municipal WWTP [b] Applications | Industrial WWTP Applications |
|---|---|---|---|---|
| **DEMON®** | hybrid (flocs and granules) | SBR with hydrocyclone | 34 | 8 |
| **Cleargreen™** | suspended | SBR | 2 | 0 |
| **NAS®** | suspended | 4 stages | 1 | 4 |
| **PN/AA (Eawag)** | suspended | SBR | 5 | 0 |
| **PANDA+** | suspended | 2 stages | 2 | 0 |
| **ANAMMOX®** | granular | 2 stages (IC [c])<br>1 stage (Continuous) | 5 | 16 |
| **ELAN®** | granular | SBR | 1 | 1 |
| **DeAmmon®** | biofilm | MBBR | 3 | 0 |
| **ANITA™MOX** | biofilm | MBBR | 5 | 1 |
| **TERRA-N®** | biofilm | 1 or 2 stages SBR | 4 | 0 |
| **OLAND** | biofilm | RBC | 3 | 3 |
| **Total** | | | 65 | 33 |

[a]All the technologies use one stage configuration, unless the contrary is indicated

[b]The anammox based process is treating the reject water from the anaerobic sludge digester.

[c]The 2 stages Anammox process combined SHARON and internal circulator (IC)-ANAMMOX process.

DEMON: DE-amMONnification; Cleargreen: Cyclic Low Energy Ammonium Removal; NAS: New Activated Sludge; PN/AA: Partial Nitritation and Anaerobic Ammonia Oxidation; PANDA: Partial Augmented Nitritation Denitritation Alkalinity recovery; ANAMMOX: Anaerobic AMmonium Oxidation; ELAN: ELiminación Autótrofa de Nitrógeno (autotrophic nitrogen removal); DeAmmon: DeAmmonification; OLAND: Oxygen-Limited Autotrophic Nitrification-Denitrification. SBR: Sequencing Batch Reactor; MBBR: Moving Bed BioReactor; RBC: Rotating Biological Contactor.

(adapted from Lackner et al. (2014))

between 6 to 12 hours. The length of the mini cycles is controlled based on the pH value. While the ammonia oxidation causes a decrease of the pH during the aeration period, the anammox process increases it during the non-aerated one. The aeration starts when the pH value reaches the upper threshold (7.02) and finishes when it reaches the lower threshold (7.01), then the pH value varies in a narrow range of 0.01 units (Wett, 2007). However, in practice in the WWTP of Plettenberg (Germany) NOB proliferation appeared, which could not be removed by applying this strategy and involved the implementation of an additional control based on the modification of the length of the aerobic and anoxic phases. This strategy intends to profit from the longer response time of NOB (lag phase) than AOB to restore their activity after being exposed to anoxic periods (Jardin & Hennerkes, 2012).

## ClearGreen® Process

The ClearGreen® process has been developed by Suez Environnement. It is a SBR system provided with continuous monitoring (ammonium, nitrite, nitrate and DO) and dynamic regulation of the aeration flow. The reactor operates in cycles of around 8 hours in accordance with a patented procedure comprising: stage 1 (feeding, mixing, aerobic period and anoxic period), stage 2 (settling) and stage 3 (withdrawal)

During the stage 1 several sub-cycles of feeding, aerobic and anoxic periods take place repeatedly before the beginning of the following stages 2 and 3. The feeding period takes place in a short time. Then the aerobic period, where the DO concentration is controlled at levels of 0.3-0.8 mg $O_2$/L, lasts roughly the 60% of the reaction time. During the aerobic period the production of nitrite by AOB is higher than its consumption by the anammox bacteria, consequently it is accumulated in the system. Then, in the anoxic period, which lasts around the 40% of the reaction time, the anammox bacteria consume this nitrite together with the remaining ammonia. The alternation of aerobic/anoxic periods allows reaching efficiencies of nitrogen removal of around 90%, probably thanks to the nitrate removal that occurs during the anoxic periods.

The sub-cycles with aerated and non-aerated periods help to ensure alternating conditions favorable for the development of AOB, followed by conditions more favorable to anammox bacteria. In this way, the NOB are suppressed by creating an inhibitory environment through controlled DO and free ammonia concentrations (McQuarrie, Johnson, Lu, & Zhao, 2015).

## NAS® Process

This process was developed by Hydroitalia-Colsen and named as New Activated Sludge (NAS®) system. It consists of an activated sludge process where a fraction of the biomass is constituted by anammox bacteria. NAS® applications are often combined with membrane bio-reactor (MBR) systems in order to obtain a higher final effluent quality (Colsen, n.d). An industrial experience of NAS® technology has been described by Desloover *et al.* (2011) as a floc-based nitrogen removal process based on the control of sludge retention time (SRT) and DO levels. This process consists of the combination of the batch-fed partial nitritation, anammox, denitrification and nitrification reactors in a four-stage configuration plant with internal recycling lines. High nitrogen removal percentages (95%) were achieved treating digested industrial wastewater, containing 264 mg N/L and 1154 mg COD/L and operated at an overall SRT of 46 d. However, high $N_2O$ (a powerful green house) emissions were recorded being necessary to reduce them in a 50% to reach a $CO_2$ neutral footprint facility.

## Partial Nitritation/Anammox (PN/AA) Process from EAWAG

The Eawag (Switzerland) developed a one stage partial nitritation/anammox process carried out in a SBR. The cycle length (between 6-10 h) is controlled by the evolution of the ammonia concentration (measured online) and it comprises five different phases: feeding, aeration, mixing, sedimentation and discharge. The feeding can be done at the beginning of the cycle or during the aeration phase being the latter strategy the preferred one. The volumetric air supply needs to be controlled and it is linked to the AOB activity in order to allow the simultaneous partial nitritation and anammox process to take place whereas NOB activity is suppressed, resulting in DO concentrations lower than 0.1 mg $O_2$/L.

The washout of the NOB is achieved by imposing a sludge age of around 45 days by means of the application of sludge purges. After two years of stable operation, nitrate build up (due to NOB development) was observed in one of the two parallel reactors at Zürich WWTP due to the supply of extra aeration periods. The aeration control has been identified as a key parameter of the process as complete anammox inhibition was observed at 0.2 mg $O_2$/L (Joss et al., 2011). There are already five full scale municipal WWTPs in operation at the moment, all of them located in Switzerland.

## PANDA+ Process

The PANDA+ process was developed by Aquaconsult Ingenieur GmbH in collaboration with the Institute for Water Quality and Waste Management (ISAH). This was initially developed as a two or more stages process and named PANDA (Partial Augmented Nitritation-Denitritation with Alkalinity recovery), where ammonia is oxidized to nitrite (partial nitritation) and then denitritation is promoted. The addition of an external carbon source was required to enhance the denitritation process. Two full scale plants were operated with this technology and then they were converted into a partial nitritation-anammox system named PANDA+, without the necessity to add external carbon source. The system comprised a partial nitritation aerated reactor and an anoxic mixed unit for anammox process. The NOB suppression was based on SRT control and free ammonia inhibition.

# Technologies Based on Granular Biomass

## ANAMMOX® Process

The first ANAMMOX® reactor designed and built in 2001 was conceived to treat the effluent from the SHARON process (Abma et al., 2007b; Mulder, van Loosdrecht, Hellinga, & van Kempen, 2001), and operated as a two-units system. However, since 2006 the company, Paques, designed the system as a one stage configuration due to the lower investment costs compared to the two stages configuration.

The anammox bacteria developed by Paques tend to form granules that can be easily separated by gravity. Due to the high settling velocity of granules the type of reactors used are internal circulators (IC), similar to the ones used for the anaerobic treatment of highly loaded effluents, with a three phase separation system in the upper zone. The separation device serves to capture the anammox granules and to retain them inside the reactor, achieving a high solid retention time, whereas the flocculent bacteria are washed out from the system.

1. **Two Stage ANAMMOX® Process**: The first stage consists of a SHARON reactor (Hellinga et al., 1998) to convert 50% of the fed ammonia to nitrite operated as described previously. The flocculent AOB generated in the SHARON process enter the ANAMMOX® reactor, and they are later washed out as the system is designed to only retain the granular anammox biomass. This two stage system is able to remove high nitrogen loads up to 7 kg N/($m^3$·d) (van der Star et al., 2007).

The two stage process was first implemented at industrial level in 2002 in the WWTP of Dokhaven (Rotterdam), where it achieved ammonia conversions of 80% treating nitrogen loads of 1.2 kg N/($m^3$·d) (van Dongen et al., 2001). The advantage of this configuration is the fact that the two main biological reactions (aerobic ammonia oxidation and anammox process) occur in separate reactors allowing a better control of each process individually. However, it presents as disadvantage the large total volume of the system, due to the requirement of the SHARON system and therefore the investment costs are high.

2. **ANAMMOX® Process in a Single Stage - CANON Process:** The combination of the partial nitritation and anammox processes in a single unit was developed by Paques as the CANON process (acronym of Completely Autotrophic Nitrogen removal Over Nitrite). For the appropriated operation of the system, two conditions are required: i) to keep high SRT values for the development of

anammox bacteria; and ii) to avoid the proliferation of microorganisms that compete with anammox bacteria for the nitrite (heterotrophs denitrifiers and, especially, NOB). The first condition is achieved by the use of a separation device inside the reactor. The second condition can be achieved by controlling parameters such as pH, dissolved oxygen, and redox potential, although it has been proved that the system remains stable by simply controlling the aeration intensity (Abma, Driessen, Haarhuis, & van Loosdrecht, 2010). The process is performed in a reactor continuously fed and aerated. The ammonium and nitrite concentrations in the effluent are monitored and they serve to regulate the aeration intensity and avoid the NOB proliferation (Abma et al., 2010).

One of the benefits of this technology is its compactness as large biomass concentrations can be accumulated in a single reactor. The disadvantage that could be attributed to this configuration is the necessity of keeping a balance between partial nitritation and anammox processes, which is not complicated given the good operational experience of the industrial scale systems (Abma et al., 2010; Lackner et al., 2014; Langone et al., 2014). In fact, the one stage configuration is the most used option to carry out the ANAMMOX® process and they are already around 20 full scale installations in operation (there are only three in the two stage configuration) and most of them treating industrial effluents. However, this configuration allows treating high nitrogen loads but with low nitrogen removal efficiency, since the effluent nitrogen concentrations are commonly higher than in other configurations.

### ELAN® Process

The ELAN® process (Spanish acronym for ELiminaci6n Autotr6fica de Nitr6geno-autotrophic nitrogen removal) was developed by FCC Aqualia in Spain. It is based on the use of a SBR with granular biomass to carry out the nitrogen removal by means of the partial nitritation-anammox processes in a single stage. This system is operated in cycles of 3 or 6 h where the feeding and the aeration are continuously supplied to the reactor during the most of the cycle duration (90-95%). The ELAN® process operates with DO concentrations between 0.8 and 4 mg $O_2$/L and mechanical stirring. One of the main operational parameters of this technology is the short period of settling time imposed, with the aim to promote the wash out of flocculent biomass and to avoid the accumulation of NOB. To maintain the stability of the process the controlled parameter is the ratio of consumed alkalinity/conductivity value. The ELAN® process is able to treat loads between 0.8 and 1.0 kg N/($m^3$·d) (Morales et al., 2015a; Vazquez-Padin et al., 2014). Two industrial scale plants were started in 2015, treating the supernatant from the anaerobic sludge co-digestion (treating municipal and meat processing wastes), and anaerobically predigested effluent from a fish canning factory, respectively.

## Technologies Based on Biofilms

In the biofilm systems the biomass grows adhered to a carrier material, allowing that its retention capacity does not depend on the settling characteristics of the biomass and achieving high biomass concentrations. For these reasons, the use of biofilm systems is adequate in the case of anammox based processes.

## DeAmmon® Process

The DeAmmon® process is based on the use of Moving Bed BioReactors (MBBR) where the biomass attachment is promoted on a specifically designed carrier material. This process was developed by Purac/Läckeby AB (Sweden), in collaboration with the University of Hannover (ISAH) in Germany. It consists of one or two systems of three reactors with a previous decanter. Reactors can be operated in serial or parallel, by means of flexibly connected pipes that allow rearranged them. To promote the biomass retention and growth, Kaldnes (AnoxKaldnes/Veolia) type K1 (500 $m^2$ of active surface per $m^3$) carriers are used, to support the biofilm, which occupy up to 40% of the reactor volume. Reactors are provided with deflector screens to retain the carriers inside the reactors. A recirculation from the third to the first reactor can be necessary to avoid problems of ammonia inhibition in the case that the feeding contains high ammonia concentrations.

The oxygen is supplied using thick bubble diffusers located at the bottom of each reactor, by intermittent aeration with the aim to promote the occurrence of the ammonia oxidation to nitrite and the anammox process. The imposed aerobic/anoxic periods limit the accumulation of nitrite in the liquid phase and restrict the growth of NOB. The length of the aerobic and anoxic periods are of 20-50 minutes and 10-20 minutes, respectively (Plaza, Stridh, Örnmark, Kanders, & Trela, 2011), keeping a DO concentration of 3 mg $O_2$/L during the aerobic periods. The operation at higher DO concentrations is not recommended to avoid the potential growth of NOB and limit the inhibition of anammox bacteria. During the non-aerated periods mixing is achieved by mechanical stirring. No pH control is applied with a typical interval of 7.3 to 7.7. With the aim of optimizing the process on-line conductivity measurements are performed and used to make adjustments to the process operational conditions as the length of the aerobic/anoxic phases (WERF, 2014a).

## ANITA™Mox Process

AnoxKaldnes/Veolia developed a one stage MBBR system under the commercial name of ANITA™Mox. The main difference compared to the DeAmmon® process is that the former uses a support material with a larger specific surface (BiofilmChip™, 1200 $m^2/m^3$) that allows to obtain nitrogen removal rates of 1.2 kg N/($m^3$·d). The support material occupies 40% (v/v) of the reactor volume (Christensson, Ekström, Chan, Le Vaillant, & Lemaire, 2013; Lemaire et al., 2011). On the other hand, the ANITA™Mox process is continuously aerated to keep the DO concentration controlled between 0.5 and 1.5 mg $O_2$/L. The DO concentration is adjusted as a function of the ammonia and nitrate levels inside the reactor (measured on line) to limit the growth of NOB. Due to the use of continuous aeration, the mechanical stirring is not necessary.

## Terra-N® Process

The Terra-N® process was developed by Clariant/Sud-Chemie AG (Munich, Germany). In spite of plastic carriers this process uses bentonite for the growth of the biofilm. The process is designed like a SBR operated as one or two stages in serial system provided with a settler (WERF, 2014a). The bentonite particles have an average size of 25-45 μm and a specific surface of 60 $m^2$/g bentonite. The bentonite is added at a concentration of 10- 12 g/L in the ammonia oxidation reactor (in the two stages system) or in the SBR (in the one stage system). The addition of bentonite in the anammox reactor in the two stages system is not necessary due to the great ability of the anammox bacteria to grow as granular biomass.

The aeration is provided by fine bubble diffusers. In the one stage SBR configuration the aeration is carried out in intermittent mode, being the aerobic/anoxic period length ratio adjusted as a function of the applied ammonia load. To keep the solids in suspension, mechanical stirrers are used during the SBR anoxic phases and in the anammox reactor of the two stages configuration. With the SBR design, a loading rate up to 1.5 kg N/m$^3$·d is possible, with nitrogen removal efficiencies of 80-90% (WERF, 2014a).

### Biodiscs: OLAND Process

The Oxygen-Limited Autotrophic Nitrification-Denitrification (OLAND) process was developed at Ghent University in Belgium, and was initially described as a mixed culture of nitrifying bacteria (Kuai & Verstraete, 1998), alluding to the capacity of some cultures of aerobic nitrifiers to denitrify and convert ammonium to dinitrogen gas under microaerophilic conditions (Philips, Wyffels, Sprengers, & Verstraete, 2002). Afterwards a more detailed examination of the mixed community biofilm, the reactions were attributed to both aerobic ammonium oxidizing bacteria and anaerobic ammonium oxidizing bacteria (Windey, De Bo, & Verstraete, 2005). OLAND process was developed on lab-scale rotating biological contactor (RBC) (Courtens et al., 2014; Pynaert, Smets, Beheydt, & Verstraete, 2004; Windey et al., 2005), where the DO concentration was regulated by the rotating speed and the immersion level of the discs. The NOB suppression can be achieved by sufficiently low dissolved oxygen (DO) levels (e.g. 0.3 mg $O_2$/L) and inhibitory concentrations of free ammonia (0.08-0.82 mg $NH_3$-N/L) (De Clippeleir et al., 2011; Courtens et al., 2014).

However, nowadays OLAND process is usually referred to as a one-stage nitritation-anammox process irrespectively from the reactor configuration (Vlaeminck et al., 2012). Several OLAND type treatments have been developed to treat medium and high-strength nitrogen wastewaters (<0.2 g N/L) such as sewage sludge digestate, landfill leachate, and digestates from specific industrial streams and concentrated black water with nitrogen removal rates between 0.4-1.2 g N/L·d (De Clippeleir et al., 2011). Different reactor types have been used, not only biofilm reactors but also immobilized and suspended biomass reactors (De Clippeleir et al., 2011; Vlaeminck et al., 2012).

## INDUSTRIAL APPLICATIONS

### Full Scale Experiences

Anammox based process have been successfully applied at mesophilic temperatures to treat wastewater with high nitrogen concentrations, like the supernatant of anaerobic digesters. In spite of technologies based on these processes being mainly implemented in municipal WWTPs (around the 75% of total installations), there are already more than 30 full scale installations treating industrial wastewater (Table 3). The one stage configuration is the more frequently applied, the DEMON® in the municipal facilities and the ANAMMOX® at the industrial sector.

The full scale anammox based processes applied to treat industrial wastewater known by the authors of this chapter are summarized in Table 3. Among all the patented technologies, Paques with its tech-

*Table 3. Full-scale installations treating industrial effluents*

| Industrial Wastewater | Technology | Location | Reactor Volume ($m^3$) | Design Load | |
|---|---|---|---|---|---|
| | | | | kg N/d | kg N/($m^3$·d) |
| **Food Processing Industry** | | | | | |
| Yeast | DEMON® | Alltech, RS | --- | 2400 | --- |
| Yeast | ANAMMOX® | Binzhou, CN | 500 | 1000 | 2.0 |
| Starch | ANAMMOX® | China | | 7000 | 2.0 |
| Starch/MSG | ANAMMOX® | Shandong, CN | 4300 | 6090 | |
| MSG | ANAMMOX® | Tongliao I, CN | 6600 | 11000 | 2.0 |
| MSG | ANAMMOX® | Tongliao II, CN | 4100 | 9000 | 2.0 |
| MSG | ANAMMOX® | Wulumuq, CN | 5400 | 10710 | 2.0 |
| Dairy | DEMON® | Rickenbach, DE | --- | 50 | --- |
| Sweetener | ANAMMOX® | Yixing, CN | 1600 | 2180 | --- |
| Potato | ANAMMOX® | Olburgen, NL | 600 | 1200 | 1.2-1.8 |
| Potato | NAS® | Bergen op Zoom, NL | 7920 (3 stages +settler) | 720 | 0.09 |
| Potato | NAS® | Budrio, IT | 1509 (3 stages +MBR) | 344 | 0.23 |
| Potato | NAS® | Kruiningen, NL | 15300 (4 stages +settler) | 325 | 0.02 |
| Meat processing | ANAMMOX® | Son, NL | 3000 | 6000 | 2.0 |
| Meat processing | ANAMMOX® | Brazil | --- | 400 | 2.0 |
| Meat processing | NAS® | Vion, NL | --- | 800 | --- |
| Fish canning | ELAN® | Catoira, SP | 250 | 84 | 0.3 |
| Distillery | ANAMMOX® | Shaoxing, CN | 560 | 900 | 2.0 |
| Distillery | ANAMMOX® | Poland | 900 | 1460 | --- |
| Winery | ANAMMOX® | China | --- | 1200 | 2.0 |
| Tannery | CIRCOX®- ANAMMOX® | Lichtenvoorde, NL | 164/100 | 325 | 0.45 |
| **Agricultural Industry** | | | | | |
| Digestate slaughterhouse | NAS® | Boxtel, NL | 5310 (3 stage, last as MBR) | 888 | 0.17 |
| Fertilizer production | RBC | NL | --- | 150 | --- |
| Digested pig manure | PN/AA Abengoa Water | Chiclana, SP | 5 | --- | 0.1 |
| **Other** | | | | | |
| Biogas production | DEMON® | Etappi Oy, FI | --- | 990 | --- |
| Biogas production | DEMON® | Etappi Oy II, FI | --- | 800 | --- |
| Leachate | DEMON® | Lavant, AT | --- | 45 | --- |
| Leachate | DEMON® | Trento, IT | --- | 700 | --- |
| Leachate | RBC | Kölliken, CH | 33 | --- | 0.4 |
| Leachate | RBC | Mechernich, DE | 80 | --- | 0.6 |
| Leachate | RBC | Pitsea, GB | 240 | --- | 1.7 |
| Leachate | One stage+Activated carbon | Emscherbruch, DE | --- | --- | 3 |
| Leachate | SNAD | Taiwan | 384 | --- | --- |
| Leachate and reject water | ANITA™Mox | Holbæk, DK | 600 | 120 | --- |
| Semiconductor | Biofilm CSTR/ ANAMMOX® | Mie, JP | 300/58 | 220 | 1.0-3.2 |

MSG: monosodium glutamate.
(adapted from Lackner et al. (2014)).

nology ANAMMOX® has been applied as preferred technology in the industry and 16 out of its 21 total installations are located in industrial facilities being more than 50% of the total industrial installations.

As an example, the monosodium glutamate (MSG) production plant of Tangliao (China) requires special mention. In this plant 18,900 $m^3$/d of anaerobically pretreated wastewater (with ammonia concentrations over 600 mg N/L) is treated in a one stage ANAMMOX® reactor. This is probably the largest ammonium load treated in one single factory in the world, and the total nitrogen ammonia removal efficiency of the 6,600 $m^3$ ANAMMOX® reactor reached 95% in less than one month. The plant has been upgraded with the construction of a second reactor and now has a design capacity to treat 20,000 kg N/d (Paques, n.d). The MSG is widely used as flavor enhancer in food products in Southeast Asia being one of major fermentation industries in China. The MSG production wastewater, after primary treatment, is characterized by high chemical oxygen demand (2-5 g/L), ammonia (1.5-2 g/L), sulfate (5-5.5 g/L) and low pH (5.5-6.5). Nitrogen removal of this wastewater was traditionally carried by air stripping with low removal efficiencies. However, the feasibility of its treatment throughout anammox based process was successfully assessed (Shen et al., 2012) and now several full scale plants are treating MSG wastewater (Table 3).

The first full-scale nitritation and anaerobic ammonium oxidation process for the treatment of an inorganic wastewater produced in a semiconductor factory has been installed and operated in Japan (Tokutomi, Yamauchi, Nishimura, Yoda, & Abma, 2011). The wastewater supplied a load of 137-220 kg N/$m^3$·d as ammonia and has to be pretreated to remove fluorine and calcium before being submitted to biological treatment. Then the nitrogen content is removed by means of a two-stage system where the conversion rate of ammonia in the partial nitritation was in the range of 0.27-0.48 kg N/$m^3$·d and the nitrogen removal rate of the anammox reactor (operated with ammonia limitation) in the range 1.04-3.29 kg N/$m^3$·d. The effluent obtained had a concentration of 100-150 mg N/L as nitrite and nitrate which is denitrified in the following unit. A denitrification post-treatment is required to fulfill the discharge limit of 8 mg N/L in the semiconductor factory.

Despite of the fact that the presence of salts has been widely reported as inhibitory to the anammox bacteria it was also proved that the anammox inhibition could be reduced by acclimatization. The specie *Kuenenia stuttgartiensis* was the anammox like bacteria most used in the salinity inhibition tests and the adaptation of other anammox species should also be proved (Dapena-Mora et al., 2007; Jin, Yang, Yu, & Zheng, 2012). The application of the ELAN® process (with *Brocadia fulgida* as the predominant anammox specie) has been tested to treat the effluent of a fish canning industry with successful results after achieving the biomass adaptation to the presence of salt. The accomplishment of the legal discharge limit (100 mg N/L) was proven, and a full-scale plant was started to be constructed on 2015 (Val del Rio et al., 2015). This full scale ELAN® reactor has a volume of 250 $m^3$ and a capacity to treat 270 $m^3$/d. In comparison with the existing nitrification-denitrification process under operation in the fish canning industry, the ELAN® technology presents the following advantages: ten times less volume necessary for nitrogen removal (important for this type of industries located near the coast and with land limitations); 98% of reduction in the waste sludge produced; 85% less energy required for aeration; energy produced in the anaerobic digester doubles its amount (since the bypass of organic matter for denitrification is not needed); and the bulking and foam problems are reduced due to the use of granular sludge. The OPEX of the ELAN® system is expected to be lower than 20% compared to that corresponding to the conventional nitrification-denitrification process.

## Future Applications to Industrial Effluents

Up to date, the anammox based processes have been already applied to remove nitrogen from several types of wastewater (Table 4) with different degree of application. The application to manure digestate is among them one of the most recent applications. This wastewater is characterized by containing high ammonium, residual organic matter and inorganic salts like potassium. Previous obtained results from laboratory studies indicated the feasibility of the anammox process for the treatment of this type of effluents (Figueroa, Vázquez-Padín, Mosquera-Corral, Campos, & Méndez, 2012; Scaglione et al., 2014; Scaglione et al., 2015). Based on the results obtained by Figueroa et al. (2012) a full scale plant of 5 $m^3$, fed with pig slurry (after solid separation and anaerobic digestion), where the partial nitritation-anammox processes took place in a single stage was operated in Chiclana (Spain) by Abengoa Water. After inoculation of the plant the nitrogen removal rate increased from 0.017 kg $N/(m^3 \cdot d)$ to 0.063 kg $N/(m^3 \cdot d)$ in 60 days which indicated a successful start-up in terms of anammox bacteria enrichment, demonstrating the potential of this process for the treatment of pig slurry.

As the composition of industrial effluents significantly differs depending on the industrial sector of origin, specific research is required to guarantee the appropriated operation of the anammox process applied to each type of wastewater. Various laboratory and pilot scale research works are currently under development to evaluate its performance when treating industrial wastewater different from those already tested (integrated gasification combined cycle, coal chemical industry, optoelectronics, etc.). In these cases, the potential inhibitory effects of the organic matter over anammox bacteria needs to be studied as this is one of the limiting factors to extend this process to further applications. This inhibition is mainly attributed to the higher affinity for the nitrite of the denitrifying than the anammox bacteria (Tomar, Gupta, & Mishra, 2015).

Recent studies tested the potential of applying the anammox based process on even traditionally more complex wastewater such as refinery wastewater (Table 4). At laboratory scale promising results have

*Table 4. Applications of the anammox based process under development*

| Wastewater Source | Reactor | Volume (L) | T (°C) | mg N/L | NLR (kg $N/m^3 \cdot d$) | Removal (%) | Reference |
|---|---|---|---|---|---|---|---|
| Swine slurry (diluted) | PN/AA | 1.5 | 18-24 | 300 | 0.46 | 75 | Figueroa et al., 2012 |
| Digested piggery/ poultry | Anammox | 3 | 34-36 | 1000-2000 | 1 | 91 | Scaglione et al., 2014; Scaglione et al., 2015 |
| IGCC | Sharon-anammox | 2 | 35 | >500 | 0.35 | 72 | Milia et al., 2015a; Milia et al., 2015b |
| Coal chemical industry | Anammox | 1 | | | 0.062 | | Toh and Ashbolt 2002 |
| Optoelectronics | PN/AA | 18 | 17.5-36 | >3000 | 0.5-0-9 | 89 | Daverey et al., 2013 |
| Film transistor liquid crystal display | Anammox | 20 | 38-40 | 300 | 0.1-0.5 | 90 | Lin et al., 2011 |

IGCC: Integrated gasification combined cycle.

been obtained treating wastewater produced by the integrated gasification combined cycle (IGCC) and characterized by a high content of formate and ammonium, a high C/N ratio (up to 1.1 g/g), high pH (around 10) and the presence of toxic substances like cyanides and phenols. A two stage approach with progressively adaptation of the biomass on both SHARON and anammox reactors were required due to the acute toxicity of this wastewater (Milia, Perra, Cappai, & Carucci, 2015; Milia, Perra, Tocco, & Carucci, 2015). However, in this case the adaptation period lasted 15 months (Toh & Ashbolt, 2002).

In the last decades, the increasing demand of electronic devices has triggered the rapid development of the technological sector such as optoelectronics and semiconductor industries. The generated wastewater from such industries is complex and hazardous. Due to its inorganic origin, they do not contain organic matter; therefore, the biological nitrogen removal is limited to the application of anammox based process. Total nitrogen removal of 89% has been obtained in a 18 L CANON laboratory reactor treating optoelectronics wastewater (3.7 g $NH_4^+$-N/L) supplemented with mineral medium and after a pH adjustment from 9.7 to 7.8-8.0 (Daverey et al., 2013). The main nitrogen sources of thin film transistor liquid crystal display wastewater are the tetramethylammonium hydroxide (TMAH) and the monoenthanolamine (MEA), used also in the natural gas industry. This industrial wastewater was treated by the anammox process and total nitrogen removal percentages up to 90%, with applied loads of 100-500 mg N/(L·d), was achieved (Lin et al., 2011).

## Anammox Microbiology in Reactors Treating Industrial Effluents

The treatment of industrial effluents, characterized by the presence of toxic compounds, involves the development of microbial consortia comprising anammox bacteria of different genera able to be active in such extreme environments.

Anammox enriched cultures, obtained treating wastewater with high nitrogen concentrations and high temperatures, allowed for the identification of five different anammox genera (*Kuenenia, Brocadia, Anammoxoglobus, Jettenia and Scalindua*), being the *Brocadia* and *Kuenenia* the most widely found. *Scalindua* bacteria were only found in marine environments and at higher salinity concentration the higher is its abundance (Oshiki et al. 2016). Morevoer, *Scalindua* is also characterized by having a lower temperature range for growth compared with the other anammox bacteria. However, conditions to favor the growth of one genus over another are still unclear. As an example van der Star et al. (2008) initially hypothesized that *Brocadia* are more sensitive to nitrite inhibition, however, these bacteria have been found in full scale reactors operated with high nitrite levels. Furthermore, this genus grows on preference at short SRT values (higher growth rate). On the other hand, they observed that *Kuenenia* possess higher substrate affinity and its presence quickly decreases when nitrite concentration increases. Therefore, according to these physiological characteristics, *Brocadia* are r-strategist while *Kuenenia* are k-strategist.

Tokutomi et al. (2011) operated a full scale partial nitritation-anammox system (2 stages) with an excess of nitrite in a semiconductor factory. After 273 days of operation they found that the anammox genus present was *Kuenenia*, supporting the hypothesis that in excess of nitrite this genus is more abundant. *Kuenenia* was also the most abundant anammox genera found in the treatment of landfill leachate (at laboratory scale with two stage system) and potato processing effluents (at full scale with NAS® technology) (Desloover et al., 2011; Li et al., 2014).

Shen et al. (2012) enriched three different anammox reactors to treat monosodium glutamate (MSG) industrial wastewater. In this study, they used as inoculum three sludge samples collected from different WWTP: landfill leachate, municipal and MSG facilities. They observed that *Kuenenia* was the dominant anammox genera in all cultures and achieved the best results in terms of activity in the case of the reactor inoculated with sludge from the MSG industrial WWTP. *Brocadia fulgida* and *Anammoxglobus* were also detected in the case of the other reactors (inoculated with landfill leachate and municipal sludge), whereas in MSG culture only *Kuenenia* organisms were detected.

Moreover, it is known that anammox bacteria are inhibited by salt. It was proved that *Kuenenia* can adapt to high salinity concentrations, but it is still unclear the salinity inhibition threshold of other species (Jin et al., 2012). It was reported that *Scalindua* bacteria lost its activity at salt concentrations lower than 1.5%. *Jettenia* is considered a freshwater genus as it is highly inhibited by salt (saline $IC_{50}$ of 68 mM) (Oshiki et al., 2016).

A shift in the anammox predominant species occurred in the long term reactor operation indicating that each genus could have specific niches. A SNAD laboratoy scale reactor treating opto-electronic wastewater was seed with sludge taken from a full scale plant treating landfill leachate having a 99% of similarity with *Kuenenia* and 93% with *Brocadia,* without *Scalindua* data (Wang et al., 2010). However, the microbial community evolved towards *Scalindua* despite being still *Kuenenia* the most abundant (Daverey, Su, Huang, & Lin, 2012). These results are contradictory with the ones that stated that *Scalindua* is only present in salty conditions. There is no information about the NaCl concentration in the mineral media, but it has high $CaCl_2$ concentrations (300 mg/L) therefore the presence of chloride might be the responsible.

Besides the anaerobic ammonium oxidation, the anammox bacteria can also oxidize organic matter such as formate, acetate, propionate, methylamine, ferrous iron and hydrogen. The specific activities of nitrate reduction through respiratory ammonification are in general much lower than the anammox activity. It was found that *Brocadia* had the higher metabolic capability to organic matter oxidation (Oshiki et al., 2016). Among the *Brocadia* genus three different species were detected: *B. anammoxidans, B. fulgida and B. sinica* being this last one the less abundant. Figueroa et al. (2012) detected *Brocadia fulgida* in a laboratory scale single stage partial nitritation-anammox reactor treating pig slurry digestate. These authors proved that *Brocadia fulgida* completely outcompeted after 4 months the *Brocadia anammoxidans* in the presence of organic matter. Scaglione et al. (2015) also treated the liquid fraction of the co-digested piggery/poultry manure with an enriched culture of *Brocadia* cells in the anammox reactor (two stage system), but no information about the specie or species predominant is available. Moreover, several laboratory scale studies have shown the successful application of the anammox based process to treat wastewater with relatively high organic carbon to ammonia ratios. Jenni, Vlaeminck, Morgenroth, & Udert (2014) found that, in the case of the dominance of *Brocadia fulgida*, successfully competes with heterotrophic bacteria and other anammox bacteria (with COD/N content up to 0.27-1.4 g COD/g N) without compromising the system stability being a high SRT the critical parameter. *Brocadia fulgida* outcompetes other anammox genera in presence of the organic matter.

## COUPLING ANAMMOX PROCESS TO PHOSPHORUS RECOVERY

Phosphorus is an essential element for life and is a key fertilizer compound for high yields in crops. The main source of phosphorus is the phosphate rock, but this source is not renewable and has a time of exhaustion estimated from 50 to 400 years (Bradford-Hartke, Lant, & Leslie, 2012; Cordell, Rosemarin, Schröder, & Smit, 2011; Reijnders, 2014). Therefore, there is increasing interest in the development of industrial scale systems that are technically and economically viable for the recovery of phosphorus from waste streams generated by human activities like wastewater effluents from urban and industry (food, agricultural, etc.).

The most common method of P recovery from wastewater is through the precipitation of struvite ($MgNH_4PO_4 \cdot 6H_2O$) (de-Bashan & Bashan, 2004). Struvite by mass is 44% crystal water, 39% phosphate, 10% magnesium and 7% ammonium. Its precipitation requires the presence of three ionic species, magnesium ($Mg^{2+}$), ammonium ($NH_4^+$) and orthophosphate ($PO_4^{3-}$) in an alkaline solution (pH between 9.0-9.5), resulting in crystals with a molar ratio of 1:1:1. The conditions required to achieve the struvite precipitation make its recovery not always economically feasible. Among them, the addition of an external source of alkalinity, to adjust the pH, or the low phosphorus concentration in the wastewater, are the most critical conditions. In this sense, the most appropriate way to achieve the struvite precipitation is to use anaerobic digestion effluents (Kataki, West, Clarke, & Baruah, 2016).

The anaerobic digestion is widely applied to treat industrial wastewater in order to reduce the organic matter content and valorize it as biogas. However, during the anaerobic digestion of wastes from the agrofood industry ammonium and phosphate are released from the proteins to the liquid medium while the alkalinity increases. The combination of the anammox based processes, for nitrogen removal, and the struvite precipitation, for phosphorus recovery, could be an adequate post-treatment of these anaerobic digester effluents.

Recent studies proposed two possible configurations for N and P removal after the anaerobic digestion:

1. Struvite precipitation and then the application of anammox process and
2. Anammox process followed by struvite precipitation.

The Company Paques (The Netherlands) developed a technology for struvite recovery called Phospaq™. In the Phospaq™ reactor struvite is produced by addition magnesium oxide (MgO) and air supply that besides provide mixing, also regulate the pH by $CO_2$ stripping. The reactor is equipped with a patented internal separator to increase retention of struvite precipitates and minimize flushing of fine particles. This patented system can recover up to 80% of the $PO_4$-P (Remy, Driessen, Hendrickx, & Haarhuis, 2013). The process is completed by the Anammox® technology, which removes around 90% of the ammonia via partial nitritation-anammox processes. Phospaq™ and Anammox® reactors are in use in The Netherlands (Olburgen) and the UK (Stoke Bardolph).

In Olburgen the potato processing wastewater from Aviko Company is treated in the municipal WWTP based on anaerobic UASB technology. Since 2006 an integrated nutrients removal approach by Paques has been implemented saving more than 1.5 million €/year on discharge costs and achieving a phosphorous and ammonium removal over 80% and 90%, respectively. The UASB effluent (1,600 kg COD/d, 1,000 kg $NH_4$-N/d, 200 kg $PO_4^{-3}$-P/day) is treated in a Phospaq™ reactor where 70% of the residual organic matter and 80% of phosphorous are removed and 400 tons/year of struvite suitable as fertilizer are obtained. Then, high nitrogen removal efficiency is accomplished (over 90%) in the ANAMMOX® reactor (Paques, n.d).

The company Colsen installed a struvite recovery unit (ANPHOS®) to precipitate P from digested effluent of a potato processing industry. The basic principle of this system is the wastewater aeration that using $CO_2$ stripping induces an increase in pH. The effluent of the ANPHOS® reactor is treated in a NAS® system, achieving a total nitrogen removal efficiency of 95%, yielding dischargeable effluent qualities (<10 mg N/L and 30 mg P/L) with polishing denitrification (Desloover et al., 2011).

The Company FCC Aqualia (Spain) developed a system that combines first the partial nitritation-anammox process (ELAN® technology) and then a cristalizer for struvite precipitation (Crutchik et al., 2015). The struvite crystallization plant is composed by a two phase fluidized bed reactor coupled in series to a settler. The effluent from the anaerobic digester fed to the crystallization system was characterized by a phosphate concentration of around 68 mg P/L. Industrial grade magnesium hydroxide was used as a cheap magnesium and alkali source. Struvite pellets were successfully obtained in the crystallization system during the experimental operation with a recovery percentage of around 68%. In the future, the plant it will be fed with the anaerobic digester supernatant after the nitrogen removal in the ELAN® unit, in order to favor the efficiency of the struvite recovery. It is expected that the consumption of reagents during the struvite precipitation will be 20% lower due to the consumption of alkalinity in the previous ELAN® process.

## CONCLUSION

Despite the anammox process being discovered in the early 90's its application at full scale is nowadays a success to treat anaerobic digester effluents from municipal and industrial facilities. The significant number of different technologies developed by water sector companies shows the commercial interest of the anammox process and it is an evidence of this success. The type of industrial effluents treated by anammox based process is very wide: food processing industries, agricultural wastes, landfill leachates, etc. While effluents from organic nature have been so far the most numerous treated, the fast development of the electronic industry is opening new possibilities for the anammox process uses. A prior wastewater characterization to determine the presence of toxic compounds such as heavy metals, salt, pharmaceuticals, organic matter or complex matrix is required. Anammox activity batch test using the raw industrial wastewater can arose a preliminary idea of its treatability. However, anammox bacteria have shown that they are able to adapt to toxic compounds reaching high efficiencies, therefore long term studies would be more adequate.

## ACKNOWLEDGMENT

This work was funded by the Spanish Government and FEDER through GRANDSEA (CTM2014-55397-JIN), FISHPOL (CTQ2014-55021-R) and REDNOVEDAR (CTQ2014-51693 REDC) projects, and by the Chilean Government through FONDECYT (1150285). The authors from the USC belong to CRETUS Strategic Partnership (AGRUP2015/02) and the Galician Competitive Research Group (GRC 2013-032), financed by the Galician Government.

## REFERENCES

Abma, W., Driessen, W., Haarhuis, R., & van Loosdrecht, M. C. M. (2010). Upgrading of sewage treatment plant by sustainable and cost-effective separate treatment of industrial wastewater. *Water Science and Technology*, *61*(7), 1715–1722. doi:10.2166/wst.2010.977 PMID:20371929

Abma, W., Schultz, C., Mulder, J. M., & van Loosdrecht, M. C. M. (2007). The advance of Anammox. *Water21, 9*(1), 36-37.

Abma, W., Schultz, C. E., Mulder, J. W., van der Star, W. R. L., Strous, M., Tokutomi, T., & van Loosdrecht, M. C. M. (2007). Full-scale granular sludge Anammox process. *Water Science and Technology, 55*(8-9), 27–33. doi:10.2166/wst.2007.238 PMID:17546966

Ahn, Y. H. (2006). Sustainable nitrogen elimination biotechnologies: A review. *Process Biochemistry, 41*(8), 1709–1721. doi:10.1016/j.procbio.2006.03.033

Bradford-Hartke, Z., Lant, P., & Leslie, G. (2012). Phosphorus recovery from centralised municipal water recycling plants. *Chemical Engineering Research & Design, 90*(1), 78–85. doi:10.1016/j.cherd.2011.08.006

Chen, H., Liu, S., Yang, F., Xue, Y., & Wang, T. (2009). The development of simultaneous partial nitrification, ANAMMOX and denitrification (SNAD) process in a single reactor for nitrogen removal. *Bioresource Technology, 100*(4), 1548–1554. doi:10.1016/j.biortech.2008.09.003 PMID:18977657

Christensson, M., Ekström, S., Chan, A. A., Le Vaillant, E., & Lemaire, R. (2013). Experience from start-ups of the first ANITA Mox Plants. *Water Science and Technology, 67*(12), 2677–2684. doi:10.2166/wst.2013.156 PMID:23787303

*Colsen: Water Energy & Environment*. (n.d.). Retrieved May 2016 from www.colsen.nl

Cordell, D., Rosemarin, A., Schröder, J. J., & Smit, A. L. (2011). Towards global phosphorus security: A systems framework for phosphorus recovery and reuse options. *Chemosphere, 84*(6), 747–758. doi:10.1016/j.chemosphere.2011.02.032 PMID:21414650

Courtens, E. N. P., Boon, N., De Clippeleir, H., Berckmoes, K., Mosquera, M., Seuntjens, D., & Vlaeminck, S. E. (2014). Control of nitratation in an oxygen-limited autotrophic nitrification/denitrification rotating biological contactor through disc immersion level variation. *Bioresource Technology, 155*, 182–188. doi:10.1016/j.biortech.2013.12.108 PMID:24457304

Crutchik, D., Morales, N., Vázquez-Padín, J. R., Rogalla, R., Garrido, J. M., & Lema, J. M. (2015). *Enhancement Of Struvite Crystallization With Magnesium Hydroxide: Results Obtained At Full Scale.* Paper presented at the 12th IWA Leading Edge Conference on Water and Wastewater Technologies, Hong Kong.

Dapena-Mora, A., Fernandez, I., Campos, J. L., Mosquera-Corral, A., Mendez, R., & Jetten, M. S. M. (2007). Evaluation of activity and inhibition effects on Anammox process by batch tests based on the nitrogen gas production. *Enzyme and Microbial Technology, 40*(4), 859–865. doi:10.1016/j.enzmictec.2006.06.018

Daverey, A., Su, S. H., Huang, Y. T., Chen, S. S., Sung, S., & Lin, J. G. (2013). Partial nitrification and anammox process: A method for high strength optoelectronic industrial wastewater treatment. *Water Research, 47*(9), 2929–2937. doi:10.1016/j.watres.2013.01.028 PMID:23548564

Daverey, A., Su, S. H., Huang, Y. T., & Lin, J. G. (2012). Nitrogen removal from opto-electronic wastewater using the simultaneous partial nitrification, anaerobic ammonium oxidation and denitrification (SNAD) process in sequencing batch reactor. *Bioresource Technology, 113*, 225–231. doi:10.1016/j.biortech.2011.12.004 PMID:22237171

de-Bashan, L. E., & Bashan, Y. (2004). Recent advances in removing phosphorus from wastewater and its future use as fertilizer (1997–2003). *Water Research, 38*(19), 4222–4246. doi:10.1016/j.watres.2004.07.014 PMID:15491670

De Clippeleir, H., Yan, X., Verstraete, W., & Vlaeminck, S. (2011). OLAND is feasible to treat sewage-like nitrogen concentrations at low hydraulic residence times. *Applied Microbiology and Biotechnology, 90*(4), 1537–1545. doi:10.1007/s00253-011-3222-6 PMID:21465304

Desloover, J., De Clippeleir, H., Boeckx, P., Du Laing, G., Colsen, J., Verstraete, W., & Vlaeminck, S. E. (2011). Floc-based sequential partial nitritation and anammox at full scale with contrasting $N_2O$ emissions. *Water Research, 45*(9), 2811–2821. doi:10.1016/j.watres.2011.02.028 PMID:21440280

Fernandez, I., Vazquez-Padin, J. R., Mosquera-Corral, A., Campos, J. L., & Mendez, R. (2008). Biofilm and granular systems to improve Anammox biomass retention. *Biochemical Engineering Journal, 42*(3), 308–313. doi:10.1016/j.bej.2008.07.011

Figueroa, M., Vázquez-Padín, J. R., Mosquera-Corral, A., Campos, J. L., & Méndez, R. (2012). Is the CANON reactor an alternative for nitrogen removal from pre-treated swine slurry? *Biochemical Engineering Journal, 65*, 23–29. doi:10.1016/j.bej.2012.03.008

Fux, C., & Siegrist, H. (2004). Nitrogen removal from sludge digester liquids by nitrification/denitrification or partial nitritation/anammox: Environmental and economical considerations. *Water Science and Technology, 50*(10), 19–26. PMID:15656291

Hellinga, C., Schellen, A. A. J. C., Mulder, J. W., van Loosdrecht, M. C. M., & Heijnen, J. J. (1998). The sharon process: An innovative method for nitrogen removal from ammonium-rich waste water. *Water Science and Technology, 37*(9), 135–142. doi:10.1016/S0273-1223(98)00281-9

Jardin, N., & Hennerkes, J. (2012). Full-scale experience with the deammonification process to treat high strength sludge water - a case study. *Water Science and Technology, 65*(3), 447–455. doi:10.2166/wst.2012.867 PMID:22258674

Jenni, S., Vlaeminck, S. E., Morgenroth, E., & Udert, K. M. (2014). Successful application of nitritation/anammox to wastewater with elevated organic carbon to ammonia ratios. *Water Research, 49*, 316–326. doi:10.1016/j.watres.2013.10.073 PMID:24355291

Jetten, M. S. M., Niftrik, L., Strous, M., Kartal, B., Keltjens, J. T., & Op den Camp, H. J. M. (2009). Biochemistry and molecular biology of anammox bacteria. *Critical Reviews in Biochemistry and Molecular Biology, 44*(2-3), 65–84. doi:10.1080/10409230902722783 PMID:19247843

Jin, R.-C., Yang, G.-F., Yu, J.-J., & Zheng, P. (2012). The inhibition of the Anammox process: A review. *Chemical Engineering Journal, 197*, 67–79. doi:10.1016/j.cej.2012.05.014

Joss, A., Derlon, N., Cyprien, C., Burger, S., Szivak, I., Traber, J., & Morgenroth, E. (2011). Combined nitritation-anammox: Advances in understanding process stability. *Environmental Science & Technology*, *45*(22), 9735–9742. doi:10.1021/es202181v PMID:21981764

Kataki, S., West, H., Clarke, M., & Baruah, D. C. (2016). *Phosphorus recovery as struvite from farm, municipal and industrial waste: Feedstock suitability, methods and pre-treatments*. Waste Management. doi:10.1016/j.wasman.2016.01.003

Khin, T., & Annachhatre, A. P. (2004). Novel microbial nitrogen removal processes. *Biotechnology Advances*, *22*(7), 519–532. doi:10.1016/j.biotechadv.2004.04.003 PMID:15262315

Kuai, L. P., & Verstraete, W. (1998). Ammonium removal by the oxygen-limited autotrophic nitrification-denitrification system. *Applied and Environmental Microbiology*, *64*(11), 4500–4506. PMID:9797314

Lackner, S., Gilbert, E. M., Vlaeminck, S. E., Joss, A., Horn, H., & van Loosdrecht, M. C. M. (2014). Full-scale partial nitritation/anammox experiences – An application survey. *Water Research*, *55*, 292–303. doi:10.1016/j.watres.2014.02.032 PMID:24631878

Langone, M., Yan, J., Haaijer, S. C. M., Op den Camp, H. J. M., Jetten, M. S. M., & Andreottola, G. (2014). Coexistence of nitrifying, anammox and denitrifying bacteria in a sequencing batch reactor. *Frontiers in Microbiology*, *5*, 28. doi:10.3389/fmicb.2014.00028 PMID:24550899

Lemaire, R., Liviano, I., Esktröm, S., Roselius, C., Chauzy, J., Thornberg, D., & Deleris, S. (2011). *1-Stage Deammonification MBBR Process for Reject Water Sidestream Treatment: Investigation of Start-Up Strategy and Carriers Design*. Paper presented at the Nutrient Recovery and Management, Miami, FL.

Li, H., Zhou, S., Ma, W., Huang, P., Huang, G., Qin, Y., & Ouyang, H. (2014). Long-term performance and microbial ecology of a two-stage PN–ANAMMOX process treating mature landfill leachate. *Bioresource Technology*, *159*, 404–411. doi:10.1016/j.biortech.2014.02.054 PMID:24681301

Liang, Y., Li, D., Zhang, X., Zeng, H., Yang, Z., & Zhang, J. (2014). Microbial characteristics and nitrogen removal of simultaneous partial nitrification, anammox and denitrification (SNAD) process treating low C/N ratio sewage. *Bioresource Technology*, *169*, 103–109. doi:10.1016/j.biortech.2014.06.064 PMID:25036337

Lin, H.-L., Chen, B.-K., Hsia, H.-P., Yang, Y.-F., Chao, Y.-C., & Cheng, S.-S. (2011). Use of two-stage biological process in treating thin film transistor liquid crystal display wastewater of tetramethylammonium hydroxide. *Sustainable Environment Research*, *21*(3), 155–160.

Lotti, T., Kleerebezem, R., Lubello, C., & van Loosdrecht, M. C. M. (2014). Physiological and kinetic characterization of a suspended cell anammox culture. *Water Research*, *60*, 1–14. doi:10.1016/j.watres.2014.04.017 PMID:24813505

McQuarrie, J., Johnson, C., Lu, T., & Zhao, H. (2015). Sidestream Deammonification. In W. E. Federation (Ed.), *Shortcut Nitrogen Removal - Nitrite Shunt and Deammonification*. Alexandria: Water Environment Federation.

Milia, S., Perra, M., Cappai, G., & Carucci, A. (2015a). SHARON process as preliminary treatment of refinery wastewater with high organic carbon-to-nitrogen ratio. *Desalination and Water Treatment*, 1–9. doi:10.1080/19443994.2015.1087341

Milia, S., Perra, M., Tocco, G., & Carucci, A. (2015b). *The anammox process as the second step for the treatment of ammonium rich refinery wastewater with high Corg/N ratio*. Paper presented at the 14th Conference on Environmental Science and Technology, Rhodes, Greece.

Molinuevo, B., García, M. C., Karakashev, D., & Angelidaki, I. (2009). Anammox for ammonia removal from pig manure effluents: Effect of organic matter content on process performance. *Bioresource Technology*, *100*(7), 2171–2175. doi:10.1016/j.biortech.2008.10.038 PMID:19097886

Morales, N., Val del Rio, A., Vazquez-Padin, J. R., Gutierrez, R., Fernandez-Gonzalez, R., Icaram, P., & Mosquera-Corral, A. (2015a). Influence of dissolved oxygen concentration on the start-up of the anammox-based process: ELAN. *Water Science and Technology*, *72*(4), 520–527. doi:10.2166/wst.2015.233 PMID:26247749

Morales, N., Val del Río, A., Vázquez-Padín, J. R., Méndez, R., Mosquera-Corral, A., & Campos, J. L. (2015b). Integration of the Anammox process to the rejection water and main stream lines of WWTPs. *Chemosphere*, *140*, 99–105. doi:10.1016/j.chemosphere.2015.03.058 PMID:25890586

Mosquera-Corral, A., González, F., Campos, J. L., & Méndez, R. (2005). Partial nitrification in a SHARON reactor in the presence of salts and organic carbon compounds. *Process Biochemistry*, *40*(9), 3109–3118. doi:10.1016/j.procbio.2005.03.042

Mulder, J. W., van Loosdrecht, M. C. M., Hellinga, C., & van Kempen, R. (2001). Full-scale application of the SHARON process for treatment of rejection water of digested sludge dewatering. *Water Science and Technology*, *43*(11), 127–134. PubMed

O'Shaughnessy, M., Sizemore, J., Musabyimana, M., Sanjines, P., Murthy, S., Wett, B., & Pallansch, K. (2008). Operations and process control of the Deammonification (DEMON) process as a sidestream option for nutrient removal. Proceedings of the Water Environment Federation, 2008(9), 6333-6348. doi:10.2175/193864708788809743

Oshiki, M., Satoh, H., & Okabe, S. (2016). Ecology and physiology of anaerobic ammonium oxidizing bacteria. *Environmental Microbiology*. doi:10.1111/1462-2920.13134

Paques. (n.d.). Retrieved May 2016 from www.paques.nl

Philips, S., Wyffels, S., Sprengers, R., & Verstraete, W. (2002). Oxygen-limited autotrophic nitrification/denitrification by ammonia oxidisers enables upward motion towards more favourable conditions. *Applied Microbiology and Biotechnology*, *59*(4-5), 557–566. doi:10.1007/s00253-002-1059-8 and PubMed

Plaza, E., Stridh, S., Örnmark, J., Kanders, L., & Trela, J. (2011). Swedish Experience of the Deammonification Process in a Biofilm System. Proceedings of the Water Environment Federation, 2011(1), 1067-1079. doi:10.2175/193864711802867397

Pynaert, K., Smets, B. F., Beheydt, D., & Verstraete, W. (2004). Start-up of autotrophic nitrogen removal reactors via sequential biocatalyst addition. *Environmental Science & Technology*, *38*(4), 1228–1235. doi:10.1021/es030081+ and PubMed

Reijnders, L. (2014). Phosphorus resources, their depletion and conservation, a review. *Resources, Conservation and Recycling*, *93*, 32–49. doi:10.1016/j.resconrec.2014.09.006

Remy, M., Driessen, W., Hendrickx, T., & Haarhuis, R. (2013). *Recovery of phosphorus by formation of struvite with the PhospaqTM process.* Paper presented at the 18th European Biosolids and Organic Resources Conference.

Scaglione, D., Ficara, E., Corbellini, V., Teli, A., Canziani, R., & Malpei, F. (2014). *Treatibiliy of digested piggery/poultry manure by anammox bacteria*. Paper presented at the 2nd International Water Association Specialized Conference EcoTechnologies for Sewage Treatment Plants, Verona, Italy.

Scaglione, D., Ficara, E., Corbellini, V., Tornotti, G., Teli, A., Canziani, R., & Malpei, F. (2015). Autotrophic nitrogen removal by a two-step SBR process applied to mixed agro-digestate. *Bioresource Technology*, *176*, 98–105. doi:10.1016/j.biortech.2014.11.019 PMID:25460989

Shen, L.-, Hu, A.-, Jin, R.-, Cheng, D.-, Zheng, P., Xu, X.-y., & Hu, B.-. (2012). Enrichment of anammox bacteria from three sludge sources for the startup of monosodium glutamate industrial wastewater treatment system. *Journal of Hazardous Materials*, *199–200*, 193–199. doi:10.1016/j.jhazmat.2011.10.081 PMID:22104770

Strous, M., Heijnen, J. J., Kuenen, J. G., & Jetten, M. S. M. (1998). The sequencing batch reactor as a powerful tool for the study of slowly growing anaerobic ammonium-oxidizing microorganisms. *Applied Microbiology and Biotechnology*, *50*(5), 589–596. doi:10.1007/s002530051340

Toh, S., & Ashbolt, N. (2002). Adaptation of anaerobic ammonium-oxidising consortium to synthetic coke-ovens wastewater. *Applied Microbiology and Biotechnology*, *59*(2-3), 344–352. doi:10.1007/s00253-002-1007-7 PMID:12111169

Tokutomi, T., Yamauchi, H., Nishimura, S., Yoda, M., & Abma, W. (2011). Application of the Nitritation and Anammox Process into Inorganic Nitrogenous Wastewater from Semiconductor Factory. *Journal of Environmental Engineering*, *137*(2), 146–154. doi:10.1061/(ASCE)EE.1943-7870.0000303

Tomar, S., Gupta, S. K., & Mishra, B. K. (2015). A novel strategy for simultaneous removal of nitrogen and organic matter using anaerobic granular sludge in anammox hybrid reactor. *Bioresource Technology*, *197*, 171–177. doi:10.1016/j.biortech.2015.08.057 PMID:26335285

Val del Rio, A., Pedrouso, A., Gutiérrez, A., Campos, J. L., Mosquera-Corral, A., Méndez, R., . . . Rogalla, F. (2015). *Scale up of an anammox based process (ELAN) for the treatment of fish canning effluents.* Paper presented at the 12th IWA Leading Edge Conference on Water and Wastewater Technologies, Hong Kong, China.

van de Graaf, A. A., de Bruijn, P., Robertson, L. A., Jetten, M. S. M., & Kuenen, J. G. (1996). Autotrophic growth of anaerobic ammonium-oxidizing micro-organisms in a fluidized bed reactor. *Microbiology*, *142*(8), 2187–2196. doi:10.1099/13500872-142-8-2187 van der Star, W. R. L., Abma, W. R., Blommers, D., Mulder, J. W., Tokutomi, T., Strous, M., & van Loosdrecht, M. C. M. (2007). Startup of reactors for anoxic ammonium oxidation: Experiences from the first full-scale anammox reactor in Rotterdam. *Water Research, 41*(18), 4149–4163. doi:10.1016/j.watres.2007.03.044 PMID:17583763

van der Star, W. R. L., Miclea, A. I., van Dongen, U. G. J. M., Muyzer, G., Picioreanu, C., & van Loosdrecht, M. C. M. (2008). The membrane bioreactor: A novel tool to grow anammox bacteria as free cells. *Biotechnology and Bioengineering, 101*(2), 286–294. doi:10.1002/bit.21891 PMID:18421799

van Dongen, U., Jetten, M. S. M., & van Loosdrecht, M. C. M. (2001). The SHARON((R))-Anammox((R)) process for treatment of ammonium rich wastewater. *Water Science and Technology, 44*(1), 153–160. PMID:11496667

Van Hulle, S. W. H., Vandeweyer, H. J. P., Meesschaert, B. D., Vanrolleghem, P. A., Dejans, P., & Dumoulin, A. (2010). Engineering aspects and practical application of autotrophic nitrogen removal from nitrogen rich streams. *Chemical Engineering Journal, 162*(1), 1–20. doi:10.1016/j.cej.2010.05.037

Van Loosdrecht, M. C. M., & Jetten, M. S. M. (1998). Microbiological conversions in nitrogen removal. *Water Science and Technology, 38*(1), 1–7. doi:10.1016/S0273-1223(98)00383-7

Varas, R., Guzmán-Fierro, V., Giustinianovich, E., Behar, J., Fernández, K., & Roeckel, M. (2015). Startup and oxygen concentration effects in a continuous granular mixed flow autotrophic nitrogen removal reactor. *Bioresource Technology, 190*, 345–351. doi:10.1016/j.biortech.2015.04.086 PMID:25965951

Vázquez-Padín, J., Fernádez, I., Figueroa, M., Mosquera-Corral, A., Campos, J.-L., & Méndez, R. (2009a). Applications of Anammox based processes to treat anaerobic digester supernatant at room temperature. *Bioresource Technology, 100*(12), 2988–2994. doi:10.1016/j.biortech.2009.01.028 PMID:19246192

Vazquez-Padin, J. R., Figueroa, M., Fernandez, I., Mosquera-Corral, A., Campos, J. L., & Mendez, R. (2009b). Post-treatment of effluents from anaerobic digesters by the Anammox process. *Water Science and Technology, 60*(5), 1135–1143. doi:10.2166/wst.2009.421 PMID:19717899

Vazquez-Padin, J. R., Morales, N., Gutierrez, R., Fernandez, R., Rogalla, F., Barrio, J. P., & Mendez, R. (2014). Implications of full-scale implementation of an anammox-based process as post-treatment of a municipal anaerobic sludge digester operated with co-digestion. *Water Science and Technology: A Journal of the International Association on Water Pollution Research, 69*(6), 1151-1158. doi:10.2166/wst.2013.795

Vlaeminck, S. E., De Clippeleir, H., & Verstraete, W. (2012). Microbial resource management of one-stage partial nitritation/anammox. *Microbial Biotechnology, 5*(3), 433–448. doi:10.1111/j.1751-7915.2012.00341.x PMID:22452819

Wang, C.-C., Lee, P.-H., Kumar, M., Huang, Y.-T., Sung, S., & Lin, J.-G. (2010). Simultaneous partial nitrification, anaerobic ammonium oxidation and denitrification (SNAD) in a full-scale landfill-leachate treatment plant. *Journal of Hazardous Materials, 175*(1–3), 622–628. doi:10.1016/j.jhazmat.2009.10.052 PMID:19913994

WERF. (2014a). Deammonification. In *Water Envrionment Research Foundation (WERF).* Retrieved from http://www.werf.org

WERF. (2014b). *Nutrient Challenge*. Deammonification.

Wett, B. (2007). Development and implementation of a robust deammonification process. *Water Science and Technology*, *56*(7), 81–88. doi:10.2166/wst.2007.611 PMID:17951871

Windey, K., De Bo, I., & Verstraete, W. (2005). Oxygen-limited autotrophic nitrification-denitrification (OLAND) in a rotating biological contactor treating high-salinity wastewater. *Water Research*, *39*(18), 4512–4520. doi:10.1016/j.watres.2005.09.002 PMID:16225902

Zhang, Z., Li, Y., Chen, S., Wang, S., & Bao, X. (2012). Simultaneous nitrogen and carbon removal from swine digester liquor by the Canon process and denitrification. *Bioresource Technology*, *114*, 84–89. doi:10.1016/j.biortech.2012.03.006 PMID:22520218

## KEY TERMS AND DEFINITIONS

**Anaerobic Digestion:** Biodegradation process performed by different types of microorganism under the absence of oxygen comprising different stages: hydrolysis, acidogenesis, acetogenesis and methanogenesis.

**Anammox:** Anaerobic ammonium oxidation.

**Partial Nitritation:** Oxidation of half of the ammonium to nitrite by ammonium oxidizing bacteria.

**Struvite:** Type of phosphate mineral containing magnesium and ammonium, which formula is $NH_4MgPO_4{\cdot}6H_2O$.

# Chapter 11
# Treatment of Landfill Leachate by Anammox Process

**Grzegorz Cema**
*Silesian University of Technology, Poland*

**Adam Sochacki**
*Silesian University of Technology, Poland & Czech University of Life Sciences, Czech Republic*

## ABSTRACT

*In most cases, the anammox process is used for nitrogen removal from reject water coming from dewatering of digested sludge. However, there are more industrial streams suitable for treatment by partial nitritation/anammox process. The landfill leachate may be a good example of such wastewater. Generally, landfilling is the most used solution for treatment of urban solid wastes. The problem with landfill leachate production and management is one of the most important issues associated with the sanitary landfills. These streams are highly contaminated wastewater with a complex mixture of organic and inorganic compounds and characterized by a high ammonia content and low biodegradable organic fraction matter. The objective of this chapter is the short characteristic of landfill leachate and a short review of its treatment methods with special focus on nitrogen removal by partial nitritation/anammox process.*

## INTRODUCTION

In Europe, the total amount of landfilled municipal waste has been diminishing, due to European Commission Council Directive (Council Directive 1999/31/EC). Nevertheless, landfilling of the municipal waste is still a common practice in the waste management system in some European countries and the rest of the world because it is the simplest and cheapest method of waste disposal. Even discontinuation of landfilling in the EU will not instantaneously eliminate the problem related to landfill leachates because even closed and capped landfills will produce these wastewater streams. Such waste management option has several negative impacts on the environment. The major potential environmental impact related to landfill leachate generation is pollution of groundwater and surface water (Kjeldsen et al., 2002). The problem with landfill leachate production and management is one of the most important issues associated with the sanitary landfills. These streams are highly contaminated wastewater with a complex mixture of organic and inorganic compounds and are characterized by high ammonia concentrations and low

DOI: 10.4018/978-1-5225-1037-6.ch011

biodegradable organic fraction matter content. High loads of refractory compounds, the low value of $BOD_5$ (five days Biochemical Oxygen Demand) to COD (Chemical Oxygen Demand) ratio and high concentration of nitrogen cause that this kind of wastewater is complicated and expensive for treatment, as well as requires various processes to be effective (Kaczorek & Ledakowicz, 2006). The changes in waste management caused by the EU directives resulted in reduced volume of leachate produced per tonne of waste landfilled and thus increased leachate strength (Brennan et al., 2015).

Both physical/chemical methods and the biological process can be used to treat landfill leachates. In spite of stable treatment effects, and preferable adaptability to the changes of wastewater quantity and quality, physical/chemical methods have several shortcomings: odors, air pollution, high chemical costs and high energy consumption. The main reason for the selection of a biological process for nitrogen removal is the lower cost compared to the physicochemical methods. Conventional biological nitrogen removal with the use of nitrification/denitrification is probably the most efficient process to eliminate nitrogen from leachate; however, the biodegradable carbon to nitrogen ratio necessitates the use of external carbon source for complete denitrification. Additionally, the nitrification process requires high energy input for aeration. For these reasons, it was important to obtain technology that facilitates enhanced nitrogen removal under the conditions of low carbon availability and with low energy requirements for nitrification. During the last two decades, the possibility of treatment of highly loaded streams with low C to N ratio by partial nitritation and the anammox process has been tested and demonstrated. However, up to now, there are still very few studies reporting the use of anammox process to treat landfill leachate.

## CHARACTERISTICS OF LANDFILL LEACHATE

Landfill leachate is generated by the rainwater infiltrating through the landfill site, the surface and groundwaters flowing into the site as well as water present in the wastes and generated as a result of organic matter degradation. The quality and the quantity of leachate is mainly related to: chemical and physical properties of waste, volumes of infiltrating water, environmental conditions, the landfill age and also storage and reclamation techniques (Christensen et al., 2001; Kulikowska & Klimiuk 2008; Müller et al., 2015).

Generally, the discharge from landfill depends on rainfall, evaporation losses and redistribution in time of infiltrating water in the soil cover and in the waste deposited. The water infiltrating into the site may be transpired to the atmosphere, evaporated from the surface soil, stored in cover soil or percolate downwards through the cover (Peyton & Schroeder 1993; Bengtsson et al., 1994). The water balance on the landfill site can be summarized as follows (Blakey, 1992):

$$L = P - R - \Delta U_s - ET - \Delta U_W \quad (1)$$

Where:L – leachate production, P – precipitation, R – surface run-off, ΔUs – change in soil moisture storage, ET – actual evaporative losses from the bare-soil/evapotranspiration losses from a vegetated surface, ΔUw – change in moisture content of the refuse components.

Leachate is known as wastewater whose treatment is very complicated and expensive as its composition is variable from different landfills and, moreover, its composition is changing with the landfill age. For this reason, the treatment methods have not been unified so far (Kaczorek & Ledakowicz 2006; Kulinowska & Klimiuk 2008). The components present in the leachate usually have different origins. However, four main groups of compounds may be distinguished (Alkalay et al., 1998; Christensen et al.,

2001; Kjeldsen et al., 2002): dissolved organic matter, inorganic macro-components, heavy metals and xenobiotic organic compounds. Generally, it is stated that low $BOD_5$/COD ratio, the presence of heavy metals and high concentration of nitrogen compounds are the most important problems associated with landfill leachate treatment (Kaczorek & Ledakowicz 2006; Renou et al., 2008).

As landfill leachate contains numerous categories of contaminants, like organic compounds, ammonia, refractory macro and micropollutants and heavy metals, it may be toxic to the aquatic environment. Additionally, the leachate composition is unstable, and furthermore the toxicity of leachate is a consequence of its composition as well as their synergistic or antagonistic effect and different physical-chemical properties (Alkalay et al., 1998; Bloor & Banks, 2005; Kalka, 2012; Marttinen et al., 2002; Słomczyńska & Słomczynski, 2004; Ribé et al., 2012). Thus, the physicochemical methods of leachate characterization should be accompanied with the toxicity tests to get more specific information about the potential environmental impact (Kalka, 2012; Oliveira et al., 2014). Research conducted by Ribé et al. (2012) showed that even treated landfill leachate may still have an acute toxic effect on aquatic organisms. Kalka (2012) studied the change of toxicity of landfill leachate in the function of landfill age in combined treatment with municipal wastewater. The leachate was mixed with the municipal wastewater in different ratios (1 and 10%) and then treated in a lab-scale A2O system (Anaerobic-Anoxic-Oxic activated sludge system). The toxicity of raw and treated wastewater was determined using toxicity tests with *Daphnia magna, Thamnocephalus platyurus, Vibrio fischeri* and *Raphidocelis subcapitata*. The toxicity of this wastewater declined during treatment, however still it was harmful to all tested organisms. It was stated, that the concentration of landfill should be only 0.05% not to be harmful to aquatic organisms (Kalka, 2012).

It is also very important that the composition of landfill leachate is not stable and undergoes changes with time. These changes in time are related to the seasonal changes and what is even more important with the age of landfill. Generally, the leachate may be characterized with its biodegradability which is decreasing with landfill age (Kulikowska & Klimiuk, 2008, Lo, 1996, Renou et al., 2008, Yao, in press). In young landfills (usually up to 5 years) there is an acidogenic phase and as a consequence, the leachate contains large amounts of readily biodegradable carbon with volatile fatty acids (VFA) as the main product (Renou et al., 2008; Słomczyńska & Słomczyński, 2004; Welander et al., 1997). Generally, young landfills are characterized by the pH in the range of 3.7 – 6.5 (as a result of presents carboxylic acid and bicarbonate) and high biodegradability as the $BOD_5$/COD ratio ranges from 0.2 to 0.7. The $BOD_5$/COD ratio of leachates decreases from 0.7 to 0.1 and even less with the aging of landfills, thus becoming less biodegradable. Additionally, the pH for old landfills rises to 8.0-8.5 (Kulikowska & Klimiuk, 2008, Renou et al., 2008; Słomczyńska & Słomczyński, 2004). The landfill leachate may be characterized by the basic standard parameters: COD, BOD, BOD/COD, suspended solids, pH, ammonium nitrogen, total Kjeldahl nitrogen (TKN) and heavy metals (Renou et al., 2008). Table 1 summarizes the composition of landfill leachate in terms of standard wastewater parameters based on different studies.

The landfill leachate contains a high concentration of ammonium nitrogen which can vary from less than 100 mg/l in low-strength leachate to over 5000 mg/l and more for high-strength leachate. The ammonium nitrogen, especially in the form of free ammonia is toxic to the aquatic environment and can also disturb the biological treatment of landfills leachate, so it should be removed from this wastewater stream. However, the $BOD_5/N\text{-}NH_4$ ratio for young leachate is usually larger than 1, conversely, for older leachate this ratio is very often lower than 1 (Słomczyńska & Słomczyński, 2004). As a result of high nitrogen concentration and, additionally, unfavorable nitrogen to biodegradable organic carbon ratio the nitrogen removal from landfill leachate is complicated and there is no standard method.

*Table 1. The chemical composition of leachate (all values are in mg/l; pH unitless)*

| COD | $BOD_5$ | $NH_4$-N | $NO_2$-N | $NO_3$-N | TN | pH | Reference |
|---|---|---|---|---|---|---|---|
| 2200±100 | 100±20 | 2000±100 | 0.4±0.1 | 0.5±0.1 | 2030±12 | 8.0±05 | Miao et al., 2014 |
| 757 | 105 | 362 | n.d. | n.d. | 397 | n.d. | Kulikowska & Klimiuk, 2004 |
| 896±36.8 | 106±5.26 | 786±15.2 | n.d. | n.d. | 834±36.8 | 8.1±0.14 | Kulikowska, 2012 |
| 2180±252 | n.d. | 1170±220 | 1±4 | 21±2 | n.d. | 8.58±0.12 | Li et al., 2014 |
| 2301 | 461 | 627 | 41.17 | n.d. | n.d. | 7.95 | Mojiri et al., 2016 |
| 4425-4860 | 433-588 | 790-821 | n.d. | n.d. | 920-977 | 8.3 – 9.0 | Ferraz et al., 2014 |
| 4295±520 | 2086±358 | 1750±89 | n.d. | n.d. | 1980±108 | 7.9±0.2 | Huang et al., 2014 |
| 1703±393.5 | n.d. | 1972±408.2 | n.d. | n.d. | 2117±426 | 8.5±0.3 | Liang & Liu, 2008 |
| 3921±1264 | 123±88.9 | 5975±213 | 0.0±0.0 | <2 | 6317±251 | 8.54±0.12 | Gabarró et al., 2012 |
| 4357±692 | 810±278 | 3772±956 | 0.2±0.5 | 1.4±3.2 | 4058±987 | 8.11±0.2 | Ganigué et al., 2009 |
| 9500 | 4515 | 2000 | n.d. | n.d. | 2500 | 7.5 | Lina et al., 2009 |
| 3203±1176 | 2296±872 | 1000±337.8 | n.d. | 17.7±23.3 | 1264±616.4 | n.d. | Quant et al., 2009 |
| 2766±528 | 485±141 | 1895±527 | n.d | n.d | 2045±489 | n.d | Fudala-Książek et al., 2012 |
| 1219 | n.d. | 800.8 | 0.14 | 5.7 | 957.84 | n.d. | Białowiec et al., 2012 |
| 7511 | 1436 | 500 | n.d. | n.d. | n.d. | 8.22 | Li et al., 1999 |
| 3006±552 | 451±105 | 2189±317 | n.d. | n.d | n.d | 8.4±0.1 | Oulego et al., 2016 |
| 6709±1863 | 2076±619 | 2813±441 | n.d. | n.d. | n.d. | 8.5±0.2 | Oulego et al., 2016 |
| 6560±226 | 3326±565 | 1740±83 | n.d. | n.d. | 2021±97 | 8.2±0.1 | Müller et al., 2015 |

## LANDFILL LEACHATE TREATMENT METHODS REVIEW

The landfill leachate treatment technologies can be based on biological, physical-chemical or integrated methods (Wiszniowski et al., 2006a). The treatability of landfill leachate depends on its composition and characteristics (Kurniawan et al. 2006). The application of biological processes for the treatment of landfill leachates is a challenging task due to their complex matrix and high toxicity. Consequently, some physical-chemical methods are needed either as a pretreatment step or a polishing step (Wiszniowski et al., 2006a). Renou et al. (2008) proposed that the treatment methods for landfill leachate can be classified into three groups:

1. Leachate transfer including recycling and combined treatment with municipal wastewater,
2. Biodegradation: aerobic and anaerobic processes,
3. Chemical and physical methods:
    a. Chemical oxidation,
    b. Adsorption,
    c. Chemical precipitation,
    d. Coagulation/flocculation,
    e. Sedimentation/flotation,

f. Air stripping,
g. Struvite precipitation, and
h. Membrane processes.

## Combined Treatment with Municipal Wastewater

Both co-treatment and recycling of landfill leachate were common methods used for the indirect treatment of this waste stream (Renou et al., 2008). The application of these methods was detrimental to the treatment efficiency at the municipal wastewater treatment plants and was found to inhibit methanogenesis in the mound (Renou et al., 2008). Quant et al. (2009) investigated the co-treatment of municipal wastewater with increasing, from 1 to 5% (vol.), quantity of landfill leachate in the A2O system. It was possible to achieve high $BOD_5$ and COD removal exceeding 90%. However, there was also a significant increase in nitrogen concentration in the effluent. Additionally, the addition of leachate caused a significant drop in phosphorus removal. Çeçen & Çakiroĝlu (2001) studied the influence of landfill leachate added to domestic wastewater in a volumetric ratio from 5 to 20%. They observed extensive nitrification, however the nitrite accumulation was also observed. It was as a result of an increase in the initial $NH_4$-N concentration and the same the increase of free ammonia concentration what caused the nitrite build-up. Ferraz et al. (2014) used pilot-scale submerged aerobic biofilter for co-treatment of leachate (pretreated by air stripping) with domestic wastewater and observed high efficiency of this treatment method with an optimum volumetric ration of the leachate of 2%. However, in other study, Kalka (2012) showed that even 1% (vol.) of leachate addition to municipal wastewater can cause acute toxicity to aquatic organisms. Noteworthy, the accepted maximum volume of leachate to be co-treated at a municipal wastewater treatment plant is based on the annual influent hydraulic loading and is regulated by the respective authorities (e.g. 4% in Ireland) (Brennan et al., 2015). In a recent article, Brennan et al. (2015) pointed out that increasingly stringent wastewater treatment plant discharge limits may disrupt sustainability of co-treatment of landfill leachate with municipal wastewater. This is also because high leachate production coincides with high maximum hydraulic loading at municipal treatment plants as suggested by Brennan et al. (2015).

## Biological Treatment

Biological methods for landfill leachate treatment are commonly used for the treatment of leachate containing a high concentration of BOD. This is a reasonable treatment especially for the young leachate (high biodegradable carbon to nitrogen content). In the case of stabilized leachate, problems with achieving permitted COD levels due to recalcitrant characteristics of organic carbon in the leachate may be encountered. Another problem is related to the toxicity of high concentration of ammonia nitrogen. Generally, the biological nitrification and denitrification are probably the most common methods for elimination of nitrogen from leachate mainly because they are well-known processes. Problems associated with nitrification and denitrification of landfill leachate are associated with ammonia toxicity, low temperature during the winter time and lack of biodegradable organic carbon, especially in mature leachate. Hoilijoki et al. (1999) studied the nitrification process of landfill leachate at low temperature. They tested three different temperatures: 10, 7 and 5°C in two kinds of activated sludge reactors. One was with the addition of plastic carrier material (ASC) and the second one without (AS). In both reactors complete nitrification was achieved at temperature 10°C, at 7°C the nitrification efficiency was decreased to 93% and finally at 5°C in the ASC reactor the complete nitrification was achieved after reducing the nitrogen load from 0.023 g $NH_4$-N/g MLVSS to 0.01 g $NH_4$-N/g MLVSS. At the same time, in the AS reactor the nitrification efficiency was only 61%. Also, Isaka et al. (2007) conducted nitrification at 10°C using nitrifying bacteria immobilized in a polyethylene glycol gel carrier material. The stable ammonium oxidation was maintained for more than one year. However, due to the low affinity of immobilized nitrifying bacteria for ammonium, in the effluent from the reactor, the ammonium nitrogen at low concentration was detected. Moreover, the nitrite oxidizing bacteria were inhibited by the free ammonia present in the reactor. Welander et al. (1997) investigated nitrification and denitrification of landfill leachate at temperature varied from 10 to 26°C. They proved that in such condition it is possible to sustain complete nitrification and denitrification using suspended carrier biofilm reactor. In the case of the denitrification process the carbon to nitrogen ratio is one of the most important issues. Generally, in landfill leachate there is a deficiency of biodegradable carbon, thus the artificial carbon source should be added. Usually acetic acid (Welander et al., 1997), sodium acetate (Kaczorek & Ledakowicz, 2006) or methanol (Kulikowska & Klimiuk, 2004) are used as a carbon source for denitrification.

An interesting and low-cost technology for leachate treatment is the use of constructed wetlands (CWs) (Białowiec et al., 2012; Kadlec & Zmarthie, 2010; Nivala et al., 2007; Yalcuk & Urgulu, 2009; Miksch et al., 2015). CWs have been shown to be very effective in the treatment of landfill leachate and additionally they have low operational and maintenance requirements (Yalcuk & Urgulu, 2009). The main types of CWs that could be used for the on-site treatment of landfill leachate are free water surface systems, and subsurface flow systems with either vertical or horizontal flow (Kadlec &Zmarthie, 2010). Ammonia can be efficiently removed in vertical flow CW due to their elevated oxygen transfer rate. Horizontal flow CWs and free water surface CWs are more effective for denitrification and may follow vertical flow CWs in order to remove nitrates and some poorly biodegradable organic compounds and residual metals via multiple-mechanism pathway including biotic and abiotic processes (Kadlec & Zmarthie, 2010). Due to the carbon deficiency of the wastewater treated in vertical flow CWs it may be necessary to add external carbon source for denitrification in the subsequent treatment steps unless the carbon generated in the CWs is sufficient (Kadlec & Wallace, 2009; Rustige & Nolde, 2007). The combination of several low-energy treatment steps offered a very high mass reduction of ammonia (99.5%) and several other contaminants (including metals and trace organics) in a well-documented system in Saginaw Township, MI, USA (Kadlec & Zmarthie, 2010). This system consists of a sedimentation basin (preceded with cascade aeration), vertical gravel beds and free water surface CWs. Białowiec et al. (2012a) showed that horizontal flow CW with reed and willow can effectively remove ammonium nitrogen. In the research performed by Nivala et al. (2007) it was shown that in spite of seasonal variation in leachate composition, and additionally in temperature, the ammonium nitrogen removal may be stable with very high efficiency exceeding 90% when horizontal flow system with forced aeration is used. Sawaittayothin and Polprasert (2007) showed that using subsurface-flow CW it is possible to remove even over 90% of total nitrogen. Kadlec and Wallace (2009) suggest, however, that multiple cells or multiple units should be used to provide very high removal efficiency. This is because of the constraint of hydraulic efficiency. Leachate quality may be detrimental to some wetland plant species, therefore species that can tolerate elevated concentrations of ammonia should be used and additionally recirculation of treated effluent may be needed to protect the plants (Kadlec & Wallace, 2009). The elevated concentrations of Fe and Mn in landfill leachate may reduce hydraulic conductivity of bed medium of subsurface flow CWs and cause clogging. According to Urbanc-Berčič (1996) the application of aerated lagoon and a sedimentation lagoon as a pretreatment step prior to subsurface flow CWs could efficiently reduce the risk of clogging. Interestingly, Sun and Austin (2006) demonstrated through mass-balance analysis that the CANON (Completely Autotrophic Nitrogen Removal Over Nitrite) process took place in a subsurface vertical flow experimental columns treating landfill leachate containing high ammonia concentration but low BOD. The overall removal of total nitrogen in one of the columns was 52% and was partly associated with CANON process. Indeed, Zhi et al. (2015) found that anammox was one of the primary processes responsible for efficient performance (76% removal of ammonia nitrogen and >81% removal of TN) of experimental tidal flow (fill-and-drain operation) CWs treating artificial wastewater (not a landfill leachate). Zhu et al. (2011) presented an integrated approach to sustain high anammox bacteria activity and biodiversity in the microcosm CW. Further studies are needed to evaluate the feasibility of using CW with enriched anammox for long-term treatment of various types of wastewater including landfill leachate.

## Physical-Chemical Treatment

The physical-chemical methods for the treatment of landfill leachate are used as a pre-treatment step, polishing step or as a main process for the removal of major contaminant (e.g. air stripping of ammonium) (Renou et al., 2016).

Air stripping is the most common method for the elimination of ammonium from various types of wastewater (Anfruns et al., 2013; Kurniawan et al., 2006; Marttinen et al., 2002; Renou et al. 2008). For this method high pH is required. Ozturk et al. (2003) studied the ammonia stripping from landfill leachate. After 2 hours of aeration at pH equal to 12 the ammonium removal efficiency was 72%, while it was only 20% at pH equal to 10 and 11. Also, Hasar et al. (2009) investigated ammonia air stripping as a pretreatment step before biological treatment. They tested four different pH values 9.0, 10.0, 11.0 and 12.0 at different air flow rates and contact time. The best results were obtained for pH 11.0 and 12.0 and were equal to 91.7% and 93.3% respectively. Guo et al. (2010) showed that at pH 11 and 16 hour contact time it is possible to remove over 96% of the ammonia. However, the drawback of this method is the necessity of pH adjustment and that the contaminated gas phase must be treated (Renou et al., 2008).

The struvite precipitation is another method used for ammonia removal from landfill leachate. Li et al. (1999) showed that ammonium may be effectively reduced as magnesium ammonium phosphate (MAP) within 15 minutes, when $MgCl_2 \cdot 6H_20$ and $Na_2HPO_4 \cdot 12H_2O$ were applied with $Mg_2^+:NH_4^+:PO_4^{3-}$ molar ratio of 1:1:1. As usually there is a lack of the $Mg_2^+$ and $PO_4^{3-}$ in the landfill leachate, and there is demand for large amounts of these compounds, Huang et al. (2014) proposed the waste phosphoric acid as phosphorus source and low cost of MgO obtained from calcinations of magnesite. The obtained ammonia nitrogen removal was 83% with Mg:N:P ratio of 3:1:1. Using low-cost of phosphate and magnesium source enabled to reduce chemical cost by 68%.

Adsorption can be used to reduce the concentrations of COD and ammonium nitrogen. Activated carbon (granular or powder) is the most commonly used adsorbent and it allows for 50-70% removal of both COD and ammonia nitrogen (Wiszniowski et al., 2006a). The aim of using activated carbon is to:

1. Remove residual heavy metals and recalcitrant organic pollutants in the final treatment steps, and
2. To provide growth support for microorganism and thus enhance biodegradation.

Other absorbents provided elimination efficiencies comparable with those achieved with activated carbon. These adsorbents are: zeolite, vermiculite, illite, keolinite, activated alumina, municipal waste incinerator bottom ash (Abbas et al., 2009) and clinoptilolite (Wang et al., 2006).

Coagulation-flocculation has been successfully used for the treatment of landfill leachate both as pre- and post-treatment step (Abbas et al., 2009). The coagulation processes served as a pre-treatment step prior to membrane bioreactor and reverse osmosis and it allowed for 36% and 46% removal of COD and suspended solids, respectively (Hasar et al., 2009). The coagulants commonly used in the treatment of landfill leachate are: aluminium sulphate (alum), ferrous sulphate, ferric chloride and ferric chlorosulphate (Wiszniowski et al., 2006a). One of the main drawbacks of coagulation is the production of large volumes of sludge and increased concentration of aluminium or iron (Abbas et al., 2009), and when coagulation is used as a pre-treatment prior to membrane processes, membrane fouling (scaling) may occur mainly due to application of lime or sodium hydroxide to control the pH (Wiszniowski et al., 2006a).

The main membrane processes applied in landfill leachate treatment are: microfiltration, ultrafiltration, nanofiltration and reverse osmosis (Renou et al., 2008). Hasar et al. (2009) stated that reverse osmosis and nanofiltration have been used alone to treat the water, whereas microfiltration and ultrafiltration have been coupled with other processes namely biological treatment and as pre-treatment for other membrane processes (Renou et al., 2008). For nanofiltration, nearly 60-70% removal of COD and 50% removal of ammonia nitrogen were widely reported, whereas reverse osmosis provided at least 98% removal of COD and 99% of heavy metals (Abbas et al., 2009). Wiszniowski et al. (2006a) identified fouling and the need to treat the retentate as the main drawbacks of the membrane processes. Also the combination of membrane processes with activated sludge technology in membrane bioreactors seems a viable technology for the treatment of wastewaters with high organic and nutrient loadings and offered removal of COD and total nitrogen as high as 90% in a single bioreactor at SRT of 30 d (Hasar et al., 2009).

Chemical oxidation (including advanced oxidation processes, which are based on the generation of highly reactive hydroxyl radicals) is widely studied and applied to remove recalcitrant organics from various types of wastewater including landfill leachate and offers high removal efficiencies for a wide range of organic pollutants (Renou et al., 2008; Miksch et al., 2015). Chemical oxidation methods can be used to fully mineralize (as a final treatment step) the target pollutants or to increase biodegradability of contaminants prior to biological treatment. Advanced oxidation processes are recommended for the treatment of industrial wastewater with relatively low COD (<5 g/L), which is due to high energy and reagents consumption (Miksch et al., 2015). Chemical oxidation techniques have been regarded as expensive, but the operation cost can be decreased when hybrid processes are used (e.g. with biological methods) and when sunlight-induced photodegradation (preferably with a catalyst) is used. Indeed, a combination of CWs with sunlight-induced photodegradation is a promising low-cost solution for treating various types of wastewater (Liu et al., 2015). Photocatalytic post-treatment (UVC with $TiO_2$ catalyst) of biologically treated leachate caused further removal of more than 80% refractory organics remaining in the leachate (Wiszniowski et al., 2006b). In a recent study, a combination of persulfate and hydrogen peroxide was applied to treat landfill leachate and 81% removal of COD and 83% removal of ammonia nitrogen was observed (Hilles et al., 2016). The treatment of landfill leachate in an $O_3/H_2O_2$-based advanced oxidation system resulted in 48% reduction of COD, enhanced biodegradability of 0.7 (from 0.1) and color removal of 94% (Tizaoui et al., 2007). Deng & Englehardt (2006) reviewed various studies related to the application of conventional Fenton process, photo-Fenton, electro-Fenton process to the treatment of landfill leachate to observe that the range of COD removal was 45-85% and that the biodegradability was enhanced up to 0.60 (as $BOD_5$/COD ratio). Despite high removal efficiency and low operating cost (in comparison to other chemical oxidation methods), Fenton treatment of landfill leachate may be limited as:

1. Some organic compounds are recalcitrant to this method,
2. Significant amount of acid needed to adjust the pH of typical leachate,
3. Slow initial kinetics at temperature less than 18.3°C (Deng & Englehardt, 2006).

Also electrochemical oxidation was proposed as an efficient method for the treatment of landfill leachate as it offered maximum removal efficiencies as high as 90% for COD, almost 100% for ammonia nitrogen and 86% for color (Deng & Englehardt, 2007). The drawbacks of this process that may somewhat limit its application are high energy consumption and potential chlorinated organics formation may limit its application (Deng & Englehardt, 2007).

## LEACHATE TREATMENT USING ANAMMOX PROCESSES

Because of requirements for high oxygen input and external carbon source addition, treating nitrogen-rich streams with nitrification/denitrification would become expensive and unsustainable. The least resources consuming pathway for the conversion of ammonium to nitrogen gas is a combination of partial nitritation and anammox processes. The main advantages of this combined processes compared to the conventional nitrification/denitrification are: low sludge production, decrease in the aeration costs by almost 60% (only half of the ammonia is oxidized to nitrite in the nitritation process without further oxidation to nitrate), and no need for external organic carbon source addition (anammox process) (De Clippeleir et al., 2011; Mulder, 2003). Furthermore, anammox bacteria oxidize ammonium under anoxic conditions with nitrite as the electron acceptor, and converse energy for $CO_2$ fixation, which resulted in 90% reduction of $CO_2$ emission in comparison to nitrification-denitrification. Additionally, the biomass yield of the anammox process is very low (0.08 kg VSS kg $NH_4$-$N^{-1}$ in comparison to 1 kg VSS kg $NH_4$-$N^{-1}$ in conventional nitrification/denitrification process) consequently, little sludge is produced. The low sludge production is another factor that contributes to the substantially lower operation costs compared to conventional denitrification systems. However, the low biomass yield also demands efficient sludge retention (Strous et al., 1997).

### Partial Nitritation

As the anammox process requires nitrite nitrogen as electron acceptor, the partial nitritation is needed as a preceding step and for its application in wastewater treatment, different set-ups are used to provide nitrite: in the single reactor systems (Van der Star et al., 2007; Cema, 2009) or in the two-reactor systems (Kartal, 2008; Vázquez-Padín et al., 2009; Cema, 2009). In the two-step process part of the ammonium nitrogen is converted to nitrite in the first reactor which is followed by an anoxic reactor where the anammox process occurs. Separation of nitritation and anammox processes in space allows flexibility and more stable process performance since both steps can be controlled separately (van Hulle et al., 2010).The partial nitritation may be achieved in several ways like using high temperature to promote ammonium oxidizing bacteria (AOB) growth over nitrite oxidizing bacteria (NOB) or by changing other parameters like pH or dissolved oxygen. Kulikowska (2012) investigated partial nitritation followed by denitrification in a two-stage SBR system treating mature landfill leachate. The operational temperature of the system was equal to 20±2°C. It was shown that for oxygen concentration between 1.2 and 1.3 mg $O_2$/l the nitrite accumulation ratio was between 32 and 37%, whereas with the decrease in the oxygen concentration to 0.8 – 0.9 mg $O_2$/l the nitrite accumulation ratio increased to 96%. Fudala-Książek et al. (2014) showed that during the co-treatment of landfill leachate with municipal landfill leachate, the volumetric addition of leachate up to 10% promoted the partial nitritation if the oxygen concentration was in the range from 0.5 to 1.5 mg $O_2$/l and free ammonia ranged from 2.01 to 35.86 mg N/l. Li et al. (2013, 2014) achieved stable partial nitritation using endpoint pH control which was necessary to get nitrite-to-ammonium ratio suitable for the anammox process. The high free ammonia and low dissolved oxygen were also necessary to inhibit the nitrite oxidation. The other possibility to achieve partial nitritation was to operate the process at high temperature (over 30°C) and with low sludge retention time to wash out nitrite oxidizers from the system (Akgul et al., 2013; Ganigué et al., 2007; Sri Shalin & Joseph, 2012). Gabarró et al. (2012) showed that for leachate with very high concentration of ammonium nitrogen (around 6000 mg N/l) it was also possible to get stable partial nitritation. At temperature 25°C the free nitrous acid was responsible for inhibition of NOB whereas at temperature equal to 35°C the combination of free ammonia and free nitrous acid was responsible for suppression of NOB.

## Anammox

The anammox process is considered as a technology specifically used for nitrogen rich wastewater with deficiency of organic carbon. This process, in combination with partial nitritation, was successfully used for the treatment of landfill leachate. The single-reactor process of partial nitritation/anammox was first observed in a rotating biological contactor (RBC) for the treatment of landfill leachate in Mechernich, Germany (Hippen and Rosenwinkel, 1997), Kölliken, Switzerland (Siegrist et al., 1998) and Pitsea, Great Britain (Schmid et al., 2003). Up to now, there is over 100 full scale applications of the anammox process (Lackner et al., 2014). Most of these installations are used for reject water treatment, however, there has been very intensive research on the application of the anammox process to landfill leachate treatment (Akgul et al. 2012; Cema et al., 2007; Miao et al., 2012; Scaglione et al., 2012; Sri Shalini & Joseph, 2012; Wang et al., 2014). Zhang & Zhou (2006) showed the possibility of nitrogen removal from landfill leachate by the anammox process performed in the Upflow Anaerobic Sludge Blanket Reactor (UASB). The leachate treated in UASB was the leachate mixture blended with leachate and effluent of SBR system. The average removal efficiency of total nitrogen was 79.6%. Cema et al. (2007) used RBC for landfill leachate treatment by the anammox process. They demonstrated that nitrogen removal by anammox process is possible at a temperature below 20°C. Scaglione et al. (2012) demonstrated the use of single-stage partial nitritation/anammox SBR with granular anammox biomass for the treatment of landfill leachate. Akgul et al. (2012) demonstrated the nitrogen removal using anammox process from young landfill. In order to remove high organic content, which can inhibit the anammox bacteria, the biodegradable carbon was removed in UASB reactor and a membrane bioreactor in sequence, and the effluent was given to the SHARON (Single reactor system for high activity Ammonium Removal Over Nitrite) reactor were partial nitrification takes place and then to the anammox reactor. In such system the COD and nitrogen removal efficiency exceeded 90%. Miao et al, (2014) demonstrated a three stage

SBR comprising: pretreatment of organic matter, nitritation and anammox for nitrogen removal from mature leachate. The nitrogen removal efficiency reached 90%.

## Heavy Metals Inhibition

Landfill leachate is characterized by high concentration of COD and ammonium nitrogen and especially in the case of old landfills by low COD to nitrogen ratio. This causes that, generally, landfill leachates are difficult to be treated by biological methods. Additionally, they can contain several heavy metals and high salinity which may lead to process inhibition. In Table 2 heavy metals concentration in landfill leachate and the $IC_{50}$ for the anammox are presented.

Kimura and Isaka (2014) studied the effect of heavy metals (Ni, Cu, Co, Zn and Mo) on anammox bacteria. Molybdenum was the most influential on the anammox activity, and the concentration higher than 0.2 mg/l strongly influences the anammox reaction. Additionally, the inhibition caused by Mo is irreversible. The 50% inhibition concentration ($IC_{50}$) of Ni, Cu, Co and Zn were 7.9, 6.5, 9.7, and 12.5 mg/l, respectively. Li et al. (2015) studied the inhibitory effect of six metals commonly present in landfill leachate (Cu, Cd, Ni, Zn, Pb and Mo). In short-term studies, Cu, Zn, Cd, and Ni were the most toxic metals with 50% inhibiting soluble concentrations of 4.2, 7.6, 11.2, and 48.6mg/l, respectively. Lead caused only 32% inhibition at the highest tested concentration (6 mg/l of soluble Pb). The $IC_{50}$ values of Cu and Zn were comparable with the results obtained by Kimura and Isaka (2014). In the research performed by Li et al (2015), Mo was not inhibitory at a soluble concentration up to 22.7 mg/l, however in contrast to studies of Kimura and Isaka (2014) this was performed in the batch test without any earlier biomass exposure to the heavy metals. Lotti et al. (2012) investigated the Co and Zn influence on the anammox sludge, and they found the 50% inhibition concentration after 24 hours exposure equal to 1.9 and 3.9 for copper and zinc, respectively. Zhang et al. (2015) studied the short and long-term effect of Zn(II) on the anammox process. The Zn concentration to 5 mg/l was tolerated by the anammox bacteria, and they were suppressed at 8 mg/l. The 50% inhibitory concentration of Zn(II) was 25 mg/l. The inhibitory effect was reversible and after short term of recovery period, the sludge characteristics were remitted to the normal. Zhang et al. (2016) evaluated the long-term effect of Cu and Zn on the anammox activity. Cu and Zn at a concentration of 1 mg/l both, and with a joint loading rate of 0.4 kg/m$^3$d did not affect the process, while the joint loading rate of 0.12 kg/m$^3$d caused it rapid collapse. Generally, the $IC_{50}$ concentrations for anammox process are quite high, however in the case of Cu, Ni, Zn their concentration in landfill leachates may be much higher than $IC_{50}$ concentration. Also, the composite inhibition research is needed.

*Table 2. Heavy metals concentration in landfill leachate and the $IC_{50}$ for the anammox*

| Heavy Metal | Concentration in Landfill Leachate | | $IC_{50}$ for Anammox | |
|---|---|---|---|---|
| | Value [mg/l] | Reference | Value [mg/l] | Reference |
| Co | 0.005-1.5 | Kjeldsen et al., 2002 | 9.7 | Kimura and Isaka 2014 |
| Cu | 0.005-10 | Kjeldsen et al., 2002 | 6.5 | Kimura and Isaka 2014 |
| Ni | 0.015-13 | Christensen et al. 2001 | 7.9 | Kimura and Isaka 2014 |
| Pb | 0.001-5 | Kjeldsen et al., 2002 | 6* | Li et al. 2015 |
| Zn | 0.03-1000 | Christensen et al. 2001 | 12.5 | Kimura and Isaka 2014 |

*32% inhibition at the highest tested concentration

## Influence of Organic Matter

Also, the presence of the organic matter in the landfill leachate could cause some problems as a result of the development of heterotrophic denitrifiers. Generally, partial nitritation/anammox process was mainly applied for wastewater with low COD to N ratio (even below 0.5). The COD requirement for nitrate denitrification is 2.9 g COD/g N-$NO_3^-$, but for nitrite denitrification COD to N ratio is decreasing to 1.71 g COD/g N-$NO_2^-$. Chamchoi et al. (2008) demonstrated the reduction of anammox activity with an increase in COD to N ratio from 0.9 to 2.0, which was caused by the concurrent denitrification. The reactors were continuously fed with synthetic wastewater and the operational temperature was 35°C. They found that the COD concentration of 300 mg/l could inactivate or even eradicate anammox biomass when the reactor is operated under low nitrogen loading rates (below 0.1 kg/$m^3$d). Similar observations were performed by Molinuevo et al. (2009) using a semi-continuous UASB reactor at 37°C, fed with synthetic wastewater and then with pig manure effluents (after UASB post digestion and after partial oxidation). By contrast, Tang et al. (2010) showed that with a high nitrogen loading rate of 13.92 kg/$m^3$d heterotrophic denitrification prevailed in the UASB (upflow anaerobic sludge blanket) reactor fed with synthetic wastewater in which high influent COD/$NO_2$ ratio of 2.92 was applied. Generally, the anammox bacteria growth is suppressed due to weaker competition for nitrite and much lower specific growth rate in comparison to heterotrophs. Recently, Janni et al. (2014) showed the application of partial nitritation/anammox with the step wise increasing COD to N ratio to 1.4. They noticed a decrease in sludge retention time (SRT) to the critical value for the anammox bacteria with increased COD loading. However, reduction of biomass loss caused the rise of anammox bacteria activity. What was interesting, the anammox population changed from mixed community to *Candidattus Brocadia fulgida* as the dominant type which is known to oxidize acetate. However, to avoid the heterotrophic bacteria growth in the anammox reactor, the two step process is recommended (Nhat et al., 2014). In such configuration, the organic carbon is oxidized in the first aerobic reactor so the second anammox reactor is protected from high organic influent and the same anammox bacteria overgrowing by heterotrophic bacteria. Nhat et al. (2014) showed the successful combination of partial nitritation in SBR reactor followed by anammox process in anammox hybrid reactor (HAR) treating old landfill leachate. The same principle of operation is in Panammox® technology which consists of two step partial nitritation-anammox process operated in two SBR reactors treating real landfill leachate. The first partial nitritation SBR reactor is operated under high temperature (35°C) with free ammonia (FA) and free nitrous acid (FNA) as key parameters for NOB suppression. The following SBR reactor was operated as anammox reactor with working temperature at 36°C. The obtained results showed that the anammox process is not inactivated by denitrification and moreover nitrates produced by anammox were removed by heterotrophic denitrification. An 85.1% of the nitrogen removal was due to the anammox process whereas the denitrification was responsible for 14.9% of nitrogen removal (Ruscalleda et al. 2008). Further study showed that the suitable influent nitrite-to-ammonium molar ratio is a crucial factor for process stability, and the lower nitrite supply may result in loss of anammox activity due to nitrite competition with denitrifiers (Ruscalleda et al., 2010). Miao et al. (2014) demonstrated nitrogen removal from landfill leachate using three stage SBR system, comprising pretreating SBR ($SBR_{pre}$), nitritation SBR and anammox SBR. The $SBR_{pre}$ was used for pretreating the biodegradable organic substances and at the same time for reducing the adverse effect of organic substances on anammox process. In this system, the overall nitrogen removal efficiency reached 90% with nitrogen loading rate at 0.81 kg/$m^3$d (Miao et al., 2014 and 2015). Lu et al. (2016) demonstrated the coupling of upflow anaerobic sludge blanked with ANITA™Mox process. The pilot study demonstrated that the combined process was capable of treating high strength leachate with high COD and nitrogen removal. The total nitrogen removal efficiency was about 70%. The system was designed and patented for young landfill leachate treatment with high organic carbon content. The high COD concentration about 45000 mg/l was first removed in UASB reactor with an efficiency about 93% and then before the anammox reactor, there was the C-stage MBBR (Moving Bed Biofilm Reactor) to treat additional 500 to 1000 mg/l of COD to protect the anammox step from high organic carbon inflow.

## Salt Inhibition

Another problem associated with the landfill leachate is high salinity considered as an inhibitory factor for biological nitrogen removal processes (Dapena-Mora et al., 2010; Liu et al., 2014). The high concentration of salt in the leachate mostly is chloride (250 – 4500 mg/l), Sodium (70-7700 mg/l) and Potassium (50-3700 mg/l) (Kjeldsen et al., 2002). Kartal et al. (2006) operated two SBRs for one year using seeding sludge consisting of equal amounts of marine and freshwater species. One reactor was operated with the salinity up to 30 g/l. It was shown that *Kuenenia stuttgartiensis* can adapt to high salt concentration. The same observation was later made by Liu et al. (2009). Also, Dapena-Mora et al. (2010) demonstrated that the anammox bacteria can adapt to high salinity and consequently can be used to remove nitrogen from effluents with high salinity.

Physical-chemical treatments have been found to be suitable not only for the removal of refractory substances from stabilized leachate, but also as a refining step for biologically treated leachate. Anfruns et al. (2013) studied the combination of partial nitritation anammox process with two AOP-based technologies (coagulation-flocculation + ozonation and photo-Fenton) in terms of nitrogen removal and carbon. Biological treatment prior to advanced oxidation processes allowed to reduce the amount of reagents. It was possible to achieve total nitrogen removal efficiency in the range from 87 to 89%.

Recently, the full scale application of partial nitritation and anammox treating landfill leachate was reported (Wang et al., 2010). Wang et al. (2010) demonstrated the occurrence of simultaneous partial nitrification, anammox and denitrification process (SNAD) in a single partially aerated full-scale bioreactor. The process was performed in two aeration tanks with 384 $m^3$ of working volume and 1.26 d of hydraulic resident time, which was located in Taiwan. The DO concentration in the reactor was maintained at 0.3 mg/l which promoted coexistence of anammox microorganism and denitrifiers. The reactor was operated at a temperature ranging from 30°C to 33°C.The removal efficiencies of COD and $N\text{-}NH_4$ were 28% and 80%, respectively, with the total nitrogen removal by partial nitritation/anammox as high as 68% and contribution of heterotrophic denitrification to 5% and 23% of total nitrogen and COD removal, respectively.

## CONCLUSION

The landfill leachate is very difficult to be treated due to its complex composition. So far, there has not been a standard treatment method for this type of wastewater. In last years there has been an increasing interest in using partial nitritation in combination with the anammox process for nitrogen removal from landfill leachate. Due to complex composition of landfill leachate and the presence of inhibitory substances like heavy metals, organic carbon or salts, there is a tendency to use combined treatment methods with some pretreatment – especially for organic carbon removal, and also with some post-treatment methods like constructed wetlands or with AOP-based technologies. The use of partial nitritation with combination of the anammox process gives a cost effective alternative for biological removal of nitrogen from landfill leachate. Although, some full scale applications are single stage technology (first installations used rotating biological contactors), there is rather tendency to use two-step partial nitritation/anammox technology instead of a one-step process, for example the Panammox® technology. It is mainly because the first partial nitritation reactor protects the following anammox reactor from organic carbon and some potentially toxic substances. However, further research is needed to determine the influence of very low temperatures especially with combination of some toxic substances which may be present in leachate. Also, further research is required on combined methods coupling partial nitritation/anammox process with other techniques.

## REFERENCES


Abbas, A.A., & Jingsong, G., LiuZhi, P., PanYing, Y., & Al-Rekabi, W.S. (2009). Review on landfill leachate treatments. *Journal of Applied Sciences Research*, *6*(4), 672–684.

Akgul, D., Aktan, C. K., Yapsakli, K., & Mertoglu, B. (2013). Treatment of landfill leachate using UASB-MBR-SHARON-Anammox configuration. *Biodegradation*, *24*(3), 399–412. doi:10.1007/s10532-012-9597-y PMID:23054184

Alkalay, D., Guerrero, L., Lema, J. M., Mendez, R., & Chamy, R. (1998). Review: Anaerobic treatment of municipal sanitary landfill leachates: the problem of refraktory and toxic components. *World Journal of Microbiology & Biotechnology*, *14*(3), 309–320. doi:10.1023/A:1008876107787

Anfruns, A., Gabarró, J., Gonzalez-Olmos, R., Puig, S., Balaguer, M. D., & Colprim, J. (2013). Coupling anammox and advanced oxidation-based technologies for mature landfill leachate. *Journal of Hazardous Materials*, *258-259*, 27–34. doi:10.1016/j.jhazmat.2013.04.027 PMID:23692680

Bengtsson, L., Bendz, D., Hogland, W., Rosqvist, H., & Åkesson, M. (1994). Water balance for landfills of different age. *Journal of Hydrology (Amsterdam)*, *158*(3-4), 203–217. doi:10.1016/0022-1694(94)90054-X

Białowiec, A., Davies, L., Albuquerque, A., & Randerson, P. F. (2012a). Nitrogen removal from landfill leachate in constructed wetlands with Reed reed and willow: Redox potential in the Root root zone. *Journal of Environmental Management*, *97*, 22–27. doi:10.1016/j.jenvman.2011.11.014 PMID:22325579

Białowiec, A., Davies, L., Albuquerque, A., & Randerson, P. F. (2012b). The influence of plant on nitrogen removal from landfill leachate in discontinuous subsurface horizontal batch shallow constructed wetland with recirculating subsurface horizontal flow. *Ecological Engineering*, *40*, 44–52. doi:10.1016/j.ecoleng.2011.12.011

Blakey, D. (1992). Model Prediction of Landfill Leachate Production. In T. H. Christensen, R. Cossu, & R. Stegmann (Eds.), *Landfilling of waste: leachate. Elsevier Applied Science*. London: Elsevier Science Publisher.

Bloor, M. C., & Banks, C. J. (2005). Acute and sub-lethal toxicity of landfill leachate towards two macro-invertebrates assessing the remediation potential of constructed wetlands. *Process Safety and Environmental Protection*, *83*(2), 184–190. doi:10.1205/psep.03303

Brennan, R. B., Healy, M. G., Morrison, L., Hynes, S., Norton, D., & Clifford, E. (2015). (in press). Management of landfill leachate: The Legacy of European Union Directives. *Waste Management (New York, N.Y.)*. doi:10.1016/j.wasman.2015.10.010 PMID:26481637

Çeçen, F., & Çakiroĝlu, D. (2001). Impact of landfill leachate on the co-treatment of domestic wastewater. *Biotechnology Letters*, *23*(10), 821–826. doi:10.1023/A:1010317823529

Cema, G. (2009). *Comparative study on different Anammox systems*. TRITA_LWR PhD Thesis 1053.

Cema, G., Wiszniowski, J., Żabczyński, S., Zabłocka-Godlewska, W., Raszka, A., & Surmacz-Górska, J. (2007). Biological nitrogen removal from landfill leachate by deammonification assisted by heterotrophic denitrification in a rotating biological contactor. *Water Science and Technology*, *55*(8-9), 35–42. doi:10.2166/wst.2007.239 PMID:17546967

Chamchoi, N., Nitisoravut, S., & Schidt, J. E. (2008). Inactivation of ANAMMOX communities under concurrent operation of anaerobic ammonium oxidation (ANAMMOX) and denitrification. *Bioresource Technology*, *99*(9), 3331–3336. doi:10.1016/j.biortech.2007.08.029 PMID:17911013

Christensen, T. H., Kjeldsen, P., Bjerg, P. L., Jensen, D. L., Christensen, J. B., Baun, A., & Heron, G. et al. (2001). Biogeochemistry of Landfill leachate plumes. *Applied Geochemistry*, *16*(7-8), 659–718. doi:10.1016/S0883-2927(00)00082-2

Council directive 1999/31/EC of 26 April 1999 on the landfill of waste

Dapena-Mora, A., Vázquéz-Padin, J. R., Campose, J. L., Mosquera-Corral, A., Jetten, M. S. M., & Méndez, R. (2010). Monitoring the stability of an Anammox reactor under high salinity conditions. *Biochemical Engineering Journal*, *51*(3), 167–171. doi:10.1016/j.bej.2010.06.014

De Clippeleir, H., Yan, X., Verstraete, W., & Vlaeminck, S. E. (2011). OLAND is feasible to treat sewage-like nitrogen concentrations at low hydraulic residence time. In *Proceedings of the IWA/WEF Nutrient Recovery and Management 2011*. doi:10.1007/s00253-011-3222-6

Deng, Y., & Englehardt, J. D. (2006). Treatment of landfill leachate by the Fenton process. *Water Research*, *40*(20), 3683–3694. doi:10.1016/j.watres.2006.08.009 PMID:17045628

Deng, Y., & Englehardt, J. D. (2007). Electrochemical oxidation for landfill leachate treatment. *Waste Management (New York, N.Y.)*, *27*(3), 380–388. doi:10.1016/j.wasman.2006.02.004 PMID:16632340

Ferraz, F. M., Povinelli, J., Pozzi, E., Vieira, E. M., & Trofino, J. C. (2014). Co-treatment of landfill leachate and domestic wastewater using a submerged aerobic biofilter. *Journal of Environmental Management*, *141*, 9–15. doi:10.1016/j.jenvman.2014.03.022 PMID:24762568

Fudala-Książek, S., Łuczkiewicz, A., Firobor, K., & Olańczuk-Neyman, K. (2012). Nitrogen removal via the nitrite pathway during wastewater co-treatment with ammonia-rich landfill leachates in a sequencing batch reactor. *Environmental Science and Pollution Research International*, *21*(12), 7307–7318. doi:10.1007/s11356-014-2641-1 PMID:24569868

Fudala-Książek, S., Łuczkiewicz, A., Fitobor, K., & Olańczuk-Neyman, K. (2014). Nitrogen removal via the nitrite pathway during wastewater co-treatment with ammonia-rich landfill leachated in a sequencing batch reactor. *Environmental Science and Pollution Research International*, *21*(12), 7307–7318. doi:10.1007/s11356-014-2641-1 PMID:24569868

Gabarró, J., Ganigué, R., Gich, F., Ruscalleda, M., Balaguer, M.D., & Colprim, J. (2012). Effect of temperature on AOB activity of partial nitritation SBR treating landfill leachate with extremely high nitrogen concentration. *Bioresource Technology*, *126*, 283-289.

Ganigué, R., Gabarró, J., Sànchez-Melsió, A., Ruscalleda, M., López, H., Vila, X., & Balaguer, M. D. et al. (2009). Long-term operation of a partial nitritation pilot plant treating leachate with extremely high ammonium concentration prior to an anammox process. *Bioresource Technology*, *100*(23), 5624–5632. doi:10.1016/j.biortech.2009.06.023 PMID:19577465

Ganigué, R., López, H., Balaguer, M. D., & Colprim, J. (2007). Partial ammonium oxidation to nitrite of high ammonium content urban landfill leachate. *Water Research*, *41*(15), 3317–3326. doi:10.1016/j.watres.2007.04.027 PMID:17597177

Guo, J.-S., Abbas, A. A., Chen, Y.-P., Liu, Z.-P., Fang, F., & Chen, P. (2010). Treatment of landfill leachate using a combined stripping, Fenton, SBR, and coagulation process. *Journal of Hazardous Materials*, *178*(1-3), 699–705. doi:10.1016/j.jhazmat.2010.01.144 PMID:20188464

Hasar, H., Unsal, S. A., Ipek, U., Karatas, S., Cinar, O., Yaman, C., & Kınacı, C. (2009). Stripping/flocculation/membrane bioreactor/reverse osmosis treatment of municipal landfill leachate. *Journal of Hazardous Materials*, *171*(1-3), 309–317. doi:10.1016/j.jhazmat.2009.06.003 PMID:19559524

Hilles, A. H., Amr, S. S. A., Hussein, R. A., Arafa, A. I., & El-Sebaie, O. D. (2016). Optimization of leachate treatment using persulfate/H2O2 based advanced oxidation process: case study: Deir El-Balah Landfill Site, Gaza Strip, Palestine. *Water Science and Technology*, *73*(1), 102–112. doi:10.2166/wst.2015.468 PMID:26744940

Hippen, A., Rosenwinkel, K.-H., Baumgarten, G., & Seyfried, C. (1997). Aerobic deammonification: A new experience in the treatment of waste waters. *Water Science and Technology*, *35*(10), 111–120. doi:10.1016/S0273-1223(97)00211-4

Hoilijoki, T. H., Kuttuen, R. H., & Rintala, J. A. (1999). Nitrification of anaerobically pretreated municipal landfill leachate at low temperature. *Water Research*, *34*(5), 1435–1446. doi:10.1016/S0043-1354(99)00278-X

Huang, H., Xiao, D., Zhang, Q., & Ding, L. (2014). Removal of ammonia from landfill leachate by struvite precipitation with the use of low-cost phosphate and magnesium sources. *Journal of Environmental Management*, *145*, 191–198. doi:10.1016/j.jenvman.2014.06.021 PMID:25043172

Isaka, K., Yoshie, S., Sumino, T., Inamori, Y., & Tsuneda, S. (2007). Nitrification of landfill leachate using immobilized nitrifying bacteria at low temperatures. *Biochemical Engineering Journal*, *37*(1), 49–55. doi:10.1016/j.bej.2007.03.008

Jenni, S., Vlaeminck, S. E., Morgenroth, E., & Udert, K. M. (2014). Successful application of nitritation/anammox to wastewater with elevated organic carbon to ammonia ratios. *Water Research*, *49*, 316–326. doi:10.1016/j.watres.2013.10.073 PMID:24355291

Kaczorek, K., & Ledakowicz, S. (2006). Kinetics of nitrogen removal from sanitary landfill leachate. *Bioprocess and Biosystems Engineering*, *29*(5-6), 291–304. doi:10.1007/s00449-006-0078-8 PMID:16944209

Kadlec, R. H., & Wallace, S. D. (2009). *Treatment Wetlands* (2nd ed.). Boca Raton: CRCPress.

Kadlec, R. H., & Zmarthie, L. A. (2010). Wetland treatment of leachate from a closed landfill. *Ecological Engineering*, *36*(7), 946–957. doi:10.1016/j.ecoleng.2010.04.013

Kalka, J. (2012). Landfill Leachate Toxicity Removal in Combined Treatment with Municipal Wastewater. *TheScientificWorldJournal*, *2012*, 1–7. doi:10.1100/2012/202897 PMID:22623882

Kartal, B. (2008). *Ecophysiology of the anammox bacteria*. (PhD thesis). Department of Microbiology, Radbound University Nijmegen, Nijmege.

Kartal, B., Koleva, M., Arsov, R., van der Star, W., Jetten, M. S. M., & Strous, M. (2006). Adaptation of a freshwater anammox population to high salinity wastewater. *Journal of Biotechnology*, *126*(4), 546–553. doi:10.1016/j.jbiotec.2006.05.012 PMID:16806555

Kimura, Y., & Isaka, K. (2014). Evaluation of inhibitory effects of heavy metals on anaerobic ammonium oxidation (anammox) by continuous feedings tests. *Applied Microbiology and Biotechnology*, *98*(16), 6965–6972. doi:10.1007/s00253-014-5735-2 PMID:24723296

Kjeldsen, P., Barlaz, M. A., Rooker, A. P., Baun, A., Ledin, A., & Christensen, T. H. (2002). Present and Long-Term Composition of MSW Landfill Leachate: A Review. *Critical Reviews in Environmental Science and Technology*, *32*(4), 297–336. doi:10.1080/10643380290813462

Kulikowska, D. (2012). Nitrogen removal from landfill leachate via the nitrite route. *Brazilian Journal of Chemical Engineering*, *29*(02), 211–219. doi:10.1590/S0104-66322012000200002

Kulikowska, D., & Klimiuk, E. (2004). Removal of Organics and Nitrogen from Municipal Landfill Leachate in Two-Stage SBR Reactors. *Polish Journal of Environmental Studies*, *13*(4), 389–396.

Kulikowska, D., & Klimiuk, E. (2008). The effect of landfill age on municipal leachate composition. *Bioresource Technology*, *99*(13), 5981–5985. doi:10.1016/j.biortech.2007.10.015 PMID:18060769

Kurniawan, T. A., Lo, W. H., & Chan, G. Y. (2006). Physico-chemical treatments for removal of recalcitrant contaminants from landfill leachate. *Journal of Hazardous Materials*, *B129*(1-3), 80–100. doi:10.1016/j.jhazmat.2005.08.010 PMID:16314043

Lackner, S., Gilbert, E. M., Vlaeminck, S. E., Joss, A., Horn, H., & van Loosdrech, M. C. M. (2014). Full-scale partial nitritation/anammox experiences – An application survey. *Water Research*, *55*, 292–303. doi:10.1016/j.watres.2014.02.032 PMID:24631878

Li, G., Puyol, D., Carvajal-Arroyo, J. M., Sierra-Alvarez, R., & Field, J. A. (2015). Inhibition of anaerobic ammonium oxidation by heavy metals. *Chemical Technology and Biotechnology*, *90*(5), 830–837. doi:10.1002/jctb.4377

Li, H., Zhou, S., Huang, G., & Xu, B. (2013). Partial nitritation of landfill leachate with varying influent composition under intermittent aeration conditions. *Process Safety and Environmental Protection*, *91*(4), 285–294. doi:10.1016/j.psep.2012.05.009

Li, H., Zhou, S., Huang, G., & Xu, B. (2014). Achieving stable partial nitritation using endpoint pH control in an SBR treating landfill leachate. *Process Safety and Environmental Protection*, *92*(3), 199–205. doi:10.1016/j.psep.2013.01.005

Li, X. Z., Zhao, Q. L., & Hao, X. D. (1999). Ammonium removal from landfill leachate by chemical precipitation. *Waste Management (New York, N.Y.)*, *19*(6), 409–415. doi:10.1016/S0956-053X(99)00148-8

Liang, Z., & Liu, J. (2008). Landfill leachate treatment with a novel process: Anaerobic ammonium oxidation (Anammox) combined with soil infiltration system. *Journal of Hazardous Materials*, *151*(1), 202–212. doi:10.1016/j.jhazmat.2007.05.068 PMID:17606322

Lin, W., Chengyao, P., Shujun, Z., & Yongzhen, P. (2009). Nitrogen removal via nitrite from municipal landfill leachate. *Journal of Environmental Sciences (China)*, *21*(11), 1480–1485. doi:10.1016/S1001-0742(08)62443-2 PMID:20108678

Liu, C., Yamamoto, T., Nishiyama, T., Fujii, T., & Furukawa, K. (2009). Effect of salt concentration in anammox treatment using non woven biomass carrier. *Journal of Bioscience and Bioengineering*, *107*(5), 519–523. doi:10.1016/j.jbiosc.2009.01.020 PMID:19393551

Liu, M., Liu, T., Peng, Y., Wang, S., & Xiao, H. (2014). Effect of salinity on $N_2O$ production during shortcut biological nitrogen removal from landfill leachate. *Journal of Bioscience and Bioengineering*, *117*(5), 582–590. doi:10.1016/j.jbiosc.2013.10.011 PMID:24225121

Liu, R., Zhao, Y., Doherty, L., Hu, Y., & Hao, X. (2015). A review of incorporation of constructed wetland with other treatment processes. *Chemical Engineering Journal*, *279*, 220–230. doi:10.1016/j.cej.2015.05.023

Lo, I. M.-C. (1996). Characteristics and treatment of leachate from domestic landfills. *Environment International*, *22*(4), 433–442. doi:10.1016/0160-4120(96)00031-1

Lotti, T., Cordola, M., Kleerebezem, R., Caffaz, S., Lubello, C., & van Loosdrecht, M. C. M. (2012). Inhibition effect of swine wastewater heavy metals and antibiotics on anammox activity. *Water Science and Technology*, *66*(7), 1519–1526. doi:10.2166/wst.2012.344 PMID:22864439

Lu, T., George, B., Zhao, H., & Liu, W. (2016). A case study of coupling upflow anaerobic sludge blanket (UASB) and ANITA™Mox process to treat high-strength landfill leachate. *Water Science and Technology*, *73*(3), 662–668. doi:10.2166/wst.2015.536 PMID:26877051

Marttinen, S. K., Kettunen, R. H., Sormunen, K. M., Soimasuo, R. M., & Rintala, J. A. (2002). Screening of physical-chemical methods for removal of organic material, nitrogen and toxicity from low strength landfill leachate. *Chemosphere*, *46*(6), 851–858. doi:10.1016/S0045-6535(01)00150-3 PMID:11922065

Miao, L., Wang, K., Ahu, R., Li, B., Peng, Y., & Weng, D. (2014). Advanced nitrogen removal from landfill leachate using real-time controlled three-stage sequencing batch reactor (SBR) system. *Bioresource Technology*, *159*, 258–265. doi:10.1016/j.biortech.2014.02.058 PMID:24657756

Miao, L., Wang, S., Tianhao, C., & Peng, Y. (2015). Optimization of three-stage Anammox system removing nitrogen from landfill leachate. *Bioresource Technology, 185*, 450–455. doi:10.1016/j.biortech.2015.03.032 PMID:25812997

Miksch, K., Cema, G., Corvini, P. F. X., Felis, E., Sochacki, A., Surmacz-Górska, J., & Żabczynski, S. et al. (2015). R&D priorities in the field of sustainable remediation and purification of agro-industrial and municipal wastewater. *New Biotechnology, 32*(1), 128–132. doi:10.1016/j.nbt.2013.11.002 PMID:24284044

Mojiri, A., Ziyang, L., Tajuddin, R. M., Farraji, H., & Alifar, N. (2016). Co-treatment of landfill leachate and municipal wastewater using the ZELIA/zeolite constructed wetlands system. *Journal of Environmental Management, 166*, 124–130. doi:10.1016/j.jenvman.2015.10.020 PMID:26496842

Molinuevo, B., Garcia, M. C., Karakashev, D., & Angelidaki, I. (2009). Anammox for ammonia removal from pig manure effluents: Effect of organic matter content on process performance. *Bioresource Technology, 100*(7), 2171–2175. doi:10.1016/j.biortech.2008.10.038 PMID:19097886

Mulder, A. (2003). The quest for sustainable nitrogen removal technologies. *Water Science and Technology, 48*(1), 67–75. PMID:12926622

Müller, G. T., Giacobbo, A., dos Santos Chiaramonte, E. A., Siqueira Rodrigues, M. A., Meneguzzi, A., & Bernardes, A. M. (2015). The effect of sanitary landfill leachate aging on the biological treatment and assessment of photoelectrooxidation as a pre-treatment process. *Waste Management (New York, N.Y.), 36*, 177–183. doi:10.1016/j.wasman.2014.10.024 PMID:25464941

Nhat, P. T., Biec, H. N., Mai, N. T. T., Thanh, B. X., & Dan, N. P. (2014). Application of a partial nitritation and anammox system for the old landfill leachate. *International Biodeterioration & Biodegradation, 95*, 144–150. doi:10.1016/j.ibiod.2014.05.025

Nivala, J., Hoos, M. B., Cross, C., Wallace, S., & Parkin, G. (2007). Treatment of landfill leachate using an aerated, horizontal subsurface-flow constructed wetland. *The Science of the Total Environment, 380*(1-3), 19–27. doi:10.1016/j.scitotenv.2006.12.030 PMID:17241653

Oliveira, L. F., Silva, S. M. C. P., & Martinez, C. B. R. (2014). Assessment of domestic landfill leachate toxicity to the Asian clam *Corbicula fluminea* via biomarkers. *Ecotoxicology and Environmental Safety, 103*, 17–23. doi:10.1016/j.ecoenv.2014.01.034 PMID:24561242

Oulego, P., Collado, S., Laca, A., & Diaz, M. (2016). Impact of leachate composition on the advanced oxidation treatment. *Water Research, 88*, 389–402. doi:10.1016/j.watres.2015.09.048 PMID:26517790

Ozturk, I., Altinbas, M., Koyuncu, I., Arikan, O., & Gomec-Yangin, C. (2003). Advanced phusicochemical treatment experiences on young municipal landfill leachates. *Waste Management (New York, N.Y.), 23*(5), 441–446. doi:10.1016/S0956-053X(03)00061-8 PMID:12893017

Peyton, R. L., & Schroeder, P. R. (1993). Water Balance for Landfills. In D. E. Daniel (Ed.), *Geotechnical Practice for Waste Disposal*. New York: Chapman & Hall. doi:10.1007/978-1-4615-3070-1_10

Quant, B., Olańczuk-Neyman, K., Fudala-Książek, S., Łuczkiewicz, A., Kulbat, E., & Czerwionka, K. (2009). Influence of landfill leachate on municipal wastewater treatment -model plant (WWTP-MP). In *Proceeding of the Conference: IWA 2nd Specialized Conference Nutrient Management In Wastewater Treatment Processes.*

Renou, S., Givaudan, J. G., Poulain, S., Dirassouyan, F., & Moulin, P. (2008). Landfill leachate treatment: Review and opportunity. *Journal of Hazardous Materials, 150*(3), 468–493. doi:10.1016/j.jhazmat.2007.09.077 PMID:17997033

Ribé, V., Nehrenheim, E., Odlare, M., Gustavsson, L., Berglind, R., & Forsberg, Å. (2012). Ecotoxicological assessment and evaluation of a pine bark biosorbent treatment of five landfill leachates. *Waste Management (New York, N.Y.), 32*(10), 1886–1894. doi:10.1016/j.wasman.2012.05.011 PMID:22703999

Ruscalleda, M., López, H., Ganigué, R., Puig, S., Balaguer, M. D., & Colprim, J. (2008). Heterotrophic denitrification on granular anammox SBR treating urban landfill leachate. *Water Science and Technology, 58*(9), 1749–1755. doi:10.2166/wst.2008.544 PMID:19029715

Ruscalleda, M., Puig, S., Mora, X., López, H., Ganigué, R., Balaguer, M. D., & Colprim, J. (2010). The effect of urban landfill leachate characteristics on the coexistence of anammox bacteria and heterotrophic denitrifiers. *Water Science and Technology, 61*(4), 1065–1071. doi:10.2166/wst.2010.610 PMID:20182087

Rustige, H., & Nolde, E. (2007). Nitrogen elimination from landfill leachates using an extra carbon source in subsurface flow constructed wetlands. *Water Science and Technology, 56*(3), 125. doi:10.2166/wst.2007.506 PMID:17802847

Sawaittayothin, V., & Polprasert, C. (2007). Nitrogen mass balance and microbial analysis of constructed wetlands treating municipal landfill leachate. *Bioresource Technology, 98*(3), 565–570. doi:10.1016/j.biortech.2006.02.002 PMID:16546377

Scaglione, D., Ruscalleda, M., Ficara, E., Balaguer, M. D., & Colprim, J. (2012). Response to high nitrite concentrations of anammox biomass from two SBR fed on synthetic wastewater and landfill leachate. *Chemical Engineering Journal, 209*, 62–68. doi:10.1016/j.cej.2012.07.106

Schmid, M., Walsh, K., Webb, R., Rijpstra, W., van de Pas-Schoonen, K., Verbruggen, M., & Strous, M. et al. (2003). Candidatus "Scalindua brodae", sp. nov., Candidatus "Scalindua wagneri", sp. nov., Two New Species Anaerobic Ammonium Oxidizing Bacteria. *Systematic and Applied Microbiology, 26*(4), 529–538. doi:10.1078/072320203770865837 PMID:14666981

Siegrist, H., Reithaar, S., Koch, G., & Lais, P. (1998). Nitrogen loss in a nitrifying rotating contactor treating ammonium-rich wastewater without organic carbon. *Water Science and Technology, 38*(8-9), 241–248. doi:10.1016/S0273-1223(98)00698-2

Słomczyńska, B., & Słomczyński, T. (2004). Physico-Chemical and Toxicological Characteristics of Leachates from MSW Landfills. *Polish Journal of Environmental Studies, 13*(6), 627–637.

Sri Shalini, S., & Joseph, K. (2012). Nitrogen management in landfill leachate: Application of SHARON, ANAMMOX and combined SHARON-ANAMMOX process. *Waste Management (New York, N.Y.), 32*(12), 2385–2400. doi:10.1016/j.wasman.2012.06.006 PMID:22766438

Strous, M., van Gerven, E., Zheng, P., Kuenen, J. G., & Jetten, M. S. M. (1997). Ammonium removal from concentrated waste streams with the anaerobic ammonium oxidation (ANAMMOX) process in different reactor configurations. *Water Research*, *31*(8), 1955–1962. doi:10.1016/S0043-1354(97)00055-9

Sun, G., & Austin, D. (2007). Completely autotrophic nitrogen-removal over nitrite in lab-scale constructed wetlands: Evidence from a mass balance study. *Chemosphere*, *68*(6), 1120–1128. doi:10.1016/j.chemosphere.2007.01.060 PMID:17349669

Tang, C.-, Zheng, P., Wang, C.-, & Mahmood, Q. (2010). Suppression of anaerobic ammonium oxidizers under high organic content in high-rate Anammox UASB reactor. *Bioresource Technology*, *101*(6), 1762–1768. doi:10.1016/j.biortech.2009.10.032 PMID:19900808

Tizaoui, C., Bouselmi, L., Mansouri, L., & Ghrabi, A. (2007). Landfill leachate treatment with ozone and ozone/hydrogen peroxide systems. *Journal of Hazardous Materials*, *140*(1), 316–324. doi:10.1016/j.jhazmat.2006.09.023 PMID:17034936

Urbanc-Berčič, O. (1996). Constructed wetlands for the treatment of landfill leachates: The Slovenian experience. *Wetlands Ecology and Management*, *4*(3), 189–197. doi:10.1007/BF01879237

van der Star, W., Abma, W. R., Blommers, D., Mulder, J. W., Tokutomi, T., Strous, M., & Van Loosdrecht, M. C. M. et al. (2007). Startup of reactor for anoxic ammonium oxidation: Experiences from the first full-scale anammox reactor in Rotterdam. *Water Research*, *41*(18), 4149–4163. doi:10.1016/j.watres.2007.03.044 PMID:17583763

Van Hulle, S., Vandeweyer, B. D., Meesschaert, P. A., Vanrolleghem, P., & Dumoulin, A. (2010). Engineering aspects and practical application of autotrophic nitrogen removal from nitrogen rich streams. *Chemical Enginering Journal*, *162*(1), 1–20. doi:10.1016/j.cej.2010.05.037

Vázquez-Padín, J., Fernádez, I., Figueroa, M., Mosquera-Corrala, A., Campos, J.-L., & Méndez, R. (2009). Applications of Anammox next term based processes to treat anaerobic digester supernatant at room previous temperature. *Bioresource Technology*, *100*(12), 2988–2994. doi:10.1016/j.biortech.2009.01.028 PMID:19246192

Wang, C.-C., Lee, P.-H., Kumar, M., Huang, Y.-T., Sung, S., & Lin, J.-G. (2010). Simultaneous partial nitrification, anaerobic ammonium oxidation and denitrification (SNAD) in a full-scale landfill-leachate treatment plant. *Journal of Hazardous Materials*, *175*(1-3), 622–628. doi:10.1016/j.jhazmat.2009.10.052 PMID:19913994

Wang, Y., Liu, S., Xu, Z., Han, T., Chuan, S., & Zhu, T. (2006). Ammonia removal from leachate solution using natural Chinese clinoptilolite. *Journal of Hazardous Materials*, *136*(3), 735–740. doi:10.1016/j.jhazmat.2006.01.002 PMID:16469437

Welander, U., Henrysson, T., & Welander, T. (1997). Biological nitrogen removal from municipal landfill leachate in a pilot scale suspended carrier biofilm process. *Water Research*, *32*(5), 1564–1570. doi:10.1016/S0043-1354(97)00351-5

Wiszniowski, J., Robert, D., Surmacz-Gorska, J., Miksch, K., & Weber, J. V. (2006a). Landfill leachate treatment methods: A review. *Environmental Chemistry Letters*, *4*(1), 51–61. doi:10.1007/s10311-005-0016-z

Wiszniowski, J., Robert, D., Surmacz-Gorska, J., Miksch, K., & Weber, J. V. (2006b). Leachate detoxification by combination of biological and $TiO_2$-photocatalytic processes. *Water Science and Technology*, *53*(3), 181–190. doi:10.2166/wst.2006.091 PMID:16605031

Yalcuk, A., & Urgulu, A. (2009). Comparison of horizontal and vertical constructed wetland system for landfill leachate treatment. *Bioresource Technology*, *100*(9), 2521–2526. doi:10.1016/j.biortech.2008.11.029 PMID:19157867

Yao, P. (in press). Perspectives on technology for landfill leachate treatment. *Arabian Journal of Chemistry*.

Zhang, H.-, & Zhou, S.-. (2006). Treating leachate mixture with anaerobic ammonium oxidation technology. *Journal of Central South University of Technology*, *14*(6), 663–667. doi:10.1007/s11771-006-0005-9

Zhang, Q.-Q., Zhang, Z.-Z., Guo, Q., Wang, J.-J., Wang, H.-Z., & Jin, R.-C. (2015). Analyzing the revolution of anaerobic ammonium oxidation (anammox) performance and sludge characteristics under zinc inhibition. *Applied Microbiology and Biotechnology*, *99*(7), 3221–3232. doi:10.1007/s00253-014-6205-6 PMID:25398286

Zhang, Z.-Z., Zhang, Q.-Q., Xu, J.-J., Shi, Z.-J., Guo, Q., Jiang, X.-Y., & Jin, R.-C. et al. (2016). Long-term effects of heavy metals and antibiotics on granule-based anammox process: Granule property and performance evolution. *Applied Microbiology and Biotechnology*, *100*(5), 2417–2427. doi:10.1007/s00253-015-7120-1 PMID:26549238

Zhi, W., Yuan, L., Ji, G., & He, C. (2015). Enhanced long-term nitrogen removal and its quantitative molecular mechanism in tidal flow constructed wetlands. *Environmental Science & Technology*, *49*(7), 4575–4583. doi:10.1021/acs.est.5b00017 PMID:25781063

Zhu, G., Wang, S., Feng, X., Fan, G., Jetten, M. S., & Yin, C. (2011). Anammox bacterial abundance, biodiversity and activity in a constructed wetland. *Environmental Science & Technology*, *45*(23), 9951–9958. doi:10.1021/es202183w PMID:21981702

## KEY TERMS AND DEFINITIONS

**Ammonia Air Stripping:** The process in which the ammonia in high concentration is stripped by aeration or air flow.

**Anammox:** Anaerobic Ammonium Oxidation is a microbial process performed by Anammox bacteria belonging to planctomycetes group. In this process the ammonium nitrogen is anaerobicallu oxidized to dinitrogen with nitrites as electron acceptor.

**Denitrification:** The reduction of nitrate nitrogen to nitrogen gas performed by heterotrophic bacteria.

**Landfill Leachate:** A liquid that passing the waste deposited in a landfill. Once in contact with solid wastes, the perocating water becomes polluted.

**Nitrification:** The microbial process performed by chemolitotrophic bacteria in which the ammonium nitrogen is oxidized to nitrate nitrogen.

**Nitritation:** In this process the nitrification process is limited to its first step – ammonium oxidation to nitrite nitrogen.

**Struvite:** Magnesium ammonium phosphate is a white crystalline substance with formula $NH_4MgPO_46H_2O$.

# Compilation of References

91/676/EEC. EEC. (1991). *Nitrates Directive. European Union.*

Aage, H. K., Andersen, B. L., Blom, A., & Jense, I. (1997). The solubility of struvite. *Journal of Radioanalytical and Nuclear Chemistry, 223*(1-2), 213–215. doi:10.1007/BF02223387

Abbas, A.A., & Jingsong, G., LiuZhi, P., PanYing, Y., & Al-Rekabi, W.S. (2009). Review on landfill leachate treatments. *Journal of Applied Sciences Research, 6*(4), 672–684.

Abdullah, N., Yuzir, A., Curtis, T. P., Yahya, A., & Ujang, Z. (2013). Characterization of aerobic granular sludge treating high strength agro-based wastewater at different volumetric loadings. *Bioresource Technology, 127*, 181–187. doi:10.1016/j.biortech.2012.09.047 PMID:23131639

Abma, W., Schultz, C., Mulder, J. M., & van Loosdrecht, M. C. M. (2007). The advance of Anammox. *Water21, 9*(1), 36-37.

Abma, W. R., Driessen, W., Haarhuis, R., & van Loosdrecht, M. C. M. (2010). Upgrading of sewage treatment plant by sustainable and cost-effective separate treatment of industrial wastewater. *Water Science and Technology, 61*(7), 1715–1722. doi:10.2166/wst.2010.977 PMID:20371929

Abma, W., Schultz, C. E., Mulder, J. W., van der Star, W. R. L., Strous, M., Tokutomi, T., & van Loosdrecht, M. C. M. (2007). Full-scale granular sludge Anammox process. *Water Science and Technology, 55*(8-9), 27–33. doi:10.2166/wst.2007.238 PMID:17546966

Adav, S. S., Lee, D.-J., & Ren, N. Q. (2007). Biodegradation of pyridine using aerobic granules in the presence of phenol. *Water Research, 41*(13), 2903–2910. doi:10.1016/j.watres.2007.03.038 PMID:17499839

Adav, S. S., Lee, D.-J., Show, K.-Y., & Tay, J.-H. (2008). Aerobic granular sludge: Recent advances. *Biotechnology Advances, 26*(5), 411–423. doi:10.1016/j.biotechadv.2008.05.002 PMID:18573633

Aguado, D., Montoya, T., Ferrer, J., & Seco, A. (2006). Relating ions concentration variations to conductivity variations in a sequencing batch reactor operated for enhanced biological phosphorus removal. *Environmental Modelling & Software, 21*(6), 845–851. doi:10.1016/j.envsoft.2005.03.004

Ahlgren, S., Baky, A., Bernesson, S., Nordberg, A., Norén, O., & Hansson, P. A. (2008). Ammonium nitrate fertiliser production based on biomass –Environmental effects from a life cycle perspective. *Bioresource Technology, 99*(17), 8034–8041. doi:10.1016/j.biortech.2008.03.041 PMID:18440225

Ahn, Y. H. (2006). Sustainable nitrogen elimination biotechnologies: A review. *Process Biochemistry, 41*(8), 1709–1721. doi:10.1016/j.procbio.2006.03.033

Aiken, A. M., Peyton, B. M., Apel, W. A., & Petersen, J. N. (2003). Heavy metal-induced inhibition of Aspergillus niger nitrate reductase: Applications for rapid contaminant detection in aqueous samples. *Analytica Chimica Acta*, *480*(1), 131–142. doi:10.1016/S0003-2670(02)01598-2

Akgul, D., Aktan, C. K., Yapsakli, K., & Mertoglu, B. (2013). Treatment of landfill leachate using UASB-MBR-SHARON-Anammox configuration. *Biodegradation*, *24*(3), 399–412. doi:10.1007/s10532-012-9597-y PMID:23054184

Alkalay, D., Guerrero, L., Lema, J. M., Mendez, R., & Chamy, R. (1998). Review: Anaerobic treatment of municipal sanitary landfill leachates: the problem of refraktory and toxic components. *World Journal of Microbiology & Biotechnology*, *14*(3), 309–320. doi:10.1023/A:1008876107787

Almeida, J. S., Reis, M. A. M., & Carrondo, M. J. T. (1995). Competition between nitrate and nitrite reduction in denitrification by Pseudomonas fluorescens. *Biotechnology and Bioengineering*, *46*, 474–484. PMID:18623340

Almeida, M. G., Silveira, C. M., Guigliarelli, B., Bertrand, P., Moura, J. J., Moura, I., & Leger, C. (2007). A needle in a haystack: The active site of the membrane-bound complex cytochrome c nitrite reductase. *FEBS Letters*, *581*(2), 284–288. doi:10.1016/j.febslet.2006.12.023 PMID:17207484

Álvarez, J. A., Ruíz, I., & Soto, M. (2008). Anaerobic digesters as a pretreatment for constructed wetlands. *Ecological Engineering*, *33*(1), 54–67. doi:10.1016/j.ecoleng.2008.02.001

Åmand, L., Laurell, C., Stark-Fujii, K., Thunberg, A., & Carlsson, B. (2014). Lessons learnt from evaluating full-scale ammonium. *Water Science and Technology*, *69*(7), 1573. doi:10.2166/wst.2014.061 PMID:24718353

Amorim, C. L., Maia, A. S., Mesquita, R. B. R., Rangel, A. O. S. S., van Loosdrecht, M. C. M., Tiritan, M. E., & Castro, P. M. L. (2014). Performance of aerobic granular sludge in a sequencing batch bioreactor exposed to ofloxacin, norfloxacin and ciprofloxacin. *Water Research*, *50*, 101–113. doi:10.1016/j.watres.2013.10.043 PMID:24361707

Amor, L., Eiroa, M., Kennes, C., & Veiga, M. C. (2005). Phenol biodegradation and its effect on the nitrification process. *Water Research*, *39*(13), 2915–2920. doi:10.1016/j.watres.2005.05.019 PMID:15998531

Andrade, A., & Schuiling, R. D. (2001). The chemistry of struvite crystallization. *Mineralogical Journal (Ukraine)*, *23*, 37–46.

Andreoli, C. V., Pegorini, E. S., Fernandes, F., & Santos, H. F. (2007). Land application of sewage sludge. In V. Cleverson, M. Von Sperling, & F. Fernandes (Eds.), *Sludge Treatment and Disposal*. London, UK: IWA Publishing.

Anfruns, A., Gabarró, J., Gonzalez-Olmos, R., Puig, S., Balaguer, M. D., & Colprim, J. (2013). Coupling anammox and advanced oxidation-based technologies for mature landfill leachate. *Journal of Hazardous Materials*, *258-259*, 27–34. doi:10.1016/j.jhazmat.2013.04.027 PMID:23692680

Anthonisen, A. C., Loehr, R. C., Prakasam, T. B. S., & Srinath, E. G. (1976). Inhibition of nitrification by ammonium and nitrous-acid. *Journal - Water Pollution Control Federation*, *48*, 835–852. PMID:948105

Anthonisen, A. C., Loehr, R. C., Prakasam, T. B. S., & Srinath, E. G. (1976). Inhibitions of nitrification by ammonia and nitrous-acid. *Journal - Water Pollution Control Federation*, *48*(5), 835–852. PMID:948105

An, Y., Li, T., Jin, Z., Dong, M., & Li, Q. (2010). Nitrate degradation and kinetic analysis of the denitrification system composed of nanoparticles and hydrogenotrophic bacteria. *Desalination*, *252*(1-3), 71–74. doi:10.1016/j.desal.2009.10.023

Arheimer, B., & Wittgren, H. B. (1994). Modelling the effects of wetlands on regional nitrogen transport. *Ambio*, *23*(6), 378–386.

Arias, C.A., Brix, H., Kilian, R.M. (2011). *Beplantede filteranlæg til miljøeffektiv spildevandsrensning i det åbne land.* Article Teknik og Miljø.

Arias, C. A., Del Bubba, M., & Brix, H. (2001). Phosphorus removal by sands for use as media in subsurface flow constructed reed beds. *Water Research, 35*(5), 1159–1168. doi:10.1016/S0043-1354(00)00368-7 PMID:11268836

Arrigo, K. R. (2005). Marine microorganisms and global nutrient cycles. *Nature, 437*(7057), 349–355. doi:10.1038/nature04159 PMID:16163345

Arrojo, B., Mosquera-Corral, A., Garrido, J. M., & Méndez, R. (2004). Aerobic granulation with industrial wastewater in sequencing batch reactors. *Water Research, 38*(14-15), 3389–3399. doi:10.1016/j.watres.2004.05.002 PMID:15276756

Aslan, S., & Dahab, M. (2008). Nitritation and denitritation of ammonium-rich wastewater using fluidized-bed biofilm reactors. *Journal of Hazardous Materials, 156*(1-3), 56–63. doi:10.1016/j.jhazmat.2007.11.112 PMID:18206304

Austin, D., Maciolek, D., Davis, B., & Wallace, S. (2006). Damköhler number design method to avoid plugging of tidal flow constructed wetlands by heterotrophic biofilms. *10th International Conference on Wetland Systems for Water Pollution Control.*

Baker, R. W. (2004). *Membrane Technology and Applications.* Wiley. doi:10.1002/0470020393

Bannert, A., Bogen, C., Esperschütz, J., Koubová, A., Buegger, F., Fischer, D., & Schloter, M. et al. (2012). Anaerobic oxidation of methane in grassland soils used for cattle husbandry. *Biogeosciences, 9*(10), 3891–3899. doi:10.5194/bg-9-3891-2012

Banning N. C., Maccarone L. D., Fisk L. M., & Murphy D. V. (2015). Ammonia-oxidising bacteria not archaea dominate nitrification activity in semi-arid agricultural soil. *Scientific Reports.* doi: 10.1038/srep11146

Barat, R., Montoya, T., Borrás, L., Ferrer, J., & Seco, A. (2008). Interactions between calcium precipitation and the polyphosphate- accumulating bacteria metabolism. *Water Research, 42*(13), 3415–3424. doi:10.1016/j.watres.2008.05.003 PMID:18538819

Barat, R., Montoya, T., Seco, A., & Ferrer, J. (2011). Modelling biological and chemically induced precipitation of calcium phosphate in enhanced biological phosphorus removal systems. *Water Research, 45*(12), 3744–3752. doi:10.1016/j.watres.2011.04.028 PMID:21565382

Barthe, P., Chaugny, M., Roudier, S., & Delgado, L. (2015). *REF BREF. Best Available Techniques (BAT) Reference Document for the Refining of Mineral Oil and Gas.* Retrieved from The European IPPC Bureau Joint Research Centre website: http://eippcb.jrc.ec.europa.eu/reference/BREF/REF_BREF_2015.pdf

Bassin, J. P., Pronk, M., Muyzer, G., Kleerebezem, R., Dezotti, M., & van Loosdrecht, M. C. M. (2011). Effect of elevated salt concentrations on the aerobic granular sludge process: Linking microbial activity with microbial community structure. *Applied and Environmental Microbiology, 77*(22), 7942–7953. doi:10.1128/AEM.05016-11 PMID:21926194

Batchelor, B., & Lawrence, A. W. (1978). A kinetic model for autotrophic denitrification using elemental sulphur. *Water Research, 12*(12), 1075–1084. doi:10.1016/0043-1354(78)90053-2

Batstone, D. J., Amerlinck, Y., Ekama, G., Goel, R., Grau, P., Johnson, B., & Volcke, E. et al. (2012). Towards a generalised physicochemical framework. *Water Science and Technology, 66*(6), 1147–1161. doi:10.2166/wst.2012.300 PMID:22828290

Batstone, D. J., Keller, J., Angelidaki, I., Kalyuzhnyi, S. V., Pavlostathis, S. G., Rozzi, A., & Vavilin, V. A. et al. (2001). The IWA Anaerobic Digestion Model No 1 (ADM1). *Water Science and Technology, 45*(10), 65–73. PMID:12188579

Beman, J. M., Roberts, K. J., Wegley, L., Rohwer, F., & Francis, C. A. (2007). Distribution and diversity of archaeal ammonia monooxygenase genes associated with corals. *Applied and Environmental Microbiology*, *73*(17), 5642–5647. doi:10.1128/AEM.00461-07 PMID:17586663

Ben Moussa, S., Tlili, M. M., Batis, N., & Ben Amor, M. (2011). Influence of temperature on struvite precipitation by CO2-deagassing method. *Crystal Research and Technology*, *46*(3), 255–260. doi:10.1002/crat.201000571

Bengtsson, L., Bendz, D., Hogland, W., Rosqvist, H., & Åkesson, M. (1994). Water balance for landfills of different age. *Journal of Hydrology (Amsterdam)*, *158*(3-4), 203–217. doi:10.1016/0022-1694(94)90054-X

Beristain-Cardoso, R., Texier, A. C., Sierra-Álvarez, R., Field, J., Razo-Flores, E., & Gomez, J. (2009). Effect of initial sulfide concentration on phenol oxidation under denitrifying conditions. *Chemosphere*, *74*(2), 200–205. doi:10.1016/j.chemosphere.2008.09.066 PMID:18990426

Beristain, R., Sierra-Alvarez, R., Rowlette, P., Flores, E. R., Gomez, J., & Field, J. A. (2006). Sulfide oxidation under chemolithoauthotrophic denitrifying conditions. *Biotechnology and Bioengineering*, *95*(6), 1148–1157. doi:10.1002/bit.21084 PMID:16807929

Bernet, N., Sanchez, O., Cesbron, D., Steyer, J. P., & Delgenes, J. P. (2005). Modeling and control of nitrite accumulation in a nitrifying biofilm reactor. *Biochemical Engineering Journal*, *24*(2), 173–183. doi:10.1016/j.bej.2005.02.002

Bernhard, A. (2010). The Nitrogen Cycle: Processes, Players, and Human Impact. *Nature Education Knowledge*, *3*(10), 25.

Beun, J. J., Hendriks, A., van Loosdrecht, M. C. M., Morgenroth, E., Wilderer, P. A., & Heijnen, J. J. (1999). Aerobic granulation in a sequencing batch reactor. *Water Research*, *33*(10), 2283–2290. doi:10.1016/S0043-1354(98)00463-1 PMID:11831218

Bhattacharjee, A. S., Motlagh, A. M., Jetten, M. S. M., & Goel, R. (2016). Methane dependent denitrification- from ecosystem to laboratory-scale enrichment for engineering applications. *Water Research*, *99*, 244–252. doi:10.1016/j.watres.2016.04.070 PMID:27176548

Bhatti, A. R., Alvi, A., & Chaudhry, G. R. (2000). Evidence on the presence of two distinct alkaline phosphatases in Serratia marcescens. *FEMS Microbiology Letters*, *182*(1), 131–135. doi:10.1111/j.1574-6968.2000.tb08886.x PMID:10612744

Białowiec, A., Davies, L., Albuquerque, A., & Randerson, P. F. (2012a). Nitrogen removal from landfill leachate in constructed wetlands with Reed reed and willow: Redox potential in the Root root zone. *Journal of Environmental Management*, *97*, 22–27. doi:10.1016/j.jenvman.2011.11.014 PMID:22325579

Białowiec, A., Davies, L., Albuquerque, A., & Randerson, P. F. (2012b). The influence of plant on nitrogen removal from landfill leachate in discontinuous subsurface horizontal batch shallow constructed wetland with recirculating subsurface horizontal flow. *Ecological Engineering*, *40*, 44–52. doi:10.1016/j.ecoleng.2011.12.011

Bialowiec, A., Janczukowicz, W., & Randerson, P. F. (2011). Nitrogen removal from wastewater in vertical flow constructed wetlands containing LWA/gravel layers and reed vegetation. *Ecological Engineering*, *37*(6), 897–902. doi:10.1016/j.ecoleng.2011.01.013

Bitton, G. (1994). Toxicity testing in wastewater treatment plants using microorganisms: wastewater microbiology. New York: Wiley-Liss.

Black, M., Canova, M., Rydin, S., Scalet, B. M., Roudier, S., & Delgado, L. (2013). *Best Available Techniques (BAT) Reference Document for theTanning of Hides and Skins*. Retrieved from The European IPPC Bureau Joint Research Centre website: http://eippcb.jrc.ec.europa.eu/reference/BREF/TAN_Adopted552013.pdf

Blakey, D. (1992). Model Prediction of Landfill Leachate Production. In T. H. Christensen, R. Cossu, & R. Stegmann (Eds.), *Landfilling of waste: leachate. Elsevier Applied Science*. London: Elsevier Science Publisher.

Blandin, G., Vervoort, H., Le-Clech, P., & Verliefde, A. R. D. (2016). Fouling and cleaning of high permeability forward osmosis membranes. *J Wat Proc Eng*, *9*, 161–169. doi:10.1016/j.jwpe.2015.12.007

Block, M. A., & Grossman, A. R. (1988). Identification and Purification of a Derepressible Alkaline Phosphatase from Anacystis nidulans R2. *Plant Physiology*, *86*(4), 1179–1184. doi:10.1104/pp.86.4.1179 PMID:16666051

Blomberg, L. M., Blomberg, M. R., & Siegbahn, P. E. (2006). A theoretical study on nitric oxide reductase activity in a ba(3)-type heme-copper oxidase. *Biochimica et Biophysica Acta*, *1757*(1), 31–46. doi:10.1016/j.bbabio.2005.11.004 PMID:16375849

Bloor, M. C., & Banks, C. J. (2005). Acute and sub-lethal toxicity of landfill leachate towards two macro-invertebrates assessing the remediation potential of constructed wetlands. *Process Safety and Environmental Protection*, *83*(2), 184–190. doi:10.1205/psep.03303

Böhler, M., & Liebi, C. (2012a). Reduzierter NaOH-Verbrauch bei der Ammoniak-Strippung durch den Einsatz einer vorgeschalteten CO2-Strippung—Betriebserfahrungen auf der ARA Kloten-Opfikon (In German). Reduction of base consumption in the free ammonia stripping process by means of a pre treatment by CO2-stripping—Operational experiences at WWTP Kloten/Opfikon. In *Proceedings of 8th Conference on Nitrogen Rejection Load—State of the Art*. Atemis GmbH.

Böhler, M., Büttner, S., Morales, N., Liebi, C., Schachtler, M., & Siegrist, H. (2012b). *Recovery of nutrients from ammonia rich sludge liquids and urine for the production of fertilizer by full scale air stripping or membrane stripping*. Paper presented at the IWA World Congress on Water, Climate and Energy, Dublin, Ireland.

Boiocchi, R., Mauricio-Iglesias, M., Vangsgaard, A. K., Gernaey, K. V., & Sin, G. (2015). Aeration control by monitoring the microbiological activity using fuzzy logic diagnosis and control. Application to a complete autotrophic nitrogen removal reactor. *Journal of Process Control*, *30*, 22–33. doi:10.1016/j.jprocont.2014.10.011

Bongards, M. (2001). Improving the efficiency of a wastewater treatment plant by fuzzy control and neural networks. *Water Science and Technology*, *43*(11), 189–196. PMID:11443962

Bonmatí, A. (2001). *Thermal energy uses for the improvement of pis slurry anaerobic digestion ant the recovery of useful products*. Catalonia, Spain: University of Lleida.

Bonmatí, A., Campos, E., & Flotats, X. (2003). Concentration of pig slurry by evaporation: Anaerobic digestion as the key process. *Water Science and Technology*, *48*(4), 189–194. PMID:14531439

Bonmatí, A., & Flotats, X. (2003a). Air stripping of ammonia from pig slurry: Characterisation and feasibility as a pre- or post- treatment to mesophilic anaerobic digestion. *Waste Management (New York, N.Y.)*, *53*(3), 261–272. doi:10.1016/S0956-053X(02)00144-7 PMID:12737968

Bonmatí, A., & Flotats, X. (2003b). Pig slurry concentration by vacuum evaporation: Influence of previous mesophilic anaerobic digestion process. *J Air Waste Manage*, *53*(1), 21–31. doi:10.1080/10473289.2003.10466112 PMID:12568250

Booker, N. A., Priestley, A. J., & Fraser, I. H. (1999). Struvite formation in wastewater treatments plants: Opportunities for nutrient recovery. *Environmental Technology*, *20*(7), 777–782. doi:10.1080/09593332008616874

Boon, N., Top, E. M., Verstraete, W., Siciliano, S. D., & Icrobiol, A. P. P. L. E. N. M. (2003). *Bioaugmentation as a Tool To Protect the Structure and Function of an Activated-Sludge Microbial Community against a 3-Chloroaniline Shock Load*. doi:10.1128/AEM.69.3.1511

Bouropoulos, N. Ch., & Koutsoukos, P. G. (2000). Spontaneous precipitation of struvite from aqueous solutions. *Journal of Crystal Growth*, *213*(3-4), 381–388. doi:10.1016/S0022-0248(00)00351-1

Bowers, K. E., & Westerman, P. W. (2005). Performance of cone-shaped fluidized bed struvite crystallizers in removing phosphorus from wastewater. *Transactions of the ASAE. American Society of Agricultural Engineers*, *48*(3), 1227–1234. doi:10.13031/2013.18523

Bradford-Hartke, Z., Lant, P., & Leslie, G. (2012). Phosphorus recovery from centralised municipal water recycling plants. *Chemical Engineering Research & Design*, *90*(1), 78–85. doi:10.1016/j.cherd.2011.08.006

Brady, N. C., & Weil, R. R. (2002). *The Nature and Properties of Soils* (13th ed.). Prentice Hall.

Brannan, D. K., & Caldwell, D. E. (1980). *Thermotrix thiopara*: Growth andmetabolism of a newly isolated thermophile capable of oxidizing sulfur and sulfur compounds. *Applied Microbiology and Biotechnology*, *88*, 1027–1042.

Brdjanovic, D., van Loosdrecht, M. C. M., Versteeg, P., Hooijmans, C. M., Alaerts, G. J., & Heijnen, J. J. (2000). Modelling COD, N and P removal in a full-scale WWTP Haarlem Waarderpolder. *Water Research*, *34*(3), 846–858. doi:10.1016/S0043-1354(99)00219-5

Brennan, R. B., Healy, M. G., Morrison, L., Hynes, S., Norton, D., & Clifford, E. (2015). (in press). Management of landfill leachate: The Legacy of European Union Directives. *Waste Management (New York, N.Y.)*. doi:10.1016/j.wasman.2015.10.010 PMID:26481637

Brix, H. (1997). Do macrophytes play a role in constructed treatment wetlands? *Water Science and Technology*, *35*(5), 11–17. doi:10.1016/S0273-1223(97)00047-4

Brix, H., & Arias, C. A. (2005). Danish guidelines for small-scale constructed wetland systems for onsite treatment of domestic sewage. *Water Science and Technology*, *51*(9), 1–9. PMID:16042237

Brix, H., Arias, C. A., & Del Bubba, M. (2001). Media selection for sustainable phosphorus removal in subsurface flow constructed wetlands. *Water Science and Technology*, *44*(11–12), 47–54. PMID:11804137

Brix, H., & Schierup, H. H. (1990). Soil oxygenation in constructed reed beds: The role of macrophyte and soil-atmosphere interface oxygen transport. *Water Research*, *29*(Part 2), 259–266. doi:10.1016/0043-1354(90)90112-J

Broda, E. (1977). Two kinds of lithotrophs missing in nature. *Zeitschrift fur Allgemeine Mikrobiologie*, *17*(6), 491–493. doi:10.1002/jobm.3630170611 PMID:930125

Buday, M. (1994). Reutilization of Ammonia from wastewater using cation-exchange resins. *Water Science and Technology*, *30*, 111–119.

Bueno, E., Gomez-Hernandez, N., Girard, L., Bedmar, E. J., & Delgado, M. J. (2005). Function of the Rhizobium etli CFN42 nirK gene in nitrite metabolism. *Biochemical Society Transactions*, *33*(Pt 1), 162–163. doi:10.1042/BST0330162 PMID:15667294

Bu, L., Wang, K., Zhao, Q. L., Wei, L. L., Zhang, J., & Yang, J. C. (2010). Characterization of dissolved organic matter during landfill leachate treatment by sequencing batch reactor, aeration corrosive cell-Fenton, and granular activated carbon in series. *Journal of Hazardous Materials*, *179*(1-3), 1096–1105. doi:10.1016/j.jhazmat.2010.03.118 PMID:20430522

Bumbac, C., Ionescu, I. A., Tiron, O., & Badescu, V. R. (2015). Continuous flow aerobic granular sludge reactor for dairy wastewater treatment. *Water Science and Technology*, *71*(3), 440–445. doi:10.2166/wst.2015.007 PMID:25714645

Burton, C. H. (2007). The potential contribution of separation technologies to the management of livestock manure. *Livestock Science*, *112*(3), 208–216. doi:10.1016/j.livsci.2007.09.004

Burton, C. H., & Turner, C. (2003). *Manure management. Treatment strategies for sustainable agriculture* (2nd ed.). Silsoe: Bedfors UK.

Butcher, S. S., Charlson, R. J., Oriand, G. H., & Wolfe, G. V. (1992). Global Biogeochemical Cycles. London: Academic Press (Harcourt Brace Jovanovich).

Bykov, D., & Neese, F. (2012). Reductive activation of the heme iron-nitrosyl intermediate in the reaction mechanism of cytochrome c nitrite reductase: A theoretical study. *Journal of Biological Inorganic Chemistry*, *17*(5), 741–760. doi:10.1007/s00775-012-0893-0 PMID:22454108

Cai, C., Hu, S., Guo, J., Shi, Y., Xie, G.-J., & Yuan, Z. (2015). Nitrate reduction by denitrifying anaerobic methane oxidizing microorganisms can reach a practically useful rate. *Water Research*, *87*, 211–217. doi:10.1016/j.watres.2015.09.026 PMID:26414889

Calderer, M., Marti, V., de Pablo, J., Guivernau, M., Prenafreta-Boldú, F. X., & Viñan, M. (2014). Efffects of ehnanced denitrification on hydrodinamics and microbial community structure in a soil column system. *Chemosphere*, *111*, 112–119. doi:10.1016/j.chemosphere.2014.03.033 PMID:24997907

Calheiros, C. S. C., Rangel, A. O. S. S., & Castro, P. M. L. (2007). Constructed wetland systems vegetated with different plants applied to the treatment of tannery wastewater. *Water Research*, *41*(8), 1790–1798. doi:10.1016/j.watres.2007.01.012 PMID:17320926

Calheiros, C. S. C., Rangel, A. O. S. S., & Castro, P. M. L. (2008). Evaluation of different substrates to support the growth of *Typha latifolia* in constructed wetlands treating tannery wastewater over long-term operation. *Bioresource Technology*, *99*(15), 6866–6877. doi:10.1016/j.biortech.2008.01.043 PMID:18334290

Calheiros, C., Rangel, A., & Castro, P. (2014). Constructed Wetlands for Tannery Wastewater Treatment in Portugal: Ten Years of Experience. *International Journal of Phytoremediation*, *16*(9), 859–870. doi:10.1080/15226514.2013.798622 PMID:24933889

Camargo, J. A., & Alonso, A. (2006). Ecological and toxicological effects of inorganic nitrogen pollution in aquatic ecosystems: A global assessment. *Environment International*, *32*(6), 831–849. doi:10.1016/j.envint.2006.05.002 PMID:16781774

Campos, E., Illa, J., Magrí, A., Palatsi, J., Solé-Mauri, F., & Flotats, X. (2004). *Guía de los tratamientos de deyecciones ganaderas.* Agencia de Residuos de Cataluña. Retrieved from http://www.arc-cat.net/es/altres/purins/guia/pdf/guia_dejeccions.pdf

Campos, J. L., Vázquez-Padín, J. R., Figueroa, M., Fajardo, C., Mosquera-Corral, A., & Méndez, R. (2010). Novel biological nitrogen-removal processes: applications and perspectives. In K.W. Canton (Ed.), Fluid Waste Disposal (pp. 153-181). Academic Press.

Campos, J. L., Carvalho, S., Portela, R., Mosquera-Corral, A., & Méndez, R. (2008). Kinetics of denitrification using sulphur compounds: Effects of S/N ratio, endogenous and exogenous compounds. *Bioresource Technology*, *99*(5), 1293–1299. doi:10.1016/j.biortech.2007.02.007 PMID:17399983

Can-Dogan, E., Turker, M., Dagasan, L., & Arslan, A. (2010). Sulfide removal from industrial wastewaters by lithotrophic denitrification using nitrate as an electron acceptor. *Water Science and Technology*, *62*(10), 2286–2293. doi:10.2166/wst.2010.545 PMID:21076214

Canfield, D. E., Glazer, A. N., & Falkowski, P. G. (2010). The evolution and future of Earth's nitrogen cycle. *Science*, *330*(6001), 192–196. doi:10.1126/science.1186120 PMID:20929768

Capodaglio, A.G., Hlavínek, P., & Raboni, M. (2015). Physico-chemical technology for nitrogen removal from wastewater: a review. *Ambiente & Agua-An interdisciplinary Journal of Applied Science*, 1980-1993.

Carballa, M., Moerman, W., De Windt, W., Grootaerd, H., & Vestraete, W. (2008). Strategies to optimize phosphate removal from industrial anaerobic effluents by magnesium ammonium phosphate (MAP) production. *Journal of Chemical Technology and Biotechnology (Oxford, Oxfordshire)*, *84*(1), 63–68. doi:10.1002/jctb.2006

Carballeira, T., Ruiz, I., & Soto, M. (2016). Effect of plants and surface loading rate on the treatment efficiency of shallow subsurface constructed wetlands. *Ecological Engineering*, *90*, 203–214. doi:10.1016/j.ecoleng.2016.01.038

Carucci, A., Milia, S., De Gioannis, G., & Piredda, M. (2008). Acetate-fed aerobic granular sludge for the degradation of chlorinated phenols. *Water Science and Technology*, *58*(2), 309–315. doi:10.2166/wst.2008.394 PMID:18701779

Carucci, A., Milia, S., De Gioannis, G., & Piredda, M. (2009). Acetate-fed aerobic granular sludge for the degradation of 4-chlorophenol. *Journal of Hazardous Materials*, *166*(1), 483–490. doi:10.1016/j.jhazmat.2008.11.049 PMID:19111986

Carvalheira, M., Oehmen, A., Carvalho, G., Eusébio, M., & Reis, M. A. M. (2014). The impact of aeration on the competition between polyphosphate accumulating organisms and glycogen accumulating organisms. *Water Research*, *66*, 296–307. doi:10.1016/j.watres.2014.08.033 PMID:25222333

Cavet, J. S., Borrelly, G. P., & Robinson, N. J. (2003). Zn, Cu and Co in cyanobacteria: Selective control of metal availability. *FEMS Microbiology Reviews*, *27*(2-3), 165–181. doi:10.1016/S0168-6445(03)00050-0 PMID:12829266

Çeçen, F., & Çakiroĝlu, D. (2001). Impact of landfill leachate on the co-treatment of domestic wastewater. *Biotechnology Letters*, *23*(10), 821–826. doi:10.1023/A:1010317823529

Celmer, D., Oleszkiewicz, J. A., & Cicek, N. (2008). Impact of shear force on the biofilm structure and performance of a membrane biofilm reactor for tertiary hydrogen-driven denitrification of municipal wastewater. *Water Research*, *42*(12), 3057–3065. doi:10.1016/j.watres.2008.02.031 PMID:18396310

Cema, G. (2009). *Comparative study on different Anammox systems*. TRITA_LWR PhD Thesis 1053.

Cema, G., Wiszniowski, J., Żabczyński, S., Zabłocka-Godlewska, W., Raszka, A., & Surmacz-Górska, J. (2007). Biological nitrogen removal from landfill leachate by deammonification assisted by heterotrophic denitrification in a rotating biological contactor. *Water Science and Technology*, *55*(8-9), 35–42. doi:10.2166/wst.2007.239 PMID:17546967

Cerrillo, M., Palatsi, J., Comas, J., Vicens, J., & Bonmatí, A. (2015). Struvite precipitation as a technology to be integrated in a manure anaerobic digestion treatment plant – removal efficiency, crystal characterization and agricultural assessment. *Journal of Chemical Technology and Biotechnology (Oxford, Oxfordshire)*, *90*(6), 1135–1143. doi:10.1002/jctb.4459

Cervantes, F. J. (2009). *Environmental Technologies to Treat Nitrogen Pollution: Principles and Engineering*. IWA Publishing.

Cervantes, F. J., de la Rosa, D. A., & Gómez, J. (2001). Nitrogen removal from wastewaters at low C/N ratios with ammonium and acetate as electron donors. *Bioresource Technology*, *79*(2), 165–170. doi:10.1016/S0960-8524(01)00046-3 PMID:11480925

Cervantes, F. J., Meza-Escalante, E. R., Texier, A. C., & Gómez, J. (2009). Kinetic limitations during the simultaneous removal of p-cresol and sulfide in a denitrifying process. *Journal of Industrial Microbiology & Biotechnology*, *36*(11), 1417–1424. doi:10.1007/s10295-009-0628-6 PMID:19672641

Cervantes, F. J., Monroy, O., & Gómez, J. (1999). Influence of ammonium on the performance of a denitrifying culture under heterotrophic conditions. *Applied Biochemistry and Biotechnology*, *81*(1), 13–21. doi:10.1385/ABAB:81:1:13 PMID:15304781

Chamchoi, N., Nitisoravut, S., & Schidt, J. E. (2008). Inactivation of ANAMMOX communities under concurrent operation of anaerobic ammonium oxidation (ANAMMOX) and denitrification. *Bioresource Technology*, *99*(9), 3331–3336. doi:10.1016/j.biortech.2007.08.029 PMID:17911013

Chang, C. C., Tseng, S. K., & Huang, H. K. (1999). Hydrogenotrophic denitrification with immobilized *Alcalogenes eutrophus* for drinking water treatment. *Bioresource Technology*, *69*(1), 53–58. doi:10.1016/S0960-8524(98)00168-0

Chang, Y., Wu, S., Zhang, T., Mazur, R., Pang, C., & Dong, R. (2014). Dynamics of nitrogen transformation depending on different operational strategies in laboratory-scale tidal flow constructed wetlands. *The Science of the Total Environment*, *487*, 49–56. doi:10.1016/j.scitotenv.2014.03.114 PMID:24768911

Chen, C., Ho, K. L., Liu, F. C., Ho, M., Wang, A., Ren, N., & Lee, D. J. (2013). Autotrophic and heterotrophic denitrificataion by a newly isolated strain *Pseudomonas* sp. C27. *Bioresource Technology*, *145*, 351–356. doi:10.1016/j.biortech.2012.12.027 PMID:23434385

Chen, G. (2004). Electrochemical technologies in wastewater treatment. *Separation and Purification Technology*, *38*(1), 11–41. doi:10.1016/j.seppur.2003.10.006

Cheng, K. Y., Kaksonen, A. H., & Cord-Ruwisch, R. (2013). Ammonia recycling enables sustainable operation of bioelectrochemical systems. *Bioresource Technology*, *143*, 25–31. doi:10.1016/j.biortech.2013.05.108 PMID:23774293

Chen, H., Liu, S., Yang, F., Xue, Y., & Wang, T. (2009). The development of simultaneous partial nitrification, ANAMMOX and denitrification (SNAD) process in a single reactor for nitrogen removal. *Bioresource Technology*, *100*(4), 1548–1554. doi:10.1016/j.biortech.2008.09.003 PMID:18977657

Chen, L. A., Carbonell, R. G., & Serad, G. A. (1999). Recovery of proteins and other biological compounds using fibrous materials: I. Adsorption by salt addition. *Journal of Chemical Technology and Biotechnology (Oxford, Oxfordshire)*, *74*(8), 733–739. doi:10.1002/(SICI)1097-4660(199908)74:8<733::AID-JCTB111>3.0.CO;2-4

Chen, Y., Randall, A. A., & McCue, T. (2004). The efficiency of enhanced biological phosphorus removal from real wastewater affected by different ratios of acetic to propionic acid. *Water Research*, *38*(1), 27–36. doi:10.1016/j.watres.2003.08.025 PMID:14630100

Chen, Y., Wen, Y., Zhou, Q., & Vymazal, J. (2014). Effects of plant biomass on nitrogen transformation in subsurface-batch constructed wetlands: A stable isotope and mass balance assessment. *Water Research*, *63*, 158–167. doi:10.1016/j.watres.2014.06.015 PMID:25000198

Chiumenti, A., da Borso, F., Chiumenti, R., Teri, F., & Segantin, P. (2013). Treatment of digestate from a co-digestion biogas plant by means of vacuum evaporation: Tests for process optimization and environmental sustainability. *Waste Management (New York, N.Y.)*, *33*(6), 1339–1344. doi:10.1016/j.wasman.2013.02.023 PMID:23562449

Chiu, Z. C., Chen, M. Y., Lee, D. J., Wang, C. H., & Lai, J. Y. (2007). Oxygen diffusion in active layer of aerobic granule with step change in surrounding oxygen levels. *Water Research*, *41*(4), 884–892. doi:10.1016/j.watres.2006.11.035 PMID:17224172

Christensen, T. H., Kjeldsen, P., Bjerg, P. L., Jensen, D. L., Christensen, J. B., Baun, A., & Heron, G. et al. (2001). Biogeochemistry of Landfill leachate plumes. *Applied Geochemistry*, *16*(7-8), 659–718. doi:10.1016/S0883-2927(00)00082-2

Christensson, M., Ekström, S., Chan, A. A., Le Vaillant, E., & Lemaire, R. (2013). Experience from start-ups of the first ANITA Mox Plants. *Water Science and Technology*, *67*(12), 2677–2684. doi:10.2166/wst.2013.156 PMID:23787303

Claus, G., & Kutzner, J. (1985). Physiology and kinetics of autotrophic denitrification by Thiobacillus denitrificans. *Applied Microbiology and Biotechnology*, *22*, 283–288.

*Clean Water Act of 1972, 33 U.S.C. § 1251 et seq.* (2002). Retrieved from http://epw.senate.gov/water.pdf

Cohen, Y., & Kirchmann, H. (2004). Increasing the pH of wastewater to high levels with different gases CO2 stripping. *Water, Air, and Soil Pollution*, *159*(1), 265–275. doi:10.1023/B:WATE.0000049185.69759.b0

*Colsen: Water Energy & Environment*. (n.d.). Retrieved May 2016 from www.colsen.nl

Comeau, Y., Hall, K. J., Hancock, R. E. W., & Oldham, W. K. (1986). Biochemical model for enhanced biological phosphorus removal. *Water Research*, *20*(12), 1511–1521. doi:10.1016/0043-1354(86)90115-6

Commission Directive 98/15/EC of 27 February 1998 amending Council Directive 91/271/EEC with respect to certain requirements established in Annex I thereof (1998) OJ L67/29.

Cooney, E. L., Booker, N. A., Shallcross, D. C., & Stevens, G. W. (1999). Ammonia removal from wastewater using natural Australian zeolite. I. Characterization of the zeolite. *Separation Science and Technology*, *34*(12), 2307–2327. doi:10.1081/SS-100100774

Cordell, D., Rosemarin, A., Schröder, J. J., & Smit, A. L. (2011). Towards global phosphorus security: A systems framework for phosphorus recovery and reuse options. *Chemosphere*, *84*(6), 747–758. doi:10.1016/j.chemosphere.2011.02.032 PMID:21414650

Corominas, L., Larsen, H. F., Flores-Alsina, X., & Vanrolleghem, P. A. (2013). Including life cycle assessment for decision-making in controlling wastewater nutrient removal systems. *Journal of Environmental Management*, *128*, 759–767. doi:10.1016/j.jenvman.2013.06.002 PMID:23856224

Correa, J. A., Castilla, J. C., Ramírez, M., Varas, M., Lagos, N., Vergara, S., & Brown, M. T. et al. (1999). Copper, copper mine tailings and their effect on marine algae in Northern Chile.[journal article]. *Journal of Applied Phycology*, *11*(1), 57–67. doi:10.1023/A:1008027610826

Corsino, S. F., Campo, R., Di Bella, G., Torregrossa, M., & Viviani, G. (2015). Cultivation of granular sludge with hypersaline oily wastewater. *International Biodeterioration & Biodegradation*, *105*, 192–202. doi:10.1016/j.ibiod.2015.09.009

Corsino, S. F., Capodici, M., Morici, C., Torregrossa, M., & Viviani, G. (2016). Simultaneous nitritation-denitritation for the treatment of high-strength nitrogen in hypersaline wastewater by aerobic granular sludge. *Water Research*, *88*, 329–336. doi:10.1016/j.watres.2015.10.041 PMID:26512811

Council directive 1999/31/EC of 26 April 1999 on the landfill of waste

Council Directive 76/160/EEC of 8 December 1975 concerning the quality of bathing water (1976) OJ L31/1.

Council Directive 76/464/EEC of 4 May 1976 on pollution caused by certain dangerous substances discharged into the aquatic environment of the Community (1976) OJ L129/23.

Council Directive 91/271/EEC of 21 May 1991 concerning urban waste-water treatment (1991) OJ L135/40.

Council Directive 91/676/EEC of 12 December 1991 concerning the protection of waters against pollution caused by nitrates from agricultural sources (1991) OJ L375/1.

Courtens, E. N. P., Boon, N., De Clippeleir, H., Berckmoes, K., Mosquera, M., Seuntjens, D., & Vlaeminck, S. E. (2014). Control of nitratation in an oxygen-limited autotrophic nitrification/denitrification rotating biological contactor through disc immersion level variation. *Bioresource Technology*, *155*, 182–188. doi:10.1016/j.biortech.2013.12.108 PMID:24457304

Crutchik, D., Morales, N., Vázquez-Padín, J. R., Rogalla, R., Garrido, J. M., & Lema, J. M. (2015). *Enhancement Of Struvite Crystallization With Magnesium Hydroxide: Results Obtained At Full Scale.* Paper presented at the 12th IWA Leading Edge Conference on Water and Wastewater Technologies, Hong Kong.

Crutchik, D., & Garrido, J. M. (2011). Struvite crystallization versus amorphous magnesium and calcium phosphate precipitation during the treatment of a saline industrial wastewater. *Water Science and Technology*, *64*(12), 2460–2467. doi:10.2166/wst.2011.836 PMID:22170842

Crutchik, D., & Garrido, J. M. (2016). Kinetics of the reversible reaction of struvite crystallisation. *Chemosphere*, *154*, 567–572. doi:10.1016/j.chemosphere.2016.03.134 PMID:27085317

Cuervo-López, F. M., Martínez-Hernández, S., Texier, A. C., & Gómez, J. (2009). Denitrification for wastewater treatment. In F. Cervantes (Ed.), *Environmental Technologies to treat nitrogen pollution. Principles and Engineering* (pp. 41–65). International Water Association Publishing.

Cuervo-López, F., Martínez Hernández, S., Texier, A. C., & Gómez, J. (2009). Denitrification for Wastewater Treatment. In F. J. Cervantes (Ed.), *Environmental Technologies to Treat Nitrogen Pollution: Principles and Engineering* (pp. 23–65). London: IWA Publication.

Cui, L., Feng, J., Ouyang, Y., & Deng, P. (2012). Removal of nutrients from septic effluent with re-circulated hybrid tidal flow constructed wetland. *Ecological Engineering*, *46*(0), 112–115. doi:10.1016/j.ecoleng.2012.06.003

Cui, M., Ma, A., Qi, H., Zhuang, X., & Zhuang, G. (2015). Anaerobic oxidation of methane: An «active» microbial process. *Microbiology Open*, *4*(1), 1–11. doi:10.1002/mbo3.232 PMID:25530008

Cunha, C. A., Macieira, S., Dias, J. M., Almeida, G., Goncalves, L. L., Costa, C., & Romao, M. J. et al. (2003). Cytochrome c nitrite reductase from Desulfovibrio desulfuricans ATCC 27774. The relevance of the two calcium sites in the structure of the catalytic subunit (NrfA). *The Journal of Biological Chemistry*, *278*(19), 17455–17465. doi:10.1074/jbc.M211777200 PMID:12618432

Cydzik-Kwiatkowska, A., & Wojnowska-Baryła, I. (2011). Nitrifying granules cultivation in a sequencing batch reactor at a low organics-to-total nitrogen ratio in wastewater. *Folia Microbiologica*, *56*(3), 201–208. doi:10.1007/s12223-011-0037-x PMID:21618047

Daiber, A., Shoun, H., & Ullrich, V. (2005). Nitric oxide reductase (P450nor) from Fusarium oxysporum. *Journal of Inorganic Biochemistry*, *99*(1), 185–193. doi:10.1016/j.jinorgbio.2004.09.018 PMID:15598501

Daims, H., Lebedeva, E. V., Pjevac, P., Han, P., Herbold, C., Albertsen, M., & Wagner, M. et al. (2015). Complete nitrification by Nitrospira bacteria. *Nature*, *528*(7583), 504–509. http://www.nature.com/nature/journal/v528/n7583/abs/nature16461.html#supplementary-information. doi:10.1038/nature16461 PMID:26610024

Dai, Y., Jiang, Y., & Su, H. (2015). Influence of an aniline supplement on the stability of aerobic granular sludge. *Journal of Environmental Management*, *162*, 115–122. doi:10.1016/j.jenvman.2015.05.017 PMID:26233584

Daniels, R. P. (2007). Tannery effluent and reed beds: Working with nature. *The Journal of American Leather Chemists Association*, *102*(8), 248–253.

Dapena-Mora, A., Fernandez, I., Campos, J. L., Mosquera-Corral, A., Mendez, R., & Jetten, M. S. M. (2007). Evaluation of activity and inhibition effects on Anammox process by batch tests based on the nitrogen gas production. *Enzyme and Microbial Technology*, *40*(4), 859–865. doi:10.1016/j.enzmictec.2006.06.018

Dapena-Mora, A., van Hulle, S. W. H., Campos, J. L., Mendez, R., Vanrolleghem, P. A., & Jetten, M. (2004). Enrichment of Anammox biomass from municipal activated sludge: Experimental and modelling results. *Journal of Chemical Technology and Biotechnology (Oxford, Oxfordshire)*, *79*(12), 1421–1428. doi:10.1002/jctb.1148

Dapena-Mora, A., Vázquéz-Padin, J. R., Campose, J. L., Mosquera-Corral, A., Jetten, M. S. M., & Méndez, R. (2010). Monitoring the stability of an Anammox reactor under high salinity conditions. *Biochemical Engineering Journal*, *51*(3), 167–171. doi:10.1016/j.bej.2010.06.014

Dasgupta, P. K., & Dong, S. (1986). Solubility of ammonia in liquid water and generation of trace levels of standard gaseous ammonia. *Atmospheric Environment*, *20*(3), 565–570. doi:10.1016/0004-6981(86)90099-5

Daverey, A., Su, S. H., Huang, Y. T., Chen, S. S., Sung, S., & Lin, J. G. (2013). Partial nitrification and anammox process: A method for high strength optoelectronic industrial wastewater treatment. *Water Research*, *47*(9), 2929–2937. doi:10.1016/j.watres.2013.01.028 PMID:23548564

Daverey, A., Su, S. H., Huang, Y. T., & Lin, J. G. (2012). Nitrogen removal from opto-electronic wastewater using the simultaneous partial nitrification, anaerobic ammonium oxidation and denitrification (SNAD) process in sequencing batch reactor. *Bioresource Technology*, *113*, 225–231. doi:10.1016/j.biortech.2011.12.004 PMID:22237171

Day, D. F., & Ingram, J. M. (1975). In vitro studies of an alkaline phosphatase-cell wall complex from Pseudomonas aeruginosa. *Canadian Journal of Microbiology*, *21*(1), 9–16. doi:10.1139/m75-002 PMID:803858

De Clippeleir, H., Yan, X., Verstraete, W., & Vlaeminck, S. (2011). OLAND is feasible to treat sewage-like nitrogen concentrations at low hydraulic residence times. *Applied Microbiology and Biotechnology*, *90*(4), 1537–1545. doi:10.1007/s00253-011-3222-6 PMID:21465304

de Kreuk, M. K. (2006). *Aerobic granular sludge: scaling up a new technology*. Delft, The Netherlands: Technical University Delft.

de Kreuk, M. K., Heijnen, J. J., & van Loosdrecht, M. C. M. (2005). Simultaneous COD, nitrogen and phosphate removal by aerobic granular sludge. *Biotechnology and Bioengineering*, *90*(6), 761–769. doi:10.1002/bit.20470 PMID:15849693

de Kreuk, M. K., Kishida, N., Tsuneda, S., & van Loosdrecht, M. C. M. (2010). Behavior of polymeric substrates in an aerobic granular sludge system. *Water Research*, *44*(20), 5929–5938. doi:10.1016/j.watres.2010.07.033 PMID:20817210

de la Varga, D., Díaz, A., Ruiz, I., & Soto, M. (2013a). Avoiding clogging in constructed wetlands by using anaerobic digesters as pre-treatment. *Ecological Engineering*, *52*, 262–269. doi:10.1016/j.ecoleng.2012.11.005

de la Varga, D., Ruíz, I., Álvarez, J. A., & Soto, M. (2015). Methane and carbon dioxide emissions from constructed wetlands receiving anaerobically pretreated sewage. *The Science of the Total Environment*, *538*, 824–833. doi:10.1016/j.scitotenv.2015.08.090 PMID:26342902

de la Varga, D., Ruiz, I., & Soto, M. (2013b). Winery wastewater treatment in subsurface constructed wetlands with different bed depths. *Water, Air, and Soil Pollution*, *224*(4), 1–13. doi:10.1007/s11270-013-1485-5

de Vries, S., Strampraad, M. J., Lu, S., Moenne-Loccoz, P., & Schroder, I. (2003). Purification and characterization of the MQH2:NO oxidoreductase from the hyperthermophilic archaeon Pyrobaculum aerophilum. *The Journal of Biological Chemistry*, *278*(38), 35861–35868. doi:10.1074/jbc.M300857200 PMID:12799376

de-Bashan, L. E., & Bashan, Y. (2004). Recent advances in removing phosphorus from wastewater and its future use as fertilizer (1997-2003). *Water Research*, *38*(19), 4222–4246. doi:10.1016/j.watres.2004.07.014 PMID:15491670

Deligeer, F., Fukunaga, R., Kataoka, K., Yamaguchi, K., Kobayashi, K., Tagawa, S., & Suzuki, S. (2002). Spectroscopic and functional characterization of Cu-containing nitrite reductase from Hyphomicrobium denitrificans A3151. *Journal of Inorganic Biochemistry*, *91*(1), 132–138. doi:10.1016/S0162-0134(02)00442-7 PMID:12121770

Deng, Y., & Englehardt, J. D. (2006). Treatment of landfill leachate by the Fenton process. *Water Research*, *40*(20), 3683–3694. doi:10.1016/j.watres.2006.08.009 PMID:17045628

Deng, Y., & Englehardt, J. D. (2007). Electrochemical oxidation for landfill leachate treatment. *Waste Management (New York, N.Y.), 27*(3), 380–388. doi:10.1016/j.wasman.2006.02.004 PMID:16632340

Desloover, J., De Clippeleir, H., Boeckx, P., Du Laing, G., Colsen, J., Verstraete, W., & Vlaeminck, S. E. (2011). Floc-based sequential partial nitritation and anammox at full scale with contrasting N2O emissions. *Water Research, 45*(9), 2811–2821. doi:10.1016/j.watres.2011.02.028 PMID:21440280

Desmidt, E., Ghyselbrecht, K., Zhang, Y., Pinoy, L., van der Bruggen, B., Verstraete, W., & Meesschaert, B. et al. (2015). Global phosphorus scarcity and full-scale P-recovery techniques: A review. *Critical Reviews in Environmental Science and Technology, 45*(4), 336–384. doi:10.1080/10643389.2013.866531

Devisscher, M., Bogaert, H., Bixio, D., Van de Velde, J., & Thoeye, C. (2002). Feasibility of automatic chemicals dosage control – a full-scale evaluation. *Water Science and Technology, 45*(4-5), 445–452. PMID:11936665

Di Bella, G., & Torregrossa, M. (2013). Simultaneous nitrogen and organic carbon removal in aerobic granular sludge reactors operated with high dissolved oxygen concentration. *Bioresource Technology, 142*, 706–713. doi:10.1016/j.biortech.2013.05.060 PMID:23751809

Di Bella, G., & Torregrossa, M. (2014). Aerobic Granular Sludge for Leachate Treatment. *Chemical Engineering Transactions, 38*, 493–498. doi:10.3303/CET1438083

Dias, V. N., Canseiro, C., Gomes, A. R., Correia, B., & Bicho, C. (2006). Constructed wetlands for wastewater treatment in Portugal: a global review.*10th International Conference on Wetland Systems for Water Pollution Control.*

DiMuro, J. L., Guertin, F. M., Helling, R. K., Perkins, J. L., & Romer, S. (2014). A financial and environmental analysis of constructed wetlands for industrial wastewater treatment. *Journal of Industrial Ecology, 18*(5), 631–640. doi:10.1111/jiec.12129

Directive 2000/60/EC of the European Parliament and of the Council of 23 October 2000 establishing a framework for Community action in the field of water policy (2000) OJ L327/1.

Directive 2006/11/EC of the European Parliament and of the Council of 15 february 2006 on pollution caused by certain dangerous substances discharged into the aquatic environment of the community (2006) OJ L64/52.

Directive 2006/118/EC of the European Parliament and of the Council of 12 December 2006 on the protection of groundwater against pollution and deterioration (Daughter to 2000/60/EC) (2006) OJ L372/19.

Directive 2006/7/EC of the European Parliament and of the Council of 15 February 2006 concerning the management of bathing water quality and repealing Directive 76/160/EEC (2006) OJ L64/37.

Directive 2008/56/EC of the European Parliament and of the Council of 17 June 2008 establishing a framework for community action in the field of marine environmental policy (Marine Strategy Framework Directive) (2008) OJ L164/19.

Directive 2008/98/EC of the European Parliament and of the Council of 19 November 2008 on waste and repealing certain Directives (2008) OJ L312/3.

Directive 2010/75/EU of the European Parliament and of the Council of 24 November 2010 on industrial emissions (integrated pollution prevention and control) (2010) OJ L334/17.

Dockhorn, T. (2009). About the economy of phosphorus recovery. In *Proceedings of the International Conference on Nutrient Recovery from Wastewater Streams*. Vancouver, Canada: IWA Publishing.

Dong, X., & Tollner, E. W. (2003). Evaluation of anammox and denitrification during anaerobic digestion of poultry manure. *Bioresource Technology, 86*(2), 139–145. doi:10.1016/S0960-8524(02)00154-2 PMID:12653278

Doyle, J., & Parsons, S. (2002). Struvite formation, control and recovery. *Water Research*, *36*(16), 3925–3940. doi:10.1016/S0043-1354(02)00126-4 PMID:12405401

Drizo, A., Frost, C. A., Smith, K. A., & Grace, J. (1997). Phosphate and ammonium removal by constructed wetlands with horizontal subsurface flow, using shale as a substrate. *Water Science and Technology*, *35*(5), 95–102. doi:10.1016/S0273-1223(97)00057-7

Dunne, E. J., Culleton, N., O'Donovan, G., Harrington, R., & Olsen, A. E. (2005). An integrated constructed wetland to treat contaminants and nutrients from dairy farmyard dirty water. *Ecological Engineering*, *24*(3), 221–234. doi:10.1016/j.ecoleng.2004.11.010

Duque, A. F. (2011). *Fluorophenol: Performance of biofilm and aerobic granular sludge reactors*. Universidade Católica Portuguesa.

Duque, A. F., Bessa, V. S., Carvalho, M. F., de Kreuk, M. K., van Loosdrecht, M. C. M., & Castro, P. M. L. (2011). 2-Fluorophenol degradation by aerobic granular sludge in a sequencing batch reactor. *Water Research*, *45*(20), 6745–6752. doi:10.1016/j.watres.2011.10.033 PMID:22060965

Eggers, E., Dirkzwager, A. H., & Van der Honing, H. (1991). Full-scale experiences with phosphate crystallization in a crystalactor. *Water Science and Technology*, *23*, 819–824.

Egli, K., Fanger, U., Alvarez, P. J. J., Siegrist, H., van der Meer, J. R., & Zehnder, A. J. B. (2001). Enrichment and characterization of an anammox bacterium from a rotating biological contactor treating ammonium-reach leachate. *Archives of Microbiology*, *175*(3), 198–207. doi:10.1007/s002030100255 PMID:11357512

Ekama, G. A., & Wentzel, M. C. (2004). Modelling inorganic material in activated sludge systems. *Water S.A.*, *30*(2), 153–174. doi:10.4314/wsa.v30i2.5060

El Diwani, G., Rafie, S., Ibiari, N., & El-Aila, H. (2007). Recovery of ammonia nitrogen from industrial wastewater treatment as struvite slow releasing fertilizer. *Desalination*, *214*(1-3), 200–214. doi:10.1016/j.desal.2006.08.019

Elvira, C., Goicoechea, M., Sampedro, L., Mato, S., & Nogales, R. (1996). Bioconversion of solid paperpulp mill sludge by earthworms. *Bioresource Technology*, *57*(2), 173–177. doi:10.1016/0960-8524(96)00065-X

Ensign, S. A., Hyman, M. R., & Arp, D. J. (1993). In vitro activation of ammonia monooxygenase from Nitrosomonas europaea by copper. *Journal of Bacteriology*, *175*(7), 1971–1980. PMID:8458839

Environmental Protection Agency. (2016). *State Development of Numeric Criteria for Nitrogen and Phosphorus Pollution.* Retrieved from https://www.epa.gov/nutrient-policy-data/state-development-numeric-criteria-nitrogen-and-phosphorus-pollution

EPA. (1987). *Process design manual for phosphorus removal. US EPA Proceedings 625/1-87/001*. Washington, DC: Environmental Protection Agency.

EPA. (2010). *Nutrient control design manual. US EPA Proceedings 600/R-10/100*. Washington, DC: Environmental Protection Agency.

EPA-820-F-15-096. (2015). *A compilation of cost data associated with the impacts and control of nutrient pollution.* Retrieved from https://www.epa.gov/sites/production/files/2015-04/documents/nutrient-economics-report-2015.pdf

EPA-821-R-04-011. (2004). *Technical Development Document for the Final Effluent Limitations Guidelines and Standards for the Meat and Poultry Products Point Source Category (40 CFR 432)*. Retrieved January 15, 2016, from http://nepis.epa.gov/Exe/ZyPDF.cgi/P1001576.PDF?Dockey=P1001576.PDF

EPA-823-R-07-002. (2007). *Bilogical Nutrient Removal Processes and Costs*. Retrieved from https://www.epa.gov/sites/production/files/documents/criteria_nutrient_bioremoval.pdf

Epsztein, R., Beliavski, M., Tarre, S., & Green, M. (2016). High-rate hydrogenotrophic denitrification in a pressurized reactor. *Chemical Engineering Journal*, *286*, 578–584. doi:10.1016/j.cej.2015.11.004

Ergas, S. J., & Reuss, A. F. (2001). Hydrogenotrophic denitrification of drinking water using a hollow fibre membrane bioreactor. *Journal of Water Supply: Research & Technology - Aqua*, *50*, 161–171.

Erisman, J. W., Sutton, M. A., Galloway, J. N., Klimont, Z., & Winiwarter, W. (2008). How a century of ammonia synthesis changed the world. *Nature Geoscience*, *1*(10), 636–639. doi:10.1038/ngeo325

Ettwig, K. F., Butler, M. K., Le Paslier, D., Pelletier, E., Mangenot, S., Kuypers, M. M. M., & Strous, M. et al. (2010). Nitrite-driven anaerobic methane oxidation by oxygenic bacteria. *Nature*, *464*(7288), 543–548. doi:10.1038/nature08883 PMID:20336137

EU-BAT. (2006). *Reference document on best available techniques for the manufacture of large volume inorganic chemicals – ammonia, acids and fertilizers*. Sevilla, Spain: Integrated Pollution Prevention and Control. European Comissions.

European Commission. (2003). *CWW BREF. Integrated Pollution Prevention and Control, Reference Document on Best Available Techniques in Common Waste Water and Waste Gas Treatment / Management Systems in the Chemical Sector*. Retrieved from The European IPPC Bureau Joint Research Centre website: http://eippcb.jrc.ec.europa.eu/reference/BREF/cww_bref_0203.pdf

European Environment Agency. (2008). *Water quality and pollution by nutriets*. Retrieved from http://www.eea.europa.eu/themes/water/status-and-monitoring/state-of-surface-waters/water-quality-and-pollution-by-nutrients

European Environment Agency. (2016a). *European Pollutant Release and Transfer Register (E-PRTR)*. Retrieved January 15, 2016, from http://prtr.ec.europa.eu/#/home

European Environment Agency. (2016b). *Emission intensity of manufacturing industries in Europe*. Retrieved January 15, 2016, from http://www.eea.europa.eu/data-and-maps/indicators/emission-intensity-of-manufacturing-industries-1/assessment

Fajardo, C., Mora, M., Fernández, I., Mosquera-Corral, A., Campos, J. L., & Méndez, R. (2014). Cross effect of temperatura, pH and free ammonia on autotrophic denitrification process with sulphide as electron donor. *Chemosphere*, *97*, 10–15. doi:10.1016/j.chemosphere.2013.10.028 PMID:24216266

Fajardo, C., Mosquera-Corral, A., Campos, J. L., & Méndez, R. (2012). Autotrophic denitrification with sulphide in a sequencing batch reactor. *Journal of Environmental Management*, *113*, 552–556. doi:10.1016/j.jenvman.2012.03.018 PMID:22704254

Fajardo, C., Mosquera-Corral, A., Campos, J. L., & Méndez, R. (2013). Post-treatment of fish canning effluents by sequential nitrification and autotrophic denitrification processes. *Process Biochemistry*, *48*(9), 1368–1374. doi:10.1016/j.procbio.2013.06.005

Fangueiro, D., Coutinhob, J., Borges, L., & Vasconcelosa, E. (2015). Recovery efficiency of nitrogen from liquid and solid fractions of pig slurry obtained using different separation technologies. *Journal of Plant Nutrition and Soil Science*, *178*(2), 229–236. doi:10.1002/jpln.201400261

Fangueiro, D., Ribeiroa, H., Vasconcelosa, E., Coutinhob, J., & Cabrala, F. (2009). Treatment by acidification followed by solid-liquid separation affects slurry and slurry fraction scomposition and their potential of N mineralization. *Bioresource Technology*, *100*(20), 4914–4917. doi:10.1016/j.biortech.2009.04.032 PMID:19464881

FAO. (2011, June). *Current world fertilizer trends and outlook to2015*. Retrieved from http://www.fao.org/3/a-av252e.pdf

Fay, A. W., Blank, M. A., Lee, C. C., Hu, Y., Hodgson, K. O., Hedman, B., & Ribbe, M. W. (2010). Characterization of isolated nitrogenase FeVco. *Journal of the American Chemical Society*, *132*(36), 12612–12618. doi:10.1021/ja1019657 PMID:20718463

Fernandes, A. T., Damas, J. M., Todorovic, S., Huber, R., Baratto, M. C., Pogni, R., & Martins, L. O. et al. (2010). The multicopper oxidase from the archaeon Pyrobaculum aerophilum shows nitrous oxide reductase activity. *The FEBS Journal*, *277*(15), 3176–3189. doi:10.1111/j.1742-4658.2010.07725.x PMID:20597980

Fernandez, I., Vazquez-Padin, J. R., Mosquera-Corral, A., Campos, J. L., & Mendez, R. (2008). Biofilm and granular systems to improve Anammox biomass retention. *Biochemical Engineering Journal*, *42*(3), 308–313. doi:10.1016/j.bej.2008.07.011

Fernández, N., Gómez, R., Amils, R., Sierra-Alvarez, R., Field, J. A., & Sanz, J. L. (2006). Microbial and structural aspects of granular sludge from autotrophic denitrifying reactors. *Water Science and Technology*, *54*(2), 11–17. doi:10.2166/wst.2006.480 PMID:16939078

Fernández, N., Sierra-Alvarez, R., Field, J. A., Amils, R., & Sanz, J. L. (2008). Microbial community dynamics in a chemolithotrophic denitrification reactor inoculated with methanogenic granular sludge. *Chemosphere*, *70*(3), 462–474. doi:10.1016/j.chemosphere.2007.06.062 PMID:17689587

Ferraz, F. M., Povinelli, J., Pozzi, E., Vieira, E. M., & Trofino, J. C. (2014). Co-treatment of landfill leachate and domestic wastewater using a submerged aerobic biofilter. *Journal of Environmental Management*, *141*, 9–15. doi:10.1016/j.jenvman.2014.03.022 PMID:24762568

Figueras, M. J., Polo, F., Inza, I., & Guarro, J. (1997). Past, present and future perspectives of the EU Bathing Water Directive. *Marine Pollution Bulletin*, *34*(3), 148–156. doi:10.1016/S0025-326X(97)00107-0

Figueroa, M., Val del Rio, A., Campos, J. L., Méndez, R., & Mosquera-Corral, A. (2015). Filamentous bacteria existence in aerobic granular reactors. *Bioprocess and Biosystems Engineering*, *38*(5), 841–851. doi:10.1007/s00449-014-1327-x PMID:25533039

Figueroa, M., Val Del Río, A., Campos, J. L., Mosquera-Corral, A., & Méndez, R. (2011). Treatment of high loaded swine slurry in an aerobic granular reactor. *Water Science and Technology*, *63*(9), 1808–1814. doi:10.2166/wst.2011.381 PMID:21902017

Figueroa, M., Vázquez-Padín, J. R., Mosquera-Corral, A., Campos, J. L., & Méndez, R. (2012). Is the CANON reactor an alternative for nitrogen removal from pre-treated swine slurry? *Biochemical Engineering Journal*, *65*, 23–29. doi:10.1016/j.bej.2012.03.008

Filipe, C. D. M., Daigger, G. T., & Grady, C. P. L. Jr. (2001). Stoichiometry and kinetics of acetate uptake under anaerobic conditions by an enriched culture of phosphorus-accumulating organisms at different pHs. *Biotechnology and Bioengineering*, *76*(1), 32–43. doi:10.1002/bit.1023 PMID:11400104

Filippelli, G. M. (2002). The Global Phosphorus Cycle. *Reviews in Mineralogy and Geochemistry*, *48*(1), 391–425. doi:10.2138/rmg.2002.48.10

Filippelli, G. M. (2008). The Global Phosphorus Cycle: Past, Present, and Future. *Elements*, *4*(2), 89–95. doi:10.2113/GSELEMENTS.4.2.89

Flere, J. M., & Zhang, T. C. (1999). Nitrate removal with sulphur-limestone autotrophic denitrification process. *Journal of Environmental Engineering*, *125*(8), 721–729. doi:10.1061/(ASCE)0733-9372(1999)125:8(721)

Flores-Alsina, X., Gernaey, K. V., & Jeppsson, U. (2012). Benchmarking biological nutrient removal in wastewater treatment plants: Influence of mathematical model assumptions. *Water Science and Technology*, *65*(8), 1496–1505. doi:10.2166/wst.2012.039 PMID:22466599

Flores-Alsina, X., Kazadi-Mbama, C., Solon, K., Vrecko, D., Tait, S., Batstone, D., & Gernaey, K. V. et al. (2015). A plant wide aqueous phase chemistry module describing pH variations and ion speciation/pairing in wastewater treatment models. *Water Research*, *85*, 255–265. doi:10.1016/j.watres.2015.07.014 PMID:26342179

Flores-Alsina, X., Solon, K., Kazadi-Mbamba, C., Tait, S., Jeppsson, U., Gernaey, K. V., & Batstone, D. J. (2016). Modelling phosphorus (P), sulfur (S) and iron (Fe) interactions for dynamic simulations of anaerobic digestion processes. *Water Research*, *95*, 370–382. doi:10.1016/j.watres.2016.03.012 PMID:27107338

Flotats, X., Foged, H. L., Bonmati, A., Palatsi, J., Magri, A., & Schelde, K. M. (2011). *Manure Processing technologies.* Technical Report No. II concerning "Manure Processing Activities in Europe" to the European Commission, Directorate-General Environment. Project reference: ENV.B.1/ETU/2010/0007. Retrieved from http://hdl.handle.net/2117/18944

Flotats, X., Bonmatí, A., Fernandez, B., & Magrí, A. (2009). Manure treatment technologies: On-farm versus centralized strategies NE of Spain as case study. *Bioresource Technology*, *100*(22), 5516–5526. doi:10.1016/j.biortech.2008.12.050 PMID:19231173

Foged, H. L., Flotats, X., Bonmati, A., Palatsi, J., & Magri, A. (2011). End and by-products from livestock manure processing - general types, chemical composition, fertilising quality and feasibility for marketing. Technical Report No. III concerning "Manure Processing Activities in Europe" to the European Commission. Directorate-General Environment. Retrieved from http://hdl.handle.net/2117/18945

Foglar, L., & Briški, F. (2003). Wastewater denitrification process-the influence of methanol and kinetic analysis. *Process Biochemistry*, *39*(1), 95–103. doi:10.1016/S0032-9592(02)00318-7

Fomina, M., & Gadd, G. M. (2014). Biosorption: Current perspectives on concept, definition and application. *Bioresource Technology*, *160*, 3–14. doi:10.1016/j.biortech.2013.12.102 PMID:24468322

Fowler, D., Coyle, M., Skiba, U., Sutton, M. A., Cape, J. N., Reis, S., & Voss, M. et al. (2013). The global nitrogen cycle in the twenty-first century. *Philosophical Transactions of the Royal Society of London. Series B, Biological Sciences*, *368*(1621). doi:10.1098/rstb.2013.0164

Franca, R. D. G., Vieira, A., Mata, A. M. T., Carvalho, G. S., Pinheiro, H. M., & Lourenço, N. D. (2015). Effect of an azo dye on the performance of an aerobic granular sludge sequencing batch reactor treating a simulated textile wastewater. *Water Research*, *85*, 327–336. doi:10.1016/j.watres.2015.08.043 PMID:26343991

Francis, C. A., Beman, J. M., & Kuypers, M. M. M. (2007). New processes and players in the nitrogen cycle: The microbial ecology of anaerobic and archaeal ammonia oxidation. *The ISME Journal*, *1*(1), 19–27. doi:10.1038/ismej.2007.8 PMID:18043610

Frear, C., Liao, W., Ewing, T., & Chen, S. (2011). Evaluation of co-digestion at a commercial dairy anaerobic digester. *Clean-Soil, Air. Water (Basel)*, *39*, 697–704.

Frijlink, M. J., Abee, T., Laanbroek, H. J., de Boer, W., & Konings, W. N. (1992). The bioenergetics of ammonia and hydroxylamine oxidation in *Nitrosomonas europaea* at acid and alkaline pH. *Archives of Microbiology*, *157*(2), 194–199. doi:10.1007/BF00245290

Fudala-Książek, S., Łuczkiewicz, A., Firobor, K., & Olańczuk-Neyman, K. (2012). Nitrogen removal via the nitrite pathway during wastewater co-treatment with ammonia-rich landfill leachates in a sequencing batch reactor. *Environmental Science and Pollution Research International*, *21*(12), 7307–7318. doi:10.1007/s11356-014-2641-1 PMID:24569868

Fukumoto, Y., & Haga, K. (2004). Advanced treatment of swine wastewater by electrodialysis with a tubular ion exchange membrane. *Animal Science Journal*, *75*(5), 479–485. doi:10.1111/j.1740-0929.2004.00216.x

Fux, C., & Siegrist, H. (2004). Nitrogen removal from sludge digester liquids by nitrification/denitrification or partial nitritation/anammox: Environmental and economical considerations. *Water Science and Technology*, *50*(10), 19–26. PMID:15656291

Gabarró, J., Ganigué, R., Gich, F., Ruscalleda, M., Balaguer, M.D., & Colprim, J. (2012). Effect of temperature on AOB activity of partial nitritation SBR treating landfill leachate with extremely high nitrogen concentration. *Bioresource Technology*, *126*, 283-289.

Gadekar, S., Nemati, M., & Hill, G. A. (2006). Batch and continuous bio-oxidation of sulphide by *Thiomicrospira* sp. CVO: Reaction kinetics and stoichiometry. *Water Research*, *40*(12), 2436–2446. doi:10.1016/j.watres.2006.04.007 PMID:16730776

Galloway, J. N., Aber, J. D., Erisman, J. W., Seitzinger, S. P., Howarth, R. W., Cowling, E. B., & Cosby, B. J. (2003). The nitrogen cascade. *Bioscience*, *53*(4), 341–356. doi:10.1641/0006-3568(2003)053[0341:TNC]2.0.CO;2

Galloway, J. N., Schlesinger, W. H., Levy, H. II, Michaels, A., & Schnoor, J. L. (1995). Nitrogen fixation: Anthropogenic enhancement-environmental response. *Global Biogeochemical Cycles*, *9*(2), 235–252. doi:10.1029/95GB00158

Ganesh, R., Balaji, G., & Ramanujam, R. A. (2006). Biodegradation of tannery wastewater using sequencing batch reactor– Respirometric assessment. *Bioresource Technology*, *97*(15), 1815–1821. doi:10.1016/j.biortech.2005.09.003 PMID:16230010

Ganigué, R., Gabarró, J., Sànchez-Melsió, A., Ruscalleda, M., López, H., Vila, X., & Balaguer, M. D. et al. (2009). Long-term operation of a partial nitritation pilot plant treating leachate with extremely high ammonium concentration prior to an anammox process. *Bioresource Technology*, *100*(23), 5624–5632. doi:10.1016/j.biortech.2009.06.023 PMID:19577465

Ganigué, R., López, H., Balaguer, M. D., & Colprim, J. (2007). Partial ammonium oxidation to nitrite of high ammonium content urban landfill leachate. *Water Research*, *41*(15), 3317–3326. doi:10.1016/j.watres.2007.04.027 PMID:17597177

Ganigué, R., Volcke, E. I. P., Puig, S., Balaguer, M. S., & Colprim, J. (2012). Impact of influent characteristics on a partial nitritation SBR treating high nitrogen load wastewater. *Bioresource Technology*, *111*, 62–69. doi:10.1016/j.biortech.2012.01.183 PMID:22405759

Gao, D., Liu, L., Liang, H., & Wu, W.-M. (2011). Aerobic granular sludge: Characterization, mechanism of granulation and application to wastewater treatment. *Critical Reviews in Biotechnology*, *31*(2), 137–152. doi:10.3109/07388551.2010.497961 PMID:20919817

Gao, J. F., Zhang, Q., Su, K., & Wang, J. H. (2010). Competitive biosorption of Yellow 2G and Reactive Brilliant Red K-2G onto inactive aerobic granules: Simultaneous determination of two dyes by first-order derivative spectrophotometry and isotherm studies. *Bioresource Technology*, *101*(15), 5793–5801. doi:10.1016/j.biortech.2010.02.091 PMID:20303747

Gao, J., Zhang, Q., Su, K., Chen, R., & Peng, Y. (2010). Biosorption of Acid Yellow 17 from aqueous solution by non-living aerobic granular sludge. *Journal of Hazardous Materials*, *174*(1-3), 215–225. doi:10.1016/j.jhazmat.2009.09.039 PMID:19783368

Garcia-González, M. C., & Vanotti, M. B. (2015). Recovery of ammonia from swine manure using gas-permeable membranes: Effect of waste strength and pH. *Waste Management (New York, N.Y.)*, *38*, 455–461. doi:10.1016/j.wasman.2015.01.021 PMID:25687948

Garuti, G., Giordano, A., & Pirozzoli, F. (2001). Full-scale ANANOX system performance. *Water S.A.*, *27*, 189–197.

Geary, P. M., & Moore, J. A. (1999). Suitability of a Treatment Wetland for Dairy Wastewaters. *Water Science and Technology*, *40*(3), 179–185. doi:10.1016/S0273-1223(99)00464-3

Gentry, T., Rensing, C., & Pepper, I. (2004). New Approaches for Bioaugmentation as a Remediation Technology. *Critical Reviews in Environmental Science and Technology*, *34*(5), 447–494. doi:10.1080/10643380490452362

Gernaey, K. V., & Jørgensen, S. B. (2004). Benchmarking combined biological phosphorus and nitrogen removal wastewater treatment processes. *Control Engineering Practice*, *12*(3), 357–373. doi:10.1016/S0967-0661(03)00080-7

Gevertz, D., Telanj, A. J., Voordouw, G., & Jenneman, G. E. (2000). Isolation and characterization of strains CVO and FWKO B, two novel nitrate reducing, sulfide-oxidizing bacteria isolated from oil field brine. *Applied and Environmental Microbiology*, *66*(6), 2491–2501. doi:10.1128/AEM.66.6.2491-2501.2000 PMID:10831429

Ghafari, S., Hasan, M., & Aroua, M. K. (2008). Bio-electrochemical removal of nitrate from water and wastewater-A review. *Bioresource Technology*, *99*(10), 3965–3974. doi:10.1016/j.biortech.2007.05.026 PMID:17600700

Ghafari, S., Hasan, M., & Aroua, M. K. (2009). Effect of carbon dioxide and bicarbonate as inorganic carbon sources on growth and adaptation of hydrogenotrophic denitrifying bacteria. *Journal of Hazardous Materials*, *162*(2-3), 1507–1513. doi:10.1016/j.jhazmat.2008.06.039 PMID:18639979

Ghafari, S., Hasan, M., & Aroua, M. K. (2010). A kinetic study of autohydrogenotrophic denitrification at the optimum pH and sodium bicarbonate dose. *Bioresource Technology*, *101*(7), 2236–2242. doi:10.1016/j.biortech.2009.11.068 PMID:20015639

Gharibi, H., Mahvi, A. H., Chehrazi, M., Sheikhi, R., & Hosseini, R. (2010). Phosphorous removal from wastewater effluent using electro-coagulation by aluminum and iron plates. *Analytical and Bioanalytical Chemistry*, *2*, 165–177.

Giesen, A. (1999). Crystallization process enables environmental friendly phosphate removal at low costs. *Environmental Technology*, *20*(7), 769–775. doi:10.1080/09593332008616873

Giesen, A., de Bruin, L. M. M., Niermans, R. P., & van der Roest, H. F. (2013). Advancements in the application of aerobic granular biomass technology for sustainable treatment of wastewater. *Water Practice and Technology*, *8*(1), 47–54. doi:10.2166/wpt.2013.007

Gilch, S., Vogel, M., Lorenz, M. W., Meyer, O., & Schmidt, I. (2009). Interaction of the mechanism-based inactivator acetylene with ammonia monooxygenase of Nitrosomonas europaea. *Microbiology*, *155*(1), 279–284. doi:10.1099/mic.0.023721-0 PMID:19118368

Ginestet, P., Audic, J. M., Urbain, V., & Block, J. C. (1998). Estimation of nitrifying bacterial activities by measuring oxygen uptake in the presence of the metabolic inhibitors allylthiourea and azide. *Applied and Environmental Microbiology*, *64*, 2266–2268. PMID:9603846

Gisvold, B., Odegaard, H., & Follesdal, M. (2000). Enhanced clay aggregate filters. *Water Science and Technology*, *41*, 409–416.

Gleeson, D. B., Clipson, N., Melville, K., Gadd, G. M., & McDermott, F. P. (2005). Characterization of fungal community structure on a weathered pegmatitic granite. *Microbial Ecology*, *50*(3), 360–368. doi:10.1007/s00248-005-0198-8 PMID:16328654

Gobi, K., & Vadivelu, V. M. (2014). Aerobic dynamic feeding as a strategy for in situ accumulation of polyhydroxyalkanoate in aerobic granules. *Bioresource Technology*, *161*, 441–445. doi:10.1016/j.biortech.2014.03.104 PMID:24725384

Gobi, K., & Vadivelu, V. M. (2015). Polyhydroxyalkanoate recovery and effect of in situ extracellular polymeric substances removal from aerobic granules. *Bioresource Technology*, *189*, 169–176. doi:10.1016/j.biortech.2015.04.023 PMID:25889804

Goldman, S., Hecht, K., Eisenberg, H., & Mevarech, M. (1990). Extracellular Ca2(+)-dependent inducible alkaline phosphatase from extremely halophilic archaebacterium Haloarcula marismortui. *Journal of Bacteriology*, *172*(12), 7065–7070. PMID:2123861

Gong, N., Chen, C., Xie, L., Chen, H., Lin, X., & Zhang, R. (2005). Characterization of a thermostable alkaline phosphatase from a novel species Thermus yunnanensis sp. nov. and investigation of its cobalt activation at high temperature. *Biochimica et Biophysica Acta*, *1750*(2), 103–111. doi:10.1016/j.bbapap.2005.05.007 PMID:15955749

González-Blanco, G., Cuervo-López, F., Cervantes, F. J., Beristain-Cardoso, R., & Gómez, J. (2013). Nitrite as oxidizing power for p-cresol removal using a denitrifying consortium: Kinetic study. *Journal of Chemical Technology and Biotechnology (Oxford, Oxfordshire)*, *88*(12), 2176–2180. doi:10.1002/jctb.4082

González-Domenech, C. M., Béjar, V., Martínez-Checa, F., & Quesada, E. (2008). *Halomonas nitroreducens* sp. nov., a novel nitrate- and nitrite-reducing species. *International Journal of Systematic and Evolutionary Microbiology*, *58*(4), 872–876. doi:10.1099/ijs.0.65415-0 PMID:18398185

González-Ponce, R., & García López, D. M. E. (2007). Evaluation of struvite as fertilizer: A comparison with traditional P sources. *Agrochimica*, *51*, 301–308.

González-Ponce, R., López-de-Sá, E. G., & Plaza, C. (2009). Lettuce Response to Phosphorus Fertilization with Struvite Recovered from Municipal Wastewater. *HortScience*, *44*, 426–430.

Gorra, R., Freppaz, M., Zanini, E., & Scalenghe, R. (2014). Mountain dairy wastewater treatment with the use of a 'irregularly shaped' constructed wetland (Aosta Valley, Italy). *Ecological Engineering*, *73*, 176–183. doi:10.1016/j.ecoleng.2014.09.013

Grady, C. P. L., Gujer, W., Henze, M., Marais, G. V. R., & Matsuo, T. (1986). A model for single-sludge wastewater treatment systems. *Water Science and Technology*, *16*(6), 47–61.

Grand View Research. (2016). *Ammonium Sulfate Market Analysis By Product (Solid Ammonium Sulfate, Liquid Ammonium Sulfate), By Application (Fertilizers, Feed Additives, Industrial Use) And Segment Forecasts To 2022*. Grand View Research.

Grüter, H., Matter, M., Oehlmann, K. H., & Hicks, M. D. (1990). Drying of sewage sludge- and important step in waste disposal. *Water Science and Technology*, *22*(12), 57–63.

Guerrero, J. (2013). *Improving EBPR stability in WWTPs aiming at simultaneous carbon and nutrient removal: from modeling studies to experimental validation.* (PhD Thesis). Universitat Autònoma de Barcelona, Spain.

Guerrero, J., Guisasola, A., & Baeza, J. A. (2011b). The nature of the carbon source rules the competition between PAO and denitrifiers in systems for simultaneous biological nitrogen and phosphorus removal. *Water Research*, *45*(16), 4793–4802. doi:10.1016/j.watres.2011.06.019 PMID:21774957

Guerrero, J., Guisasola, A., & Baeza, J. A. (2014). A novel control strategy for efficient biological phosphorus removal with carbon-limited wastewaters. *Water Science and Technology*, *70*(4), 691–697. doi:10.2166/wst.2014.280 PMID:25116500

Guerrero, J., Guisasola, A., & Baeza, J. A. (2015). Controlled crude glycerol dosage to prevent EBPR failures in C/N/P removal WWTPs. *Chemical Engineering Journal*, *271*, 114–127. doi:10.1016/j.cej.2015.02.062

Guerrero, J., Guisasola, A., Comas, J., Rodríguez-Roda, I., & Baeza, J. A. (2011a). Multi-criteria selection of optimum WWTP control setpoints base on microbiology-related failures, effluent quality and operating costs. *Chemical Engineering Journal*, *188*, 23–29. doi:10.1016/j.cej.2012.01.115

Guerrero, J., Guisasola, A., Vilanova, R., & Baeza, J. A. (2011c). Improving the performance of a WWTP control system by model-based setpoint optimization. *Environmental Modelling & Software*, *26*(4), 492–497. doi:10.1016/j.envsoft.2010.10.012

Guisasola, A., Petzet, S., Baeza, J. A., Carrera, J., & Lafuente, J. (2007). Inorganic carbon limitations on nitrification: Experimental assessment and modelling. *Water Research*, *41*(2), 277–286. doi:10.1016/j.watres.2006.10.030 PMID:17169397

Guisasola, A., Vargas, M., Marcelino, M., Lafuente, J., Casas, C., & Baeza, J. A. (2007b). On-line monitoring of the enhanced biological phosphorus removal process using respirometry and titrimetry. *Biochemical Engineering Journal*, *35*(3), 371–379. doi:10.1016/j.bej.2007.02.001

Gu, J. D., Qiu, W., Koenig, A., & Fan, Y. (2004). Removal of high $NO_3^-$ concentration in saline water through autotrophic denitrification by the bacterium *Thiobacillus denitrificans* strain MP. *Water Science and Technology*, *49*(5-6), 105–112. PMID:15137413

Gujer, W., Henze, M., Mino, T., Matsuo, T., Wentzel, M. C., & Marais, G. R. (1995). The Activated Sludge Model No. 2: Biological phosphorus removal. *Water Science and Technology*, *31*(2), 1–11. doi:10.1016/0273-1223(95)00175-M

Gujer, W., Henze, M., Takahashi, M., & van Loosdrecht, M. C. M. (1999). Activated Sludge Model No. 3. *Water Science and Technology*, *29*(1), 183–193. doi:10.1016/S0273-1223(98)00785-9

Guo, H., Chen, C., Lee, D. J., Wang, A., Gao, D., & Ren, N. (2014). Couple carbon, sulfur and nitrogen cycles of mixotrophic growth of *Pseudomonas* sp. C27 under denitrifying sulfide removal conditions. *Bioresource Technology*, *171*, 120–126. doi:10.1016/j.biortech.2014.08.035 PMID:25189517

Guo, J. S., Abbas, A. A., Chen, Y. P., Liu, Z. P., & Chen, P. (2010). Treatment of landfill leachate using a combined stripping, Fenton, SBR, and coagulation process. *Journal of Hazardous Materials*, *178*(1-3), 699–705. doi:10.1016/j.jhazmat.2010.01.144 PMID:20188464

Guo, Q., Xing, B.-S., Li, P., Xu, J.-L., Yang, C.-C., & Jin, R.-C. (2015). Anaerobic ammonium oxidation (anammox) under realistic seasonal temperature variations: Characteristics of biogranules and process performance. *Bioresource Technology*, *192*, 765–773. doi:10.1016/j.biortech.2015.06.049 PMID:26111630

Guo, X., & Jin, X. (2014). Purification of UF-treated anaerobically digested manure wastewater by two-pass reverse osmosis. *Desalination and Water Treatment*, *52*(16-18), 3027–3034. doi:10.1080/19443994.2013.797671

Haaker, H., & Klugkist, J. (1987). The bioenergetics of electron transport to nitrogenase. *FEMS Microbiology Letters*, *46*(1), 57–71. doi:10.1111/j.1574-6968.1987.tb02452.x

Hailei, W., Guosheng, L., Ping, L., & Feng, P. (2006). The effect of bioaugmentation on the performance of sequencing batch reactor and sludge characteristics in the treatment process of papermaking wastewater. *Bioprocess and Biosystems Engineering*, *29*(5-6), 283–289. doi:10.1007/s00449-006-0077-9 PMID:16983527

Haimi, H., Mulas, M., Corona, F., & Vahala, E. (2013). Data-derived soft-sensors for biological wastewater treatment plants: An overview. *Environmental Modelling & Software*, *47*, 88–107. doi:10.1016/j.envsoft.2013.05.009

Hao, X.D., Heijnen, J.J., & van Loosdrecht, M.C.M. (2002b) Model-based evaluation of temperature and inflow variations on a partial nitrification-ANAMMOX biofilm process. *Water Research, 36*, 4839-4849.

Hao, X. D., Heijnen, J. J., & van Loosdrecht, M. C. M. (2002a). Sensitivity analysis of a biofilm model describing a one-stage completely autotrophic nitrogen removal (CANON) process. *Biotechnology and Bioengineering*, *77*(3), 266–277. doi:10.1002/bit.10105 PMID:11753935

Hao, X. D., & van Loosdrecht, M. C. M. (2004). Model-based evaluation of COD influence on a partial nitrification-Anammox biofilm (CANON) process. *Water Science and Technology*, *49*(11-12), 83–90. PMID:15303727

Haroon, M., Hu, S., Shi, Y., Imelfort, M., Keller, J., Hugenholtz, P., & Tyson, G. W. et al. (2013). Anaerobic oxidation of methane coupled to nitrate reduction in a novel archael lineage. *Nature*, *500*(7464), 567–572. doi:10.1038/nature12375 PMID:23892779

Hasar, H., Unsal, S. A., Ipek, U., Karatas, S., Cinar, O., Yaman, C., & Kınacı, C. (2009). Stripping/flocculation/membrane bioreactor/reverse osmosis treatment of municipal landfill leachate. *Journal of Hazardous Materials*, *171*(1-3), 309–317. doi:10.1016/j.jhazmat.2009.06.003 PMID:19559524

Hatzenpichler, R. (2012). Diversity, Physiology, and Niche Differentiation of Ammonia-Oxidizing Archaea. *Applied and Environmental Microbiology*, *78*(21), 7501–7510. doi:10.1128/AEM.01960-12 PMID:22923400

Hauduc, H., Guillot, S., Rieger, L., Ohtsuki, T., Shaw, A., Takács, I., & Winkler, S. (2009). Activated sludge modelling practice: An international survey. *Water Science and Technology*, *60*(8), 1943–1951. doi:10.2166/wst.2009.223 PMID:19844041

Hauduc, H., Rieger, L., Oehmen, A., van Loosdrecht, M. C. M., Comeau, Y., Héduit, A., & Guillot, S. et al. (2013). Critical review of activated sludge modeling: State of process knowledge, modeling concepts and limitations. *Biotechnology and Bioengineering*, *1*(1), 24–46. doi:10.1002/bit.24624 PMID:22886494

Hauduc, H., Takács, I., Smith, S., Szabo, A., Murthy, S., Daigger, G. T., & Spérandio, M. (2015). A dynamic physicochemical model for chemical phosphorus removal. *Water Research*, *73*, 157–170. doi:10.1016/j.watres.2014.12.053 PMID:25655322

Haug, R. T., & McCarty, P. L. (1972). Nitrification with submerged filters. *Journal - Water Pollution Control Federation*, *44*, 2086.

Hauksson, J. B., Andresson, O. S., & Asgeirsson, B. (2000). Heat-labile bacterial alkaline phosphatase from a marine Vibrio sp. *Enzyme and Microbial Technology*, *27*(1-2), 66–73. doi:10.1016/S0141-0229(00)00152-6 PMID:10862903

Have, P. J. W., & Chiappini, U. (1993). *Processing of manure surpluses.* European Conference. Environment Agriculture Stock Farming in Europe, Montova, Italy.

Häyrynen, K., Pongrácz, E., Väisänen, V., Pap, N., Mänttäri, M., Langwaldt, J., & Keiski, R. L. (2009). Concentration of ammonium and nitrate from mine water by reverse osmosis and nanofiltration. *Desalination*, *240*(1-3), 280–289. doi:10.1016/j.desal.2008.02.027

Healy, M. G., Rodgers, M., & Mulqueen, J. (2007). Treatment of dairy wastewater using constructed wetlands and intermittent sand filters. *Bioresource Technology*, *98*(12), 2268–2281. doi:10.1016/j.biortech.2006.07.036 PMID:16973357

Healy, M., & Cawley, A. M. (2002). Nutrient processing capacity of a constructed wetland in western Ireland. *Journal of Environmental Quality*, *31*(5), 1739–1747. doi:10.2134/jeq2002.1739 PMID:12371194

Hellinga, C., Schellen, A. A. J. C., Mulder, J. W., van Loosdrecht, M. C. M., & Heijnen, J. J. (1998). The sharon process: An innovative method for nitrogen removal from ammonium-rich waste water. *Water Science and Technology*, *37*(9), 135–142. doi:10.1016/S0273-1223(98)00281-9

Hellinga, C., van Loosdrecht, M. C. M., & Heijnen, J. J. (1999). Model based design of a novel process for nitrogen removal from concentrated flows. *Mathematical and Computer Modelling*, *5*(4), 351–371.

Hendrickx, T. L. G., Wang, Y., Kampman, C., Zeeman, G., Temmink, H., & Buisman, C. J. N. (2012). Autotrophic nitrogen removal from low strength waste water at low temperature. *Water Research*, *46*(7), 2187–2193. doi:10.1016/j.watres.2012.01.037 PMID:22349000

Henze, M., Grady, C. P. L., Jr., Gujer, W., Marais, G. R., & Matsuo, T. (1987). *Activated Sludge Model No. 1*. IAWQ Scientific and Technical Report No. 1.

Henze, M., Gujer, W., Mino, T., Matsuo, T., & van Loosdrecht, M. C. M. (2000). *Activated sludge models ASM1, ASM2, ASM2d and ASM3*. IWA Publishing.

Henze, M., Gujer, W., Mino, T., Matsuo, T., Wentzel, M. C., Marais, G. R., & van Loosdrecht, M. C. M. (1999). Activated sludge model No. 2D, ASM2D. *Water Science and Technology*, *39*(1), 165–182. doi:10.1016/S0273-1223(98)00829-4

Henze, M., van Loosdrecht, M., Ekama, G., & Brdjanovic, D. (2008). *Biological Wastewater Treatment*. London, UK: IWA Publishing.

He, S., Niu, Q., Ma, H., Zhang, Y., & Li, Y. Y. (2015). The Treatment Performance and the Bacteria Preservation of Anammox: A Review. *Water, Air, and Soil Pollution*, *226*(5), 163. doi:10.1007/s11270-015-2394-6

He, Y., Tao, W., Wang, Z., & Shayya, W. (2012). Effects of pH and seasonal temperature variation on simultaneous partial nitrification and anammox in free-water surface wetlands. *Journal of Environmental Management*, *110*, 103–109. doi:10.1016/j.jenvman.2012.06.009 PMID:22762912

He, Z., Geng, S., Pan, Y., Cai, C., Wang, J., Wang, L., & Hu, B. et al. (2015b). Improvement of the trace metal composition of medium for nitrite-dependent anaerobic methane oxidation bacteria: Iron (II) and copper (II) make a difference. *Water Research*, *85*, 235–243. doi:10.1016/j.watres.2015.08.040 PMID:26340061

He, Z., Geng, S., Shen, L., Lou, L., Zheng, P., Xu, X., & Hu, B. (2015a). The short- and long-term effects of environmental conditions on anaerobic methane oxidation coupled to nitrite reduction. *Water Research*, *68*, 554–562. doi:10.1016/j.watres.2014.09.055 PMID:25462761

He, Z., Geng, S., Wang, L., Cai, C., Wang, J., Liu, S., & Hu, B. et al. (2016). Improvement of mineral nutrient concentrations and pH control for the nitrite-dependent anaerobic methane oxidation process. *Separation and Purification Technology*, *162*, 148–153. doi:10.1016/j.seppur.2016.02.016

He, Z., Wang, J., Zhang, X., Cai, C., Geng, S., Zheng, P., & Hu, B. et al. (2015). Nitrogen removal from wastewater by anaerobic methane-driven denitrification in a lab-scale reactor: Heterotrophic denitrifiers associated with denitrifying methanotrophs. *Applied Microbiology and Biotechnology*, *99*(24), 10853–10860. doi:10.1007/s00253-015-6939-9 PMID:26342737

Hiatt, W. C., & Grady, C. P. L. Jr. (2008). An updated process model for carbon oxidation, nitrification, and denitrification. *Water Environment Research*, *80*(11), 2145–2156. doi:10.2175/106143008X304776 PMID:19024730

Hijosa-Valsero, M., Sidrach-Cardona, R., & Bécares, E. (2012). Comparison of interannualremoval variation of various constructed wetland types. *The Science of the Total Environment*, *430*, 174–183. doi:10.1016/j.scitotenv.2012.04.072 PMID:22647241

Hilles, A. H., Amr, S. S. A., Hussein, R. A., Arafa, A. I., & El-Sebaie, O. D. (2016). Optimization of leachate treatment using persulfate/H2O2 based advanced oxidation process: case study: Deir El-Balah Landfill Site, Gaza Strip, Palestine. *Water Science and Technology*, *73*(1), 102–112. doi:10.2166/wst.2015.468 PMID:26744940

Hippen, A., Rosenwinkel, K.-H., Baumgarten, G., & Seyfried, C. (1997). Aerobic deammonification: A new experience in the treatment of waste waters. *Water Science and Technology*, *35*(10), 111–120. doi:10.1016/S0273-1223(97)00211-4

Hjorth, M., Christensen, K. V., Christensen, M. L., & Sommer, S. G. (2010). Solid-liquid separation of animal slurry in theory and practice. A review. *Agron. Sust. Devel*, *30*(1), 153–180. doi:10.1051/agro/2009010

Hoilijoki, T. H., Kuttuen, R. H., & Rintala, J. A. (1999). Nitrification of anaerobically pretreated municipal landfill leachate at low temperature. *Water Research*, *34*(5), 1435–1446. doi:10.1016/S0043-1354(99)00278-X

Honda, H., & Kobayashi, T. (2000). Fuzzy control of bioprocess. *Journal of Bioscience and Bioengineering*, *89*(5), 401–408. doi:10.1016/S1389-1723(00)89087-8 PMID:16232768

Hood, C. R., & Randall, A. A. (2001). A biochemical hypothesis explaining the response of enhanced biological phosphorus removal biomass to organic substrates. *Water Research*, *35*(11), 2758–2766. doi:10.1016/S0043-1354(00)00573-X PMID:11456176

Howarth, R., Chan, F., Conley, D. J., Garnier, J., Doney, S. C., Marino, R., & Billen, G. (2011). Coupled biogeochemical cycles: Eutrophication and hypoxia in temperate estuaries and coastal marine ecosystems. *Frontiers in Ecology and the Environment*, *9*(1), 18–26. doi:10.1890/100008

Huang, H., Xiao, D., Zhang, Q., & Ding, L. (2014). Removal of ammonia from landfill leachate by struvite precipitation with the use of low-cost phosphate and magnesium sources. *Journal of Environmental Management*, *145*, 191–198. doi:10.1016/j.jenvman.2014.06.021 PMID:25043172

Huang, H., Xu, C., & Zhang, W. (2011). Removal of nutrients from piggery wastewater using struvite precipitation and pyrogenation technology. *Bioresource Technology*, *102*(3), 2523–2528. doi:10.1016/j.biortech.2010.11.054 PMID:21146405

Huang, J., Cai, W., Zhong, Q., & Wang, S. (2013). Influence of temperature on micro-environment, plant eco-physiology and nitrogen removal effect in subsurface flow constructed wetland. *Ecological Engineering*, *60*, 242–248. doi:10.1016/j.ecoleng.2013.07.023

Huang, L. Y., Lee, D. J., & Lai, J. Y. (2015). Forward osmosis membrane bioreactor for wastewater treatment with phosphorus recovery. *Bioresource Technology*, *198*, 418–423. doi:10.1016/j.biortech.2015.09.045 PMID:26409853

Hu, S., Zeng, R. J., Burow, L. C., Lant, P., Keller, J., & Yuan, Z. (2009). Enrichment of denitrifying anaerobic methane oxidizing microorganisms. *Environmental Microbiology Reports*, *1*(5), 377–384. doi:10.1111/j.1758-2229.2009.00083.x PMID:23765890

Hu, S., Zeng, R. J., Keller, J., Lant, P. A., & Yuan, Z. (2011). Effect of nitrate and nitrite on the selection of microorganisms in the denitrifying anaerobic methane oxidation process. *Environmental Microbiology Reports*, *3*(3), 315–319. doi:10.1111/j.1758-2229.2010.00227.x PMID:23761277

Hutnik, N., Kozik, A., Mazienczuk, A., Piotrowski, K., Wierzbowska, B., & Matynia, A. (2013). Phosphates (V) recovery from phosphorus mineral fertilizers industry wastewater by continuous struvite reaction crystallization process. *Water Research*, *47*(11), 3635–3643. doi:10.1016/j.watres.2013.04.026 PMID:23726699

Hwang, I. S., Min, K. S., Choi, E., & Yun, Z. (2005). Nitrogen removal from piggery waste used combined SHARON and ANAMMOX process. *Water Science and Technology*, *52*(10-11), 487–494. PMID:16459825

Iacopozzi, I., Innocenti, V., Marili-Libelli, S., & Giusti, E. (2007). A modified Activated Sludge Model No. 3 (ASM3) with two-step nitrification-denitrification. *Environmental Modelling & Software*, *22*(6), 847–861. doi:10.1016/j.envsoft.2006.05.009

Ibrahim, K. M. (2001). Evaluation of Jordanian Faujasite tuff by comparison with other natural and synthetic zeolites. *Environmental Geology*, *40*(4-5), 440–445. doi:10.1007/s002540000235

Ingildsen, P., Rose, C., Gernaey, K. V., Nielsen, M. K., Guidal, T., & Jacobsen, B. N. (2006). Modelling and control strategy testing of biological and chemical phosphorus removal at Avedøre WWTP. *Water Science and Technology*, *53*(4-5), 105–113.

Isaka, K., Yoshie, S., Sumino, T., Inamori, Y., & Tsuneda, S. (2007). Nitrification of landfill leachate using immobilized nitrifying bacteria at low temperatures. *Biochemical Engineering Journal*, *37*(1), 49–55. doi:10.1016/j.bej.2007.03.008

Işik, M., & Sponza, D. T. (2008). Anaerobic/aerobic treatment of a simulated textile wastewater. *Separation and Purification Technology*, *60*(1), 64–72. doi:10.1016/j.seppur.2007.07.043

Islas-Lima, S., Thalasso, F., & Gómez, J. (2004). Evidence of anoxic methane oxidation coupled to denitrification. *Water Research*, *38*(1), 13–16. doi:10.1016/j.watres.2003.08.024 PMID:14630098

Itokawa, H., Hanaki, K., & Matsuo, T. (2007). Nitrous oxide production in high-loading biological nitrogen removal process under low cod/n ratio condition. *Water Research*, *35*(3), 657–664. doi:10.1016/S0043-1354(00)00309-2 PMID:11228962

Ivanov, V., Wang, X. H., Tay, S. T. L., & Tay, J. H. (2006). Bioaugmentation and enhanced formation of microbial granules used in aerobic wastewater treatment. *Applied Microbiology and Biotechnology*, *70*(3), 374–381. doi:10.1007/s00253-005-0088-5 PMID:16091930

Jaffer, Y., Clark, T. A., Pearce, P., & Parsons, S. A. (2002). Potential phosphorus recovery by struvite formation. *Water Research*, *36*(7), 1834–1842. doi:10.1016/S0043-1354(01)00391-8 PMID:12044083

Jardin, N., & Hennerkes, J. (2012). Full-scale experience with the deammonification process to treat high strength sludge water - a case study. *Water Science and Technology*, *65*(3), 447–455. doi:10.2166/wst.2012.867 PMID:22258674

Jemaat, Z., Suárez-Ojeda, M. E., Pérez, J., & Carrera, J. (2013). Simultaneous nitritation and p-nitrophenol removal using aerobic granular biomass in a continuous airlift reactor. *Bioresource Technology*, *150*, 307–313. doi:10.1016/j.biortech.2013.10.005 PMID:24177164

Jemaat, Z., Suárez-Ojeda, M. E., Pérez, J., & Carrera, J. (2014a). Partial nitritation and o-cresol removal with aerobic granular biomass in a continuous airlift reactor. *Water Research*, *48*(1), 354–362. doi:10.1016/j.watres.2013.09.048 PMID:24140352

Jemaat, Z., Suárez-Ojeda, M. E., Pérez, J., & Carrera, J. (2014b). Sequentially alternating pollutant scenarios of phenolic compounds in a continuous aerobic granular sludge reactor performing simultaneous partial nitritation and o-cresol biodegradation. *Bioresource Technology*, *161*, 354–361. doi:10.1016/j.biortech.2014.03.071 PMID:24727695

Jenni, S., Vlaeminck, S. E., Morgenroth, E., & Udert, K. M. (2014). Successful application of nitritation/anammox to wastewater with elevated organic carbon to ammonia ratios. *Water Research*, *49*, 316–326. doi:10.1016/j.watres.2013.10.073 PMID:24355291

Jetten, M. S. M., Niftrik, L., Strous, M., Kartal, B., Keltjens, J. T., & Op den Camp, H. J. M. (2009). Biochemistry and molecular biology of anammox bacteria. *Critical Reviews in Biochemistry and Molecular Biology*, *44*(2-3), 65–84. doi:10.1080/10409230902722783 PMID:19247843

Jetten, M. S., Wagner, M., Fuerst, J., van Loosdrecht, M., Kuenen, G., & Strous, M. (2001). Microbiology and application of the anaerobic ammonium oxidation ('anammox') process. *Current Opinion in Biotechnology*, *12*(3), 283–288. doi:10.1016/S0958-1669(00)00211-1 PMID:11404106

Jiang, A., Zhang, T., Zhao, Q.-B., Li, X., Chen, S., & Frear, C. (2014). Evaluation of an integrated ammonia stripping, recovery, and biogas scrubbing system for use with anaerobically digested dairy manure. *Biosystems Engineering*, *119*, 117–126. doi:10.1016/j.biosystemseng.2013.10.008

Jiang, H. L., Tay, J. H., & Tay, S. T. L. (2002). Aggregation of immobilized activated sludge cells into aerobically grown microbial granules for the aerobic biodegradation of phenol. *Letters in Applied Microbiology*, *35*(5), 439–445. doi:10.1046/j.1472-765X.2002.01217.x PMID:12390497

Jing, C., Ping, Z., & Mahmood, Q. (2009). Simultaneous sulfide and nitrate removal in anaerobic reactor under shock loading. *Bioresource Technology*, *100*(12), 3010–3014. doi:10.1016/j.biortech.2008.12.041 PMID:19231174

Jin, R.-C., Yang, G.-F., Yu, J.-J., & Zheng, P. (2012). The inhibition of the Anammox process: A review. *Chemical Engineering Journal*, *197*, 67–79. doi:10.1016/j.cej.2012.05.014

Johansson, A. E., Gustavsson, A. M., Öquist, M. G., & Svensson, B. H. (2004). Methane emissions from a constructed wetland treating wastewater — seasonal and spatial distribution and dependence on edaphic factors. *Water Research*, *38*(18), 3960–3970. doi:10.1016/j.watres.2004.07.008 PMID:15380986

Jones, R. M., Dold, P. L., Takács, I., Chapman, K., Wett, B., Murphy, S., & O'Shaughnessy, M. (2007). Simulation for operation and control of reject water treatment processes. In *Proc. WEFTEC2007.*

Jorgensen, K., & Stoumann, J. L. (2009). Chemical and biochemical variation in animal manure solids separated using different commercial separation technologies. *Bioresource Technology*, *100*(12), 3088–3096. doi:10.1016/j.biortech.2009.01.065 PMID:19272772

Joss, A., Derlon, N., Cyprien, C., Burger, S., Szivak, I., Traber, J., & Morgenroth, E. (2011). Combined nitritation-anammox: Advances in understanding process stability. *Environmental Science & Technology*, *45*(22), 9735–9742. doi:10.1021/es202181v PMID:21981764

Jung, J. Y., Kang, S. H., Chung, Y. C., & Ahn, D. H. (2007). Factors affecting the activity of anammox bacteria during start up in the continuous culture reactor. *Water Science and Technology*, *55*(1-2), 459–468. doi:10.2166/wst.2007.023 PMID:17305171

Justin, P., & Kelly, D. P. (1978). Growth kinetics of *Thiobacillus denitrificans* in anaerobic and aerobic chemostat culture. *Journal of General Microbiology*, *107*(1), 123–130. doi:10.1099/00221287-107-1-123 PMID:24083

Ju, X., Wu, S., Zhang, Y., & Dong, R. (2014). Intensified nitrogen and phosphorus removal in a novel electrolysis-integrated tidal flow constructed wetland system. *Water Research*, *59*(0), 37–45. doi:10.1016/j.watres.2014.04.004 PMID:24784452

Kaczorek, K., & Ledakowicz, S. (2006). Kinetics of nitrogen removal from sanitary landfill leachate. *Bioprocess and Biosystems Engineering*, *29*(5-6), 291–304. doi:10.1007/s00449-006-0078-8 PMID:16944209

Kadlec, R. H., Knight, R. L., Vymazal, J., Brix, H., Cooper, P., & Haberl, R. (2000). *Constructed wetlands for pollution control – processes, performance, design and operation*. IWA Scientific and Technical Report No. 8. London, UK: IWA Publishing.

Kadlec, R. H., & Wallace, S. D. (2009). *Treatment Wetlands* (2nd ed.). Boca Raton, FL: CRC Press.

Kadlec, R. H., & Zmarthie, L. A. (2010). Wetland treatment of leachate from a closed landfill. *Ecological Engineering*, *36*(7), 946–957. doi:10.1016/j.ecoleng.2010.04.013

Kainz, H., & Gruber, G. (2006). *Wastewater treatment, sludge treatment. Lecture notes*. Austria: Technical University Graz.

Kajie, S., & Anraku, Y. (1986). Purification of a hexaheme cytochrome c552 from Escherichia coli K 12 and its properties as a nitrite reductase. *European Journal of Biochemistry*, *154*(2), 457–463. doi:10.1111/j.1432-1033.1986.tb09419.x PMID:3002798

Kalka, J. (2012). Landfill Leachate Toxicity Removal in Combined Treatment with Municipal Wastewater. *TheScientificWorldJournal*, *2012*, 1–7. doi:10.1100/2012/202897 PMID:22623882

Kampman, C., Hendrickx, T. L. G., Luesken, F. A., van Alen, T. A., Op den Camp, H. J. M., Jetten, M. S. M., & Temmink, H. et al. (2012). Enrichment of denitrifying methanotrophic bacteria for application after direct low-temperature anaerobic sewage treatment. *Journal of Hazardous Materials*, *227-228*, 164–171. doi:10.1016/j.jhazmat.2012.05.032 PMID:22657102

Kampschreur, M. J., Temmink, H., Kleerebezem, R., Jetten, M. S. M., & van Loosdrecht, M. C. M. (2009). Nitrous oxide emission during wastewater treatment. *Water Research*, *43*(17), 4093–4103. doi:10.1016/j.watres.2009.03.001 PMID:19666183

Kantawanichkul, S., Kladprasert, S., & Brix, H. (2009). Treatment of high-strength wastewater in tropical vertical flow constructed wetlands planted with *Typha angustifolia* and *Cyperus involucratus*. *Ecological Engineering*, *35*(2), 238–247. doi:10.1016/j.ecoleng.2008.06.002

Karabulut, A., Bouraoui, F., Grizzetti, B., Bidoglio, G., & Pistocchi, A. (2014). *Managing Nitrogen and Phosphorus Loads to Water Bodies: Characterisation and Solutions. Towards Macro-Regional Integrated Nutrient Management.* Retrieved from The European IPPC Bureau Joint Research Centre website: http://publications.jrc.ec.europa.eu/repository/bitstream/JRC91624/lb-na-26822-en-n.pdf

Karanasios, K. A., Michailidis, M. K., Vasiliadou, I. A., Pavlou, S., & Vayenas, D. V. (2011). Potable water hydrogenotrophic denitrification in packed-bed bioreactors coupled with a solar-electrolysis hydrogen production system. *Desalination and Water Treatment*, *33*(1-3), 86–96. doi:10.5004/dwt.2011.2614

Karanasios, K. A., Vasiliadou, I. A., Pavlou, S., & Vayenas, D. V. (2010). Hydrogenotorophic denitrification of potable water: A review. *Journal of Hazardous Materials*, *180*(1-3), 20–37. doi:10.1016/j.jhazmat.2010.04.090 PMID:20471745

Kartal, B. (2008). *Ecophysiology of the anammox bacteria.* (PhD thesis). Department of Microbiology, Radbound University Nijmegen, Nijmege.

Kartal,, B., van Niftrik,, L., Rattray,, J., van de Vossenberg,, J.L., & Schmid,, M.C., Sinninghe Damst, E. J., … Strous, M. (2008). Candidatus 'Brocadia fulgida': An autofluorescent anaerobic ammonium oxidizing bacterium. *FEMS Microbiology Ecology*, *63*, 46–55. doi:10.1111/j.1574-6941.2007.00408.x PMID:18081590

Kartal, B., de Almeida, N. M., Maalcke, W. J., Op den Camp, H. J. M., Jetten, M. S. M., & Keltjens, J. T. (2013). How to make a living from anaerobic ammonium oxidation. *FEMS Microbiology Reviews*, *37*(3), 428–461. doi:10.1111/1574-6976.12014 PMID:23210799

Kartal, B., Koleva, M., Arsov, R., van der Star, W., Jetten, M. S. M., & Strous, M. (2006). Adaptation of a freshwater anammox population to high salinity wastewater. *Journal of Biotechnology*, *126*(4), 546–553. doi:10.1016/j.jbiotec.2006.05.012 PMID:16806555

Kartal, B., Kuenen, J. G., & van Loosdrecht, M. C. M. (2010). Sewage treatment with anammox. *Science*, *328*(5979), 702–703. doi:10.1126/science.1185941 PMID:20448175

Kartal, B., Maalcke, W. J., de Almeida, N. M., Cirpus, I., Gloerich, J., Geerts, W., & Strous, M. et al. (2011). Molecular mechanism of anaerobic ammonium oxidation. *Nature*, *479*(7371), 127–130. doi:10.1038/nature10453 PMID:21964329

Kataki, S., West, H., Clarke, M., & Baruah, D. C. (2016). *Phosphorus recovery as struvite from farm, municipal and industrial waste: Feedstock suitability, methods and pre-treatments*. Waste Manage. doi:10.1016/j.wasman.2016.01.003

Kazadi Mbamaba, C., Flores-Alsina, X., Batstone, D., & Tait, S. (2015a). A generalised chemical precipitation modelling approach in wastewater treatment applied to calcite. *Water Research*, *68*, 342–353. doi:10.1016/j.watres.2014.10.011 PMID:25462741

Kazadi Mbamaba, C., Flores-Alsina, X., Batstone, D., & Tait, S. (2015b). A systematic study of multiple minerals precipitation modelling in wastewater treatment. *Water Research*, *85*, 359–370. doi:10.1016/j.watres.2015.08.041 PMID:26360229

Keener, W. K., & Arp, D. J. (1994). Transformations of aromatic compounds by *Nitrosomonas europea. Applied and Environmental Microbiology*, *60*, 1914–1920. PMID:16349282

Kee, T. C., Bay, H. H., Lim, C. K., Muda, K., & Ibrahim, Z. (2015). Development of bio-granules using selected mixed culture of decolorizing bacteria for the treatment of textile wastewater. *Desalination and Water Treatment*, *54*(1), 132–139. doi:10.1080/19443994.2013.877853

Kellerman, C., & Griebler, C. (2009). *Thiobacillus tiophillus*, sp. Nov., a chemolitoautotrophic, thiosulfate-oxidizing bacterium isolated from contaminated aquifer sediments. *International Journal of Systematic and Evolutionary Microbiology*, *59*(3), 583–588. doi:10.1099/ijs.0.002808-0 PMID:19244446

Kelly, D. P., & Wood, A. P. (2000). Reclasification of some species of *Thiobacillus* to the newly designated genera *Acidithiobacillus* gen. nov. *Halothiobacillus* gen. nov. and *Thermithiobacillus* gen. nov. *International Journal of Systematic and Evolutionary Microbiology*, *50*(2), 511–516. doi:10.1099/00207713-50-2-511 PMID:10758854

Kennedy, K. J., & Lentz, E. M. (2000). Treatment of landfill leachate using sequencing batch and continuous flow upflow anaerobic sludge blanket (UASB) reactors. *Water Research*, *34*(14), 3640–3656. doi:10.1016/S0043-1354(00)00114-7

Kern, J., & Idler, C. (1999). Treatment of domestic and agricultural wastewater by reed bed systems. *Ecological Engineering*, *12*(1), 13–25. doi:10.1016/S0925-8574(98)00051-2

Khin, T., & Annachhatre, A. P. (2004). Novel microbial nitrogen removal processes. *Biotechnology Advances*, *22*(7), 519–532. doi:10.1016/j.biotechadv.2004.04.003 PMID:15262315

Kim, H.-R., Lee, I.-S. & Bae, J.-H. (2004). Performance of a sulphur-utilizing fluidized bed reactor for post-denitrification. *Process Biotechnology*, *39*, 1591-1597.

Kim, D. G., Mishurov, M., & Kiely, G. (2010). Effect of increased N use and dry periods on N2O emission from a fertilized grassland. *Nutrient Cycling in Agroecosystems*, *88*(3), 397–410. doi:10.1007/s10705-010-9365-5

Kim, D. J., Lee, D. I., & Kellet, J. (2006). Effect of temperature and free ammonia on nitrification and nitrite accumulation in landfill leachate and analusis of itr nitrifying bacterial community by FISH. *Bioresource Technology*, *97*(3), 459–468. doi:10.1016/j.biortech.2005.03.032 PMID:15927463

Kim, E.-W., & Bae, J.-H. (2000). Alkalinity requirements and the possibility of simultaneous heterotrophic denitrification during sulphur-utilizing autotrophic denitrification. *Water Science and Technology*, *42*(3-4), 233–238.

Kim, I. S., & Son, J.-H. (2000). Impact of COD/N/S ratio on denitrification by the mixed culture of sulphate reducing bacteria and sulphur denitrifying bacteria. *Water Science and Technology*, *42*(3-4), 69–76.

Kim, J., Lim, J., & Lee, C. (2013). Quantitative real-time PCR approaches for microbial community studies in wastewater treatment systems: Applications and considerations. *Biotechnology Advances*, *31*(8), 1358–1373. doi:10.1016/j.biotechadv.2013.05.010 PMID:23747590

Kim, R. Y., Yoon, J. K., Kim, T. S., Yang, J. E., Owens, G., & Kim, K. R. (2015). Bioavailability of heavy metals in soils: Definitions and practical implementation--a critical review. *Environmental Geochemistry and Health*, *37*(6), 1041–1061. doi:10.1007/s10653-015-9695-y PMID:25841357

Kimura, K., Nakamura, M., & Watanabe, Y. (2002). Nitrate removal by a combination of elemental sulphur-based denitrification and membrane filtration. *Water Research*, *36*(7), 1758–1766. doi:10.1016/S0043-1354(01)00376-1 PMID:12044075

Kimura, Y., & Isaka, K. (2014). Evaluation of inhibitory effects of heavy metals on anaerobic ammonium oxidation (anammox) by continuous feedings tests. *Applied Microbiology and Biotechnology*, *98*(16), 6965–6972. doi:10.1007/s00253-014-5735-2 PMID:24723296

Kioussis, D. R., Wheaton, F. W., & Kofinas, P. (2000). Reactive nitrogen and phosphorus removal from aquaculture wastewater effluent using polymer hydrogels. *Aquacultural Engineering*, *23*(4), 315–332. doi:10.1016/S0144-8609(00)00058-3

Kjeldsen, P., Barlaz, M. A., Rooker, A. P., Baun, A., Ledin, A., & Christensen, T. H. (2002). Present and Long-Term Composition of MSW Landfill Leachate: A Review. *Critical Reviews in Environmental Science and Technology*, *32*(4), 297–336. doi:10.1080/10643380290813462

Kleerebezem, R., & Méndez, R. (2002). Autotrophic denitrification for combined hydrogen sulphide removal from and post-denitrification. *Water Science and Technology*, *45*(10), 349–356. PMID:12188569

Klimiuk, E., & Kulikowska, D. (2006). The influence of hydraulic retention time and sludge age on the kinetics of nitrogen removal from leachate in SBR. *Polish Journal of Environmental Studies*, *15*(2), 283–289.

Knight, R. L., Payne, V. W. E. Jr, Borer, R. E., Clarke, R. A. Jr, & Pries, J. H. (2000). Constructed wetlands for livestock wastewater management. *Ecological Engineering*, *15*(1-2), 41–55. doi:10.1016/S0925-8574(99)00034-8

Knowles, R. (1982). Denitrification. *Microbiological Reviews*, *46*(1), 43–70. PMID:7045624

Knowles, R. (2005). Denitrifiers associated with methanotrophs and their potential impact on the nitrogen cycle. *Ecological Engineering*, *24*(5), 441–446. doi:10.1016/j.ecoleng.2005.01.001

Kocaturk, I., & Erguder, T. H. (2015). Investigation of the use of aerobic granules for the treatment of sugar beet processing wastewater. *Environmental Technology*, *36*(20), 2577–2587. doi:10.1080/09593330.2015.1039070 PMID:25851439

Koch, F. A., Mavinic, S., Yonemitsu, N., & Britton, A. T. (2009). *Fluidized bed wastewater treatment*. United States patent US 7622047 B2.

Koch, G., Egli, K., van der Meer, J.R., & Siegrist, H. (2000). Mathematical modeling of autotrophic denitrification in a nitrifying biofilm of a rotating biological contactor. *Water Science & Technology*, *41*, 191-198.

Kodama, Y., & Watanabe, K. (2004). *Sulfuricurvum kujiense* gen. nov. sp. nov., a facultative anaerobic chelithoautotrophic, sulfur oxidyinz bacterium isolated from an underground crude-oil storage cavity. *International Journal of Systematic and Evolutionary Microbiology*, *54*(6), 2297–2300. doi:10.1099/ijs.0.63243-0 PMID:15545474

Koenig, A., & Liu, H. (1996). Autotrophic denitrification of landfill leachate using elemental sulphur. *Water Science and Technology*, *34*(5-6), 469–476. doi:10.1016/0273-1223(96)00680-4

Koenig, A., & Liu, L. H. (2001). Kinetic model of autotrophic denitrification in sulphur packed-bed reactors. *Water Research*, *35*(8), 1969–1978. doi:10.1016/S0043-1354(00)00483-8 PMID:11337843

Koenig, A., Tong, Z., Liu, L., & Fang, H. P. (2005). Microbial community and biochemistry process in autosulfurotrophic denitrifying biofilm. *Chemosphere*, *58*(8), 1041–1047. doi:10.1016/j.chemosphere.2004.09.040 PMID:15664612

Koening, A., & Liu, L. H. (2004). Autotrophic denitrification of high-salinity wastewater using elemental sulfur: Batch tests. *Water Environment Research*, *76*(1), 37–46. doi:10.2175/106143004X141564 PMID:15058463

Komarowski, S., & Yu, Q. (1997). Ammonium ion removal from wastewater using Australian natural zeolite: Batch equilibrium and kinetic studies. *Environmental Technology*, *18*(11), 1085–1097. doi:10.1080/09593331808616628

Komori, M., & Sakakibara, Y. (2008). High-rate hydrogenotrophic denitrification in a fluidized-bed biofilm reactor using solid-polymer-electrolyte membrane electrode (SPEME). *Water Science and Technology*, *58*(7), 1441–1446. doi:10.2166/wst.2008.725 PMID:18957757

Kong, Q., Zhi-Bin, W., Shu, L., & Miao, M.-S. (2015). Characterization of the extracellular polymeric substances and microbial community of aerobic granulation sludge exposed to cefalexin. *International Biodeterioration & Biodegradation*, *102*, 375–382. doi:10.1016/j.ibiod.2015.04.020

Konieczny, K., Kwiecińska, A., & Gworek, B. (2011). The recovery of water from slurry produced in high density livestock farming with the use of membrane processes. *Separation and Purification Technology*, *80*(3), 490–498. doi:10.1016/j.seppur.2011.06.002

Koops, H. P. & Pommerening-Röser, A. (2001). Distribution and ecophysiology of the nitrifying bacteria emphasizing cultured species. *FEMS Microbiology Ecology*, *37*, 1–9. doi:10.1111/j.1574-6941.2001.tb00847.x

Koparal, A. S., & Öğütveren, Ü. B. (2002). Removal of nitrate from water by electroreduction and electrocoagulation. *Journal of Hazardous Materials*, *89*(1), 83–94. doi:10.1016/S0304-3894(01)00301-6 PMID:11734348

Kourti, I., Corden, C., Grebot, B., Green, C., Cherrier, V., Pooley, M., . . . Claero, J. (2015). *Service contract for assessing the potential emission reductions deliverated by BAT conclusions adopted under the directive on industrial emissions (EID)*. Retrieved from https://circabc.europa.eu/sd/a/44aaf4c4-d716-4f02-91ab-a526b07ee6b7/Final%20report_20150501.pdf

Krishnakumar, B., & Manilal, V. B. (1999). Bacterial oxidation of sulphide under denitrifying condition. *Biotechnology Letters*, *21*(5), 437–440. doi:10.1023/A:1005584920800

Kuai, L. P., & Verstraete, W. (1998). Ammonium removal by the oxygen-limited autotrophic nitrification-denitrification system. *Applied and Environmental Microbiology*, *64*(11), 4500–4506. PMID:9797314

Kuai, L., & Verstraete, W. (1999). Autotrophic denitrification with elemental sulphur in small-scale wastewater treatment facilities. *Environmental Technology*, *20*(2), 201–209. doi:10.1080/09593332008616809

Kuba, T., Murnleitner, E., van Loosdrecht, M. C. M., & Heijen, J. J. (1996). A metabolic model for biological phosphorus removal by denitrifying organisms. *Biotechnology and Bioengineering*, *52*(6), 685–695. doi:10.1002/(SICI)1097-0290(19961220)52:6<685::AID-BIT6>3.3.CO;2-M PMID:18629947

Kucuk, O. S., Sengul, F., & Kapdan, I. K. (2003). Removal of ammonia from tannery effluents in a reed bed constructed wetland. *Water Science and Technology*, *48*(11–12), 179–186. PMID:14753534

Kulikowska, D. (2012). Nitrogen removal from landfill leachate via the nitrite route. *Brazilian Journal of Chemical Engineering*, *29*(02), 211–219. doi:10.1590/S0104-66322012000200002

Kulikowska, D., & Klimiuk, E. (2004). Removal of Organics and Nitrogen from Municipal Landfill Leachate in Two-Stage SBR Reactors. *Polish Journal of Environmental Studies*, *13*(4), 389–396.

Kulikowska, D., & Klimiuk, E. (2008). The effect of landfill age on municipal leachate composition. *Bioresource Technology*, *99*(13), 5981–5985. doi:10.1016/j.biortech.2007.10.015 PMID:18060769

Kumar, M., & Lin, J. G. (2010). Co-existence of anammox and denitrification for simultaneous nitrogen and carbon removal—Strategies and issues. *Journal of Hazardous Materials*, *178*(1-3), 1–9. doi:10.1016/j.jhazmat.2010.01.077 PMID:20138428

Kurniawan, T. A., Lo, W. H., & Chan, G. Y. (2006). Physico-chemical treatments for removal of recalcitrant contaminants from landfill leachate. *Journal of Hazardous Materials*, *B129*(1-3), 80–100. doi:10.1016/j.jhazmat.2005.08.010 PMID:16314043

Kurt, M., Dunn, J. I., & Bourne, R. J. (1987). Biological denitrification of drinking water using autotrophic organisms with $H_2$ in a fluidized-bed biofilm reactor. *Biotechnology and Bioengineering*, *29*(4), 493–501. doi:10.1002/bit.260290414 PMID:18576478

Kuypers, M. M. M., Sliekers, A. O., Lavik, G., Schmid, M., Jørgensen, B. B., Kuenen, J. G., & Jetten, M. S. M. et al. (2003). Anaerobic ammonium oxidation by anammox bacteria in the Black Sea. *Nature*, *422*(6932), 608–611. doi:10.1038/nature01472 PMID:12686999

Labrenz, M., Jost, G., Pohl, C., Beckmann, S., Martens-Habbena, W., & Jurgens, K. (2005). Impact of different in-vitro electron donor/acceptor conditions on potential chemolithoautotrophic communities from marine pelagic redoxiclines. *Applied and Environmental Microbiology*, *71*(11), 6664–6672. doi:10.1128/AEM.71.11.6664-6672.2005 PMID:16269695

Lacasa, E., Cañizares, P., Sáez, C., Fernández, F. J., & Rodrigo, M. A. (2011). Removal of nitrates from groundwater by electrocoagulation. *Chemical Engineering Journal*, *171*(3), 1012–1017. doi:10.1016/j.cej.2011.04.053

Lacasa, E., Cañizares, P., Sáez, C., Martínez, F., & Rodrigo, M. A. (2013). Modelling and cost evaluation of electrocoagulation processes for the removal of anions from water. *Separation and Purification Technology*, *107*, 219–227. doi:10.1016/j.seppur.2013.01.035

Lackner, S., Gilbert, E. M., Vlaeminck, S. E., Joss, A., Horn, H., & Van Loosdrecht, M. C. M. (2014). Full-scale partial nitritation/anammox experiences - an application survey. *Water Research*, *55*, 292–303. doi:10.1016/j.watres.2014.02.032 PMID:24631878

Lackner, S., Terada, A., & Smets, B. F. (2008). Heterotrophic activity compromises autotrophic nitrogen removal in membrane-aerated biofilms: Results of a modeling study. *Water Research*, *42*(4-5), 1102–1112. doi:10.1016/j.watres.2007.08.025 PMID:17915280

Lahav, O., & Green, M. (2009). Ammonium removal from primary and secondary effluents using a bioregenerated ion-exchange process. *Water Science and Technology*, *42*, 179–185.

Laitinen, N., Luonsi, A., & Vilen, J. (2006). Landfill leachate treatment with sequencing batch reactor and membrane bioreactor. *Desalination*, *191*(1-3), 86–91. doi:10.1016/j.desal.2005.08.012

Lalucat, J., Bennasar, A., Bosch, R., García-Valdés, E., & Palleroni, N. J. (2006). Biology of *Pseudomonas stutzeri*. *Microbiology and Molecular Biology Reviews*, *70*(2), 510–547. doi:10.1128/MMBR.00047-05 PMID:16760312

Lam, P., & Kuypers, M. M. M. (2011). Microbial nitrogen cycling processes in oxygen minimum zones. *Annual Review of Marine Science*, *3*(1), 317–345. doi:10.1146/annurev-marine-120709-142814 PMID:21329208

Lam, Y., & Nicholas, D. J. (1969). A nitrite reductase with cytochrome oxidase activity from Micrococcus denitrificans. *Biochimica et Biophysica Acta*, *180*(3), 459–472. doi:10.1016/0005-2728(69)90025-5 PMID:4309366

Langone, M., Yan, J., Haaijer, S. C. M., Op den Camp, H. J. M., Jetten, M. S. M., & Andreottola, G. (2014). Coexistence of nitrifying, anammox and denitrifying bacteria in a sequencing batch reactor. *Frontiers in Microbiology*, *5*, 28. doi:10.3389/fmicb.2014.00028 PMID:24550899

Lanham, A. B., Oehmen, A., Saunders, A. M., Carvalho, G., Nielsen, P. H., & Reis, M. A. M. (2014). Metabolic modelling of full-scale enhanced biological phosphorus removal sludge. *Water Research*, *66*, 283–295. doi:10.1016/j.watres.2014.08.036 PMID:25222332

Lansing, S., & Martin, J. F. (2006). Use of an ecological treatment system (ETS) for removal of nutrients from dairy wastewater. *Ecological Engineering*, *28*(3), 235–245. doi:10.1016/j.ecoleng.2006.04.006

Laureni, M., Palatsi, J., Llovera, M., & Bonmatí, A. (2013). Influence of pig slurry characteristics on ammonia stripping efficiencies and quality of the recovered ammonium-sulfate solution. *Journal of Chemical Technology and Biotechnology (Oxford, Oxfordshire)*, *88*(9), 1654–1662. doi:10.1002/jctb.4016

Laureni, M., Weissbrodt, D. G., Szivak, I., Robin, O., Nielsen, J. L., Morgenroth, E., & Joss, A. (2015). Activity and growth of anammox biomass on aerobically pre-treated municipal wastewater. *Water Research*, *80*, 325–336. doi:10.1016/j.watres.2015.04.026 PMID:26024830

Le Corre, K. S., Valsami-Jones, E., Hobbs, P., & Parsons, A. (2009). Phosphorus recovery from wastewater by struvite crystallization: A review. *Environmental Science & Technology*, *39*(6), 433–477. doi:10.1080/10643380701640573

Le Corre, K. S., Valsami-Jones, E., Hobbs, P., & Parsons, S. A. (2005). Impact of calcium on struvite crystal size, shape and purity. *Journal of Crystal Growth*, *283*(3-4), 514–522. doi:10.1016/j.jcrysgro.2005.06.012

Ledda, C., Schievano, A., Salati, S., & Adani, F. (2013). Nitrogen and water recovery from animal slurries by a new integrated ultrafiltration, reverse osmosis and cold stripping process: A case study. *Water Research*, *47*(16), 6157–6166. doi:10.1016/j.watres.2013.07.037 PMID:23948560

Lee, J., Lee, K., Park, K., & Maeng, S. (2010). Hydrogenotrophic denitrification in a packed bed reactor: Effects of hydrogen-to-water flow rate ratio. *Bioresource Technology*, *101*(11), 3940–3947. doi:10.1016/j.biortech.2010.01.022 PMID:20144861

Lee, K. C., & Rittmann, B. E. (2000). A novel hollow-fibre membrane biofilm reactor for autohydrogenotrophic denitrification of drinking water. *Water Science and Technology*, *41*, 219–226.

Lee, K. C., & Rittmann, B. E. (2002). Applying a novel autoydrogenotrophic hollow-fiber membrane biofilm reactor for denitrification of drinking water. *Water Research*, *36*(8), 2040–2052. doi:10.1016/S0043-1354(01)00425-0 PMID:12092579

Lee, S. I., Weon, S. Y., Lee, C. W., & Koopman, B. (2003). Removal of nitrogen and phosphate from wastewater by addition of bittern. *Chemosphere*, *51*(4), 265–271. doi:10.1016/S0045-6535(02)00807-X PMID:12604078

Lekuona, A. (2004). Planta de tratamiento de purines de Egiluze. RETEMA. *Revista Técnica de Medio Ambiente.*, *103*, 20–24.

Lemaire, R., Liviano, I., Esktröm, S., Roselius, C., Chauzy, J., Thornberg, D., & Deleris, S. (2011). *1-Stage Deammonification MBBR Process for Reject Water Sidestream Treatment: Investigation of Start-Up Strategy and Carriers Design.* Paper presented at the Nutrient Recovery and Management, Miami, FL.

Lemos, P. C., Serafim, L. S., Santos, M. M., Reis, M. A. M., & Santos, H. (2003). Metabolic pathways for propionate utilisation by phosphorus accumulating organisms in activated sludge:13 C labelling and in vivo nuclear magnetic resonance. *Applied and Environmental Microbiology*, *69*(1), 241–251. doi:10.1128/AEM.69.1.241-251.2003 PMID:12514001

Levasseur, P. (2004). *Traitement des effluents procins. Guide Paractique des Procédés*. Institut Technique du Porc.

Liang, Y., Li, D., Zhang, X., Zeng, H., Yang, Z., & Zhang, J. (2014). Microbial characteristics and nitrogen removal of simultaneous partial nitrification, anammox and denitrification (SNAD) process treating low C/N ratio sewage. *Bioresource Technology*, *169*, 103–109. doi:10.1016/j.biortech.2014.06.064 PMID:25036337

Liang, Z., & Liu, J. (2008). Landfill leachate treatment with a novel process: Anaerobic ammonium oxidation (Anammox) combined with soil infiltration system. *Journal of Hazardous Materials*, *151*(1), 202–212. doi:10.1016/j.jhazmat.2007.05.068 PMID:17606322

Li, C., Wu, S., & Dong, R. (2015a). Dynamics of organic matter, nitrogen and phosphorus removal and their interactions in a tidal operated constructed wetland. *Journal of Environmental Management*, *151*, 310–316. doi:10.1016/j.jenvman.2015.01.011 PMID:25585144

Lide, D. (Ed.). (1993). *CRC Handbook of Chemistry and Physics. A Ready-reference Book of Chemical and Physical Data*. Boca Raton, USA: CRC Press.

Li, G., Puyol, D., Carvajal-Arroyo, J. M., Sierra-Alvarez, R., & Field, J. A. (2015). Inhibition of anaerobic ammonium oxidation by heavy metals. *Chemical Technology and Biotechnology*, *90*(5), 830–837. doi:10.1002/jctb.4377

Li, H., Zhou, S., Huang, G., & Xu, B. (2013). Partial nitritation of landfill leachate with varying influent composition under intermittent aeration conditions. *Process Safety and Environmental Protection*, *91*(4), 285–294. doi:10.1016/j.psep.2012.05.009

Li, H., Zhou, S., Huang, G., & Xu, B. (2014). Achieving stable partial nitritation using endpoint pH control in an SBR treating landfill leachate. *Process Safety and Environmental Protection*, *92*(3), 199–205. doi:10.1016/j.psep.2013.01.005

Li, H., Zhou, S., Ma, W., Huang, P., Huang, G., Qin, Y., & Ouyang, H. (2014). Long-term performance and microbial ecology of a two-stage PN–ANAMMOX process treating mature landfill leachate. *Bioresource Technology*, *159*, 404–411. doi:10.1016/j.biortech.2014.02.054 PMID:24681301

Li, J. P., Healy, M. G., Zhan, X., Norton, D., & Rodgers, M. (2008). Effect of aeration rate on nutrient removal from slaughterhouse wastewater in intermittently aerated sequencing batch reactors. *Water, Air, and Soil Pollution*, *192*(1-4), 251–261. doi:10.1007/s11270-008-9652-9

Li, J., Ding, L.-B., Cai, A., Huang, G.-X., & Horn, H. (2014). Aerobic sludge granulation in a full-scale sequencing batch reactor. *BioMed Research International*. doi:10.1155/2014/268789 PMID:24822190

Li, L., He, C., Ji, G., Zhi, W., & Sheng, L. (2015b). Nitrogen removal pathways in a tidal flow constructed wetland under flooded time constraints. *Ecological Engineering*, *81*, 266–271. doi:10.1016/j.ecoleng.2015.04.073

Li, M., & Gu, J. D. (2011). Advances in methods for detection of anaerobic ammonium oxidizing (anammox) bacteria. *Applied Microbiology and Biotechnology*, *90*(4), 1241–1252. doi:10.1007/s00253-011-3230-6 PMID:21476137

Lin, H., Gan, J., Rajendran, A., Rodrigues Reis, C. E., & Hu, B. (2015). Phosphorus Removal and Recovery from Digestate after Biogas Production. In Biofuels - Status and Perspective (pp. 517-546). Academic Press.

Lin, H.-L., Chen, B.-K., Hsia, H.-P., Yang, Y.-F., Chao, Y.-C., & Cheng, S.-S. (2011). Use of two-stage biological process in treating thin film transistor liquid crystal display wastewater of tetramethylammonium hydroxide. *Sustainable Environment Research*, *21*(3), 155–160.

Lin, W., Chengyao, P., Shujun, Z., & Yongzhen, P. (2009). Nitrogen removal via nitrite from municipal landfill leachate. *Journal of Environmental Sciences (China)*, *21*(11), 1480–1485. doi:10.1016/S1001-0742(08)62443-2 PMID:20108678

Lin, Y. M., Bassin, J. P., & van Loosdrecht, M. C. M. (2012). The contribution of exopolysaccharides induced struvites accumulation to ammonium adsorption in aerobic granular sludge. *Water Research*, *46*(4), 986–992. doi:10.1016/j.watres.2011.11.072 PMID:22209260

Lin, Y. M., Nierop, K. G. J., Girbal-Neuhauser, E., Adriaanse, M., & van Loosdrecht, M. C. M. (2015). Sustainable polysaccharide-based biomaterial recovered from waste aerobic granular sludge as a surface coating material. *Sustainable Materials and Technologies*, *4*, 24–29. doi:10.1016/j.susmat.2015.06.002

Lin, Y. M., Sharma, P. K., & van Loosdrecht, M. C. M. (2013). The chemical and mechanical differences between alginate-like exopolysaccharides isolated from aerobic flocculent sludge and aerobic granular sludge. *Water Research*, *47*(1), 57–65. doi:10.1016/j.watres.2012.09.017 PMID:23084341

Lin, Y. M., Wang, L., Chi, Z. M., & Liu, X. Y. (2008). Bacterial Alginate Role in Aerobic Granular Bio-particles Formation and Settleability Improvement. *Separation Science and Technology*, *43*(7), 1642–1652. doi:10.1080/01496390801973805

Lin, Y., de Kreuk, M., van Loosdrecht, M. C. M., & Adin, A. (2010). Characterization of alginate-like exopolysaccharides isolated from aerobic granular sludge in pilot-plant. *Water Research*, *44*(11), 3355–3364. doi:10.1016/j.watres.2010.03.019 PMID:20417539

Liu, B., Giannis, A., Zhang, J., Chang, V. W. C., & Wang, J. Y. (2013). Characterization of induced struvite formation from source-separated urine using seawater and brine as magnesium sources. *Chemosphere*, *93*(11), 2738–2747. doi:10.1016/j.chemosphere.2013.09.025 PMID:24134888

Liu, C., Yamamoto, T., Nishiyama, T., Fujii, T., & Furukawa, K. (2009). Effect of salt concentration in anammox treatment using non woven biomass carrier. *Journal of Bioscience and Bioengineering*, *107*(5), 519–523. doi:10.1016/j.jbiosc.2009.01.020 PMID:19393551

Liu, H., Li, G., Li, X., & Chen, J. (2008). Molecular characterization of bacterial community in aerobic granular sludge stressed by pentachlorophenol. *Journal of Environmental Sciences (China)*, *20*(10), 1243–1249. doi:10.1016/S1001-0742(08)62216-0 PMID:19143350

Liu, H., & Yoo, C. (2011). Performance assessment of cascade controllers for nitrate control in a wastewater treatment process. *Korean Journal of Chemical Engineering*, *28*(3), 667–673. doi:10.1007/s11814-010-0442-x

Liu, L., Sheng, G.-P., Li, W.-W., Tong, Z.-H., Zeng, R. J., & Liu, J.-X. et al. (2011). Cultivation of aerobic granular sludge with a mixed wastewater rich in toxic organics. *Biochemical Engineering Journal*, *57*, 7–12. doi:10.1016/j.bej.2011.07.005

Liu, M., Liu, T., Peng, Y., Wang, S., & Xiao, H. (2014). Effect of salinity on $N_2O$ production during shortcut biological nitrogen removal from landfill leachate. *Journal of Bioscience and Bioengineering*, *117*(5), 582–590. doi:10.1016/j.jbiosc.2013.10.011 PMID:24225121

Liu, R., Zhao, Y., Doherty, L., Hu, Y., & Hao, X. (2015). A review of incorporation of constructed wetland with other treatment processes. *Chemical Engineering Journal*, *279*, 220–230. doi:10.1016/j.cej.2015.05.023

Liu, S. T., Yang, F. L., Xue, Y., Gong, Z., Chen, H. H., Wang, T., & Su, Z. C. (2008). Evaluation of oxygen adaptation and identification of functional bacteria composition for anammox consortium in non-woven biological rotating contactor. *Bioresource Technology*, *99*(17), 8273–8279. doi:10.1016/j.biortech.2008.03.006 PMID:18439820

Liu, W., Zhang, J., Jin, Y., Zhao, X., & Cai, Z. (2015). Adsorption of Pb (II), Cd (II) and Zn (II) by extracellular polymeric substances extracted from aerobic granular sludge : Efficiency of protein. *Journal of Environmental Chemical Engineering*, *3*(2), 1223–1232. doi:10.1016/j.jece.2015.04.009

Liu, X., Chen, Y., Zhang, X., Jiang, X., Wu, S., & Shen, J. et al. (2015). Aerobic granulation strategy for bioaugmentation of a sequencing batch reactor (SBR) treating high strength pyridine wastewater. *Journal of Hazardous Materials*, *295*, 153–160. doi:10.1016/j.jhazmat.2015.04.025 PMID:25897697

Liu, Y. (2003). A general model for biosorption of Cd2+, Cu2+ and Zn2+ by aerobic granules. *Journal of Biotechnology, 102*(3), 233–239. doi:10.1016/S0168-1656(03)00030-0 PMID:12730006

Liu, Y., & Chen, J. (2014). *Phosphorus Cycle Reference Module in Earth Systems and Environmental Sciences*. Elsevier.

Liu, Y.-Q., Liu, Y., & Tay, J.-H. (2004). The effects of extracellular polymeric substances on the formation and stability of biogranules. *Applied Microbiology and Biotechnology, 65*(2), 143–148. doi:10.1007/s00253-004-1657-8 PMID:15197510

Liu, Y.-Q., Tay, J.-H., Ivanov, V., Moy, B. Y.-P., Yu, L., & Tay, S. T.-L. (2005). Influence of phenol on nitrification by microbial granules. *Process Biochemistry, 40*(10), 3285–3289. doi:10.1016/j.procbio.2005.03.018

Liu, Y., Qin, L., & Yang, S. F. (2007). *Microbial granulation technology for nutrient removal wastewater*. New York: Nova Science Publisher.

Liu, Y., & Xu, H. (2007). Equilibrium, thermodynamics and mechanisms of Ni2+ biosorption by aerobic granules. *Biochemical Engineering Journal, 35*(2), 174–182. doi:10.1016/j.bej.2007.01.020

Liu, Y., Yang, S.-F., Tan, S.-F., Lin, Y.-M., & Tay, J.-H. (2002). Aerobic granules: A novel zinc biosorbent. *Letters in Applied Microbiology, 35*(6), 548–551. doi:10.1046/j.1472-765X.2002.01227.x PMID:12460442

Liu, Y., Yang, S.-F., Xu, H., Woon, K.-H., Lin, Y.-M., & Tay, J.-H. (2003). Biosorption kinetics of cadmium(II) on aerobic granular sludge. *Process Biochemistry, 38*(7), 997–1001. doi:10.1016/S0032-9592(02)00225-X

Li, X. Z., Zhao, Q. L., & Hao, X. D. (1999). Ammonium removal from landfill leachate by chemical precipitation. *Waste Management (New York, N.Y.), 19*(6), 409–415. doi:10.1016/S0956-053X(99)00148-8

Li, X., & Zhang, R. (2002). Aerobic treatment of dairy wastewater with sequencing batch reactor systems. *Bioprocess and Biosystems Engineering, 25*(2), 103–109. doi:10.1007/s00449-002-0286-9 PMID:14505010

Li, Y., Zou, J., Zhang, L., & Sun, J. (2014). Aerobic granular sludge for simultaneous accumulation of mineral phosphorus and removal of nitrogen via nitrite in wastewater. *Bioresource Technology, 154*, 178–184. doi:10.1016/j.biortech.2013.12.033 PMID:24388958

Lizarralde, I., Fernandez-Arevalo, T., Brouckaert, C., Vanrolleghem, P. A., Ikumi, D., Ekama, D., & Grau, P. et al. (2015). A new general methodology for incorporating physico-chemical transformations into multiphase wastewater treatment process models. *Water Research, 74*, 239–256. doi:10.1016/j.watres.2015.01.031 PMID:25746499

Lo, I. M.-C. (1996). Characteristics and treatment of leachate from domestic landfills. *Environment International, 22*(4), 433–442. doi:10.1016/0160-4120(96)00031-1

Lopez-Vazquez, C. M., Oehmen, A., Hooijmans, C. M., Brdjanovic, D., Gijzen, H. J., Yuan, Z., & van Loosdrecht, M. C. M. (2009). Modelling the PAO-GAO competition: Effects of carbon source, pH and temperature. *Water Research, 43*(2), 450–462. doi:10.1016/j.watres.2008.10.032 PMID:19022471

Lotito, A. M., De Sanctis, M., Di Iaconi, C., & Bergna, G. (2014). Textile wastewater treatment: Aerobic granular sludge vs activated sludge systems. *Water Research, 54*, 337–346. doi:10.1016/j.watres.2014.01.055 PMID:24583525

Lotti, T., Cordola, M., Kleerebezem, R., Caffaz, S., Lubello, C., & van Loosdrecht, M. C. M. (2012). Inhibition effect of swine wastewater heavy metals and antibiotics on anammox activity. *Water Science and Technology, 66*(7), 1519–1526. doi:10.2166/wst.2012.344 PMID:22864439

Lotti, T., Kleerebezem, R., Lubello, C., & van Loosdrecht, M. C. M. (2014). Physiological and kinetic characterization of a suspended cell anammox culture. *Water Research, 60*, 1–14. doi:10.1016/j.watres.2014.04.017 PMID:24813505

Luesken, F. A., van Alen, T. A., van der Biezen, E., Frijters, C., Toonen, G., Kampman, C., & Jetten, M. S. M. et al. (2011). Diversity and enrichment of nitrite-dependent anaerobic methane oxidizing bacteria from wastewater sludge. *Environmental Biotechnology*, *92*(4), 845–854. doi:10.1007/s00253-011-3361-9 PMID:21667086

Lu, H., Oehmen, A., Virdis, B., & Yuan, Z. (2006). Obtaining highly enriched cultures of Candidatus Accumulibacter phosphates through alternating carbon sources. *Water Research*, *40*(20), 3838–3848. doi:10.1016/j.watres.2006.09.004 PMID:17070894

Lukat, P., Rudolf, M., Stach, P., Messerschmidt, A., Kroneck, P. M., Simon, J., & Einsle, O. (2008). Binding and reduction of sulfite by cytochrome c nitrite reductase. *Biochemistry*, *47*(7), 2080–2086. doi:10.1021/bi7021415 PMID:18201106

Luo, J., Hao, T., Wei, L., Mackey, H. R., Lin, Z., & Chen, G. H. (2014). Impact of influent COD/N ratio on disintegration of aerobic granular sludge. *Water Research*, *62*, 127–135. doi:10.1016/j.watres.2014.05.037 PMID:24950459

Luo, M., Guo, Y. C., Deng, J. Y., Wei, H. P., Zhang, Z. P., Leng, Y., & Zhou, Y. F. et al. (2010). Characterization of a monomeric heat-labile classical alkaline phosphatase from Anabaena sp. PCC7120. *Biochemistry (Moscow)*, *75*(5), 655–664. doi:10.1134/S0006297910050172 PMID:20632947

Lu, T., George, B., Zhao, H., & Liu, W. (2016). A case study of coupling upflow anaerobic sludge blanket (UASB) and ANITA™Mox process to treat high-strength landfill leachate. *Water Science and Technology*, *73*(3), 662–668. doi:10.2166/wst.2015.536 PMID:26877051

Lv, T., Wu, S., Hong, H., Chen, L., & Dong, R. (2013). Dynamics of nitrobenzene degradation and interactions with nitrogen transformations in laboratory-scale constructed wetlands. *Bioresource Technology*, *133*, 529–536. doi:10.1016/j.biortech.2013.02.003 PMID:23455225

Ma, B., Wang, S., Cao, S., Miao, Y., Jia, F., Du, R., & Peng, Y. (2016). Biological nitrogen removal from sewage via anammox: Recent advances. *Bioresource Technology*, *200*, 981–990. doi:10.1016/j.biortech.2015.10.074 PMID:26586538

MacGregor, B. J., Biddle, J. F., Siebert, J. R., Staunton, E., Hegg, E. L., Matthysse, A. G., & Teske, A. (2013). Why orange Guaymas Basin Beggiatoa spp. are orange: Single-filament-genome-enabled identification of an abundant octaheme cytochrome with hydroxylamine oxidase, hydrazine oxidase, and nitrite reductase activities. *Applied and Environmental Microbiology*, *79*(4), 1183–1190. doi:10.1128/AEM.02538-12 PMID:23220958

Machado, V. C., Gabriel, D., Lafuente, J., & Baeza, J. A. (2009). Cost and effluent quality controllers design based on the relative gain array for a nutrient removal WWTP. *Water Research*, *43*(20), 5129–5141. doi:10.1016/j.watres.2009.08.011 PMID:19815250

Machado, V. C., Lafuente, J., & Baeza, J. A. (2015). Model-based control structure design of a full-scale WWTP under the retrofitting process. *Water Science and Technology*, *71*(11), 1661–1671. doi:10.2166/wst.2015.140 PMID:26038931

Madsen, E. L. (2011). Microorganisms and their roles in fundamental biogeochemical cycles. *Current Opinion in Biotechnology*, *22*(3), 456–464. doi:10.1016/j.copbio.2011.01.008 PMID:21333523

Mahmood, Q., Zheng, P., Cai, J., Wu, D., Hu, B., & Li, J. (2007). Anoxic sulfide biooxidation using nitrite as electron acceptor. *Journal of Hazardous Materials*, *147*(1-2), 249–256. doi:10.1016/j.jhazmat.2007.01.002 PMID:17261349

Mahmood, Q., Zheng, P., Hayat, Y., Islam, E., Wu, D., & Ren-cun, J. (2008). Effect of pH on anoxic sulfide oxidizing reactor performance. *Bioresource Technology*, *99*(8), 3291–3296. doi:10.1016/j.biortech.2007.07.006 PMID:17702569

Ma, J., Quan, X., Yang, Z., & Li, A. (2012). Biodegradation of a mixture of 2,4-dichlorophenoxyacetic acid and multiple chlorophenols by aerobic granules cultivated through plasmid pJP4 mediated bioaugmentation. *Chemical Engineering Journal*, *181-182*, 144–151. doi:10.1016/j.cej.2011.11.041

Maltais-Landry, G., Maranger, R., Brisson, J., & Chazarenc, F. (2009). Greenhouse gas production and efficiency of planted and artificially aerated constructed wetlands. *Environmental Pollution*, *157*(3), 748–754. doi:10.1016/j.envpol.2008.11.019 PMID:19110349

Mañas, A., Biscans, B., Spérandio, M., Angela, M., Béatrice, B., & Mathieu, S. (2011). Biologically induced phosphorus precipitation in aerobic granular sludge process. *Water Research*, *45*(12), 3776–3786. doi:10.1016/j.watres.2011.04.031 PMID:21616518

Manconi, I., Carucci, A., & Lens, P. (2007). Combined removal of sulfur compounds and nitrate by autotrophic denitrification in bioaugmented activated sludge system. *Biotechnology and Bioengineering*, *98*(3), 551–560. doi:10.1002/bit.21383 PMID:17724757

Manconi, I., Carucci, A., Lens, P., & Rossetti, S. (2006). Simultaneous biological removal of sulphide and nitrate by autotrophic denitrification in an activated sludge system. *Water Science and Technology*, *53*(12), 91–99. doi:10.2166/wst.2006.410 PMID:16889245

Mander, Ü., Dotro, G., Ebie, Y., Towprayoon, S., Chiemchaisri, C., Nogueira, S. F., & Mitsch, W. J. et al. (2014). Greenhouse gas emission in constructed wetlands for wastewater treatment: A review. *Ecological Engineering*, *66*, 19–35. doi:10.1016/j.ecoleng.2013.12.006

Mander, Ü., Kuusemets, V., Lõhmus, K., Mauring, T., Teiter, S., & Augustin, J. (2003). Nitrous oxide, dinitrogen and methane emission in a subsurface flow constructed wetland. *Water Science and Technology*, *48*(5), 135–142. PMID:14621157

Manga, J., Ferrer, J., Garcia-Usach, F., & Seco, A. (2001). A modification to the Activated Sludge Model No. 2 based on the competition between phosphorus-accumulating organisms and glycogen-accumulating organisms. *Water Science and Technology*, *43*(11), 161–171. PMID:11443958

Mansell, B. O., & Schroeder, E. D. (2002). Hydrogenotrophic denitrification in a microporous membrane bioreactor. *Water Research*, *36*(19), 4683–4690. doi:10.1016/S0043-1354(02)00197-5 PMID:12448509

Mantovi, P., Marmiroli, M., Maestri, E., Tagliavini, S., Piccinni, S., & Marmiroli, N. (2003). Application of a horizontal subsurface flow constructed wetland on treatment of dairy parlor wastewater. *Bioresource Technology*, *88*(2), 85–94. doi:10.1016/S0960-8524(02)00291-2 PMID:12576000

Marhuenda-Egea, F. C., Piera-Velazquez, S., Cadenas, C., & Cadenas, E. (2001). Kinetic regulation of an alkaline p-nitrophenylphosphate phosphatase from Halobacterium salinarum in low water system by Mn2+ and monovalent cations. *FEMS Microbiology Letters*, *198*(2), 111–115. doi:10.1111/j.1574-6968.2001.tb10627.x PMID:11430399

Marks, A. L., Luthy, R. G., & Diweker, U. M. (1994). Semi-continuous evaporation model for leachate treatment process evaluation. *Environment and Progress*, *13*(4), 278–289. doi:10.1002/ep.670130419

Marques, A. P. G. C., Duque, A. F., Bessa, V. S., Mesquita, R. B. R., Rangel, A. O. S. S., & Castro, P. M. L. (2013). Performance of an aerobic granular sequencing batch reactor fed with wastewaters contaminated with Zn2+. *Journal of Environmental Management*, *128*, 877–882. doi:10.1016/j.jenvman.2013.06.052 PMID:23880431

Martínez-Gutiérrez, E., González-Márquez, H., Martínez-Hernández, S., Texier, A.-C., Cuervo-López, F. M., & Gómez, J. (2012). Effect of phenol and acetate addition on 2-chlorophenol consumption by a denitrifying sludge. *Environmental Technology*, *33*(12), 1375–1382. doi:10.1080/09593330.2011.627882 PMID:22856312

Martínez-Hernández, S., Olguín, E. J., Gómez, J., & Cuervo-López, F. M. (2009). Acetate enhances the specific consumption rate of toluene under denitrifying conditions. *Archives of Environmental Contamination and Toxicology*, *57*(4), 679–687. doi:10.1007/s00244-009-9321-z PMID:19387525

Martinez, J., Burton, C. H., Sneath, R. W., & Farrent, J. W. (1995). A study of the potential contribution of sedimentation to aerobic treatment processes for pig slurry. *J. Agric. Ng. Res*, *61*, 8–96.

Marttinen, S. K., Kettunen, R. H., Sormunen, K. M., Soimasuo, R. M., & Rintala, J. A. (2002). Screening of physical-chemical methods for removal of organic material, nitrogen and toxicity from low strength landfill leachate. *Chemosphere*, *46*(6), 851–858. doi:10.1016/S0045-6535(01)00150-3 PMID:11922065

Masi, F., Conte, G., Martinuzzi, N., & Pucci, B. (2002). *Winery high organic content wastewater treated by constructed wetland in mediterranean climate.* IWA 8th International Conference on Wetland Systems for Water Pollution Control, Arusha, Tanzania.

Masi, F., Rochereau, J., Troesch, S., Ruiz, I., & Soto, M. (2014). Wineries wastewater treatment by constructed wetlands: A review. *Water Science and Technology*, *71*(8), 1113–1127. doi:10.2166/wst.2015.061 PMID:25909720

Masse, L., Massé, D. I., & Pellerinb, Y. (2007). The use of membranes for the treatment of manure: A critical literature review. *Biosystems Engineering*, *98*(4), 371–380. doi:10.1016/j.biosystemseng.2007.09.003

Matsumiya, Y., Yamasita, T., & Nawamura, Y. (2000). Phosphorus removal from sidestreams by crystallization of magnesium-ammonium-phosphate using seawater. *Journal of the Chartered Institution of Water and Environmental Management*, *14*(4), 291–296. doi:10.1111/j.1747-6593.2000.tb00263.x

Maurer, M., Abramovich, D., Siegrist, H., & Gujer, W. (1999). Kinetics of biologically induced phosphorus precipitation in wastewater treatment. *Water Research*, *33*(2), 484–493. doi:10.1016/S0043-1354(98)00221-8

Maurer, M., & Gujer, W. (1995). Monitoring of microbial phosphorus release in batch experiments using electric conductivity. *Water Research*, *29*(11), 2613–2617. doi:10.1016/0043-1354(95)00146-C

Maurer, M., Pronk, W., & Larsen, T. A. (2006). Treatment processes for source-separated urine. *Water Research*, *40*(17), 3151–3166. doi:10.1016/j.watres.2006.07.012 PMID:16949123

Maurer, M., Schwegler, P., & Larsen, T. A. (2003). Nutrient in urine: Energetic aspects of removal and recovery. *Water Science and Technology*, *48*(1), 37–46. PMID:12926619

Mauricio-Iglesias, M., Vangsgaard, A. K., Gernaey, K. V., Smets, B. F., & Sin, G. (2015). A novel control strategy for single-stage autotrophic nitrogen removal in SBR. *Chemical Engineering Journal*, *260*, 64–73. doi:10.1016/j.cej.2014.08.069

Ma, Y.-L., Yang, B.-L., & Zhao, J.-L. (2006). Removal of $H_2S$ by *Thiobacillus denitrificans* immobilized on different matrices. *Bioresource Technology*, *97*(16), 2041–2046. doi:10.1016/j.biortech.2005.09.023 PMID:16289624

McAdam, E. J., & Judd, S. J. (2006). A review of membrane bioreactor potential for nitrate removal from drinking water. *Destilation*, *196*, 135–148.

McQuarrie, J., Johnson, C., Lu, T., & Zhao, H. (2015). Sidestream Deammonification. In W. E. Federation (Ed.), *Shortcut Nitrogen Removal - Nitrite Shunt and Deammonification.* Alexandria: Water Environment Federation.

McSwain, B. S., Irvine, R., Hausner, M., & Wilderer, P. A. (2005). Composition and Distribution of Extracellular Polymeric Substances in Aerobic Flocs and Granular Sludge. *Applied and Environmental Microbiology*, *71*(2), 1051–1057. doi:10.1128/AEM.71.2.1051-1057.2005 PMID:15691965

Meers, E., Tack, F. M. G., Tolpe, I., & Michels, E. (2008). Application of full-scale constructed wetland for tertiary treatment of piggery manure: Monitoring results. *Water, Air, and Soil Pollution*, *193*(1-4), 15–24. doi:10.1007/s11270-008-9664-5

Meijer, S. C. F. (2004). *Theoretical and practical aspects of modeling activated sludge processes*. (PhD Thesis). Delft University of Technology, The Netherlands.

Meinhold, J., Arnold, E., & Isaacs, S. (1999b). Effect of nitrite on anoxic phosphate uptake in biological phosphorus removal activated sludge. *Water Research*, *33*(8), 1871–1883. doi:10.1016/S0043-1354(98)00411-4

Meinhold, J., Filipe, C. D. M., Digger, G. T., & Isaacs, S. (1999a). Characterization of the denitrifying fraction of phosphate accumulating organisms in biological phosphate removal. *Water Science and Technology*, *39*(1), 31–42. doi:10.1016/S0273-1223(98)00773-2

Metcalf, E., & Eddy, H. (2003). *Wastewater engineering: Treatment, disposal and reuse*. New York: Ed. Mc Graw-Hill.

Metcalf, E., & Eddy, H. (2007). *Wastewater treatment*. New York: Ed. Mc Graw-Hill.

Meza-Escalante, E. R., Anne-Claire, T., Cuervo-López, F., Gómez, J., & Cervantes, F. (2008). Inhibition of sulfide on the simultaneous removal of nitrate and p-cresol by a denitrifying sludge. *Journal of Chemical Technology and Biotechnology (Oxford, Oxfordshire)*, *83*(3), 372–377. doi:10.1002/jctb.1823

Miao, L., Wang, K., Ahu, R., Li, B., Peng, Y., & Weng, D. (2014). Advanced nitrogen removal from landfill leachate using real-time controlled three-stage sequencing batch reactor (SBR) system. *Bioresource Technology*, *159*, 258–265. doi:10.1016/j.biortech.2014.02.058 PMID:24657756

Miao, L., Wang, S., Tianhao, C., & Peng, Y. (2015). Optimization of three-stage Anammox system removing nitrogen from landfill leachate. *Bioresource Technology*, *185*, 450–455. doi:10.1016/j.biortech.2015.03.032 PMID:25812997

Mielczarek, A. T., Kragelund, C., Eriksen, P. S., & Nielsen, P. H. (2012). Population dynamics of filamentous bacteria in Danish wastewater treatment plants with nutrient removal. *Water Research*, *46*(12), 3781–3795. doi:10.1016/j.watres.2012.04.009 PMID:22608099

Mielczarek, A. T., Nguyen, H. T. T., Nielsen, J. L., & Nielsen, P. H. (2013). Population dynamics of bacteria involved in enhanced biological phosphorus removal in Danish wastewater treatment plants. *Water Research*, *47*(4), 1529–1544. doi:10.1016/j.watres.2012.12.003 PMID:23317522

Miksch, K., Cema, G., Corvini, P. F. X., Felis, E., Sochacki, A., Surmacz-Górska, J., & Żabczynski, S. et al. (2015). R&D priorities in the field of sustainable remediation and purification of agro-industrial and municipal wastewater. *New Biotechnology*, *32*(1), 128–132. doi:10.1016/j.nbt.2013.11.002 PMID:24284044

Milia, S., Perra, M., Tocco, G., & Carucci, A. (2015b). *The anammox process as the second step for the treatment of ammonium rich refinery wastewater with high Corg/N ratio*. Paper presented at the 14th Conference on Environmental Science and Technology, Rhodes, Greece.

Milia, S., Perra, M., Cappai, G., & Carucci, A. (2015a). SHARON process as preliminary treatment of refinery wastewater with high organic carbon-to-nitrogen ratio. *Desalination and Water Treatment*, 1–9. doi:10.1080/19443994.2015.1087341

Mills, H. J., Martínez, R. J., Story, S., & Sobecky, P. A. (2004). Identification of members of of the metabolically active microbial populations associated with *Beggiatoa* species mat communities from Gulf of Mexico cold-deep sediments. *Applied and Environmental Microbiology*, *7*(9), 5447–5458. doi:10.1128/AEM.70.9.5447-5458.2004 PMID:15345432

Mino, T., Arun, V., Tsuzuki, Y., & Matsuo, T. (1987). Effect of phosphorus accumulation on acetate metabolism in the biological phosphorus removal process. In R. Ramadori (Ed.), *Biological phosphate removal from wastewaters* (pp. 27–38). Oxford, UK: Pergamon Press. doi:10.1016/B978-0-08-035592-4.50009-2

Mino, T., van Loosdrecht, M. C. M., & Heijnen, J. J. (1998). Microbiology and biochemistry of enhanced biological phosphate removal process. *Water Research*, *32*(11), 3193–3207. doi:10.1016/S0043-1354(98)00129-8

Mo, H., Oleszkiewicz, J.A., Cicek, N., Rezania, B. (2005). Incorporating membrane gas diffusion into a membrane bioreactor for hydrogenotrophic denitrification of groundwater. *Water Science and Technology*, *51*, 357-364.

Mobarry, B., Wagner, M., Urbain, V., Rittmann, B., & Stahl, D. (1996). Phylogenetic probes for analyzing abundance and spatial organization of nitrifying bacteria. *Applied and Environmental Microbiology*, *62*(6), 2156–2162. PMID:8787412

Modin, O., Fukushi, K., Nakajima, F., & Yamamoto, K. (2010). Nitrate removal and biofilm characteristics in methanotrophic membrane biofilm reactors with various gas supply regimes. *Water Research*, *44*(1), 85–96. doi:10.1016/j.watres.2009.09.009 PMID:19781736

Modin, O., Fukushi, K., & Yamamoto, K. (2007). Denitrification with methane as external carbon source. *Water Research*, *41*(12), 2726–2738. doi:10.1016/j.watres.2007.02.053 PMID:17433401

Moerman, W., Carball, M., Vandekerckhove, A., Derycke, D., & Verstraete, W. (2009). Phosphate removal in agro-industry: Pilot- and full-scale operational considerations of struvite crystallization. *Water Research*, *43*(7), 1887–1892. doi:10.1016/j.watres.2009.02.007 PMID:19264340

Mojiri, A., Ziyang, L., Tajuddin, R. M., Farraji, H., & Alifar, N. (2016). Co-treatment of landfill leachate and municipal wastewater using the ZELIA/zeolite constructed wetlands system. *Journal of Environmental Management*, *166*, 124–130. doi:10.1016/j.jenvman.2015.10.020 PMID:26496842

Molinuevo, B., García, M. C., Karakashev, D., & Angelidaki, I. (2009). Anammox for ammonia removal from pig manure effluents: Effect of organic matter content on process performance. *Bioresource Technology*, *100*(7), 2171–2175. doi:10.1016/j.biortech.2008.10.038 PMID:19097886

Møller, H. B., Lund, I., & Sommer, S. G. (2000). Solid-liquid separation of livestock slurry: Efficiency and cost. *Bioresource Technology*, *74*(3), 223–229. doi:10.1016/S0960-8524(00)00016-X

Møller, H. B., Sommer, S. G., & Ahring, B. K. (2002). Separation efficiency and particle size distribution in relation to manure type and storage conditions. *Bioresource Technology*, *85*(2), 189–196. doi:10.1016/S0960-8524(02)00047-0 PMID:12227544

Moon, H. S., Ahn, K.-H., Lee, S., Nam, K., & Kim, J. Y. (2004). Use of autotrophic sulphur-oxidizers to remove nitrate from bank filtrate in a permeable reactive barrier system. *Environmental Pollution*, *129*(3), 499–507. doi:10.1016/j.envpol.2003.11.004 PMID:15016470

Moon, H. S., Chang, S. W., Nam, K., Choe, J., & Kim, J. Y. (2006). Effect of reactive media composition and co-contaminants on sulfur-based autotrophic denitrification. *Environmental Pollution*, *144*(3), 802–807. doi:10.1016/j.envpol.2006.02.020 PMID:16632130

Moon, H. S., Shin, D. Y., Nam, K., & Kim, J. Y. (2008). A long-term performance test on an autotrophic denitrification column for application as a permeable reactive barrier. *Chemosfere*, *73*(5), 723–728. doi:10.1016/j.chemosphere.2008.06.065 PMID:18722639

Morales, N., Boehler, M. A., Buettner, S., Liebi, C., & Siegrist, H. (2013). Recovery of N and P from Urine by Struvite Precipitation Followed by Combined Stripping with Digester Sludge Liquid at Full Scale. *Water (Basel)*, *5*(3), 1262–1278. doi:10.3390/w5031262

Morales, N., Val del Rio, A., Vazquez-Padin, J. R., Gutierrez, R., Fernandez-Gonzalez, R., Icaram, P., & Mosquera-Corral, A. (2015a). Influence of dissolved oxygen concentration on the start-up of the anammox-based process: ELAN. *Water Science and Technology*, *72*(4), 520–527. doi:10.2166/wst.2015.233 PMID:26247749

Morales, N., Val del Río, A., Vázquez-Padín, J. R., Méndez, R., Mosquera-Corral, A., & Campos, J. L. (2015b). Integration of the Anammox process to the rejection water and main stream lines of WWTPs. *Chemosphere*, *140*, 99–105. doi:10.1016/j.chemosphere.2015.03.058 PMID:25890586

Mora, M., Lopez, R. L., Gamisans, X., & Gabriel, D. (2014). Coupling respirometry ans titrimetry for the characterization of the biological activity of SO-NR consortium. *Chemical Engineering Journal*, *251*, 111–115. doi:10.1016/j.cej.2014.04.024

Moreira, I. S., Amorim, C. L., Ribeiro, A. R., Mesquita, R. B. R., Rangel, A. O. S. S., & van Loosdrecht, M. C. M. et al. (2015). Removal of fluoxetine and its effects in the performance of an aerobic granular sludge sequential batch reactor. *Journal of Hazardous Materials*, *287*, 93–101. doi:10.1016/j.jhazmat.2015.01.020 PMID:25625631

Morse, G. K., Brett, S. W., Guy, J. A., & Lester, J. N. (1998). Review: Phosphorus removal and recovery technologies. *The Science of the Total Environment*, *212*(1), 69–81. doi:10.1016/S0048-9697(97)00332-X

Moscatelli, G., & Fabbri, C. (2008). Strippaggio, tecnica efficient per l'abbattimento dell'azoto. *L'informatore Agrario*, *18*, 45–50.

Mosquera-Corral, A., González, F., Campos, J. L., & Méndez, R. (2005). Partial nitrification in a SHARON reactor in the presence of salts and organic carbon compounds. *Process Biochemistry*, *40*(9), 3109–3118. doi:10.1016/j.procbio.2005.03.042

Mosteo, R., Ormad, P., Mozas, E., Sarasa, J., & Ovelleiro, J. L. (2006). Factorial experimental design of winery wastewater treatment by heterogeneous photo-Fenton process. *Water Research*, *40*(8), 1561–1568. doi:10.1016/j.watres.2006.02.008 PMID:16574187

Mosteo, R., Sarasa, J., Ormad, M., & Ovelleiro, J. (2008). Sequential Solar Photo-Fenton-Biological System for the Treatment of Winery Wastewater. *Journal of Agricultural and Food Chemistry*, *56*(16), 7333–7338. doi:10.1021/jf8005678 PMID:18642841

Muda, K., Aris, A., Salim, M. R., Ibrahim, Z., Yahya, A., & van Loosdrecht, M. C. M. et al. (2010). Development of granular sludge for textile wastewater treatment. *Water Research*, *44*(15), 4341–4350. doi:10.1016/j.watres.2010.05.023 PMID:20580402

Mulder, A. (2003). The quest for sustainable nitrogen removal technologies. *Water Science and Technology*, *48*, 67–75. PMID:12926622

Mulder, A., Vandegraaf, A. A., Robertson, L. A., & Kuenen, J. G. (1995). Anaerobic ammonium oxidation discovered in a denitrifying fluidized-bed reactor. *FEMS Microbiology Ecology*, *16*(3), 177–183. doi:10.1111/j.1574-6941.1995.tb00281.x

Mulder, J. W., van Loosdrecht, M. C. M., Hellinga, C., & van Kempen, R. (2001). Full-scale application of the SHARON process for treatment of rejection water of digested sludge dewatering. *Water Science and Technology*, *43*(11), 127–134. PubMed

Müller, G. T., Giacobbo, A., dos Santos Chiaramonte, E. A., Siqueira Rodrigues, M. A., Meneguzzi, A., & Bernardes, A. M. (2015). The effect of sanitary landfill leachate aging on the biological treatment and assessment of photoelectrooxidation as a pre-treatment process. *Waste Management (New York, N.Y.)*, *36*, 177–183. doi:10.1016/j.wasman.2014.10.024 PMID:25464941

Mutlu, A. G., Vangsgaard, A. K., Sin, G., & Smets, B. F. (2013). An operation protocol for facilitating start-up of single-stage autotrophic nitrogen removing reactors based on process stoichiometry. *Water Science and Technology*, *68*(3), 514–521. doi:10.2166/wst.2013.157 PMID:23925177

Myhre, G., Shindell, D., Bréon, F.-M., Collins, W., Fuglestvedt, J., Huang, J., . . . Zhang, H. (2013). Anthropogenic and Natural Radiative Forcing. In T. F. Stocker, D. Qin, G.-K. Plattner, M. Tignor, S. K. Allen, J. Boschung, . . . Midgley, P. M. (Eds.), Climate Change 2013: The Physical Science Basis. Contribution of Working Group I to the Fifth Assessment Report of the Intergovernmental Panel on Climate Change. Cambridge, UK: Cambridge University Press. Available from http://www.ipcc.ch/pdf/assessment-report/ar5/wg1/WG1AR5_Chapter08_FINAL.pdf

Nancharaiah, Y. V., Joshi, H. M., Hausner, M., & Venugopalan, V. P. (2008). Bioaugmentation of aerobic microbial granules with Pseudomonas putida carrying TOL plasmid. *Chemosphere*, *71*(1), 30–35. doi:10.1016/j.chemosphere.2007.10.062 PMID:18076969

Nancharaiah, Y. V., Joshi, H. M., Krishna Mohan, T. V., Venugopalan, V. P., & Narasimhan, S. V. (2008). Formation of aerobic granules in the presence of a synthetic chelating agent. *Environmental Pollution*, *153*(1), 37–43. doi:10.1016/j.envpol.2007.11.017 PMID:18166252

Neczaj, E., Okoniewska, E., & Kacprzak, M. (2005). Treatment of landfill leachate by sequencing batch reactor. *Desalination*, *185*(1-3), 357–362. doi:10.1016/j.desal.2005.04.044

Nelson, N. O., Mikkelsen, R. L., & Hesterberg, D. L. (2003). Struvite precipitation in anaerobic swine lagoon liquid: effect of pH and Mg:P ratio and determination of rate constant. *Bioresource Technology*, *89*(3), 229–236. doi:10.1016/S0960-8524(03)00076-2 PMID:12798112

Newman, J. M., & Clausen, J. C. (2000). Seasonal performance of a wetland constructed to process milkhouse wastewater in Connecticut. *Ecological Engineering*, *14*(1), 181–198.

Nguyen, M. L., & Tabber, C. C. (1998). Ammonium removal from wastewater using natural new Zealand zeolites. *New Zealand Journal of Agricultural Research*, *41*(3), 427–446. doi:10.1080/00288233.1998.9513328

Nhat, P. T., Biec, H. N., Mai, N. T. T., Thanh, B. X., & Dan, N. P. (2014). Application of a partial nitritation and anammox system for the old landfill leachate. *International Biodeterioration & Biodegradation*, *95*, 144–150. doi:10.1016/j.ibiod.2014.05.025

Ni, B. J., & Yuan, Z. (2015). Recent Advances in Mathematical Modeling of Nitrous Oxides Emissions from Wastewater Treatment Processes. *Water Research*, *87*, 336–346. doi:10.1016/j.watres.2015.09.049 PMID:26451976

Nielsen, S. (2015). *Sludge Treatment in Reed Beds System Technology: Theory, design and application.* IRSTEA 2015 6-8th July Workshop program, Lyon, France.

Niewersch, C., Battaglia Bloch, A. L., Yüce, S., Melin, T., & Wessling, M. (2014). Nanofiltration for the recovery of phosphorus: Development of a mass transport model. *Desalination*, *346*, 70–78. doi:10.1016/j.desal.2014.05.011

Nikolic-Hughes, I., O'Brien, P. J., & Herschlag, D. (2005). Alkaline phosphatase catalysis is ultrasensitive to charge sequestered between the active site zinc ions. *Journal of the American Chemical Society*, *127*(26), 9314–9315. doi:10.1021/ja051603j PMID:15984827

Nishimura, F., Somiya, I., Tsuno, H., & Iwabu, H. (1996). Development of a combined bac and Bz reactor for removal of nitrogen in wastewater from sludge drying process. *Water Science and Technology*, *34*(1-2), 145–151. doi:10.1016/0273-1223(96)00504-5

Nitrates Action Programme. (2015). *Nitrates Action Programme and Phosphorus regulations 2015-2018.* Department of Agriculture and Rural Development. Retrieved from: https://www.daera-ni.gov.uk/sites/default/files/publications/doe/water-policy-nitrates-action-programme-201518-2015.PDF

Nivala, J., Hoos, M. B., Cross, C., Wallace, S., & Parkin, G. (2007). Treatment of landfill leachate using an aerated, horizontal subsurface-flow constructed wetland. *The Science of the Total Environment*, *380*(1-3), 19–27. doi:10.1016/j.scitotenv.2006.12.030 PMID:17241653

Nkoa, R. (2014). Agricultural benefits and environmental risks of soil fertilization with anaerobic digestates: A review. *Agron Sustain Dev*, *34*(2), 473–492. doi:10.1007/s13593-013-0196-z

Noroi, K. (2014). Nitrate-dependent anaerobic methane oxidation in a freshwater sediment. *Geochimica et Cosmochimica Acta*, *132*, 141–150. doi:10.1016/j.gca.2014.01.032

Nowak, O., Svardal, K., & Schweighofer, P. (1995). The dynamic behaviour of nitrifying activated sludge systems influenced by inhibiting wastewater compounds. *Water Science and Technology*, *31*(2), 115–124. doi:10.1016/0273-1223(95)00185-P

Oehmen, A., Carvalho, G., Lopez-Vazquez, C. M., van Loosdrecht, M. C. M., & Reis, M. A. M. (2010b). Incorporating microbial ecology into the metabolic modelling of polyphosphate accumulating organisms and glycogen accumulating organisms. *Water Research*, *44*(17), 4992–5004. doi:10.1016/j.watres.2010.06.071 PMID:20650504

Oehmen, A., Lemos, P. C., Carvalho, G., Yuan, Z., Keller, J., Blackhall, L. L., & Reis, M. A. M. (2007). Advances in enhanced biological phosphorus removal: From micro to macro scale. *Water Research*, *41*(11), 2271–2300. doi:10.1016/j.watres.2007.02.030 PMID:17434562

Oehmen, A., Lopez-Vazquez, C. M., Carvalho, G., Reis, M. A. M., & van Loosdrecht, M. C. M. (2010a). Modelling the population dynamics and metabolic diversity of organisms relevant in anaerobic/anoxic/aerobic enhanced biological phosphorus removal processes. *Water Research*, *44*(15), 4473–4486. doi:10.1016/j.watres.2010.06.017 PMID:20580055

Oehmen, A., Saunders, A. M., Vives, M. T., Yuan, Z., & Keller, J. (2006). Competition between polyphosphate and glycogen accumulating organisms in enhance biological phosphorus removal systems with acetate and propionate as carbon sources. *Journal of Biotechnology*, *123*(1), 22–32. doi:10.1016/j.jbiotec.2005.10.009 PMID:16293332

Oehmen, A., Yuan, Z. G., Blackall, L. L., & Keller, J. (2005). Comparison of acetate and propionate uptake by polyphosphate accumulating organisms and glycogen accumulating organisms. *Biotechnology and Bioengineering*, *91*(2), 162–168. doi:10.1002/bit.20500 PMID:15892052

Oehmen, A., Zeng, R. J., Yuan, Z., & Keller, J. (2005). Anaerobic metabolism of propionate by polyphosphate-accumulating organisms in enhanced biological phosphorus removal systems. *Biotechnology and Bioengineering*, *91*(1), 43–53. doi:10.1002/bit.20480 PMID:15880463

Oenema, O., Bleeker, A., Braathen, N. A., Budňáková, M., Bull, K., Čermák, P., . . . Winiwarter, W. (2011). Nitrogen in current European policies. In The European Nitrogen Assessment. Cambridge, UK: Cambridge University Press. doi:10.1017/CBO9780511976988.007

Ohlinger, K. N., Young, T. M., & Schroeder, E. D. (1999). Kinetics effects on preferential struvite accumulation in wastewater. *Journal of Environmental Engineering*, *125*, 730–737. doi:10.1061/(ASCE)0733-9372(1999)125:8(730)

Oh, S. E., Bum, M. S., Yoo, Y. B., Zubair, A., & Kim, I. S. (2002). Nitrate removal by simultaneous sulphur utilizing autotrophic and heterotrophic denitrification under different organic and alkalinity condition: Batch experiments. *Water Science and Technology*, *47*(1), 237–244. PMID:12578200

Oh, S.-E., Kim, K.-S., Choi, H.-C., Cho, J., & Kim, I. S. (2000). Kinetics and physiological characteristics of autotrophic denitrifying sulphur bacteria. *Water Research & Technology*, *42*(3-4), 959–968.

Oji, Y., Hamano, T., Ryoma, Y., Miki, Y., Wakiuchi, N., & Okamoto, S. (1985). Purification and Characterization of NADH-Nitrate Reductase from Leaves of 2-Row Barley, and Its Activity as Affected by Some Metabolites. *Journal of Plant Physiology*, *119*(3), 247–256. doi:10.1016/S0176-1617(85)80184-X

Oliveira, L. F., Silva, S. M. C. P., & Martinez, C. B. R. (2014). Assessment of domestic landfill leachate toxicity to the Asian clam *Corbicula fluminea* via biomarkers. *Ecotoxicology and Environmental Safety*, *103*, 17–23. doi:10.1016/j.ecoenv.2014.01.034 PMID:24561242

Olsson, G. (2011). Automation development in water and wastewater systems. *Environmental Engineering Research*, *16*(4), 197–200.

Olsson, G. (2012). ICA and me – A subjective review. *Water Research*, *46*(6), 1585–1624. doi:10.1016/j.watres.2011.12.054 PMID:22284982

Orentlicher, M. (2012). Overview of nitrogen removal technologies and application/use of associated end products. In *Proceedings of Got Manure Tradeshow and Conference*. New York: Cornell University.

Orentlicher, M., Brown, S., & Hall, C. (2009). Centrate ammonia reduction with ARP®: pilot data with synthetic and actual wastewaters. In *Proceedings of WEF/WERF Nutrient Removal Specialty Conference*. New York: Water Environment Federation.

O'Shaughnessy, M., Sizemore, J., Musabyimana, M., Sanjines, P., Murthy, S., Wett, B., & Pallansch, K. (2008). Operations and process control of the Deammonification (DEMON) process as a sidestream option for nutrient removal. Proceedings of the Water Environment Federation, 2008(9), 6333-6348. doi:10.2175/193864708788809743

Oshiki, M., Satoh, H., & Okabe, S. (2016). Ecology and physiology of anaerobic ammonium oxidizing bacteria. *Environmental Microbiology*. doi:10.1111/1462-2920.13134

Ostace, G. S., Baeza, J. A., Guerrero, J., Guisasola, A., Cristea, V. M., Agachi, P. S., & Lafuente, J. (2013). Development and economic assessment of different WWTP control strategies for optimal simultaneous removal of carbon, nitrogen and phosphorus. *Computers & Chemical Engineering*, *53*, 164–177. doi:10.1016/j.compchemeng.2013.03.007

Othman, I., Anuar, A. N., Ujang, Z., Rosman, N. H., Harun, H., & Chelliapan, S. (2013). Livestock wastewater treatment using aerobic granular sludge. *Bioresource Technology*, *133*, 630–634. doi:10.1016/j.biortech.2013.01.149 PMID:23453799

Oulego, P., Collado, S., Laca, A., & Diaz, M. (2016). Impact of leachate composition on the advanced oxidation treatment. *Water Research*, *88*, 389–402. doi:10.1016/j.watres.2015.09.048 PMID:26517790

Ozturk, I., Altinbas, M., Koyuncu, I., Arikan, O., & Gomec-Yangin, C. (2003). Advanced phusico-chemical treatment experiences on young municipal landfill leachates. *Waste Management (New York, N.Y.)*, *23*(5), 441–446. doi:10.1016/S0956-053X(03)00061-8 PMID:12893017

Pambrun, V., Paul, E., & Sperandio, M. (2006). Modeling the partial nitrification in sequencing batch reactor for biomass adapted to high ammonium concentrations. *Biotechnology and Bioengineering*, *95*(1), 120–131. doi:10.1002/bit.21008 PMID:16715534

Panda, S. K., & Choudhury, S. (2005). Changes in nitrate reductase activity and oxidative stress response in the moss Polytrichum commune subjected to chromium, copper and zinc phytotoxicity. *Brazilian Journal of Plant Physiology*, *17*(2), 191–197. doi:10.1590/S1677-04202005000200001

Paques. (n.d.). Retrieved May 2016 from www.paques.nl

Park, H. I., Choi, Y. J., & Pak, D. (2005a). Autohydrogenotrophic denitrifying microbial community in a glass beads biofilm reactor. *Biotechnology Letters*, *27*(13), 949–953. doi:10.1007/s10529-005-7654-x PMID:16091891

Park, H. I., Kim, D. K., Choi, Y., & Pak, D. (2005b). Nitrate reduction using an electrode as direct electron donor in a biofilm-electrode reactor. *Process Biochemistry*, *40*(10), 3383–3388. doi:10.1016/j.procbio.2005.03.017

Park, J. Y., & Yoo, Y. J. (2009). Biological nitrate removal in industrial wastewater treatment: Which electron donor we can choose. *Applied Microbiology and Biotechnology*, *82*(3), 415–429. doi:10.1007/s00253-008-1799-1 PMID:19148639

Park, S., & Bae, W. (2009). Modeling kinetics of ammonium oxidation and nitrite oxidation under simultaneous inhibition by free ammonia and free nitrous acid. *Process Biochemistry*, *44*(6), 631–640. doi:10.1016/j.procbio.2009.02.002

Patyk, A., & Reinhardt, G. A. (1997). *Düngemittel – Energie- und Stoffstrombilanzen.* Vieweg. doi:10.1007/978-3-663-08014-5

Pei-Dong, . (2002). Greenhouse gas emissions from a constructed wetland for municipal sewage treatment. *Journal of Environmental Sciences (China)*, *14*(1), 27–33. PMID:11887314

Peña-Calva, A., Olmos, D. A., Viniegra, G. G., Cuervo, L. F., & Gómez, J. (2004). Denitrification in presence of benzene, toluene, and *m*-xylene. *Applied Biochemistry and Biotechnology*, *119*(3), 195–208. doi:10.1007/s12010-004-0002-0 PMID:15591614

Peng, Y., & Zhu, G. (2006). Biological nitrogen removal with nitrification and denitrification via nitrite pathway. *Applied Microbiology and Biotechnology*, *73*(1), 15–26. doi:10.1007/s00253-006-0534-z PMID:17028876

Pepper, I., Gerba, C., & Gentry, T. (2015). *Environmental Microbiology* (3rd ed.). San Diego, CA: Academic Press.

Pérez, R., Galí, A., Dosta, J., & Mata-Álvarez, J. (2007). Biological nitrogen removal (BNR) using sulfides for autotrophic denitrification in a sequencing batch reactor (SBR) to treat reject water. *Industrial & Engineering Chemistry Research*, *46*(21), 6646–6649. doi:10.1021/ie061496h

Perry, R. H. (1992). *Perry's Chemical Engineers' Handbook* (6th ed.). McGraw Hill.

Petersen, J., Fisher, K., & Lowe, D. J. (2008). Structural basis for VO2+ inhibition of nitrogenase activity (A): 31P and 23Na interactions with the metal at the nucleotide binding site of the nitrogenase Fe protein identified by ENDOR spectroscopy. *Journal of Biological Inorganic Chemistry*, *13*(4), 623–635. doi:10.1007/s00775-008-0360-0 PMID:18351402

Petersen, J., Mitchell, C. J., Fisher, K., & Lowe, D. J. (2008). Structural basis for VO(2+)-inhibition of nitrogenase activity: (B) pH-sensitive inner-sphere rearrangements in the 1H-environment of the metal coordination site of the nitrogenase Fe-protein identified by ENDOR spectroscopy. *Journal of Biological Inorganic Chemistry*, *13*(4), 637–650. doi:10.1007/s00775-008-0364-9 PMID:18386081

Petruccioli, M., Duarte, J. C., Eusebio, A., & Federici, F. (2002). Aerobic treatment of winery wastewater using a jet-loop activated sludge reactor. *Process Biochemistry*, *37*(8), 821–829. doi:10.1016/S0032-9592(01)00280-1

Peyton, R. L., & Schroeder, P. R. (1993). Water Balance for Landfills. In D. E. Daniel (Ed.), *Geotechnical Practice for Waste Disposal.* New York: Chapman & Hall. doi:10.1007/978-1-4615-3070-1_10

Philips, S., Wyffels, S., Sprengers, R., & Verstraete, W. (2002). Oxygen-limited autotrophic nitrification/denitrification by ammonia oxidisers enables upward motion towards more favourable conditions. *Applied Microbiology and Biotechnology*, *59*(4-5), 557–566. doi:10.1007/s00253-002-1059-8 and PubMed

Picioreau, C., van Loosdrecht, M. C. M., & Heijnen, J. J. (1997). Modelling the effect of oxygen concentration on nitrite accumulation in a biofilm airlift suspension reactor. *Water Science and Technology*, *36*(1), 147–156. doi:10.1016/S0273-1223(97)00347-8

Pijuan, M. (2004). *Effect of different carbon sources and continous aeration on EBPR process*. (PhD Thesis). Universitat Autònoma de Barcelona, Spain.

Pijuan, M., Casas, C., & Baeza, J. A. (2009). Polyhydroxyalkanoate synthesis using different carbon sources by two enhanced biological phosphorus removal microbial communities. *Process Biochemistry*, *44*(1), 97–105. doi:10.1016/j.procbio.2008.09.017

Pijuan, M., Ye, L., & Yuan, Z. (2010). Free nitrous acid inhibition on the aerobic metabo-lism of poly-phosphate accumulating organisms. *Water Research*, *44*(20), 6063–6072. doi:10.1016/j.watres.2010.07.075 PMID:20709350

Pinnekamp, J., & Friedrich, H. (2006). *Membrane technology for waste water treatment. Municipal water and waste management* (Vol. 2). FiW-Verlag.

Plaza, E., Stridh, S., Örnmark, J., Kanders, L., & Trela, J. (2011). Swedish Experience of the Deammonification Process in a Biofilm System. Proceedings of the Water Environment Federation, 2011(1), 1067-1079. doi:10.2175/193864711802867397

Prasad, M. N. V. (2001). *Metals in the Environment: Analysis by Biodiversity*. Taylor & Francis.

Prodouz, K. N., & Garrett, R. H. (1981). Neurospora crassa NAD(P)H-nitrite reductase. Studies on its composition and structure. *The Journal of Biological Chemistry*, *256*(18), 9711–9717. PMID:6457037

Pronk, M., Abbas, B., Al-zuhairy, S. H. K., Kraan, R., Kleerebezem, R., & van Loosdrecht, M. C. M. (2015). Effect and behaviour of different substrates in relation to the formation of aerobic granular sludge. *Applied Microbiology and Biotechnology*, *99*(12), 5257–5268. doi:10.1007/s00253-014-6358-3 PMID:25616527

Pronk, M., de Kreuk, M. K., de Bruin, B., Kamminga, P., Kleerebezem, R., & van Loosdrecht, M. C. M. (2015a). Full scale performance of the aerobic granular sludge process for sewage treatment. *Water Research*, *84*, 207–217. doi:10.1016/j.watres.2015.07.011 PMID:26233660

Pronk, W., Palmquist, H., Biebow, M., & Boller, M. (2006). Nanofiltration for the separation of pharmaceuticals from nutrients in source-separated urine. *Water Research*, *40*(7), 1405–1412. doi:10.1016/j.watres.2006.01.038 PMID:16530802

Prosser, J. I. (1989). Autotrophic nitrification in bacteria. *Advances in Microbial Physiology*, *30*, 125–181. doi:10.1016/S0065-2911(08)60112-5 PMID:2700538

Puñal, A., Rodríguez, J., Franco, A., Carrasco, E. F., Roca, E., & Lema, J. M. (2001). Advanced monitoring and control of anaerobic wastewater treatment plants: Diagnosis and supervision by a fuzzy-based expert system. *Water Science and Technology*, *43*(7), 191–198. PMID:11385847

Purtschert, I., Siegrist, H., & Gujer, W. (1996). Enhanced denitrification with methanol at WWTP Zürich-Werdhölzli. *Water Science and Technology*, *33*(12), 117–126. doi:10.1016/0273-1223(96)00465-9

Pynaert, K., Smets, B. F., Beheydt, D., & Verstraete, W. (2004). Start-up of autotrophic nitrogen removal reactors via sequential biocatalyst addition. *Environmental Science & Technology*, *38*(4), 1228–1235. doi:10.1021/es030081+ PMID:14998042

Quant, B., Olańczuk-Neyman, K., Fudala-Książek, S., Łuczkiewicz, A., Kulbat, E., & Czerwionka, K. (2009). Influence of landfill leachate on municipal wastewater treatment -model plant (WWTP-MP). In *Proceeding of the Conference: IWA 2nd Specialized Conference Nutrient Management In Wastewater Treatment Processes*.

Quan, X., Ma, J., Xiong, W., & Wang, X. (2015). Bioaugmentation of half-matured granular sludge with special microbial culture promoted establishment of 2,4-dichlorophenoxyacetic acid degrading aerobic granules. *Bioprocess and Biosystems Engineering*, *38*(19), 1081–1090. doi:10.1007/s00449-014-1350-y PMID:25567414

Quan, X., Tang, H., & Ma, J. (2011). Effects of gene augmentation on the removal of 2,4-dichlorophenoxyacetic acid in a biofilm reactor under different scales and substrate conditions. *Journal of Hazardous Materials*, *185*(2-3), 689–695. doi:10.1016/j.jhazmat.2010.09.073 PMID:20951494

Quan, X., Tang, H., Xiong, W., & Yang, Z. (2010). Bioaugmentation of aerobic sludge granules with a plasmid donor strain for enhanced degradation of 2, 4-dichlorophenoxyacetic acid. *Journal of Hazardous Materials*, *179*(1-3), 1136–1142. doi:10.1016/j.jhazmat.2010.04.002 PMID:20430519

Quist-Jensen, C. A., Macedonio, F., & Drioli, E. (2015). Membrane technology for water production in agriculture: Desalination and wastewater reuse. *Desalination*, *364*, 17–32. doi:10.1016/j.desal.2015.03.001

Raboni, M., Torretta, V., Viotti, P., & Urbini, G. (2013). Experimental plant for the chemico-physical treatment of groundwater polluted by MSW leachate, with ammonia recovery. *Revista Ambiente & Agua*, *8*(3), 22–32.

Raghoebarsing, A. A., Pol, A., van de Pas-Schoonen, K. T., Smolders, A. J. P., Ettwig, K. F., Rijpstra, W. I. C., & Strous, M. et al. (2006). A microbial consortium couples anaerobic methane oxidation to denitrification. *Nature*, *440*(7086), 918–921. doi:10.1038/nature04617 PMID:16612380

Rahaman, M., Mavinic, D. S., Meikleham, A., & Ellis, N. (2014). Modeling Phosphorus Removal and Recovery from Anaerobic Digester Supernatant through Struvite Crystallization in a Fluidized Bed Reactor. *Water Research*, *51*, 1–10. doi:10.1016/j.watres.2013.11.048 PMID:24384559

Rahman, M., Liu, Y. H., Kwag, J.-H., & Ra, C. S. (2011). Recovery of struvite from animal wastewater and its nutrient leaching loss in soil. *Journal of Hazardous Materials*, *186*(2-3), 2026–2030. doi:10.1016/j.jhazmat.2010.12.103 PMID:21236571

Rahman, M., Salleh, M. A. M., Rashid, U., Ahsan, A., Hossain, M. M., & Ra, C. S. (2014). Production of slow release crystal fertilizer from wastewaters through struvite crystallization - A review. *Arabian Journal of Chemistry*, *7*(1), 139–155. doi:10.1016/j.arabjc.2013.10.007

Rajapakse, J. P., & Scutt, J. E. (1999). Denitrification with natural gas and various new growth media. *Water Research*, *33*(18), 3723–3734. doi:10.1016/S0043-1354(99)00088-3

Ramírez, C. A., & Worrell, E. (2006). Feeding fossil fuels to the soil. An analysis of energy embedded and technological learning in the fertilizer industry. *Resources, Conservation and Recycling*, *46*, 75–93.

Ramos, C., Suárez-Ojeda, M. E., & Carrera, J. (2015). Long-term impact of salinity on the performance and microbial population of an aerobic granular reactor treating a high-strength aromatic wastewater. *Bioresource Technology*, *198*, 844–851. doi:10.1016/j.biortech.2015.09.084 PMID:26457833

Ramos, C., Suárez-Ojeda, M. E., & Carrera, J. (2016a). Biodegradation of a high-strength wastewater containing a mixture of ammonium, aromatic compounds and salts with simultaneous nitritation in an aerobic granular reactor. *Process Biochemistry*, *51*(3), 399–407. doi:10.1016/j.procbio.2015.12.020

Ramos, C., Suárez-Ojeda, M. E., & Carrera, J. (2016b). Long-term performance and stability of a continuous granular airlift reactor treating a high-strength wastewater containing a mixture of aromatic compounds. *Journal of Hazardous Materials*, *303*, 154–161. doi:10.1016/j.jhazmat.2015.10.031 PMID:26530892

Rasmussen, T., Brittain, T., Berks, B. C., Watmough, N. J., & Thomson, A. J. (2005). Formation of a cytochrome c-nitrous oxide reductase complex is obligatory for N2O reduction by Paracoccus pantotrophus. *Dalton Transactions (Cambridge, England)*, (21): 3501–3506. doi:10.1039/b501846c PMID:16234931

Records, A., & Sutherland, K. (2001). *Decanter Centrifuge Handbook*. Amsterdam: Elsevier.

Reijnders, L. (2014). Phosphorus resources, their depletion and conservation, a review. *Resources, Conservation and Recycling*, *93*, 32–49. doi:10.1016/j.resconrec.2014.09.006

Remus, R., Aguado, M. A., Roudier, S., & Delgado, L. (2013). *IS BREF. Best Available Techniques (BAT) Reference Document for Iron and Steel Production*. Retrieved from The European IPPC Bureau Joint Research Centre website: http://eippcb.jrc.ec.europa.eu/reference/BREF/IS_Adopted_03_2012.pdf

Remy, M., Driessen, W., Hendrickx, T., & Haarhuis, R. (2013). *Recovery of phosphorus by formation of struvite with the PhospaqTM process*. Paper presented at the 18th European Biosolids and Organic Resources Conference.

Renou, S., Givaudan, J. G., Poulain, S., Dirassouyan, F., & Moulin, P. (2008). Landfill leachate treatment: Review and opportunity. *Journal of Hazardous Materials*, *150*(3), 468–493. doi:10.1016/j.jhazmat.2007.09.077 PMID:17997033

Reyes-Ávila, J., Flores-Razo, E., & Gómez, J. (2004). Simultaneous biological removal of nitrogen, carbon and sulfur by denitrification. *Water Research*, *38*(14-15), 3313–3321. doi:10.1016/j.watres.2004.04.035 PMID:15276748

Rezaei, L. S., Ayati, B., & Ganjidoust, H. (2012). Cultivation of aerobic granules in a novel configuration of sequencing batch airlift reactor. *Environmental Technology*, *33*(19-21), 2273–2280. doi:10.1080/09593330.2012.665490 PMID:23393968

Rezania, B., Cicek, N., & Olezkiewics, J. A. (2005). Kintics of hydrogen-dependent denitrification under varying pH and temperature conditions. *Biotechnology and Bioengineering*, *92*(7), 900–906. doi:10.1002/bit.20664 PMID:16116656

Rezania, B., Oleszkiewicz, J. A., & Cicek, N. (2006). Hydrogen-driven denitrification of wastewater in an anaerobic submerged membrane bioreactor: Potential for water reuse. *Water Science and Technology*, *54*(11), 207–214. doi:10.2166/wst.2006.844 PMID:17302322

Rezania, B., Oleszkiewicz, J. A., & Cicek, N. (2007). Hidrogen-dependent denitrification of water in an anaerobic submerged membrane bioreactor coupled with a novel hydrogen delivery system. *Water Research*, *41*(5), 1074–1080. doi:10.1016/j.watres.2006.11.016 PMID:17207833

Rhee, G. Y., & Fuhs, G. W. (1978). Wastewater denitrification with one carbon compounds as energy source. *Journal - Water Pollution Control Federation*, *50*(9), 2111–2119.

Ribé, V., Nehrenheim, E., Odlare, M., Gustavsson, L., Berglind, R., & Forsberg, Å. (2012). Ecotoxicological assessment and evaluation of a pine bark biosorbent treatment of five landfill leachates. *Waste Management (New York, N.Y.)*, *32*(10), 1886–1894. doi:10.1016/j.wasman.2012.05.011 PMID:22703999

Rieger, L., Koch, G., Kühni, M., Gujer, W., & Siegrist, H. (2001). The EAWAG bio-P module for activated sludge model No. 3. *Water Research*, *35*(16), 3887–3903. doi:10.1016/S0043-1354(01)00110-5 PMID:12230171

Rittman, B. E., Mayer, B., Westerhoff, P., & Edwards, M. (2011). Capturing the lost phosphorus[Review]. *Chemosphere*, *84*(6), 846–853. doi:10.1016/j.chemosphere.2011.02.001 PMID:21377188

Rodriguez, D. C., Pino, N., & Penuela, G. (2011). Monitoring the removal of nitrogen by applying a nitrification–denitrification process in a sequencing batch reactor (SBR). *Bioresource Technology*, *102*(3), 2316–2321. doi:10.1016/j.biortech.2010.10.082 PMID:21075619

Rosman, N. H., Nor Anuar, A., Othman, I., Harun, H., Sulong Abdul Razak, M. Z., & Elias, S. H. et al. (2013). Cultivation of aerobic granular sludge for rubber wastewater treatment. *Bioresource Technology*, *129*, 620–623. doi:10.1016/j.biortech.2012.12.113 PMID:23317554

Ruggieri, L., Gea, T., Artola, A., & Sanchez, A. (2008). Influence of different co-substances biochemical composition on raw sludge co-composting. *Biodegradation*, *19*(3), 403–415. doi:10.1007/s10532-007-9146-2 PMID:17674125

Ruscalleda, M., López, H., Ganigué, R., Puig, S., Balaguer, M. D., & Colprim, J. (2008). Heterotrophic denitrification on granular anammox SBR treating urban landfill leachate. *Water Science and Technology*, *58*(9), 1749–1755. doi:10.2166/wst.2008.544 PMID:19029715

Ruscalleda, M., Puig, S., Mora, X., López, H., Ganigué, R., Balaguer, M. D., & Colprim, J. (2010). The effect of urban landfill leachate characteristics on the coexistence of anammox bacteria and heterotrophic denitrifiers. *Water Science and Technology*, *61*(4), 1065–1071. doi:10.2166/wst.2010.610 PMID:20182087

Rustige, H., & Nolde, E. (2007). Nitrogen elimination from landfill leachates using an extra carbon source in subsurface flow constructed wetlands. *Water Science and Technology*, *56*(3), 125. doi:10.2166/wst.2007.506 PMID:17802847

Rybicki, S. (1997). *Phosphorus removal from wastewater: A Literature Review*. Advanced Wastewater Treatment. Report No 1. Stockholm, Sweden: Swedish Reports.

SA BREF. (2005). *Reference Document on Best Available Techniques in the Slaughterhouses and Animal By-products Industries*. Retrieved January 15, 2016, from http://eippcb.jrc.ec.europa.eu/reference/BREF/sa_bref_0505.pdf

Sadr, S. M., & Saroj, D. P. (2015). *Membrane technologies for municipal wastewater treatment. In Advances in Membrane Technologies for Water Treatment: Materials, Processes and Applications* (pp. 443–463). Woodhead Publishing. doi:10.1016/B978-1-78242-121-4.00014-9

Saeed, T., Afrin, R., Muyeed, A. A., & Sun, G. (2012a). Treatment of tannery wastewater in a pilot-scale hybrid constructed wetland system in Bangladesh. *Chemosphere*, *88*(9), 1065–1073. doi:10.1016/j.chemosphere.2012.04.055 PMID:22673399

Saeed, T., & Sun, G. (2012b). A review on nitrogen and organics removal mechanisms in subsurface flow constructed wetlands: Dependency on environmental parameters, operating conditions and supporting media. *Journal of Environmental Management*, *112*, 429–448. doi:10.1016/j.jenvman.2012.08.011 PMID:23032989

Sahu, A. K., Conneely, T., Nusslein, K., & Ergas, S. J. (2009). Hydrogenotrophic denitrification and perchlorate reduction in ion exchange brines using membrane biofilm reactors. *Biotechnology and Bioengineering*, *104*(3), 483–491. doi:10.1002/bit.22414 PMID:19544384

Saito, T., Brdjanovic, D., & van Loosdrecht, M. C. M. (2004). Effect of nitrite on phosphate uptake by phosphate accumulatingorganisms. *Water Research*, *38*(17), 3760–3768. doi:10.1016/j.watres.2004.05.023 PMID:15350428

Sakurai, Y., Toda, K., & Hideo, S. (1981). Multiple Forms and Some Properties of Alkaline Phosphatase Produced by Aspergillus oryzae on Solid Medium. *Agricultural and Biological Chemistry*, *45*(9), 1959–1967. doi:10.1080/00021369.1981.10864831

Sanchez, M., Hernadez, S., Bravo, C., Antolin, J. M., & Gallardo, T. (2010). *Sustainable use of biodegradables products in agricultural fertilization, reforestation, plant production and recovery of degraded areas*. Special Orbit. International Congress about Organic Soils and Organic Matter from Waste, Valladolid, Spain. (in Spanish)

Santin, I., Pedret, C., & Vilanova, R. (2015). Applying variable dissolved oxygen set point in a two level hierarchical control structure to a wastewater treatment process. *Journal of Process Control*, *28*, 40–55. doi:10.1016/j.jprocont.2015.02.005

Sato, K., Okubo, A., & Yamazaki, S. (1998). Characterization of a multi-copper enzyme, nitrous oxide reductase, from Rhodobacter sphaeroides f. sp. denitrificans. *Journal of Biochemistry*, *124*(1), 51–54. doi:10.1093/oxfordjournals.jbchem.a022096 PMID:9644245

Savchenko, A., Wang, W., Vieille, C., & Zeikus, J. G. (2001). Alkaline phosphatase from Thermotoga neapolitana. *Methods in Enzymology*, *331*, 298–305. doi:10.1016/S0076-6879(01)31067-4 PMID:11265472

Savelieff, M. G., Wilson, T. D., Elias, Y., Nilges, M. J., Garner, D. K., & Lu, Y. (2008). Experimental evidence for a link among cupredoxins: Red, blue, and purple copper transformations in nitrous oxide reductase. *Proceedings of the National Academy of Sciences of the United States of America*, *105*(23), 7919–7924. doi:10.1073/pnas.0711316105 PMID:18535143

Sawaittayothin, V., & Polprasert, C. (2007). Nitrogen mass balance and microbial analysis of constructed wetlands treating municipal landfill leachate. *Bioresource Technology*, *98*(3), 565–570. doi:10.1016/j.biortech.2006.02.002 PMID:16546377

Scaglione, D., Ficara, E., Corbellini, V., Teli, A., Canziani, R., & Malpei, F. (2014). *Treatibiliy of digested piggery/poultry manure by anammox bacteria*. Paper presented at the 2nd International Water Association Specialized Conference EcoTechnologies for Sewage Treatment Plants, Verona, Italy.

Scaglione, D., Ficara, E., Corbellini, V., Tornotti, G., Teli, A., Canziani, R., & Malpei, F. (2015). Autotrophic nitrogen removal by a two-step SBR process applied to mixed agro-digestate. *Bioresource Technology*, *176*, 98–105. doi:10.1016/j.biortech.2014.11.019 PMID:25460989

Scaglione, D., Ruscalleda, M., Ficara, E., Balaguer, M. D., & Colprim, J. (2012). Response to high nitrite concentrations of anammox biomass from two SBR fed on synthetic wastewater and landfill leachate. *Chemical Engineering Journal*, *209*, 62–68. doi:10.1016/j.cej.2012.07.106

Schaffel, S. D., & Hulett, F. M. (1978). Alkaline phosphatase from Bacillus licheniformis. Solubility dependent on magnesium, purification and characterization. *Biochimica et Biophysica Acta*, *526*(2), 457–467. doi:10.1016/0005-2744(78)90137-7 PMID:718947

Schipper, W. J., Klapwijk, A., Potjer, B., Rulkens, W. H., Temmink, B. G., Kiestra, F. D. G., & Lijmbach, A. C. M. (2001). Phosphate recycling in the phosphorus industry. *Environmental Technology*, *22*(11), 1337–1345. doi:10.1080/09593330.2001.9619173 PMID:11804355

Schlesinger, W. H., & Bernhardt, E. S. (2013). Biogeochemistry. In E. S. Bernhardt (Ed.), *Biogeochemistry* (3rd ed., pp. i–ii). Boston: Academic Press.

Schmid, M. C., Maas, B., Dapena, A., van de Pas-Schoonen, K., van de Vossenberg, J., Kartal, B., & Strous, M. et al. (2005). Biomarkers for in situ detection of anaerobic ammonium-oxidizing (anammox) bacteria. *Applied and Environmental Microbiology*, *71*(4), 1677–1684. doi:10.1128/AEM.71.4.1677-1684.2005 PMID:15811989

Schmid, M., Walsh, K., Webb, R., Rijpstra, W., van de Pas-Schoonen, K., Verbruggen, M., & Strous, M. et al. (2003). Candidatus "Scalindua brodae", sp. nov., Candidatus "Scalindua wagneri", sp. nov., Two New Species Anaerobic Ammonium Oxidizing Bacteria. *Systematic and Applied Microbiology*, *26*(4), 529–538. doi:10.1078/072320203770865837 PMID:14666981

Schmidt, I., & Bock, E. (1998). Anaerobic ammonia oxidation by cell-free extracts of *Nitrosomonas eutropha*. *Antonie van Leeuwenhoek*, *73*(3), 271–278. doi:10.1023/A:1001572121053 PMID:9801772

Schuiling, R. D., & Andrade, A. (1999). Recovery of struvite from calf manure. *Environmental Technology*, *20*(7), 765–768. doi:10.1080/09593332008616872

Schuler, A. J. (2005). Diversity matters: Dynamic simulation of distributed bacterial states in suspended growth biological wastewater treatment systems. *Biotechnology and Bioengineering*, *91*(1), 62–74. doi:10.1002/bit.20484 PMID:15880520

Schütte, T., Niewersch, C., Wintgens, T., & Yüce, S. (2015). Phosphorus recovery from sewage sludge by nanofiltration in diafiltration mode. *Journal of Membrane Science*, *480*, 74–82.

Schwarzenbeck, N., Borges, J. M., & Wilderer, P. (2005). Treatment of dairy effluents in an aerobic granular sludge sequencing batch reactor. *Applied Microbiology and Biotechnology*, *66*(6), 711–718. doi:10.1007/s00253-004-1748-6 PMID:15558277

Schwarzenbeck, N., Erley, R., Mc Swain, B. S., Wilderer, P., & Irvine, R. L. (2004). Treatment of malting wastewater in a granular sludge sequencing batch reactor (SBR). *Acta Hydrochimica et Hydrobiologica*, *32*(1), 16–24. doi:10.1002/aheh.200300517

Seeger, E. M., Maier, U., Grathwohl, P., Kuschk, P., & Kaestner, M. (2013). Performance evaluation of different horizontal subsurface flow wetland types by characterization of flow behavior, mass removal and depth-dependent contaminant load. *Water Research*, *47*(2), 769–780. doi:10.1016/j.watres.2012.10.051 PMID:23200508

Seidel, K. (1961). Zur Problematik der Keim- und Pflanzengewasser. Verh Internat. Verein. *Limnology*, *14*, 1035–1039.

Señoráns, L., Barros, S., García, G., & Garrido, J. M. (2011). Phosphorus recovery by struvite crystallization in a wastewater from a frozen fish factory. *Tecnología del Agua*, *335*, 26–35. (in Spanish)

Serralta, J., Borrás, L., Blanco, C., Barat, R., & Seco, A. (2004b). Monitoring pH and electric conductivity in an EBPR sequencing batch reactor. *Water Science and Technology*, *50*(10), 145–152. PMID:15656307

Serralta, J., Ferrer, J., Borrás, L., & Seco, A. (2004a). An extension of ASM2d including pH calculation. *Water Research*, *38*(19), 4029–4038. doi:10.1016/j.watres.2004.07.009 PMID:15491650

Serrano, L., de la Varga, D., Díaz, M. A., Ruiz, I., Bondelle, R., & Soto, M. (2009). HUSB-CW treatment system for winery wastewater.*Proceedings of the 5th International Specialized Conference on Sustainable Viticulture: Winery Waste and Ecologic Impacts Management.* (pp.317-324).

Serrano, L., de la Varga, D., Ruiz, I., & Soto, M. (2011). Winery wastewater treatment in a hybrid constructed wetland. *Ecological Engineering*, *37*(5), 744–753. doi:10.1016/j.ecoleng.2010.06.038

Serrano-Silva, N., Sarria-Guzmán, Y., Dendooven, L., & Luna-Guido, M. (2014). Methanogenesis and Methanotrophy in soil: A Review. *Pedosphere*, *24*(3), 291–307. doi:10.1016/S1002-0160(14)60016-3

Shalaby, A., Nassef, E., Mubark, A., & Hussein, M. (2014). Phosphate removal from wastewater by electrocoagulation using aluminium electrodes. *American Journal of Environmental Engineering and Science*, *1*(5), 90–98.

Shamir, E., Thompson, T. L., Karpiscak, M. M., Freitas, R. J., & Zauderer, J. (2001). Nitrogen accumulation in a constructed wetland for dairy wastewater treatment. *Journal of the American Water Resources Association*, *37*(2), 315–325. doi:10.1111/j.1752-1688.2001.tb00971.x

Shen, L.-D., He, Z.-F., Zhu, Q., Chen, D.-Q., Lou, L.-P., Xu, X.-Y., Zheng, P & Hu B.L. (2012) Microbiology, ecology, and application of the nitrite-dependent anaerobic methane oxidation process. *Frontiers in Microbiology*, *3*, 269-273.

Sheng, G.-P., Yu, H.-Q., & Li, X.-Y. (2010). Extracellular polymeric substances (EPS) of microbial aggregates in biological wastewater treatment systems: A review. *Biotechnology Advances*, *28*(6), 882–894. doi:10.1016/j.biotechadv.2010.08.001 PMID:20705128

Shen, L.-, Hu, A.-, Jin, R.-, Cheng, D.-, Zheng, P., Xu, X.-y., & Hu, B.-. (2012). Enrichment of anammox bacteria from three sludge sources for the startup of monosodium glutamate industrial wastewater treatment system. *Journal of Hazardous Materials*, *199–200*, 193–199. doi:10.1016/j.jhazmat.2011.10.081 PMID:22104770

Shepher, H. L., Grismer, M. E., & Tchobanoglous, G. (2001). Treatment of high-strength winery wastewater using a subsurface-flow constructed wetland. *Water Environment Research*, *73*(4), 394–403. doi:10.2175/106143001X139434

Shimamura, M., Nishiyama, T., Shigetomo, H., Toyomoto, T., Kawahara, Y., Furukawa, K., & Fujii, T. (2007). Isolation of a multiheme protein with features of a hydrazine-oxidizing enzyme from an anaerobic ammonium-oxidizing enrichment culture. *Applied and Environmental Microbiology*, *73*(4), 1065–1072. doi:10.1128/AEM.01978-06 PMID:17172456

Shimamura, M., Nishiyama, T., Shinya, K., Kawahara, Y., Furukawa, K., & Fujii, T. (2008). Another multiheme protein, hydroxylamine oxidoreductase, abundantly produced in an anammox bacteriumbesides the hydrazine-oxidizing enzyme. *Journal of Bioscience and Bioengineering*, *105*(3), 243–248. doi:10.1263/jbb.105.243 PMID:18397776

Shin, J. H., Sang, B. I., Chung, Y. C., & Choung, Y. K. (2005). The removal of nitrogen using an autotrophic hybrid hollow-fiber membrane biofilm reactor. *Desalination*, *183*(1-3), 447–454. doi:10.1016/j.desal.2005.03.045

Shi, Y.-J., Wang, X.-H., Qi, Z., Diao, M.-H., Gao, M.-M., & Xing, S.-F. et al. (2011). Sorption and biodegradation of tetracycline by nitrifying granules and the toxicity of tetracycline on granules. *Journal of Hazardous Materials*, *191*(1-3), 103–109. doi:10.1016/j.jhazmat.2011.04.048 PMID:21570181

Shi, Y., Xing, S., Wang, X., & Wang, S. (2013). Changes of the reactor performance and the properties of granular sludge under tetracycline (TC) stress. *Bioresource Technology*, *139*, 170–175. doi:10.1016/j.biortech.2013.03.037 PMID:23660380

Show, K.-Y., Lee, D.-J., & Tay, J.-H. (2012). Aerobic granulation: Advances and challenges. *Applied Biochemistry and Biotechnology*, *167*(6), 1622–1640. doi:10.1007/s12010-012-9609-8 PMID:22383048

Siegrist, H., Laureni, M., & Udert, K. M. (2013). Transfer into the gas phase: Ammonia stripping. In T.A. Larsen, K.M. Udert, & J. Lienert (Eds.), Source Separation and Decentralization for Wastewater Management (pp. 337-350). London, UK: IWA.

Siegrist, H., Brunner, I., Koch, G., Linh, C. P., & Van Chieu, L. (1999). Reduction of biomass decay rate under anoxic and anaerobic conditions. *Water Science and Technology*, *39*(1), 129–137. doi:10.1016/S0273-1223(98)00782-3

Siegrist, H., Reithaar, S., Koch, G., & Lais, P. (1998). Nitrogen loss in a nitrifying rotating contactor treating ammonium-rich wastewater without organic carbon. *Water Science and Technology*, *38*(8-9), 241–248. doi:10.1016/S0273-1223(98)00698-2

Siemann, S., Schneider, K., Drottboom, M., & Muller, A. (2002). The Fe-only nitrogenase and the Mo nitrogenase from Rhodobacter capsulatus: A comparative study on the redox properties of the metal clusters present in the dinitrogenase components. *European Journal of Biochemistry*, *269*(6), 1650–1661. doi:10.1046/j.1432-1327.2002.02804.x PMID:11895435

Sierra-Alvarez, R., Beristan-Cardoso, R., Salazer, M., Gómez, J., Razo-Flores, E., & Field, J. A. (2007). Chemolithotrophic denitrification with elemental sulphur for groundwater treatment. *Water Research*, *41*(6), 1253–1262. doi:10.1016/j.watres.2006.12.039 PMID:17296214

Silva, C. D., Gómez, J., & Beristain-Cardoso, R. (2011). Simultaneous removal of 2-chlorophenol, phenol, p-creso and p-hydroxybenzaldehyde under nitrifying conditions: Kinetic study. *Bioresource Technology*, *102*(11), 6464–6468. doi:10.1016/j.biortech.2011.03.105 PMID:21504846

Silva, D. D., Cuervo-López, F. M., Gómez, J., & Texier, A.-C. (2011). Nitrite effect on ammonium and nitrite oxidizing processes in a nitrifying sludge. *World Journal of Microbiology & Biotechnology*, *27*(5), 1241–1245. doi:10.1007/s11274-010-0573-6

Sin, G., Kaelin, D., Kampschreur, M. J., Takács, I., Wett, B., Gernaey, K. V., & van Loosdrecht, M. C. M. et al. (2008). Modelling nitrite in wastewater treatment systems: A discussion of different modelling concepts. *Water Science and Technology*, *58*(6), 1155–1171. doi:10.2166/wst.2008.485 PMID:18845852

Sin, G., Niville, K., Bachis, G., Jiang, T., Nopens, I., van Hulle, S., & Vanrolleghem, P. A. (2007b). Nitrite effect on the phosphorus uptake activity of phosphate accumulating organisms (PAOs) in pilot-scale SBR and MBR reactors. *Water S.A., 34*(2), 249–260.

Sin, G., & Vanrolleghem, P. A. (2006). Evolution of an ASM2d-like model structure due to operational changes of an SBR process. *Water Science and Technology, 53*(12), 237–245. doi:10.2166/wst.2006.426 PMID:16889260

Sin, G., & Vanrolleghem, P. A. (2007a). Extensions to modeling aerobic carbon degradation using combined respirometric-titrimetric measurements in view of activated sludge model calibration. *Water Research, 41*(15), 3345–3358. doi:10.1016/j.watres.2007.03.029 PMID:17498768

Singer, P. C., & Stumm, W. (1970). Acid mine drainage—the rate limiting step. *Science, 167*(3921), 1121–1123. doi:10.1126/science.167.3921.1121 PMID:17829406

Słomczyńska, B., & Słomczyński, T. (2004). Physico-Chemical and Toxicological Characteristics of Leachates from MSW Landfills. *Polish Journal of Environmental Studies, 13*(6), 627–637.

Smil, V. (2000). Phosphorus in the environment: Natural flows and human interferences. *Annual Review of Energy and the Environment, 25*(1), 53–88. doi:10.1146/annurev.energy.25.1.53

Smith, D. P., Rector, T., Reid-Black, K., Hummerick, M., Strayer, R., Birmele, M., & Garland, J. L. et al. (2007). Redox control bioreactor: A unique biological water processor. *Biotechnology and Bioengineering, 99*(4), 830–845. doi:10.1002/bit.21617 PMID:17705250

Smith, L. K., Sartoris, J., Thullen, J., & Anderson, D. C. (2000). Investigation of denitrification rates in an ammonia-dominated treatment wetland. *Wetlands, 20*(4), 684–696. doi:10.1672/0277-5212(2000)020[0684:IODRIA]2.0.CO;2

Smith, R. L., Ceazan, M. L., & Brooks, M. H. (1994). Autotrophic, hydrogen oxidizing, denitrifying bacteria in groundwater, potential agents for bioremediation of nitrate contamination. *Applied and Environmental Microbiology, 60*(6), 949–955. PMID:16349284

Smith, R. L., Ceazan, M. L., & Brooks, M. H. (1994). Autotrophic, hydrogen-oxidizing, denitrifying bacteria in groundwater, potential agents for bioremediation of nitrate contamination. *Applied and Environmental Microbiology, 60*(6), 1949–1955. PMID:16349284

Smolders, G. J. F., van der Meij, J., van Loosdrecht, M. C. M., & Heijnen, J. J. (1994a). Model of the anaerobic metabolism of the biological phosphorus removal process: Stoichiometry and pH influence. *Biotechnology and Bioengineering, 43*(6), 461–470. doi:10.1002/bit.260430605 PMID:18615742

Smolders, G. J. F., van der Meij, J., van Loosdrecht, M. C. M., & Heijnen, J. J. (1994b). Stoichiometric model of the aerobic metabolism of the biological phosphorus removal proces. *Biotechnology and Bioengineering, 44*(7), 837–848. doi:10.1002/bit.260440709 PMID:18618851

Snyder, S. W., & Hollocher, T. C. (1987). Purification and some characteristics of nitrous oxide reductase from Paracoccus denitrificans. *The Journal of Biological Chemistry, 262*(14), 6515–6525. PMID:3032972

Soares, M. I. M. (2002). Denitrification of groundwater with elemental sulphur. *Water Research, 36*(5), 1392–1395. doi:10.1016/S0043-1354(01)00326-8 PMID:11905445

Sollo, F. W., Mueller, H. F., & Larson, T. E. (1976). Denitrification of wastewater effluents with methane. *Journal - Water Pollution Control Federation, 48*(7), 1840–1842.

Solon, K., Flores-Alsina, X., Kazadi Mbamaba, C., Gernaey, K. V., Tait, S., Batstone, D., & Jeppsson, U. et al. (2015). Effects of ion strength and ion pairing on plant wide modelling on anaerobic digester. *Water Research*, *70*, 235–245. doi:10.1016/j.watres.2014.11.035 PMID:25540837

Sotres, A., Cerrillo, M., Viñas, M., & Bonmatí, A. (2015). Nitrogen recovery from pig slurry in a two-chambered bioelectrochemical system.[PubMed]. *Bioresource Technology*, *194*, 373–382.

Spagni, A., Buday, J., Ratini, P., & Bortone, G. (2001). Experimental considerations on monitoring ORP, pH, conductivity and dissolved oxygen in nitrogen and phosphorus biological removal processes. *Water Science and Technology*, *43*(11), 197–204. PMID:11443963

Spagni, A., & Marsili-Libelli, S. (2009). Nitrogen removal via nitrite in a sequencing batch reactor treating sanitary landfill leachate. *Bioresource Technology*, *100*(2), 609–614. doi:10.1016/j.biortech.2008.06.064 PMID:18707876

Sri Shalini, S., & Joseph, K. (2012). Nitrogen management in landfill leachate: Application of SHARON, ANAMMOX and combined SHARON-ANAMMOX process. *Waste Management (New York, N.Y.)*, *32*(12), 2385–2400. doi:10.1016/j.wasman.2012.06.006 PMID:22766438

Stare, A., Vrecko, D., Hvala, N., & Strmcnik, S. (2007). Comparison of control strategies for nitrogen removal in an activated sludge process in terms of operating costs: A simulation study. *Water Research*, *41*(9), 2004–2014. doi:10.1016/j.watres.2007.01.029 PMID:17346768

Stein, L. Y., & Arp, D. J. (1998). Loss of ammonia monooxygenase activity in *Nitrosomonas europaea* upon exposure to nitrite. *Applied and Environmental Microbiology*, *64*, 4098–4102. PMID:9758853

Stein, L. Y., & Klotz, M. G. (2016). The nitrogen cycle. *Current Biology*, *26*(3), R94–R98. doi:10.1016/j.cub.2015.12.021 PMID:26859274

Strous, M., Fuerst, J. A., Kramer, E. H., Logemann, S., Muyzer, G., van de Pas-Schoonen, K. T., & Jetten, M. S. M. et al. (1999). Missing lithotroph identified as new planctomycete. *Nature*, *400*(6743), 446–449. doi:10.1038/22749 PMID:10440372

Strous, M., Heijnen, J. J., Kuenen, J. G., & Jetten, M. S. M. (1998). The sequencing batch reactor as a powerful tool for the study of slowly growing anaerobic ammonium-oxidizing microorganisms. *Applied Microbiology and Biotechnology*, *50*(5), 589–596. doi:10.1007/s002530051340

Strous, M., Kuenen, J. G., & Jetten, M. S. M. (1999). Key physiology of anaerobic ammonium oxidation. *Applied and Environmental Microbiology*, *65*, 3248–3250. PMID:10388731

Strous, M., Pelletier, E., Mangenot, S., Rattei, T., Lehner, A., Taylor, M. W., & Le Paslier, D. et al. (2006). Deciphering the evolution and metabolism of an anammox bacterium from a community genome. *Nature*, *440*(7085), 790–794. doi:10.1038/nature04647 PMID:16598256

Strous, M., van Gerven, E., Kuenen, J. G., & Jetten, M. S. M. (1997). Effects of aerobic and microaerobic conditions on anaerobic ammonium-oxidizing (Anammox) sludge. *Applied and Environmental Microbiology*, *63*(3), 2446–2448. PMID:16535633

Strous, M., van Gerven, E., Zheng, P., Kuenen, J. G., & Jetten, M. S. M. (1997). Ammonium removal from concentrated waste streams with the anaerobic ammonium oxidation (ANAMMOX) process in different reactor configurations. *Water Research*, *31*(8), 1955–1962. doi:10.1016/S0043-1354(97)00055-9

Stumm, W., & Morgan, J. J. (1981). *Aquatic Chemistry. An Introduction Emphasizing Chemical Equilibria in Natural Waters* (2nd ed.). New York: Wiley-Interscience.

Suarez, S., Lema, J. M., & Omil, F. (2010). Removal of pharmaceutical and personal care products (PPCPs) under nitrifying and denitrifying conditions. *Water Research*, *44*(10), 3214–3224. doi:10.1016/j.watres.2010.02.040 PMID:20338614

Sublette, K. L., & Sylvester, N. D. (1997). Oxidation of hydrogen sulphide by mixed cultures of *Thiobacillus denitrificans* and heterotrophs. *Biotechnology and Bioengineering*, *29*(6), 759–761. doi:10.1002/bit.260290614 PMID:18576512

Suharti, H., Heering, H. A., & de Vries, S. (2004). NO reductase from Bacillus azotoformans is a bifunctional enzyme accepting electrons from menaquinol and a specific endogenous membrane-bound cytochrome c551. *Biochemistry*, *43*(42), 13487–13495. doi:10.1021/bi0488101 PMID:15491156

Suhr, M., Klein, G., Kourti, I., Rodrigo, M., Giner, G., Roudier, S., & Delgado, L. (2015). *PP BREF. Best Available Techniques (BAT) Reference Document for the Production of Pulp, Paper and Board.* Retrieved from The European IPPC Bureau Joint Research Centre website: http://eippcb.jrc.ec.europa.eu/reference/BREF/PP_revised_BREF_2015.pdf

Sukalyan, S., Tabish, N., & Jeffrey, J. (2015). Nitrogen and Phosphorus Recovery from Wastewater. *Water Pollution*, *1*(3), 155–166.

Sun, G. J., Zhao, Y., Allen, S., & Cooper, D. (2006). Generating "tide" in pilot-scale constructed wetlands to enhance agricultural wastewater treatment. *Engineering in Life Sciences*, *6*(6), 560–565. doi:10.1002/elsc.200620156

Sun, G., & Austin, D. (2007). Completely autotrophic nitrogen-removal over nitrite in lab-scale constructed wetlands: Evidence from a mass balance study. *Chemosphere*, *68*(6), 1120–1128. doi:10.1016/j.chemosphere.2007.01.060 PMID:17349669

Sun, G., Zhao, Y., & Allen, S. (2005). Enhanced removal of organic matter and ammoniacal-nitrogen in a column experiment of tidal flow constructed wetland system. *Journal of Biotechnology*, *115*(2), 189–197. doi:10.1016/j.jbiotec.2004.08.009 PMID:15607237

Sun, X.-F., Wang, S.-G., Liu, X.-W., Gong, W.-X., Bao, N., Gao, B.-Y., & Zhang, H.-Y. (2008). Biosorption of Malachite Green from aqueous solutions onto aerobic granules: Kinetic and equilibrium studies. *Bioresource Technology*, *99*(9), 3475–3483. doi:10.1016/j.biortech.2007.07.055 PMID:17855080

Sutton, M. A., Bleeker, A., Howard, C. M., Bekunda, M., Grizzett, B., de Vries, W., . . . Zhang, F. S. (2013). Our Nutrient World: The challenge to produce more food and energy with less pollution. In *Global Overview on Nutrient Management.* Centre for Ecology and Hydrology, Edinburgh on behalf of the Global Partnership on Nutrient Management and the International Nitrogen Initiative. Retrieved from: http://www.unep.org/gpa/documents/publications/ONW.pdf

Suzuki, Y., Mizutani, Y., Tsuji, T., Ohtani, N., Takano, K., Haruki, M., & Kanaya, S. et al. (2005). Gene cloning, overproduction, and characterization of thermolabile alkaline phosphatase from a psychrotrophic bacterium. *Bioscience, Biotechnology, and Biochemistry*, *69*(2), 364–373. doi:10.1271/bbb.69.364 PMID:15725663

Tan, A. S., & Worobec, E. A. (1993). Isolation and characterization of two immunochemically distinct alkaline phosphatases from Pseudomonas aeruginosa. *FEMS Microbiology Letters*, *106*(3), 281–286. doi:10.1111/j.1574-6968.1993.tb05977.x PMID:8454193

Tandukar, M., Pavlostathis, S. G., & Cervantes, F. J. (2009). Autotrophic denitrification for the removal of nitrogenous and sulphurous contaminants from wastewaters. In F. Cervantes (Ed.), *Environmental Technologies to treat nitrogen pollution. Principles and Engineering* (pp. 319–371). International Water Association Publishing.

Tang, C.-, Zheng, P., Wang, C.-, & Mahmood, Q. (2010). Suppression of anaerobic ammonium oxidizers under high organic content in high-rate Anammox UASB reactor. *Bioresource Technology*, *101*(6), 1762–1768. doi:10.1016/j.biortech.2009.10.032 PMID:19900808

Tang, Y., Zhou, C., Van Ginkel, S. W., Ontiveros-Valencia, A., Shin, J., & Rittmann, B. E. (2012). Hydrogen permeability of the hollow fibers used in $H_2$-based membrane biofilm reactors. *Journal of Membrane Science*, *407–408*, 176–183. doi:10.1016/j.memsci.2012.03.040

Tang, Y., Zhou, C., Ziv-El, M., & Rittmann, B. E. (2011). A pH-control model for heterotrophic and hydrogen-based autotrophic denitrification. *Water Research*, *45*(1), 232–240. doi:10.1016/j.watres.2010.07.049 PMID:20705316

Tanner, C. (2001). Plants as ecosystem engineers in subsurface-flow treatment wet-lands. *Water Science and Technology*, *44*(11–12), 9–17. PMID:11804163

Tanner, C. C., Adams, D. D., & Downes, M. T. (1997). Methane emissions from constructed wetlands treating agricultural wastewaters. *Journal of Environmental Quality*, *26*(4), 1056–1062. doi:10.2134/jeq1997.00472425002600040017x

Tanner, C. C., Clayton, J. S., & Upsdell, M. P. (1995). Effect of loading rate and planting on treatment of dairy farm wastewaters in constructed wetlands I. Removal of oxygen demand, suspended solids and faecal coliforms. *Water Research*, *29*(1), 17–26. doi:10.1016/0043-1354(94)00139-X

Tao, W., Fattah, K. P., & Huchzermeier, M. P. (2016). Struvite recovery from anaerobically digested dairy manure: A review of application potential and hindrances. *Journal of Environmental Management*, *169*, 46–57. doi:10.1016/j.jenvman.2015.12.006 PMID:26720329

Tayà, C. (2013). *Facing current EBPR bottlenecks in view of full-scale implementation.* (PhD. Thesis). Universitat Autònoma de Barcelona, Spain.

Tay, S. T.-L., Moy, B. Y.-P., Jiang, H.-L., & Tay, J.-H. (2005). Rapid cultivation of stable aerobic phenol-degrading granules using acetate-fed granules as microbial seed. *Journal of Biotechnology*, *115*(4), 387–395. doi:10.1016/j.jbiotec.2004.09.008 PMID:15639100

Tay, S. T., Zhuang, W.-Q., & Tay, J.-H. (2005). Start-up, microbial community analysis and formation of aerobic granules in a tert-butyl alcohol degrading sequencing batch reactor. *Environmental Science & Technology*, *39*(15), 5774–5780. doi:10.1021/es050278x PMID:16124314

Tchamango, S., Nanseu-Njiki, C. P., Ngameni, E., Hadjiev, D., & Darchen, A. (2010). Treatment of dairy effluents by electrocoagulation using aluminium electrodes. *The Science of the Total Environment*, *408*(4), 947–952. doi:10.1016/j.scitotenv.2009.10.026 PMID:19900696

Tchobanoglous, G., Burton, F. L., & Stensel, H. D. (2003). *Wastewater engineering: Treatment and reuse*. Boston: Ed. Mc Graw-Hill.

Terada, A., Kaku, S., Matsumoto, S., & Tsuneda, S. (2006). Rapid autohydrogenotrophic denitrification by a membrane biofilm reactor equipped with a fibrous support around a gas-permeable membrane. *Biochemical Engineering Journal*, *31*(1), 84–91. doi:10.1016/j.bej.2006.06.004

Terada, A., Lackner, S., Tsuneda, S., & Smets, B. F. (2007). Redox-stratification controlled biofilm (ReSCoBi) for completely autotrophic nitrogen removal: The effect of co- versus counter-diffusion on reactor performance. *Biotechnology and Bioengineering*, *97*(1), 40–51. doi:10.1002/bit.21213 PMID:17013935

Teraguchi, S., & Hollocher, T. C. (1989). Purification and some characteristics of a cytochrome c-containing nitrous oxide reductase from Wolinella succinogenes. *The Journal of Biological Chemistry*, *264*(4), 1972–1979. PMID:2536696

Texier, A. C., & Gomez, J. (2002). Tolerance of nitrifying sludge to different concentrations of *p*-cresol. *Biotechnology Letters*, *40*(4), 321–332. doi:10.1023/A:1014057907953

Texier, A. C., & Gomez, J. (2007). Simultaneous nitrification and p-cresol oxidation in a nitrifying sequencing batch reactor. *Water Research*, *41*(2), 315–322. doi:10.1016/j.watres.2006.10.017 PMID:17126873

Texier, A. C., Gomez, J., & Cuervo-López, F. M. (2013). Inhibitory, toxic and structure effects of toluene on microbial consortia involved in wastewater treatment. In M. Palminteri (Ed.), *Toluene: Chemical Properties, Application and Toxicology* (pp. 93–124). Nova Science Publishers, Inc.

Thalasso, F., Vallecillo, A., García-Encina, P., & Fdz-Polanco, F. (1997). The use of methane as a sole carbon source for wastewater denitrification. *Water Research*, *31*(1), 55–60. doi:10.1016/S0043-1354(96)00228-X

Thomas, M., Wright, P., Blackall, L., Urbain, V., & Keller, J. (2003). Optimisation of Nossa BNR plant to improve prformance and reduced operating costs. *Water Science and Technology*, *47*(12), 141–148. PMID:12926681

Thörn, M., & Sörensson, F. (1996). Variation of nitrous oxide formation in the denitrification basin in a wastewater treatment plant with nitrogen removal. *Water Research*, *30*(6), 1543–1547. doi:10.1016/0043-1354(95)00327-4

Thürlimann, C. M., Dürrenmatt, D. J., & Villez, K. (2015). Evaluation of Qualitative Trend Analysis as a Tool for Automation. *Computer Aided Chemical Engineering*, *37*, 2531–2536. doi:10.1016/B978-0-444-63576-1.50116-3

Tiedje, J. M. (1988). *Biology of anaerobic microorganisms* (A. J. B. Zehnder, Ed.). Wiley.

Tikhonova, T. V., Slutsky, A., Antipov, A. N., Boyko, K. M., Polyakov, K. M., Sorokin, D. Y., & Popov, V. O. et al. (2006). Molecular and catalytic properties of a novel cytochrome c nitrite reductase from nitrate-reducing haloalkaliphilic sulfur-oxidizing bacterium Thioalkalivibrio nitratireducens. *Biochimica et Biophysica Acta*, *1764*(4), 715–723. doi:10.1016/j.bbapap.2005.12.021 PMID:16500161

Tilche, A., Bortone, G., Malaspina, F., Piccinini, S., & Stante, L. (2001). Biological nutrient removal in a full-scale SBR treating piggery wastewater: Results and modelling. *Water Science and Technology*, *43*(3), 363–371. PMID:11381929

Timur, H., & Ozturk, I. (1999). Anaerobic sequencing batch reactor treatment of landfill leachate. *Water Research*, *33*(15), 3225–3230. doi:10.1016/S0043-1354(99)00048-2

Tirado, R. (2008). *Dead zones. How agricultural fertilizers kill our rivers, lakes and oceans.* Retrieved from Greenpeace webpage: http://www.greenpeace.org/canada/Global/canada/report/2008/7/dead-zones.pdf

Tizaoui, C., Bouselmi, L., Mansouri, L., & Ghrabi, A. (2007). Landfill leachate treatment with ozone and ozone/hydrogen peroxide systems. *Journal of Hazardous Materials*, *140*(1), 316–324. doi:10.1016/j.jhazmat.2006.09.023 PMID:17034936

Toh, S., & Ashbolt, N. (2002). Adaptation of anaerobic ammonium-oxidising consortium to synthetic coke-ovens wastewater. *Applied Microbiology and Biotechnology*, *59*(2-3), 344–352. doi:10.1007/s00253-002-1007-7 PMID:12111169

Tokutomi, T., Yamauchi, H., Nishimura, S., Yoda, M., & Abma, W. (2011). Application of the Nitritation and Anammox Process into Inorganic Nitrogenous Wastewater from Semiconductor Factory. *Journal of Environmental Engineering*, *137*(2), 146–154. doi:10.1061/(ASCE)EE.1943-7870.0000303

Tomar, S., Gupta, S. K., & Mishra, B. K. (2015). A novel strategy for simultaneous removal of nitrogen and organic matter using anaerobic granular sludge in anammox hybrid reactor. *Bioresource Technology*, *197*, 171–177. doi:10.1016/j.biortech.2015.08.057 PMID:26335285

Torà, J. A., Lafuente, J., Baeza, J. A., & Carrera, J. (2010). Combined effect of inorganic carbon limitation and inhibition by free ammonia and free nitrous acid on ammonia oxdazing bacteria. *Bioresource Technology*, *101*(15), 6051–6058. doi:10.1016/j.biortech.2010.03.005 PMID:20356734

Torrijos, V., Trueba, A., Ruiz, I., & Soto, M. (2015). Investigating different hybrid configurations of constructed wetlands for nitrogen removal from domestic and industrial wastewaters.*Proceedings of the WETPOL 2015 International Symposium on Wetland Pollutant Dynamics and Control and the Annual Conference of the Constructed Wetland Association* (pp. 294-295).

Trautwein, K., Kühner, S., Wöhlbrand, L., Halder, T., Kuchta, K., Steinbuchel, A., & Rabus, R. (2008). Solvent Stress Response of the Denitrifying Bacterium '' *Aromatoleum aromaticum*'' Strain EbN1. *Applied and Environmental Microbiology*, *74*(8), 2267–2274. doi:10.1128/AEM.02381-07 PMID:18263750

Treusch, A. H., Leininger, S., Kietzin, A., Schuster, S. C., Klenk, H. P., & Schleper, C. (2005). Novel genes for nitrite reductase and Amo-related proteins indicate a role of uncultivated mesophilic Crenarchaeota in nitrogen cycling. *Environmental Microbiology*, *7*(12), 1985–1995. doi:10.1111/j.1462-2920.2005.00906.x PMID:16309395

Trollberg, O., Carlsson, B., & Jacobsen, E. W. (2014). Extremum seeking control of the CANON process - Existence of multiple stationary solutions. *Journal of Process Control*, *24*(2), 348–356. doi:10.1016/j.jprocont.2013.11.007

Türker, M., & Çelen, I. (2007). Removal of ammonia as struvite from anaerobic digester effluents and recycling of magnesium and phosphate. *BioresourceTechnol*, *98*(8), 1529–1534. doi:10.1016/j.biortech.2006.06.026 PMID:17084077

Turk, O., & Mavinic, D. S. (1987). Benefits of using selective inhibition to remove nitrogen from highly nitrogenous wastes. *Environmental Technology Letters*, *8*(9), 419–426. doi:10.1080/09593338709384500

Tyson, K., Busby, S., & Cole, J. (1997). Catabolite regulation of two Escherichia coli operons encoding nitrite reductases: Role of the Cra protein. *Archives of Microbiology*, *168*(3), 240–244. doi:10.1007/s002030050494 PMID:9382703

Ueno, Y., & Fujii, M. (2001). Three years experience of operating and selling recovered struvite from full-scale plant. *Environmental Technology*, *22*(11), 1373–1381. doi:10.1080/09593332208618196 PMID:11804359

Uggetti, E., Ferrer, I., Nielsen, S., Arias, C., Brix, H., & García, J. (2012). Characteristics of Biosolids from Sludge Treatment Wetlands for Agricultural Reuse. *Ecological Engineering*, *40*, 210–216. doi:10.1016/j.ecoleng.2011.12.030

Uludag-Demirer, S., Demirer, G. N., Frear, C., & Chen, S. (2008). Anaerobic digestion of dairy manure with enhanced ammonia removal. *Journal of Environmental Management*, *86*(1), 193–200. doi:10.1016/j.jenvman.2006.12.002 PMID:17257738

Urbanc-Berčič, O. (1996). Constructed wetlands for the treatment of landfill leachates: The Slovenian experience. *Wetlands Ecology and Management*, *4*(3), 189–197. doi:10.1007/BF01879237

*USDA*. (2013). Retrieved from: www.ers.usda.gov/Data/FertilizerUse/

Uygur, A., & Kargi, F. (2004). Biological nutrient removal from pre-treated landfill leachate in a sequencing batch reactor. *Journal of Environmental Management*, *71*(1), 9–14. doi:10.1016/j.jenvman.2004.01.002 PMID:15084355

Uysal, A., Yilmazel, Y. D., & Demirer, G. N. (2010). The determination of fertilizer quality of the formed struvite from effluent of a sewage sludge anaerobic digester. *Journal of Hazardous Materials*, *181*(1-3), 248–254. doi:10.1016/j.jhazmat.2010.05.004 PMID:20605682

Vaiopoulou, E., Melidis, P., & Aivasidis, A. (2005). Sulfide removal in wastewater from petrochemical industries by autotrophic denitrification. *Water Research*, *39*(17), 4101–4109. doi:10.1016/j.watres.2005.07.022 PMID:16139866

Vajralaa, N., Martens-Habbenab, W., Sayavedra-Sotoa, L. A., Schauerc, A., Bottomleyd, P. J., Stahlb, D. A., & Arpa, D. J. (2013). Hydroxylamine as an intermediate in ammonia oxidation by globally abundant marine archaea. *Proceedings of the National Academy of Sciences of the United States of America*, *110*(3), 1006–1011. doi:10.1073/pnas.1214272110 PMID:23277575

Val del Rio, A., Pedrouso, A., Gutiérrez, A., Campos, J. L., Mosquera-Corral, A., Méndez, R., . . . Rogalla, F. (2015). *Scale up of an anammox based process (ELAN) for the treatment of fish canning effluents.* Paper presented at the 12th IWA Leading Edge Conference on Water and Wastewater Technologies, Hong Kong, China.

Valverde-Pérez, B., Fuentes-Martínez, J. M., Flores-Alsina, X., Gernaey, K. V., Huusom, J. K., & Plósz, B. (in press). Gy., (2016b). Control structure design for resource recovery using the enhanced biological phosphorus removal and recovery (EBP2R) activated sludge process combined with green microalgae cultivation. *Chemical Engineering Journal*. doi:10.1016/j.cej.2016.03.021

Valverde-Pérez, B., Mauricio-Iglesias, M., & Sin, G. (2016a). Systematic design of an optimal control system for the SHARON-Anammox process. *Journal of Process Control*, *39*, 1–10. doi:10.1016/j.jprocont.2015.12.009

Valverde-Pérez, B., Ramin, E., Smets, B. F., & Plósz, B. (2015). EBP2R – An innovative enhanced biological nutrient recovery activated sludge system to produce growth medium for green microalgae cultivation. *Water Research*, *68*, 821–830. doi:10.1016/j.watres.2014.09.027 PMID:25480432

Van de Graaf, A. A., de Bruijn, P., Robertson, L. A., Jetten, M. S. M., & Kuenen, J. G. (1996). Autotrophic growth of anaerobic ammonium-oxidizing micro-organisms in a fluidized bed reactor. *Microbiology*, *142*(8), 2187–2196. doi:10.1099/13500872-142-8-2187

Van de Vossenberg, J., Woebken, D., Maalcke, W. J., Wessels, H. J. C. T., Dutilh, B. E., Kartal, B., & Jetten, M. S. M. et al. (2013). The metagenome of the marine anammox bacterium 'Candidatus Scalindua profunda' illustrates the versatility of this globally important nitrogen cycle bacterium. *Environmental Microbiology*, *15*(5), 1275–1289. doi:10.1111/j.1462-2920.2012.02774.x PMID:22568606

van der Star, W. R. L., Abma, W., Blommers, D., Mulder, J. W., Tokutomi, T., Strous, M., & van Loosdrecht, M. C. M. et al. (2007). Startup of reactors for anoxic ammonium oxidation: Experiences from the first full-scale anammox reactor in Rotterdam. *Water Research*, *41*(18), 4149–4163. doi:10.1016/j.watres.2007.03.044 PMID:17583763

van der Star, W. R. L., Miclea, A. I., van Dongen, U. G. J. M., Muyzer, G., Picioreanu, C., & van Loosdrecht, M. C. M. (2008). The membrane bioreactor: A novel tool to grow anammox bacteria as free cells. *Biotechnology and Bioengineering*, *101*(2), 286–294. doi:10.1002/bit.21891 PMID:18421799

van Dongen, U., Jetten, M. S. M., & van Loosdrecht, M. C. M. (2001). The SHARON((R))-Anammox((R)) process for treatment of ammonium rich wastewater. *Water Science and Technology*, *44*(1), 153–160. PMID:11496667

van Eekert, M., Verdoes, N., & de Buisonjé, F. (2012). *Explorative research on innovative nitrogen recovery*. Amersfoort: STOWA. Retrieved from: http://www.stowa.nl/Upload/publicaties/STOWA%202012%2051_LR.pdf

Van Hulle, S. W. H., Vandeweyer, H. J. P., Meesschaert, B. D., Vanrollenghem, P. A., Dejans, P., & Dumoulin, A. (2010). Engineering aspects and practical application of autotrophic nitrogen removal from nitrogen rich streams. *Chemical Engineering Journal*, *162*(1), 1–20. doi:10.1016/j.cej.2010.05.037

Van Hulle, S. W. H., Volcke, E. I. P., Lopez Teruel, J., Donckels, B., van Loosdrecht, M. C. M., & Vanrolleghem, P. A. (2007). Influence of temperature and pH on the kinetics of the SHARON nitritation process. *Journal of Chemical Technology and Biotechnology (Oxford, Oxfordshire)*, *82*(5), 471–480. doi:10.1002/jctb.1692

Van Kessel, M. A. H. J., Speth, D. R., Albertsen, M., Per, H., Nielsen, P. H., Op den Camp, H. J. M., & Lücker, S. et al. (2015). Complete nitrification by a single microorganism. *Nature*, *528*, 555–559. PMID:26610025

Van Loosdrecht, M. C. M., & Jetten, M. S. M. (1998). Microbiological conversions in nitrogen removal. *Water Science and Technology*, *38*(1), 1–7. doi:10.1016/S0273-1223(98)00383-7

Van Niftrik, L., & Jetten, M. S. M. (2012). Anaerobic ammonium oxidizing bacteria: Unique microorganisms with exceptional properties. *Microbiology and Molecular Biology Reviews*, *76*(3), 585–596. doi:10.1128/MMBR.05025-11 PMID:22933561

Van Oirschot, D. (2015). RietLand vertical flow wetlands: 16 years of experience in the Netherlands and Belgium. *Proceedings 12th IWA International Conference on Wetland Systems for Water Pollution Control.*

Vangsgaard, A. K., Mauricio-Iglesias, M., Gernaey, K. V., & Sin, G. (2014). Development of novel control strategies for single-stage autotrophic nitrogen removal: A process oriented approach. *Computers & Chemical Engineering*, *66*, 71–81. doi:10.1016/j.compchemeng.2014.01.017

Vangsgaard, A. K., Mauricio-Iglesias, M., Gernaey, K. V., Smets, B. F., & Sin, G. (2012). Sensitivity analysis of autotrophic N removal by a granule based bioreactor: Influence of mass transfer versus microbial kinetics. *Bioresource Technology*, *123*, 230–241. doi:10.1016/j.biortech.2012.07.087 PMID:22940324

Vangsgaard, A. K., Mauricio-Iglesias, M., Valverde-Pérez, B., Gernaey, K. V., & Sin, G. (2013). pH variation and influence in an autotrophic nitrogen removing biofilm system using an efficient numerical solution strategy. *Water Science and Technology*, *67*(11), 2608–2615. doi:10.2166/wst.2013.174 PMID:23752396

Varas, R., Guzmán-Fierro, V., Giustinianovich, E., Behar, J., Fernández, K., & Roeckel, M. (2015). Startup and oxygen concentration effects in a continuous granular mixed flow autotrophic nitrogen removal reactor. *Bioresource Technology*, *190*, 345–351. doi:10.1016/j.biortech.2015.04.086 PMID:25965951

Vasiliadou, I. A., Pavlou, S., & Vallenas, D. V. (2006). A kinetic study of hydrogenotrophic denitrification. *Process Biochemistry*, *41*(6), 1401–1408. doi:10.1016/j.procbio.2006.02.002

Vasiliadou, I. A., Pavlou, S., & Vayenas, D. V. (2009a). Dynamics of a chemostat with three competitive hydrogen oxidizing denitrifying microbial populations and their efficiency for denitrification. *Ecological Modelling*, *220*(8), 1169–1180. doi:10.1016/j.ecolmodel.2009.02.009

Vasiliadou, I. A., Pavlou, S., & Vayenas, D. V. (2009b). Experimental and modelling study of drinking water hydrogenotrophic denitrification in packed-bed reactors. *Journal of Hazardous Materials*, *165*(1-3), 812–824. doi:10.1016/j.jhazmat.2008.10.067 PMID:19054612

Vasiliadou, I. A., Siozios, S., Papadas, I. T., Bourtzis, K., Pavlou, S., & Vayenas, D. V. (2006b). Kinetics of pure cultures of hydrogen-oxidizing denitrifying bacteria and modeling of the interactions among them in mixed cultures. *Biotechnology and Bioengineering*, *95*(3), 513–525. doi:10.1002/bit.21031 PMID:16758460

Vasudevan, S., Epron, F., Lakshmi, J., Ravichandran, S., Mohan, S., & Sozhan, G. (2010). Removal of $NO^{3-}$ from Drinking Water by Electrocoagulation – An Alternate Approach. *Clean – Soil, Air. Water (Basel)*, *38*, 225–229.

Vazquez-Padin, J. R., Fernandez, R., Campos, J. L., Mosquera, A., & Mendez, R. (2014). *Method for starting up and controlling a biological process for ammonium removal by the action of autotrophic bacteria in wastewater*. European Patent, EP2740713B1

Vazquez-Padin, J. R., Morales, N., Gutierrez, R., Fernandez, R., Rogalla, F., Barrio, J. P., & Mendez, R. (2014). Implications of full-scale implementation of an anammox-based process as post-treatment of a municipal anaerobic sludge digester operated with co-digestion. *Water Science and Technology: A Journal of the International Association on Water Pollution Research*, *69*(6), 1151-1158. doi:10.2166/wst.2013.795

Vázquez-Padín, J. R., Fernández, I., Mosquera-Corral, A., Campos, J. L., & Méndez, R. (2009). The anammox process. In F. J. Cervantes (Ed.), *Environmental technologies to treat nitrogen pollution* (pp. 67–91). London: IWA Publishing.

Vazquez-Padin, J. R., Figueroa, M., Fernandez, I., Mosquera-Corral, A., Campos, J. L., & Mendez, R. (2009b). Post-treatment of effluents from anaerobic digesters by the Anammox process. *Water Science and Technology*, *60*(5), 1135–1143. doi:10.2166/wst.2009.421 PMID:19717899

Vázquez-Padín, J. R., Pozo, M. J., Jarpa, M., Figueroa, M., Franco, A., Mosquera-Corral, A., & Méndez, R. et al. (2009). Treatment of anaerobic sludge digester effluents by the CANON process in an air pulsing SBR. *Journal of Hazardous Materials*, *166*(1), 336–341. doi:10.1016/j.jhazmat.2008.11.055 PMID:19117673

Vázquez-Padín, J., Fernádez, I., Figueroa, M., Mosquera-Corral, A., Campos, J.-L., & Méndez, R. (2009a). Applications of Anammox based processes to treat anaerobic digester supernatant at room temperature. *Bioresource Technology*, *100*(12), 2988–2994. doi:10.1016/j.biortech.2009.01.028 PMID:19246192

Velthof, G. L. (2011). Synthesis of the research within the framework of the Mineral Concentrates Pilot. Wageningen, Alterra, Alterra report 2224.

Verhamme, D. T., Prosser, J. I., & Nicol, G. N. (2011). Ammonia concentration determines differential growth of ammonia-oxidising archaea and bacteria in soil microcosms. *The ISME Journal*, *5*(6), 1067–1071. doi:10.1038/ismej.2010.191 PMID:21228892

Viotti, P., & Gavasci, R. (2015). Scaling of ammonia stripping towers in the treatment of groundwater polluted by the leachate of municipal solid waste landfill: Study of the causes and the effects on the stripping performance. *Revista Ambiente & Agua*, *10*(2), 240–252.

Vitousek, P. M., Aber, J. D., Howarth, R. W., Likens, G. E., Matson, P. A., Schindler, D. W., & Tilman, D. G. et al. (1997). Human alteration of the global nitrogen cycle: Sources and consequences. *Ecological Applications*, *7*(3), 737–750. doi:10.1890/1051-0761(1997)007[0737:haotgn]2.0.co;2

Vitousek, P. M., Naylor, R., Crews, T., David, M. B., Drinkwater, L. E., Holland, E., & Zhang, F. S. et al. (2009). Nutrient Imbalances in Agricultural Development. *Science*, *324*(5934), 1519–1520. doi:10.1126/science.1170261 PMID:19541981

Vlaeminck, S. E., De Clippeleir, H., & Verstraete, W. (2012). Microbial resource management of one-stage partial nitritation/anammox. *Microbial Biotechnology*, *5*(3), 433–448. doi:10.1111/j.1751-7915.2012.00341.x PMID:22452819

Vlaeminck, S. E., Terada, A., Smets, B. F., De Clippeleir, H., Schaubroeck, T., Bolca, S., & Verstraete, W. et al. (2010). Aggregate size and architecture determine microbial activity balance for one-stage partial nitritation and Anammox. *Applied and Environmental Microbiology*, *76*(3), 900–909. doi:10.1128/AEM.02337-09 PMID:19948857

Vohla, C., Koiv, M., Bavor, H. J., Chazarenc, F., & Mander, U. (2011). Filter materials for phosphorus removal from wastewater in treatment wetlands—A review. *Ecological Engineering*, *37*(1), 70–89. doi:10.1016/j.ecoleng.2009.08.003

Volcke, E. I. P., Picioreanu, C., De Baets, B., & van Loosdrecht, M. C. M. (2010). Effect of granule size on autotrophic nitrogen removal in a granular sludge reactor. *Environmental Technology*, *31*(11), 1271–1280. doi:10.1080/09593331003702746 PMID:21046957

Volcke, E. I. P., van Loosdrecht, M. C. M., & Vanrolleghem, P. A. (2006). Controlling the nitrite: Ammonium ratio in a SHARON reactor in view of its coupling with an Anammox process. *Water Science and Technology*, *53*(4-5), 45–54. doi:10.2166/wst.2006.109 PMID:16722054

Volcke, E. I. P., van Loosdrecht, M. C. M., & Vanrolleghem, P. A. (2007). Interaction between control and design of a SHARON reactor: Economic considerations in a plant-wide (BSM2) context. *Water Science and Technology*, *56*(9), 117–125. doi:10.2166/wst.2007.607 PMID:18025739 von Schulthess, R., & Gujer, W. (1996). Release of nitrous oxide ($N_2O$) from denitrifying activated sludge: Verification and application of a mathematical model. *Water Research, 30*(3), 521–530. doi:10.1016/0043-1354(95)00204-9

Vymazal, J. (2011a). Constructed wetlands for wastewater treatment: Five decades of experience. *Environmental Science & Technology, 45*(1), 61–69. doi:10.1021/es101403q PMID:20795704

Vymazal, J. (2011b). Plants used in constructed wetlands with horizontal subsurface flow: A review. *Hydrobiologia, 674*(1), 133–156. doi:10.1007/s10750-011-0738-9

Vymazal, J. (2013). The use of hybrid constructed wetlands for wastewater treatment with special attention to nitrogen removal: A review of a recent development. *Water Research, 47*(14), 4795–4811. doi:10.1016/j.watres.2013.05.029 PMID:23866134

Vymazal, J. (2014). Constructed wetlands for treatment of industrial wastewater: A review. *Ecological Engineering, 73*, 724–751. doi:10.1016/j.ecoleng.2014.09.034

Wágner, D. S., Ramin, E., Szabo, P., Dechesne, A., & Plósz, B. (2015). *Microthrix parvicella* abundance associates with activated sludge settling velocity and rheology - Quantifying and modelling filamentous bulking. *Water Research, 78*, 121–132. doi:10.1016/j.watres.2015.04.003 PMID:25935367

Waki, M., Yasuda, T., Yokoyama, H., Hanajima, D., Ogino, A., Suzuki, K., & Tanaka, Y. et al. (2009). Nitrogen removal by co-occurring methane oxidation, denitrification, aerobic ammonium oxidation, and anammox. *Applied Microbiology and Biotechnology, 84*(5), 977–985. doi:10.1007/s00253-009-2112-7 PMID:19603162

Wang, Y. P., Houlton, B. Z., & Field, C. B. (2007). A model of biogeochemical cycles of carbon, nitrogen, and phosphorus including symbiotic nitrogen fixation and phosphatase production. *Global Biogeochemical Cycles, 21*(1). doi: 10.1029/2006gb002797

Wang, A., Du, D., Ren, N., & van Groenestijn, J. W. (2005). An innovative process of simultaneous desulfurization and denitrication by *Thiobacillus denitrificans. Journal of Environmental Science and Health, 40*(10), 1939–1949. doi:10.1080/10934520500184590 PMID:16194914

Wang, C.-C., Lee, P.-H., Kumar, M., Huang, Y.-T., Sung, S., & Lin, J.-G. (2010). Simultaneous partial nitrification, anaerobic ammonium oxidation and denitrification (SNAD) in a full-scale landfill-leachate treatment plant. *Journal of Hazardous Materials, 175*(1–3), 622–628. doi:10.1016/j.jhazmat.2009.10.052 PMID:19913994

Wang, H., He, Q., Chen, D., Wei, L., Zou, Z., Zhou, J., & Zhang, H. (2015). Microbial community in hydrogenotrophic denitrification reactor based on pyrosequencing. *Applied Microbiology and Biotechnology, 99*(24), 10829–10837. doi:10.1007/s00253-015-6929-y PMID:26286515

Wang, L., Wan, C., Lee, D.-J., Liu, X., Zhang, Y., Chen, X. F., & Tay, J.-H. (2014). Biosorption of antimony(V) onto Fe(III)-treated aerobic granules. *Bioresource Technology, 158*(V), 351–354. doi:10.1016/j.biortech.2014.02.046 PMID:24630496

Wang, L., Wan, C., Lee, D.-J., Tay, J.-H., Chen, X. F., Liu, X., & Zhang, Y. (2013). Adsorption–desorption of strontium from waters using granules. *Journal of the Taiwan Institute of Chemical Engineers, 44*(3), 454–457. doi:10.1016/j.jtice.2012.12.005

Wang, L., Zheng, P., Xing, Y., Li, W., Yang, J., & Abbas, G. et al. (2014). Effect of particle size on the performance of autotrophic nitrogen removal in the granular sludge bed reactor and microbiological mechanisms. *Bioresource Technology, 157*, 240–246. doi:10.1016/j.biortech.2014.01.116 PMID:24561629

Wang, S.-G., Liu, X.-W., Gong, W.-X., Gao, B.-Y., Zhang, D.-H., & Yu, H.-Q. (2007). Aerobic granulation with brewery wastewater in a sequencing batch reactor. *Bioresource Technology*, *98*(11), 2142–2147. doi:10.1016/j.biortech.2006.08.018 PMID:17071083

Wang, S.-G., Liu, X.-W., Zhang, H.-Y., Gong, W.-X., Sun, X.-F., & Gao, B.-Y. (2007). Aerobic granulation for 2,4-dichlorophenol biodegradation in a sequencing batch reactor. *Chemosphere*, *69*(5), 769–775. doi:10.1016/j.chemosphere.2007.05.026 PMID:17617438

Wang, S., Peng, Y., Ma, B., Wang, S., & Zhu, G. (2015). Anaerobic ammonium oxidation in traditional municipal wastewater treatment plants with low-strength ammonium loading: Widespread but overlooked. *Water Research*, *84*, 66–75. doi:10.1016/j.watres.2015.07.005 PMID:26210031

Wang, X.-H., Gai, L.-H., Sun, X.-F., Xie, H.-J., Gao, M.-M., & Wang, S.-G. (2010). Effects of long-term addition of Cu(II) and Ni(II) on the biochemical properties of aerobic granules in sequencing batch reactors. *Applied Microbiology and Biotechnology*, *86*(6), 1967–1975. doi:10.1007/s00253-010-2467-9 PMID:20169340

Wang, Y., Inamori, R., Kong, H., Xu, K., Inamori, Y., Kondo, T., & Zhang, J. (2008). Influence of plant species and wastewater strength on constructed wetland methane emissions and associated microbial populations. *Ecological Engineering*, *32*(1), 22–29. doi:10.1016/j.ecoleng.2007.08.003

Wang, Y., Liu, S., Xu, Z., Han, T., Chuan, S., & Zhu, T. (2006). Ammonia removal from leachate solution using natural Chinese clinoptilolite. *Journal of Hazardous Materials*, *136*(3), 735–740. doi:10.1016/j.jhazmat.2006.01.002 PMID:16469437

Wang, Z., Liu, L., Yao, J., & Cai, W. (2006). Effects of extracellular polymeric substances on aerobic granulation in sequencing batch reactors. *Chemosphere*, *63*(10), 1728–1735. doi:10.1016/j.chemosphere.2005.09.018 PMID:16288800

Watsuji, T. O., Takaya, N., Nakamura, A., & Shoun, H. (2003). A possible role of NADPH-dependent cytochrome P450nor isozyme in glycolysis under denitrifying conditions. *Bioscience, Biotechnology, and Biochemistry*, *67*(5), 1109–1114. doi:10.1271/bbb.67.1109 PMID:12834289

Webb, J. M., Quintã, R., Papadimitriou, S., Norman, L., Rigby, M., Thomas, D. N., & Le Vay, L. (2013). The effect of halophyte planting density on the efficiency of constructed wetlands for the treatment of wastewater from marine aquaculture. *Ecological Engineering*, *61*, 145–153. doi:10.1016/j.ecoleng.2013.09.058

Wei, D., Li, M., Wang, X., Han, F., Li, L., & Guo, J. et al. (2016). Extracellular polymeric substances for Zn (II) binding during its sorption process onto aerobic granular sludge. *Journal of Hazardous Materials*, *301*, 407–415. doi:10.1016/j.jhazmat.2015.09.018 PMID:26410269

Wei, D., Shi, L., Yan, T., Zhang, G., Wang, Y., & Du, B. (2014). Aerobic granules formation and simultaneous nitrogen and phosphorus removal treating high strength ammonia wastewater in sequencing batch reactor. *Bioresource Technology*, *171*, 211–216. doi:10.1016/j.biortech.2014.08.001 PMID:25203228

Wei, D., Wang, B., Ngo, H. H., Guo, W., Han, F., & Wang, X. et al. (2015). Role of extracellular polymeric substances in biosorption of dye wastewater using aerobic granular sludge. *Bioresource Technology*, *185*, 14–20. doi:10.1016/j.biortech.2015.02.084 PMID:25746473

Wei, D., Wang, Y., Wang, X., Li, M., Han, F., & Ju, L. et al. (2015). Toxicity assessment of 4-chlorophenol to aerobic granular sludge and its interaction with extracellular polymeric substances. *Journal of Hazardous Materials*, *289*, 101–107. doi:10.1016/j.jhazmat.2015.02.047 PMID:25710820

Welander, U., Henrysson, T., & Welander, T. (1997). Biological nitrogen removal from municipal landfill leachate in a pilot scale suspended carrier biofilm process. *Water Research*, *32*(5), 1564–1570. doi:10.1016/S0043-1354(97)00351-5

WERF. (2014a). Deammonification. In *Water Envrionment Research Foundation (WERF)*. Retrieved from http://www.werf.org

WERF. (2014b). *Nutrient Challenge*. Deammonification.

Wett, B. (2007). Development and implementation of a robust deammonification process. *Water Science and Technology*, *56*(7), 81–88. doi:10.2166/wst.2007.611 PMID:17951871

Wett, B., & Rauch, W. (2003). The role of inorganic carbon limitation in biological nitrogen removal of extremely ammonia concentrated wastewater. *Water Research*, *37*(5), 1100–1110. doi:10.1016/S0043-1354(02)00440-2 PMID:12553985

Wiesmann, U. (1994). Biological nitrogen removal from wastewater. In A. Fletcher (Ed.), *Advances in biochemical engineering biotechnology* (Vol. 51, pp. 113–154). Berlin: Spinger Verlag.

Wiesmann, U. (1994). Biological nitrogen removal from wastewater. *Advances in Biochemical Engineering/Biotechnology*, *51*, 113–154. doi:10.1007/BFb0008736 PMID:8165950

Windey, K., De Bo, I., & Verstraete, W. (2005). Oxygen-limited autotrophic nitrification-denitrification (OLAND) in a rotating biological contactor treating high-salinity wastewater. *Water Research*, *39*(18), 4512–4520. doi:10.1016/j.watres.2005.09.002 PMID:16225902

Winkler, M.-K. H. (2012). *Segregation of biomass in aerobic granular sludge*. Delft, The Netherlands: Technical University Delft.

Winogradsky, S. (1890). Recherches sur les organismes de la nitrification. *Annales de l'Institut Pasteur*, *4*, 257–275.

Wiszniowski, J., Robert, D., Surmacz-Gorska, J., Miksch, K., & Weber, J. V. (2006a). Landfill leachate treatment methods: A review. *Environmental Chemistry Letters*, *4*(1), 51–61. doi:10.1007/s10311-005-0016-z

Wiszniowski, J., Robert, D., Surmacz-Gorska, J., Miksch, K., & Weber, J. V. (2006b). Leachate detoxification by combination of biological and $TiO_2$-photocatalytic processes. *Water Science and Technology*, *53*(3), 181–190. doi:10.2166/wst.2006.091 PMID:16605031

Wojciechowski, C. L., Cardia, J. P., & Kantrowitz, E. R. (2002). Alkaline phosphatase from the hyperthermophilic bacterium T. maritima requires cobalt for activity. *Protein Science*, *11*(4), 903–911. doi:10.1110/ps.4260102 PMID:11910033

Wollast, R., Mackenzie, F. T., & Chou, L. (1993). *Interactions of C, N, P and S. Biogeochemical Cycles and Global Change*. Springer-Verlag Berlin Heidelberg. doi:10.1007/978-3-642-76064-8

Wunderlin, P., Mohn, J., Joss, A., Emmenegger, L., & Siegrist, H. (2012). Mechanisms of N2O production in biological wastewater treatment under nitrifying and denitrifying conditions. *Water Research*, *46*(4), 1027–1103. doi:10.1016/j.watres.2011.11.080 PMID:22227243

Wun-Jern, N. (2006). *Industrial Wastewater Treatment*. London, UK: Imperial College Press.

Wu, S., Kuschk, P., Brix, H., Vymazal, J., & Dong, R. (2014). Development of constructed wetlands in performance intensifications for wastewater treatment: A nitrogen and organic matter targeted review. *Water Research*, *57*(15), 40–55. doi:10.1016/j.watres.2014.03.020 PMID:24704903

Wu, S., Zhang, D., Austin, D., Dong, R., & Pang, C. (2011). Evaluation of a lab-scale tidal flow constructed wetland performance: Oxygen transfer capacity, organic matter and ammonium removal. *Ecological Engineering*, *37*(11), 1789–1795. doi:10.1016/j.ecoleng.2011.06.026

Wyckoff, H. W., Handschumacher, M., Murthy, H. M., & Sowadski, J. M. (1983). The three dimensional structure of alkaline phosphatase from E. coli. *Advances in Enzymology and Related Areas of Molecular Biology*, *55*, 453–480. PMID:6353889

Wyffels, S., van Hulle, S. W. H., Boeckx, P., Volcke, E. I. P., van Cleemput, O., Vanrolleghem, P. A., & Verstraete, W. (2004). Modeling and simulation of oxygen-limited nitritation in a membrane-assisted bioreactor (MBR). *Biotechnology and Bioengineering*, *86*(5), 531–542. doi:10.1002/bit.20008 PMID:15129436

Xia, S., Zhang, Y. H., & Zhong, F. H. (2009). A continuous stirred hydrogen-based polyvinyl chloride membrane biofilm reactor for the treatment of nitrate contaminated drinking water. *Bioresource Technology*, *100*(24), 6223–6228. doi:10.1016/j.biortech.2009.07.002 PMID:19656675

Xia, S., Zhang, Z., Zhong, F., & Zhang, J. (2011). High efficiency removal of 2-chlorophenol from drinking water by a hydrogen-based polyvinyl chloride membrane biofilm reactor. *Journal of Hazardous Materials*, *186*(2-3), 1367–1373. doi:10.1016/j.jhazmat.2010.12.023 PMID:21215515

Xia, Z., Xiao-chun, W., Zhong-lin, C., Hao, X., & Qing-fang, Z. (2014). Microbial community structure and pharmaceuticals and personal care products removal in a membrane bioreactor seeded with aerobic granular sludge. *Applied Microbiology and Biotechnology*, *99*(1), 425–433. doi:10.1007/s00253-014-5984-0 PMID:25099174

Xie, M., Shon, H. K., Gray, S. R., & Elimelech, M. (2016). Membrane-based processes for wastewater nutrient recovery: Technology, challenges, and future direction. *Water Research*, *89*, 210–221. doi:10.1016/j.watres.2015.11.045 PMID:26674549

Xu, G., Peng, J., Feng, C., Fang, F., Chen, S., Xu, Y., & Wang, X. (2015). Evaluation of simultaneous autotrophic and heterotrophic processes and bacterial community structure analysis. *Applied Microbiology and Biotechnology*, *99*(15), 6527–6536. doi:10.1007/s00253-015-6532-2 PMID:25825049

Xu, H., Liu, Y., & Tay, J.-H. (2006). Effect of pH on nickel biosorption by aerobic granular sludge. *Bioresource Technology*, *97*(3), 359–363. doi:10.1016/j.biortech.2005.03.011 PMID:15905090

Yagci, N., Insel, G., Artan, N., & Orhon, D. (2004). Modelling and calibration of phosphate and glycogen accumulating organisms competition for acetate uptake in a sequencing batch reactor. *Water Science and Technology*, *50*(6), 241–250. PMID:15537013

Yalcuk, A., & Urgulu, A. (2009). Comparison of horizontal and vertical constructed wetland system for landfill leachate treatment. *Bioresource Technology*, *100*(9), 2521–2526. doi:10.1016/j.biortech.2008.11.029 PMID:19157867

Yamamoto-Ikemoto, R., Komori, T., Nomura, M., Ide, Y., & Matsukami, T. (2000). Nitrogen removal from hydroponic culture wastewater by autotrophic denitrification using thiosulfate. *Water Science and Technology*, *42*(3-4), 369–376.

Yao, P. (in press). Perspectives on technology for landfill leachate treatment. *Arabian Journal of Chemistry*.

Yetilmezsoy, K., & Sapci-Zengin, Z. (2009). Recovery of ammonium nitrogen from the effluent of UASB treating poultry manure wastewater by MAP precipitation as a slow release fertilizer. *Journal of Hazardous Materials*, *166*(1), 260–269. doi:10.1016/j.jhazmat.2008.11.025 PMID:19097699

Yeung, S. L., Cheng, C., Lui, T. K., Tsang, J. S., Chan, W. T., & Lim, B. L. (2009). Purple acid phosphatase-like sequences in prokaryotic genomes and the characterization of an atypical purple alkaline phosphatase from Burkholderia cenocepacia J2315. *Gene*, *440*(1-2), 1–8. doi:10.1016/j.gene.2009.04.002 PMID:19376213

Yilmaz, G., Lemaire, R., Keller, J., & Yuan, Z. (2008). Simultaneous nitrification, denitrification, and phosphorus removal from nutrient-rich industrial wastewater using granular sludge. *Biotechnology and Bioengineering*, *100*(3), 529–541. doi:10.1002/bit.21774 PMID:18098318

Yi, S., Zhuang, W.-Q., Wu, B., Tay, S. T.-L., & Tay, J.-H. (2006). Biodegradation of p-nitrophenol by aerobic granules in a sequencing batch reactor. *Environmental Science & Technology*, *40*(7), 2396–2401. doi:10.1021/es0517771 PMID:16646480

Yong, M., Yongzhen, P., & Jeppsson, U. (2006). Dynamic evaluation of integrated control strategies for enhanced. *Control Engineering Practice*, *14*(11), 1269–1278. doi:10.1016/j.conengprac.2005.06.018

Yousafzai, F. K., & Eady, R. R. (2002). Dithionite reduction kinetics of the dissimilatory copper-containing nitrite reductase of Alcalegenes xylosoxidans. The SO(2)(.-) radical binds to the substrate binding type 2 copper site before the type 2 copper is reduced. *The Journal of Biological Chemistry*, *277*(37), 34067–34073. doi:10.1074/jbc.M204305200 PMID:12082116

Yuan, Z., Oehmen, A., Peng, Y., Ma, Y., & Keller, J. (2008). Sludge population optimization in biological nutrient removal wastewater treatment systems through on-line process control: A review. *Reviews in Environmental Science and Biotechnology*, *7*(3), 243–254. doi:10.1007/s11157-008-9134-y

Yuan, Z., Pratt, S., & Batstone, D. (2012). Phosphorus recovery from wastewater through microbial processes. *Current Opinion in Biotechnology*, *23*(6), 878–883. doi:10.1016/j.copbio.2012.08.001 PMID:22922003

Zahn, J. A., Arciero, D. M., Hooper, A. B., & DiSpirito, A. A. (1996). Evidence for an iron center in the ammonia monooxygenase from Nitrosomonas europaea. *FEBS Letters*, *397*(1), 35–38. doi:10.1016/S0014-5793(96)01116-7 PMID:8941709

Zalatan, J. G., Fenn, T. D., & Herschlag, D. (2008). Comparative enzymology in the alkaline phosphatase superfamily to determine the catalytic role of an active-site metal ion. *Journal of Molecular Biology*, *384*(5), 1174–1189. doi:10.1016/j.jmb.2008.09.059 PMID:18851975

Zepeda, A., Texier, A. C., & Gómez, J. (2007). Batch nitrifying cultures in presence of mixtures of benzene, toluene, and *m*-xylene. *Environmental Technology*, *28*(3), 355–360. doi:10.1080/09593332808618796 PMID:17432387

Zhang, H., He, Y., Jiang, T., & Yang, F. (2011). Research on characteristics of aerobic granules treating petrochemical wastewater by acclimation and co-metabolism methods. *Desalination*, *279*(1-3), 69–74. doi:10.1016/j.desal.2011.05.060

Zhang, H.-, & Zhou, S.-. (2006). Treating leachate mixture with anaerobic ammonium oxidation technology. *Journal of Central South University of Technology*, *14*(6), 663–667. doi:10.1007/s11771-006-0005-9

Zhang, L. L., Chen, J. M., & Fang, F. (2008). Biodegradation of methyl t-butyl ether by aerobic granules under a cosubstrate condition. *Applied Microbiology and Biotechnology*, *78*(3), 543–550. doi:10.1007/s00253-007-1321-1 PMID:18183384

Zhang, L.-L., Feng, X. X., Xu, F., Xu, S., & Cai, W. M. (2005). Biosorption of Rare Earth Metal Ion on Aerobic Granules. *Journal of Environmental Science and Health. Part A*, *40*(4), 857–867. doi:10.1081/ESE-200048290

Zhang, L., Zhang, C., Hu, C., Liu, H., Bai, Y., & Qu, J. (2015). Sulfur-based mixotrophic denitrification corresponding to different electron donors and microbial profiling in anoxic fluidized-bed membrane bioreactors. *Water Research*, *85*, 422–431. doi:10.1016/j.watres.2015.08.055 PMID:26364226

Zhang, Q., Hu, J., & Lee, D. (2016). Aerobic granular processes : Current research trends. *Bioresource Technology*, *210*, 74–80. doi:10.1016/j.biortech.2016.01.098 PMID:26873285

Zhang, Q.-Q., Zhang, Z.-Z., Guo, Q., Wang, J.-J., Wang, H.-Z., & Jin, R.-C. (2015). Analyzing the revolution of anaerobic ammonium oxidation (anammox) performance and sludge characteristics under zinc inhibition. *Applied Microbiology and Biotechnology*, *99*(7), 3221–3232. doi:10.1007/s00253-014-6205-6 PMID:25398286

Zhang, T. C., & Lampe, D. G. (1999). Sulfur: limestone autotrophic denitrification processes for treatment of nitrate-contaminated water: batch experiments. *Water Research*, *33*(3), 599–608. doi:10.1016/S0043-1354(98)00281-4

Zhang, Y., Desmidt, E., Van Looveren, A., Pinoy, L., Meesschaert, B., & Van der Bruggen, B. (2013). Phosphate separation and recovery from wastewater by novel electrodialysis. *Environmental Science & Technology*, *47*(11), 5888–5895. doi:10.1021/es4004476 PMID:23651001

Zhang, Y., Zhang, W., & Pan, B. (2015). Struvite-based phosphorus recovery from the concentrated bioeffluent by using HFO nanocomposite adsorption: Effect of solution chemistry. *Chemosphere*, *141*, 227–234. doi:10.1016/j.chemosphere.2015.07.023 PMID:26246192

Zhang, Y., Zhong, F., Xia, S., Wang, X., & Li, J. (2009). Autohydrogenotrophic denitrification of drinking water using a polyvinyl chloride hollow fiber membrane biofilm reactor. *Journal of Hazardous Materials*, *170*(1), 203–209. doi:10.1016/j.jhazmat.2009.04.114 PMID:19473764

Zhang, Z., Li, Y., Chen, S., Wang, S., & Bao, X. (2012). Simultaneous nitrogen and carbon removal from swine digester liquor by the Canon process and denitrification. *Bioresource Technology*, *114*, 84–89. doi:10.1016/j.biortech.2012.03.006 PMID:22520218

Zhang, Z.-Z., Zhang, Q.-Q., Xu, J.-J., Shi, Z.-J., Guo, Q., Jiang, X.-Y., & Jin, R.-C. et al. (2016). Long-term effects of heavy metals and antibiotics on granule-based anammox process: Granule property and performance evolution. *Applied Microbiology and Biotechnology*, *100*(5), 2417–2427. doi:10.1007/s00253-015-7120-1 PMID:26549238

Zhao, D., Liu, C., Zhang, Y., & Liu, Q. (2011). Biodegradation of nitrobenzene by aerobic granular sludge in a sequencing batch reactor (SBR). *Desalination*, *281*, 17–22. doi:10.1016/j.desal.2011.07.037

Zhao, Q., Zhang, T., Frear, C., Bowers, K., Harrison, J., & Chen, S. (2010). *Phosphorous Recovery Technology in Conjunction with Dairy Anaerobic Digestion. Climate Friendly Farming: Research Report 2010 – 001*. CSANR.

Zhao, X., Chen, Z., Wang, X., Li, J., Shen, J., & Xu, H. (2015). Remediation of pharmaceuticals and personal care products using an aerobic granular sludge sequencing bioreactor and microbial community profiling using Solexa sequencing technology analysis. *Bioresource Technology*, *179*, 104–112. doi:10.1016/j.biortech.2014.12.002 PMID:25531682

Zhao, Y., Huang, J., Zhao, H., & Yang, H. (2013). Microbial community and N removal of aerobic granular sludge at high COD and N loading rates. *Bioresource Technology*, *143*, 439–446. doi:10.1016/j.biortech.2013.06.020 PMID:23827440

Zhao, Y., Zhang, B., Feng, C., Huang, F., Zhang, P., Zhang, Z., & Sugiura, N. et al. (2012). Behavior of autotrophic denitrification and heterotrophic denitrification in an intensified biofilm-electrode reactor for nitrate-contaminated drinking water treatment. *Bioresource Technology*, *107*, 159–165. doi:10.1016/j.biortech.2011.12.118 PMID:22244955

Zheng, X., He, Y., Chen, W., Wang, M., Cao, S., Ni, M., & Chen, Y. (2016). A comparative adsorption study : 17β-estradiol onto aerobic granular sludge and activated sludge. *Environmental Technology*, *37*(1), 136–144. doi:10.1080/09593330.2015.1065004 PMID:26209151

Zheng, Y., Wang, C. X., Ge, Y., Dzakpasu, M., Zhao, Y., & Xiong, J. (2015). Effects of annual harvesting on plants growth and nutrients removal in surface-flow constructed wetlands in northwestern China. *Ecological Engineering*, *83*, 268–275. doi:10.1016/j.ecoleng.2015.06.035

Zhi, W., Yuan, L., Ji, G., & He, C. (2015). Enhanced long-term nitrogen removal and its quantitative molecular mechanism in tidal flow constructed wetlands. *Environmental Science & Technology*, *49*(7), 4575–4583. doi:10.1021/acs.est.5b00017 PMID:25781063

Zhou, D., Liu, M., Wang, J., Dong, S., Cui, N., & Gao, L. (2013). Granulation of activated sludge in a continuous flow airlift reactor by strong drag force. *Biotechnology and Bioprocess Engineering*, *18*(2), 289–299. doi:10.1007/s12257-012-0513-4

Zhou, D., Niu, S., Xiong, Y., Yang, Y., & Dong, S. (2014). Microbial selection pressure is not a prerequisite for granulation: Dynamic granulation and microbial community study in a complete mixing bioreactor. *Bioresource Technology*, *161*, 102–108. doi:10.1016/j.biortech.2014.03.001 PMID:24690580

Zhou, M., Fu, W., Gu, H., & Lei, L. (2007). Nitrate removal from groundwater by a novel three-dimensional electrode biofilm reactor. *Electrochimica Acta*, *52*(19), 6052–6059. doi:10.1016/j.electacta.2007.03.064

Zhu, G., Peng, Y., Li, B., Guo, J., Yang, Q., & Wang, S. (2008). Biological removal of nitrogen from wastewater. In D. Whitacre (Ed.), *Reviews of Environmental Contamination and Toxicology* (Vol. 192, pp. 159–195). New York: Springer. doi:10.1007/978-0-387-71724-1_5

Zhu, G., Wang, S., Feng, X., Fan, G., Jetten, M. S., & Yin, C. (2011). Anammox bacterial abundance, biodiversity and activity in a constructed wetland. *Environmental Science & Technology*, *45*(23), 9951–9958. doi:10.1021/es202183w PMID:21981702

Zhu, L., Lv, M., Dai, X., Zhou, J.-H., & Xu, X.-Y. (2013). The stability of aerobic granular sludge under 4-chloroaniline shock in a sequential air-lift bioreactor (SABR). *Bioresource Technology*, *140*, 126–130. doi:10.1016/j.biortech.2013.04.017 PMID:23685649

Zhu, L., Xu, X., Luo, W., Cao, D., & Yang, Y. (2008). Formation and microbial community analysis of chloroanilines-degrading aerobic granules in the sequencing airlift bioreactor. *Journal of Applied Microbiology*, *104*(1), 152–160. doi:10.1111/j.1365-2672.2007.03548.x PMID:17887986

Zhu, L., Xu, X., Luo, W., Tian, Z., Lin, H., & Zhang, N. (2008). A comparative study on the formation and characterization of aerobic 4-chloroaniline-degrading granules in SBR and SABR. *Applied Microbiology and Biotechnology*, *79*(5), 867–874. doi:10.1007/s00253-008-1476-4 PMID:18449538

Zuleeg, S., Wüthrich, F., Egli, C., Hürlimann, M., Grömping, M., Böhler, M., & Siegrist, H. (2012). ARA Altenrhein—Rücklaufbehandlung Variantenvergleich. ARA Altenrhein/Ingenieurbüro AG. (in German)

Zumft, W. G. (1997). Cell biology and molecular basis of denitrification. *Microbiology and Molecular Biology Reviews*, *64*, 533–616. PMID:9409151

# About the Contributors

**Ángeles Val del Río** received the Ph.D. in Chemical and Environmental Engineering in 2012 at the University of Santiago de Compostela (USC) on the topic of aerobic granular systems for wastewater treatment. Currently, she is a post-doctoral researcher in the Group of Environmental Engineering and Bioprocesses at USC. Her research activities are focused on the application of biological processes (nitrification, anammox, anaerobic digestion, etc.) operated in suspended or biofilm (aerobic granular sludge) reactors for the treatment of urban and industrial wastewater.

**José Luis Campos Gómez**, Ph.D. in Chemical Sciences, Dep. of Chemical Engineering, Univ. of Santiago de Compostela (June, 2000). His present position is teaching and research assistant at University Adolfo Ibáñez (Chile) (since 2014). During the first years (1994-2001), he has worked in the field of nitrogen removal from industrial wastewater, optimization of aquaculture systems and minimization of sludge production. From 2001, his research is focused on two main topics: aerobic granulation, in order to improve the activated sludge systems, and anammox process to remove ammonia from wastewater with a low C/N ratio. He has participated and/or supervised several projects and contracts with industries where industrial and pilot plants were applied and new technologies were developed to management and treat different kinds of wastewater and sludge.

**Anuska Mosquera-Corral**, Associate Professor at the University of Santiago de Compostela (USC) since 2007. Ph.D. in Chemical Sciences from the USC in 1998 focused on the nitrogen removal from industrial effluents. Postdoctoral researcher, at the TU Delft (The Netherlands) in 2000-2001, where she specialized in aerobic granular biomass and biofilm systems. Since 2001, in the USC, she worked extensively in aerobic granular reactors to remove organic matter (COD) and nitrogen from industrial wastewater and on the application of autotrophic denitrification (anammox, denitrification with sulphur compounds) to effluents with low COD content. Then in 2005, she began applying the molecular biology techniques (FISH and DGGE) for the identification of bacterial populations in bioreactors. In 2011, she began the research line of biopolymers (polyhydroxyalkanoates) production from liquid wastes using mixed cultures.

* * *

**Juan A. Álvarez** holds a PhD in Environmental Chemical Engineering. He has expertise in waste-water treatment, specially focused on anaerobic co-digestion and constructed wetlands. Dr. Álvarez has participated actively in about 10 projects of regional, national and international calls. These projects were based on: Urban wastewater anaerobic treatment, harvested plant effect on CW performance and post-treatment of anaerobic effluent by CW, Anaerobic-co-digestion of agro-industrial waste, aerobic and anaerobic treatment of pig manure and antibiotic effect on pig manure anaerobic treatment. He has carried out about 20 scientific contributions to important journals in WW area. He has coordinated several European projects (FP7 and H2020): SWINGS; HIGHWET; and currently INCOVER project.

**Catarina Leite Amorim** holds a degree in Biochemistry from Faculty of Sciences - University of Porto and a PhD in Biotechnology, with a specialization in Environmental Science and Engineering from Universidade Católica Portuguesa-Faculty of Biotechnology. At the present she is a post-doc researcher in the field of wastewater treatment using the aerobic granular sludge technology, in a partnership between University of Aveiro and Universidade Católica Portuguesa. Particular interests include the biodegradation/bioremediation of pollutants and Aerobic granular sludge technology and its microbial community.

**Carlos A. Arias** is a civil Engineer with a MSc. In environmental engineering and holds a PhD in science from the University of Aarhus where he is a Senior Researcher. With more than 20 years of experience in the development of eco-technological systems for the treatment of waters, wastewaters and runoff control. His research has been dedicated to the identification, development and optimization of processes involved in the removal of pollutants from waters. Efforts in the research have been devoted to improve the removal of nitrogen, phosphorous, pathogen and emergent pollutants in decentralized wastewater systems by developing innovative operational schemes, introducing the use of the intensification of constructed wetlands and reactive media for the improvement of performance. He has participated in multiple national and international projects and cooperates in knowledge and know how to transfer to less developed countries. He has more than 100 publications, including peer reviewed articles, conference proceedings and book chapters.

**Marisol Belmonte** is Assistant Professor in the Department of Environment at the University of Playa Ancha since 2013. She was a member of the Research Group of Engineering and Environmental Biotechnology of EULA-Chile Center (Universidad de Concepción). Her research activities are focused on the wastewater treatment by means of biological processes for organic matter (anaerobic and aerobic) and nitrogen removal (partial nitrification, and heterotrophic denitrification) with suspended and granular biomass systems. She has various scientific products (more than ten articles published in ISI journals, author of book chapters, a number of contributions to national and international conferences in the field of Environmental Engineering and Biotechnology) and actually is the principal investigator in a project post-doctoral (FONDECYT 3140276) financed by Chilean Government, about hydrogenotrophic denitrification in the School of Biochemistry Engineering (EIB), of the Pontificia Universidad Católica de Valparaíso, Chile.

**August Bonmatí Blasi,** PhD, Agricultural Engineer, Universidad de Lleida (UdL). He is a researcher at GIRO Joint Research Unit IRTA-UPC in the Institute of Agrifood Research and Technology (IRTA). His research is focused on the optimization and development of new treatment technologies for processing wastewater and organic wastes. Energy and products recovery is the main topic of his research.

**Paula M. L. Castro** holds a degree in Food Engineering from Universidade Católica Portuguesa – ESB (1989) and a PhD in Biochemical Engineering from University College London (1994). She is an Assistant Professor with Habilitation at UCP. She is actively involved in research in environmental science and engineering, including biodegradation of toxic organic compounds and micropollutants and biological wastewater treatment with immobilized systems, with wide experience on the technology of aerobic granular sludge combined with bioaugmentation.

**Grzegorz Cema** On the February 2010 He defended his doctoral dissertation at Royal Institute of Technology (Stockholm, Sweden) and obtained doctoral degree in Environmental Engineering. Additionally on March 2010 He was graduated with honors to doctor of engineering at Silesian University of Technology (Gliwice). 06.2012 – 03.2013 – Postdoctoral fellowship awarded in the framework of KOLUMB Program by Foundation for Polish Science in Institute for Sanitary Engineering and Waste Management (Gottfried Wilhelm Leibniz Universität Hannover, Germany). Currently - Assistant Professor at The Silesian University of Technology, Faculty of Power and Environmental Engineering.

**Miriam Cerrillo** obtained her BSc degree in Environmental Science in 2002 and MSc degree in Advanced Biotechnology in 2012 at Universitat Autònoma de Barcelona in Spain. She is also a Technical Agricultural Engineer by the University of Vic, Spain, since 2012. She is a PhD student at GIRO Joint Research Unit IRTA-UPC in the Institute of Agrifood Research and Technology (IRTA) in Spain. Her research is focused on the combination of anaerobic digestion with other technologies such as struvite precipitation and bioelectrochemical systems to improve energy and ammonia recovering.

**Francisco Cervantes** is a Professor of Environmental Biotechnology at IPICYT. He obtained his PhD in Environmental Sciences at Wageningen University (The Netherlands, 2002). Author of more than 100 peer-reviewed publications. Lettinga Award 2007.

**Jorge Gómez** obtained his PhD at Universidad de Santiago de Compostela, Spain. The author works in microbial physiology on topics of the respiratory processes, such as nitrification and denitrification for carbon and sulfur removal. He has published nearly 40 papers on these topics. At present, Dr Gómez is Professor of Department of Biotechnology at Universidad Autónoma Metropolitana, campus Iztapalapa, in Mexico.

**Flor Cuervo-López** is a Professor-Researcher at UAM-I. She obtained her PhD in Biotechnology at Universidad Autónoma Metropolitana. Author of more than 30 indexed and peer-reviewed publications.

**David de la Varga** has a PhD in Environmental Sciences (treatment wetlands for winery wastewater) and has a MsC in Environmental Engineering. He has 7 years of experience in waste water (constructed wetlands and industrial treatment plants), working in private companies and Universities. He also has worked in Health and Safety in construction during a year. In 2012 he started his own firm as General Manager of SEDAQUA (Spin-off from University of A Coruña, Spain), and he works within the field of wastewater treatment, project design, construction, operation, or related to water treatment plants based on anaerobic digestion and constructed wetlands. He has also participated in some regional, national and international research projects related with treatment wetlands.

**Anouk F. Duque** graduated in Microbiology from Escola Superior de Biotecnologia, Universidade Católica Portuguesa (ESB-UCP), Portugal, in 2003. After 3 years working in industry, she returned to academia, having participated as a researcher in several research projects. She has completed her PhD in 2012 in Environmental Sciences and Engineering from ESB-UCP. During her PhD she was at Technical University of Delft, The Netherlands, as an invited researcher in the scope of a European Project, INNOWATECH (Innovative and Integrated Technologies for the Treatment of Industrial Wastewater), for 11 months. Since 2012, she is a post-doctoral researcher at Faculdade de Ciências e Tecnologia, Universidade Nova de Lisboa (FCT-UNL), Portugal, under the topic waste treatment and valorization through polyhydroxyalkanoates production. She has been collaborating with ESB-UCP and with Unicer, which is the largest Portuguese beverages company, with a multibrand and multimarket strategy, focused on the beer and bottled water businesses. During her career, Anouk F. Duque has been a team member of 10 research projects, including 4 European projects, EcoBioCap (Ecoefficient Biodegradable Composite Advanced Packaging), SONATURA (Vapour Phase Bioreactors for Agro-non-food Industries), BIOSAP (Biotreatment of Sequentially Alternating Pollutants (SAP) in Wastewaters) and INNOWATECH. In addition, Anouk F. Duque has integrated the organizing committee of the 1st National Biotechnology Olympiads. Since 2007, Anouk F. Duque has published 16 papers in high level, peer-reviewed scientific journals (ISI Web of Science) with more than 190 citations (without self-citations), including 7 papers as first author (H-index of 8). These papers were published in many of the top journals in her field, such as Water Research (IF: 5.528) and Bioresource Technology (IF: 4.494). She has also contributed to over 30 international and national conference communications.

**Maria del Carmen Fajardo Ortiz** received the Ph.D. in Chemical and Environmental Engineering in 2011 at the University of Santiago de Compostela (USC) on the topic of autotrophic denitrification for wastewater treatment. Currently, she is professor of Universidad Autónoma Metropolitana, Mexico. Her research activities are focused on the application of biological processes (simultaneous nitrification/denitrification, acidogenesis, anaerobic digestion, etc.) for the treatment of industrial wastewater (leachates from organic composting, cheese whey, slaughterhouse, vinasses, etc.).

**Francesco Fatone**, PhD, is environmental engineer and associate professor of Chemical Engineering. He has published 57 peer-reviewed papers in international journals. He is expert of advanced treatment, phosphorus and biopolymers recovery in municipal wastewater treatment plant. He is co-chair of the second largest IWA specialist group, editor of international water related scientific journals, chair or member of the scientific committees of IWA international conferences. Francesco Fatone is author of n.2 pending patent and supervisor for application of innovative techniques for supernatant treatment. As far as international coordination and management is concerned, Francesco Fatone is co-chair of the second largest specialist group of the International Water Association. He recently organized and chaired the IWA international conference EcoTechnologies for Wastewater Treatment" and is member of the management committee and steering committee of the COST Action Water2020. Within UTILITALIA, the national association of water, waste and energy utilities, Francesco is member of the IWA governing member working group.

**Paula Carrera Fernández** has studied Chemical Engineering at the University of Santiago de Compostela. Since 2015 she is a PhD Student in the Department of Chemical Engineering at the University of Santiago de Compostela and her research is focused on aerobic granular sludge processes for wastewater treatment.

**Andrea Fra-Vázquez** studied Biology and Master in Environmental Engineering at the University of Santiago de Compostela. From 2013, she is doing her PhD in the Group of Environmental Engineering and Bioprocesses.

**Alba Pedrouso Fuentes** has studied Chemical Engineering at the University of Santiago de Compostela and later a MSc in Chemical Engineering and Bioprocess at the same university. Since 2014 she is a PhD Student in the Department of Chemical Engineering at the University of Santiago de Compostela and her research is focused on nitrogen removal from wastewater through anammox based processes.

**Elisa A. Giustinianovich** has studied Bioengineering at the University of Concepción and later a MSc in Chemical Engineering at the same university. Since 2013 she is a PhD Student in the Department of Chemical Engineering at the University of Concepción and her research is focused on nitrogen removal from wastewater through anammox based processes.

**René Kilian** received his MS in Environmental Science from Wageningen Agricultural University in Holland in 1996. Driven by an intense engagement in watersaving, wastewatertreatment and reuse of treated wastewater, he started his own company, Kilian Water, in 1997. The main focus of the company is on developing, designing and constructing constructed wetlands and willow-evaporating systems. René Kilian has participated in several projects on reseach and development of different constructed wetlands, both on EU-level and with Danish authorities. He is bringing the achieved knowledge into practice through Kilian Water, resulting in designing, establishing and delivering of more than 80% of all reedbed systems, constructed wetlands and aerated constructed wetlands in Denmark.

**Miguel Mauricio-Iglesias** received a MSc in Chemical Engineering (Univ. Santiago de Compostela) and a PhD in Food Technology at the University of Montpellier II. After working as a postdoc (2010-14) at the Technical University of Denmark, he joined the University of Santiago de Compostela as a Marie Curie postdoctoral fellow. His research focuses on applying process system engineering tools to the modelling, development, control and optimization of wastewater treatment and bioprocesses. He is member of the International Federation of Automatic Control (IFAC), involved at the Technical Committee on Biosystems and Bioprocesses.

**Ramón Méndez,** Prof Dr in Chemical Engineering Chemical Engineering Department ETSE School of Engineering H Index 35 Director of one PhD Thesis and two financed projects in this subject publishing several papers.

**Irina S. Moreira** holds a degree in Biology from Faculty of Sciences - University of Porto and a PhD in Biotechnology, with a specialization in Environmental Science and Engineering from Universidade Católica Portuguesa-Faculty of Biotechnology. At the present she is a post-doc researcher in the field of Biodegradation and Biocatalysis of aromatic compounds. Particular interests include the biodegradation/bioremediation of pollutants, metagenomics and Aerobic granular sludge technology.

**Ana Pascual** is technical R & D + i Unit Environment and specialty degree in Chemistry in Inorganic Chemistry from the University of Vigo and she has an M. Sc. in Environmental Technology from the University of Vigo. Since 2009 he has taken an active part in R&D at AIMEN in regional and international projects. Her expertise is focused on monitoring of domestic wastewater treatment plants and operation of anaerobic digestion reactors and constructed wetlands.

**Sergio Ponsa** is Associate Professor in the Department of Biosciences at University of Vic-Central University of Catalonia (UVic-UCC), Coordinator of the Research Group BETA and Director of BETA Tech. Center (UVic-UCC)-TECNIO Network (Catalan Government). Dr Ponsa has a long and convincing experience in solid waste and wastewater treatment processes, mainly working in biological technologies such as anaerobic digestion, composting, solid state fermentation, biological nutrient removal and biodrying processes. He has participated in several conferences, workshops and seminars related with biological processes for solid waste and wastewater treatment and valorisation. Furthermore Dr. Ponsa is participating in different European Networks, including the COST Action Water2020, where he is the Coordinator of the Task Group in advanced technologies for sludge valorisation. Dr. Ponsa has also a renowned experience in the promotion and development of research and technologies, since he has promoted, coordinated and participated in more than 45 R&D&I projects with companies and public administrations and promoted more than 25 international research and innovation projects in the last 5 years. He has also participated in more than 15 national competitive research projects related with biological technologies. He is co-funder of the spin-off AXIA Environmental Technologies (2013) that is developing and applying at industrial level advanced biological technologies related with the sustainable treatment of solid waste and wastewater effluents.

**Victor Riau** obtained his PhD degree in Environmental Technology and Water Management at the University of Cadiz (Spain) in 2011. In February 2012 he was hired as Postdoctoral researcher at the University of Cadiz and he worked at this position until July 2012. Then, he worked at the National Research Council (CSIC) in the framework of a private-funding project. In January 2014, he joined IRTA (Institute of Agrifood Research and Technology) where he is currently working as Researcher and Innovation Manager in the Research Programme “Integral Organic Waste Management”. His research topics are anaerobic digestion of organic wastes and wastewaters, urban and industrial wastewater treatment, biogas and nutrient removal and recovery from waste streams, among others.

**Gonzalo Ruiz-Filippi** was Associate Professor of the School of Biochemistry Engineering (EIB), of the Pontificia Universidad Católica de Valparaíso, Chile. His research was development in the biological treatment of wastewater in Chile, and its use as an alternative for the generation of non-conventional renewable energy. In his career as a researcher (over 15 years) was an outstanding professor in the scientific and academic area, in Chile as international level. This was based on his prolific scientific products (ex. Book publishing, authoring numerous articles published in ISI journals, author of book chapters, a number of contributions to national and international conferences in the field of Environmental Engineering and Biotechnology, and was a member active of the scientific networks (ex. IWA)); main and partner investigator in several research projects with national and international financing (FP7, Horizon2020); as well as numerous awards won by Dr. Ruiz Filippi in his short academic life. The research of Dr. Ruiz-Filippi was mainly in the application of anaerobic digestion to produce biogas, mainly hydrogen, which it can be used as a source of non-conventional renewable energy, and the modelling and control of anaerobic digestion process as well.

**Adam Sochacki** received his PhD in Environmental Engineering from the Silesian University of Technology (Gliwice, Poland) and from Ecole Nationale Superieure des Mines de Saint-Etienne (France) in 2013. His PhD thesis was dedicated to the polishing of electroplating wastewater in constructed wetland systems. Since 2013, at the Silesian University of Technology, he has been participating in research projects dedicated to transformation of organic micropollutants in constructed wetlands and by light-induced treatment processes. In 2016 he was hired as a research and teaching assistant at the Silesian University of Technology and as a postdoctoral researcher at the Czech University of Life Sciences Prague.

**Manuel Soto** is a Professor of Chemical Engineering and director of the Office of Environment of the University of A Coruña. Doctor by the University of Santiago de Compostela (1990), he conducted research stays at the Agricultural University of Wageningen (Netherlands). He was the research coordinator on 20 projects from public calls and contracts. He has more than a hundred publications and papers at conferences on topics of environmental science and technology. His research interests focus on anaerobic digestion and constructed wetlands for wastewater treatment, and composting and organic waste recovery. In these lines, he has an extensive experience in the application of anaerobic digestion for the treatment of very diverse types of wastewater, both in the industrial sector (fish canning industry, dairy industry, cellulose and wood industry, winery) and domestic wastewater. More recently, his interest also focused on wastewater treatment and post-treatment using constructed wetlands and on composting of organic waste.

**Estela Tapia-Venegas**, 2008 Biochemist engineer 2010 Magister in engineering sciences 2014 Doctorate in biotechnology areas of expertise related mainly to environmental biotechnology, renewable energy production, microbiology of mixed cultures and bio-refinery waste.

**Carlota Tayà** is a postdoctoral research of BETA Technological Center and Chemical Engineer. She has deep experience on wastewater treatment processes, especially biological treatments. Her PhD thesis was focused on solving the main bottlenecks on enhanced biological phosphorus removal (EBPR) together with nitrogen removal. Furthermore, she also has experience on microbial analysis for identify the different communities involved in wastewater treatment. She has participated as a researcher in one European project focused on wastewater treatment using microalgae to remove nutrients and capture CO2 from other processes. During her PhD, she has also participated in Spanish projects related to biological wastewater treatments. She is a coordinator of the research line on wastewater treatment and valorisation. Dr Tayà has also taken part in several conferences and workshops related with the project field. During her career, she has also published 8 in international peer reviewed scientific journals and has presented 3 studies in international conferences and 2 studies in national conferences.

**Javiera Toledo-Alarcon** is PhD student since 2014 at Laboratory of Environmental Biotechnology (LBE) - French National Institute for Agricultural Research (INRA), France. Currently she is working with electro-fermentation process with objective to control patterns fermentation. In 2014 she received her MS in Engineering Sciences majoring in Biochemical Engineering from Pontificia Universidad Catolica de Valparaiso, Chili. She worked with hydrogen production from dark fermentation using different treatments on inoculum. During 2012 worked with hydrogenotrophic denitrification of groundwater with high nitrate level using CSTR for obtain her Biochemical Engineering diploma.

**Daniel Valenzuela-Heredia** is bachelor in biological science from the Universidad Católica de Chile (2008). He received his PhD in Biotechnology from the Universidad Católica de Valparaíso and from the Universidad Técnica Federico Santa Maria in 2015. In 2016 he was hired by the Biotechnological Center of the University Adolfo Ibáñez as a postdoctoral researcher. The focus of his researches has been water and soil microbiology.

**Borja Valverde-Pérez** graduated as chemical engineer at the Unversidade de Santiago de Compostela, Spain, in 2012. He received his PhD in environmental engineering at the Technical University of Denmark in December 2015. He is an expert on design, modelling and control of processes for biological nutrient removal and recovery.

**Mark C. M. van Loosdrecht** is Professor in Environmental Biotechnology at Delft University of Technology, The Netherlands. He graduated from and did his PhD research at Wageningen University. His PhD topic was a combination of microbiology and colloid chemistry. He was appointed at Delft in 1988 and became Full Professor in 1998. His research is characterized by the combination of scientific understanding of complex systems and development of new processes. Dr. van Loosdrecht's scientific interests are mainly related to biofilm processes, nutrient conversion processes and the role of storage polymers in microbial ecology. In particular, he is interested in new processes related to wastewater treatment and resource recovery. His research has resulted in several processes currently applied on full scale such as the Sharon process, Anammox process and Nereda process. He is active member of the International Water Association (IWA) and past chairman of the Biofilm and the Nutrient removal specialist groups. He is Editor in Chief of Water Research. He obtained several prizes for his work, including the Lee Kuan Yew Singapore Water Prize and the IWA Grand Award. He is member of both the Royal Dutch Academy of Arts and Sciences and the Dutch as well as U.S. Academy of Engineering (AcTI and NAE). He was awarded a knighthood in the order of the Dutch Lion. He has published over 600 scientific papers, has 15 patents and has supervised over 50 PhD students.

**Dion van Oirschot**, owner of Rietland, started his career at De Twaalf Ambachten, a center for alternative technology in Boxtel the Netherlands. He wrote several guides for owner built systems and a co-authored a booklet with Sietz Leeflang: "Hoe het ook zonder riool kan" (How to do without sewerage). After starting his own firm, being the first firm specializing in treatment wetlands in the Netherlands, he has devoted his professional career to finding ecological solutions for waste water problems. His extensive knowledge and practical experience with these systems have made him an expert in implementation of treatment wetlands. His greatest strength is his creativity and versatility in taking on technical challenges.

# Index

www.ingramcontent.com/pod-product-compliance
Lightning Source LLC
LaVergne TN
LVHW060405300926
849162LV00006B/1348